Ergebnisse der Mathematik und ihrer Grenzgebiete

Volume 41

3. Folge

A Series of Modern Surveys in Mathematics

Springer
Berlin
Heidelberg
New York
Barcelona
Hong Kong
London
Milan
Paris
Singapore
Tokyo

Igor Nikolaev

Foliations on Surfaces

With 23 Figures

Springer

Igor Nikolaev
The Fields Institute for Research
in Mathematical Sciences
222 College Street
Toronto
Ontario M5T 3J1
Canada
e-mail: inikolae@fields.utoronto.ca

Library of Congress Cataloging-in-Publication Data applied for

Die Deutsche Bibliothek - CIP-Einheitsaufnahme

Nikolaev, Igor:
Foliations on surfaces / Igor Nikolaev. - Berlin ; Heidelberg ; New York ; Barcelona ; Hong Kong ; London ; Milan ; Paris ; Singapore ; Tokyo : Springer, 2001
(Ergebnisse der Mathematik und ihrer Grenzgebiete ; Folge 3,Vol. 41)
ISBN 3-540-67524-8

Mathematics Subject Classification (2000):
Primary: 57Rxx, 58Fxx
Secondary: 05Cxx, 28Fxx, 30Exx, 34Cxx, 46Lxx

ISSN 0071-1136
ISBN 3-540-67524-8 Springer-Verlag Berlin Heidelberg New York

Springer-Verlag Berlin Heidelberg New York
a member of BertelsmannSpringer Science+Business Media GmbH

Printed in Germany

Typeset by the author.
Edited and reformatted by Ludwig Feuchte, Flein, using a Springer TEX macro package.
Printed on acid-free paper SPIN 10723139 44/3142LK - 5 4 3 2 1 0

Моим дорогим, маме и папе

If you can dream – and not make dreams your master,
If you can think – and not make thoughts your aim..

Rudyard Kipling : 'If–'
("Poems. Short Stories")

We are what we think.
All that we are arises with our thoughts,
With our thoughts we make the World...

Dhammapada : Choices
("The Sayings of the Buddha")

Foreword

H. Poincaré, the founder of the qualitative theory of differential equations, was the first to realize the significance of the simple fact that the trajectories of a smooth vector field, v, determine a geometric picture called the *phase portrait* of v. From a general viewpoint, such a picture can be considered as a *locally oriented one-dimensional foliation*, $\mathcal{F}$, *with singularities*. Actually, these two geometric concepts are equivalent because for a given smooth foliation $\mathcal{F}$ one can determine the corresponding vector field v up to a change of time.

There is a vast literature devoted to the case of locally oriented foliations on two-dimensional manifolds. Let me only mention two recent texts: S. Aranson, G. Belitsky, E. Zhuzhoma, *Introduction to qualitative theory of dynamical systems on surfaces*, AMS Math. Monographs, 1996; I. Nikolaev, E. Zhuzhoma, *Flows on two-dimensional manifolds*, Lecture Notes in Math., 1705, Springer Verlag, 1999. But there exist foliations on surfaces that are non-orientable at some singular points. Indeed, consider a saddle point with an *odd* number, n, of separatrices. If $n = 1$ ($n = 3$), the singularity is called a *thorn* (*tripod*, respectively). A reader with a vivid imagination can recognize thorns and tripods even in her (his) own fingerprints. Of course, there are really interesting fields where thorns and tripods are met (for example, the theory of liquid crystals). Although the above mentioned books are devoted mainly to the locally oriented case, a few thorns and tripods flashed there on the horizon.

The present book gives an account of basic results obtained in exploration of foliations on surfaces with emphasis on the *locally non-orientable* case. It demonstrates that the general setting of not necessarily locally orientable foliations leads to a better understanding of some classical problems initially investigated by A. Cayley, G. Darboux and É. Picard. Various approaches to the theory of foliations are discussed and some unsolved problems formulated.

I invite the reader interested in geometry and analysis on surfaces to visit the refreshingly new land of locally non-orientable foliations and enjoy the world of thorns, tripods, apples, sun-sets, labyrinths and many other exotic things (including a foliation with a leaf everywhere dense in a disk).

Idel Bronshteyn

Preface

A foliated space is one of the fundamental concepts of modern geometry and topology. Given a topological space X, by a foliation, $\mathcal{F}$, one understands a partition of X into a disjoint sum of 'leaves' with a regular 'microscopic' behaviour.

For example, if X is an n-dimensional manifold, then every small part of $\mathcal{F}$ looks like a family of the parallel planes $\mathbb{R}^n = \mathbb{R}^k \oplus \mathbb{R}^{n-k}$. The global behaviour of leaves can be quite tricky and one of the goals of the theory is to understand the asymptotic properties of leaves.

The theory of foliations begins with the work of Ch. Ehresmann and G. Reeb in the 1940's. Perhaps they were inspired by the ideas of H. Poincaré, A. M. Lyapunov and G. D. Birkhoff who promoted geometric methods in the study of differential equations.

The development of the subject since the 1940s is reflected in a beautiful survey by B. Lawson, *Foliations*, Bull. Amer. Math. Soc. 80 (1974), 369-418. The advances in the area are due to D. V. Anosov, R. Bott, L. Conlon, A. Connes, A. Haefliger, G. Hector, R. Hermann, F. W. Kamber J. W. Milnor, S. P. Novikov, B. Reinhart, H. Rosenberg, P. Schweitzer, H. Seifert, I. Tamura, W. P. Thurston and P. Tondeur, to name a very few. The younger generation is represented by G. Gabai, E. Ghys, C. Godbillon, S. Hurder, J. Martinet, S. Matsumoto, P. Molino, R. Moussu, J. F. Plante, R. Sacksteder, D. Tischler, T. Tsuboi, P. Walczak and J. Vey; see the vast bibliography of the book by C. Godbillon, *Feuilletages. Étude géometriques*, Birkhäuser, Basel-Boston-Berlin, 1991.

A modest aim of the present monograph is to cover the case when X is a two-dimensional manifold (a surface).

Let k be a dimension of leaves. In the case $k = 0$ (0-dimensional leaves) the surface is foliated by points, which is trivial. The case $k = 1$ assumes that the Euler characteristic of X vanishes. In other words, only the torus and Klein bottle can carry such a foliation. This theory received a full treatment in the monograph of G. Hector and U. Hirsch, *Introduction to the Geometry of Foliations*, Parts A and B, Vieweg. 1981.

The point of view adopted in this book is that foliations can be *singular*. In other words, the dimension of leaves of $\mathcal{F}$ is not fixed and it can vary from point to point, being either 1 or 0. Moreover, until otherwise mentioned, the set Sing $\mathcal{F}$ is assumed to be finite. In this setting, by a (singular) foliation $\mathcal{F}$ on a surface X one means a partition of $X\setminus$ Sing $\mathcal{F}$ into disjoint curves (1-leaves), this partition being locally homeomorphic to a family of parallel lines. As far as the singularity set Sing $\mathcal{F}$ (0-leaves) is concerned, it must be defined separately for each class of foliation.

An immediate example of singular foliations is given by a flow v^t on X. The reader can see that 1-leaves correspond to the trajectories of v^t and singular points to the fixed points of v^t. This simplicity is false, since the majority of foliations can be given by no flow whatsoever.

There are two kinds of obstacles to turning a foliation into a flow. The local obstacle is represented by a *non-orientable singularity*, i.e. a singularity whose 'phase portrait' cannot be coherently oriented along the leaves. Many examples are known (thorns, tripods, etc).

The global obstacle is given by a *labyrinth*, i.e. a collection of leaves which are everywhere dense in the disk. The reader can take a look at Fig. 0.3 to see the idea. The obstacle is the Poincaré-Bendixson theorem which says that the long-time behaviour of trajectories in the disk is either periodic or stationary.

There are however parallels between the two theories. To encourage the reader at this point, we shall mention that eventually every foliation can be represented by a ($\mathbb{Z}_2$-symmetric) flow on a surface with involution. This procedure is called a *normalization* and will be discussed later on.

The theory of surface foliations is as old as the number theory. The problem of approximation of an irrational number using rationals is linked to the geometry of foliations. This fact was recognized by Leopold Kronecker in the middle of the 19th century. The irrational (Kronecker's) foliations on the torus have been applied to the problems of celestial mechanics by H. Poincaré, A. Denjoy and T. Cherry. The study of the geometry and topology of foliations on surfaces was undertaken by S. Kh. Aranson, T. O'Brien, C. Gardiner, V. Z. Grines, C. Gutierrez, T. Inaba, R. Langevin, G. Levitt, N. Markley, D. Neumann, Ch. Pugh, E. V. Zhuzhoma and others. The theory of measure preserving transformations linked to a surface foliation is due to V. I. Arnold, G. Forni, M. Herman, J. Hubbard, A. Katok, M. Keane, M. Kontsevich, M. Martens, H. Masur, W. de Melo, P. Mendes, E. A. Sataev, Ya. Sinai, S. van Strien, W. A. Veech, J. C. Yoccoz and A. Zorich.

Cayley-Darboux-Picard Problem and Normalization. The doctoral dissertation of H. Poincaré, *Sur les courbes défenie par les équations différentielles*, C. R. Acad. Sci. Paris 90 (1880), 673-675, was devoted to the geometric theory of singular points of differential equations. He classified all simple singular points of planar (analytic) vector fields by showing that there are only three types of such points: a simple saddle, a node and a focus (the last two types are topologically equivalent).

Around 1895 É. Picard (*Sur les points singuliers des équations différentielles du premier ordre*, Math. Annalen 46, 521-528) was looking for a similar classification of the non-orientable singular points. He was probably influenced by an earlier work of A. Cayley, *On differential equations and umbilici,* Philos. Mag. 26 (1863), 373-379, 441-452, Coll. Works: V. 6.

In 1896 G. Darboux (*Sur la forme des lignes de courbure dans la voisinage d'un ombilic,* Lecons sur la Theorie des Surfaces, IV, Note 7, Gauthier Villars, Paris) used a geometric approach to settle the same problem.

Unfortunately, from our point of view, no satisfactory theory was built at that time. Let us call the following question, "Find a regular procedure of resolving the non-orientable singularities of planar foliations" the *Cayley–Darboux–Picard problem.*

Several research teams in Brazil (V. Guinez, C. Gutierrez, J. Sotomayor), France (J. F. Mattei, M. F. Michel), Russia [1] (A. A. Kadyrov, A. G. Kuzmin) and USA (B. Smyth, F. Xavier) were involved in this problem for different reasons.

In a short note, *Singular points of the line element fields in the plane,* Izv. Akad. Nauk Resp. Moldova, Matematika 3(9), 1992, 23-29, I. U. Bronstein (Bronshteyn) and the author suggested a method leading to a solution of the Cayley-Darboux-Picard problem. The idea was to consider a double covering to the phase space of the differential equation studied by Cayley, Darboux and Picard. A surprising consequence was a short list of simple (structurally stable) singular points. These appeared to be thorns, tripods, sun-sets and apples.

We call the method a *normalization* (in the sense of Riemann) since it reduces planar foliations to the planar vector fields with a symmetry. This book gives a systematic account of the normalization method.

Final Remarks. This book is addressed to graduate students. It will be useful to the specialists in geometry and topology, ergodic theory, dynamical systems, complex analysis, differential and noncommutative geometry.

Toronto, September 2000 *Igor Nikolaev*

[1] The author learned about this problem from A. A. Kadyrov while working as a postdoctoral researcher in St. Petersburg.

Acknowledgements

This book was inspired by fruitful contacts with many people. Among the very few mentioned here are D. V. Anosov, S. Kh. Aranson, I. U. Bronshteyn (who kindly agreed to write a foreword to this book), G. A. Elliott, G. Forni, J. Guckenheimer, G. Hector, A. Katok, R. Langevin, G. Levitt, H. Masur, S. Matsumoto, C. Possani, P. Schweitzer, B. Smyth, J. Sotomayor, P. Tondeur and A. Zorich. Gilbert Levitt intensively read the early version of the manuscript, making numerous helpful suggestions. David Kerr from the Fields Institute verified both the linguistic and the mathematical quality (especially the C^*-algebra section) of the manuscript. I thank him and also Paul Bracken from the CRM for this painstaking work. Of course, the remaining mistakes and bugs belong exclusively to me.

I would like to thank Evgeny Zhuzhoma who is morally a coauthor of this book.

Chapter 16 was written by Benedetto Piccoli to whom I extend my sincere thanks. Chapter 5 was prepared with the help of materials courtesy of E. Zhuzhoma.

I am grateful to the Centre de Recherches Mathematiques and the Fields Institute for technical aid. I thank François Lalonde and the Centre Interuniversitaire de Recherche en Géométrie Différentielle et Topologie for financial support.

I would like to thank Leonie Kunz and Ute Motz from Springer-Verlag for sound help at the different stages of the project.

Finally, the most kind words should be said to Louise Huppé whose patience and support allowed me to complete this book.

Contents

Part III. Appendix

Index of Notation

ϕ^t	continuous flow
$O(x)$	orbit through the point x
$O^{\pm}(x)$	forward (backward) orbit through the point x
$\alpha(x)$	α–limit set of the point x
$\omega(x)$	ω–limit set of the point x
$\mathcal{Q}(x)$	quasiminimal set of the orbit $O(x)$
Λ	minimal set of the flow ϕ^t
(M, π^t, θ)	foliation $\mathcal{F}$ given by the flow π^t and the involution θ on the surface M
θ	involution defined by the foliation $\mathcal{F}$
$Sing\ \mathcal{F}$	set of 0–dimensional leaves of the foliation $\mathcal{F}$
$\mathcal{F}$	foliation
$TM(T^*M)$	tangent (cotangent) bundle over the manifold M
ω	differential 1–form on M
L_v	Lie derivative along the vector field v
i_v	inner product in the space of differential forms
BM	bivector bundle over the manifold M
$Susp\ d$	suspension over the interval exchange transformation d
$\mathcal{L}$	labyrinth on M
$\mathcal{F}_1 \# \mathcal{F}_2$	sum of the foliations $\mathcal{F}_1$ and $\mathcal{F}_2$
C^r	differentiability class of the foliation
$\mathcal{F}^r(M)$	space of C^r–smooth foliations on the manifold M
$\mathcal{F}_0^r(M)$	space of structurally stable (rough) foliations on the manifold M
X^R	graph X endowed with the rotation system R
$\lvert\mathcal{Q}\rvert$	number of the disjoint quasiminimal sets of the foliation $\mathcal{F}$
g	genus of the orientable surface M
p	genus of the non orientable surface N
$\mathbb{H}$	Lobachevsky half–plane endowed with the hyperbolic metric
$SL_2(\mathbb{Z})$	modular group
G	finite group
$Cay\ G$	Cayley graph of the group G
$H_1(M, \mathbb{R})$	first real homology group of the surface M
$H^1(M, \mathbb{R})$	first real cohomology group of the surface M
$\pi_1(M)$	fundamental group of the surface M

l	curve on the surface M
$\widetilde{l}$	preimage of the curve l on the universal covering to M
$\partial\mathbb{H}$	absolute
μ	invariant measure of the flow ϕ^t
AF	approximately finite C^*–algebra A
A_α	irrational rotation algebra
$\Gamma(n)$	principal congruence subgroup of the modular group
$A(\phi)$	Artin number of the minimal flow ϕ^t
$K_0(A)$	Grothendieck group of the C^*–algebra A
$\phi(z)dz^2$	quadratic differential on the surface M
$Arg\ \phi(z)$	argument of the quadratic differential $\phi(z)dz^2$
$\Omega^1(\Omega^2)$	first (the second) fundamental forms of the surface M
$\mathbb{C}^*$	Riemann sphere $\mathbb{C}\cup\{\infty\}$
$PSL(2,\mathbb{C})$	group of Möbius transformations of the complex plane $\mathbb{C}$
T_g	Teichmüller space of the Riemann surface of genus g
$\mathbb{P}_2$	complex projective plane
C	complex projective curve on $\mathbb{P}_2$
$\wp(z)$	Weierstrass function
D	divisor of the non singular projective curve C

0. Foliations on 2-Manifolds

0.1 Notations

Let M be a compact surface and let $\mathbb{R}$ be the set of real numbers. When it is not specified explicitly, all the objects are supposed to be smooth (C^∞-smooth).

Definition 0.1.1 *By a continuous flow , ϕ^t, on M we mean a continuous mapping*

$$\phi^t : M \times \mathbb{R} \to M,$$

such that for all $x \in M$ and for all $s, t \in \mathbb{R}$,

(i) $\phi^0(x) = x$,
(ii) $\phi^{s+t}(x) = \phi^s(\phi^t(x))$.

If the mapping $M \times \mathbb{R} \to M$ is r times differentiable in t at every point $x \in M$, then ϕ^t is said to be of *differentiability class* C^r.

The set $O(x) \stackrel{\text{def}}{=} \{\phi^t(x) | t \in \mathbb{R}\}$ is called an *orbit of* ϕ^t passing through the point $x \in M$. The *forward orbit* of ϕ^t through $x \in M$ is defined by $O^+(x) = \{\phi^t(x) | t \geq 0\}$. Similarly, the *backward orbit* of ϕ^t, passing through the point $x \in M$, is the set $O^-(x) = \{\phi^t(x) | t \leq 0\}$. Clearly, $O(x) = O^-(x) \cup O^+(x)$.

The long–time behaviour of every point $x \in M$, moving under the action of the flow ϕ^t, is described by its α- and ω-limit set, which is given by

$$\alpha(x) = \lim_{t \to -\infty} \phi^t(x), \quad \omega(x) = \lim_{t \to +\infty} \phi^t(x).$$

Usually, each orbit is a *proper* embedding of an open interval I into the surface M, that is, $I \hookrightarrow M$. But there are several important exceptions.

The point $x \in M$ is called an *equilibrium* ($\equiv$ *fixed point*, or *singularity)* of a flow ϕ^t, if $\phi^t(x) = x$ for all values of $t \in \mathbb{R}$. In this case the 0-dimensional orbit $O(x)$ is called a *stationary orbit.*

Let $x \in M$ and let $\phi^T(x) = x$ for some finite value of $t = T$, different from 0. By the homogeneity of the flow, all points of the orbit $O(x)$ enjoy the same property. In this case $O(x)$ is the image of an embedding $S^1 \hookrightarrow M$, and is called a *closed orbit* of the flow ϕ^t.

An orbit $O(x)$ through the point $x \in M$ is *positively recurrent* if, for every small neighborhood U of x and any positive real T, there exist $t > T$ such that

$\phi^t(x) \subset U$. *Negative recurrence* is defined similarly. The orbit $O(x)$ through the point $x \in M$ is called *recurrent* if it is both negatively and positively recurrent. Immediate examples of recurrent orbits are the stationary and the closed orbits. They are called *trivially recurrent.*

The closure of any non trivially recurrent orbit $O(x)$ is said to be a *quasi-minimal set* and is denoted by $\mathcal{Q}(x)$.

A subspace A of the surface M is called *invariant* with respect to the flow ϕ^t if $\phi^t(A) \in A$ for all points $x \in A$ in all moments of time $t \in I\!R$. This possibility is usually written as $\phi^t(A) = A$.

A non-empty set Λ lying in the surface M is a *minimal set of a flow* $\phi^t : M \times I\!R \to M$ if it is:

- invariant
- closed
- contains no smaller invariant closed subsets.

Λ is called a *trivial* minimal set if it is either an equilibrium, or a closed orbit, or else $\Lambda = T^2$, where T^2 is the two-dimensional torus. Flow ϕ^t is called *minimal* if its minimal set coincides with M.

Definition 0.1.2 *A foliation $\mathcal{F}$ of a class C^r is defined to be a triple (M, π^t, θ), where $\pi^t : M \times I\!R \to M$ is a C^r-smooth flow on M and*

$$\theta : M \to M, \qquad \theta^2 = id_M$$

is a smooth involution on M which obeys the following conditions:

(i) θ keeps the orbit structure, induced by π^t,

$$\forall x \in M, \qquad O(\theta(x)) = \theta(O(x)).$$

(ii) θ fixes a finite even number of points $W = \{p_1, ..., p_{2k}\}$ on M,

$$\theta(x) \cap x = \begin{cases} x, \text{ if } & x \in W, \\ \emptyset, \text{ if } & x \in M \backslash W \end{cases}$$

(iii) if $W = \emptyset$, then M consist of two connected components N_1 and N_2, so that $\theta : N_1 \to N_2$ is a homeomorphism between them.

The foliation $\mathcal{F} = (M, \pi^t, \theta)$ is called non-orientable if W is non-empty. If $W = \emptyset$, then $\mathcal{F}$ is said to be orientable.

Triples (M, π^t, θ) do exist. To see this, one should take any foliation with non orientable singularities in the points $W = \{p_1, ..., p_{2k}\}$ and 'lift' the foliation to a 2-fold covering surface, ramified over the set W. The covering surface, M, will have only orientable singularities, given by a flow π^t, which fits the conditions (i)-(iii) of the definition. Conversely, if a triple (M, π^t, θ) is defined, then factorizing by θ one obtains in general a non orientable foliation $\mathcal{F}(\pi^t)/\theta$ on the manifold M/θ, where $\mathcal{F}(\pi^t)$ denotes a foliation on M, induced by the flow π^t.

Proposition 0.1.1 *Let $\mathcal{F} = (M, \pi^t, \theta)$ be a non orientable foliation on a surface M. Then all points $x \in W$ are equilibria of the flow π^t. Moreover, for α- and ω-limit set of the orbit $O(x)$,*

$$\theta(\alpha(x)) = \omega(\theta(x)).$$

In other words, the forward orbit through a point x is sent by θ to the backward orbit through the point $\theta(x)$, and vice versa.

Proof. Follows from Definition 0.1.2. □

Definition 0.1.3 *Let $\mathcal{F} = (M, \pi^t, \theta)$ be a foliation of the class C^r. Then the factor foliation $\mathcal{F}/\theta$ of the factor surface M/θ is said to be C^r-smooth.*

0.2 Examples

0.2.1 Smooth Functions

Let M be a surface and $\mathbb{R}$ the real line. Suppose that $f : M \to \mathbb{R}$ is a differentiable function and df is its tangent mapping. Point $x \in M$ is called *critical point* of f if the rank of $df(x)$ is different from the maximal, that is, $df(x) = 0$. Otherwise, the point $x \in M$ is a *regular point* of f.

Let $y \in Imf$ be a real number. Consider a partition $\mathcal{F}$ of the manifold M by the level set $f^{-1}(y)$ of the function $f : M \to \mathbb{R}$.

Proposition 0.2.1 *$\mathcal{F}$ is an orientable foliation (M, π^t, θ) of the surface M and the set Sing $\mathcal{F}$ coincides with the set of critical points of the mapping $f : M \to \mathbb{R}$.*

Proof. Let $f : M \to \mathbb{R}$ be a smooth mapping on the manifold M. Denote by $v : M \to TM$ a vector field on M which satisfies the equation

$$L_v f = 0,$$

where L_v is the Lie derivative of f in the direction of vector field v. It is an easy exercise to check that v is tangent to the level set $f^{-1}(y)$, passing through the regular point $x \in M$. For the critical points $p \in M$, one has $v(p) = 0$. The flow π^t is tangent to v at all points where $v \neq 0$. One obtains a triple (M, π^t, θ) which represents an orientable foliation $\mathcal{F}$ on the surface M. □

Note that foliations $\mathcal{F}$ given by the level set of smooth functions are regular. Indeed, every leaf $f^{-1}(y)$ is properly embedded into M (i.e., the 'inner' topology of the leaves coincides with the topology of the ambient space M). In particular, $\mathcal{F}$ admits only trivial recurrent leaves.

Example 0.2.1 *Let $p : M \to B$ be a fibration of the torus over $B = S^1$ with fibre $p^{-1}(x) \cong S^1$. Parametrizing S^1 by $t \in [0, 2\pi) \subset I\!R$, one obtains a smooth function $f : T^2 \to I\!R$. Clearly, f has no critical points and its level set $f^{-1}(x)$ yields a non singular foliation $\mathcal{F}$ on the 2-dimensional torus.*

Example 0.2.2 *Let $h : T^2 \to I\!R$ be a height function $x \mapsto h(x)$ of points lying on the 2-dimensional torus T^2. The rank of dh is maximal everywhere, except at the points $\{p_1, p_2, p_3, p_4\}$, where dh vanishes. The level set $h^{-1}(y)$ is homeomorphic either to S^1 or $S^1 \cup S^1$ or $S^1 \# S^1$ or a point. The induced orientable foliation $\mathcal{F}$ has four singularities $Sing\ \mathcal{F} = \{p_1, p_2, p_3, p_4\}$ of saddle and centre types.*

0.2.2 1–Forms

Let M be a compact surface and $\omega : T^*M \to I\!R$ a C^r-smooth differential 1-form defined on M. For $r \geq 1$, one considers the class $\Omega^r(M) = \{\omega | d\omega = 0\}$ of *closed* 1-forms. The form $\omega \in \Omega^r(M)$ is *exact* if there exists a smooth function $f : M \to I\!R$ such that

$$df = \omega.$$

Since $d^2 = 0$, an exact 1-form is always closed. Conversely, if the first (*de Rham*) cohomology class $H^1(M, \mathbb{Z})$ of the manifold M vanishes, then the above equation is solvable for all $\omega \in \Omega^r(M)$.

Proposition 0.2.2 *Let ω be a closed differential 1-form on the surface M such that ω vanishes at a finite set of points $x \in M$ (which may be empty). Then ω defines an orientable foliation $\mathcal{F} = (M, \pi^t, \theta)$ on M which is area-preserving (that is, $\mathcal{F}$ preserves a volume form given on M). Moreover, if ω is exact, then $\mathcal{F}$ is induced by the level set of a smooth function $f : M \to I\!R$.*

Proof. Locally, every closed 1-form $\omega \in \Omega^r(M)$ is exact and the equation $df = \omega$ is solvable. The generating flow $\pi^t : M \times I\!R \to M$ is defined by a vector field $v : M \to TM$, which satisfies the equation $L_v f = 0$.

Let us show that π^t is volume-preserving. Denote by Ω a volume form on the manifold M. Ω is a closed 2-form defined on M, and it follows from $L_v f = 0$ that $i_v(\Omega) = df$, where i_v and d are the inner and the exterior differentiation in the space of forms. Applying the homotopy formula $L_v = d \circ i_v + i_v \circ d$ to Ω, one obtains $L_v \Omega = di_v(\Omega) = d(df) = 0$. Therefore, π^t preserves the volume form Ω.

In case ω is exact, the cohomology class $H^1(M, \mathbb{Z})$ of M is trivial and there exists a global solution $f : M \to I\!R$ of the equation $df = \omega$. Clearly, $\mathcal{F}$ is induced by the level set $f^{-1}(y)$ of a smooth function f. □

The closed 1-forms which are not exact are especially interesting. They define the orientable foliations with a non trivial structure. Let $\omega \in \Omega^r(M)$ be a closed 1-form on the surface M. Consider a mapping Per $\omega : H_1(M,\mathbb{Z}) \times H^1(M,\mathbb{Z}) \to I\!R$, defined on the simple closed curves $C \subset M$ and the closed 1-forms ω, which acts by the formula

$$(C,\omega) \mapsto \int_C \omega.$$

By the *rank* of ω one understands the rank of the mapping *Per* ω. Let M be a compact orientable surface of genus g. It is known that $0 \leq \text{Rank } \omega \leq 2g$. In particular, when *Rank* $\omega = 2g$ every leaf of $\mathcal{F}$ is dense in M.

Example 0.2.3 *Let $T^2 = S^1 \times S^1$ be a 2-dimensional torus. Points of T^2 are parametrized by (ϕ,ψ), where $\phi,\psi \in S^1$ are polar coordinates on T^2. Consider a non singular foliation $\mathcal{F}$ given by the closed 1-form $\omega = d\phi + \alpha d\psi$, where $\alpha \in I\!R$ is a real number.*

(a) *Suppose $\alpha \in \mathbb{Q}$ is rational number. Then ω induces a foliation $\mathcal{F}$, whose leaves are simple closed curves filling up the 2-dimensional torus. This is a rational foliation on T^2 and Rank $\omega = 1$.*
(b) *Let $\alpha \in I\!R \backslash \mathbb{Q}$ be irrational. Then all leaves of $\mathcal{F}$ are dense on T^2. In this case Rank $\omega = 2$, which is the maximal possible for the torus. $\mathcal{F}$ is called irrational foliation on T^2.*

0.2.3 Line Elements

Take a C^r vector field $v : M \to TM$ on the manifold M and wipe off the arrows on every $v(x)$ associated with the point $x \in M$. What one obtains will be a field of segments ($\equiv$ *line elements*), the length and the direction of which vary slowly from point to point. In fact, it is easier to think about the bivectors, i.e., pairs $(v(x), -v(x))$ of equal and opposite directed vectors associated to every $x \in M$.

Let M be a 2-dimensional manifold and let $TM = \cup_{x\in M} T_xM$ be its tangent bundle. A C^r vector field on M is a C^r mapping $v : M \to TM$ which attaches to every point $x \in M$ a unique vector $v(x) \in T_xM$. In each fibre T_xM over x, one introduces an equivalence relation $E : v(x) = \pm v(x)$, which is symmetric and transitive.

Definition 0.2.1 *Denote by $BM = \cup_{x\in M}(T_xM/E)$ a bundle of bivectors over the manifold M. By the C^r field of line elements one understands a C^r mapping $w : M \to BM$ which associates to every point $x \in M$ a pair of vectors $(v(x), -v(x))$ in T_xM.*

A point $p \in M$ of a line element field $w : M \to BM$ is called *singular* if $w = (v, -v)$ vanishes at p, that is, $||v(p)|| = 0$. Let $p \in \text{Sing } w$ be a singular point of the line element field $w : M \to BM$. Suppose that γ is a simple closed curve bounding a disc region $G \subset M$ such that $\overline{G} \cap \text{Sing } w = p$. Let $w(\gamma) = \{(v(x), -v(x)) | x \in \gamma\}$ be the restriction of w to γ. Fix a point $x_0 \in \gamma$

and fix one of two vectors in the pair $(v(x_0), -v(x_0))$. Singular point p is called *non orientable* if, moving along γ, one returns to x_0 with the opposite sign of the chosen vector. Otherwise, the point $p \in \text{Sing } w$ is called *orientable*. Clearly, the orientability of singular points does not depend on the particular choice of γ and x_0.

Proposition 0.2.3 *Let $w : M \to BM$ be a field of line elements on the surface M. Then w induces a foliation (N, π^t, θ) on a surface N such that $M = N/\theta$. Conversely, each triple (N, π^t, θ) yields a line element field on the surface $M = N/\theta$.*

Proof. Suppose $w : M \to BM$ is a line element field with $\text{Sing } w = P \cup Q$, consisting of $2k$ non orientable singularities $p \in P$ and l orientable singularities $q \in Q$ (by the index argument which is omitted here, the number of $p \in P$ is always even). Let N be a 2-fold covering surface for the surface M, with the ramification of index 2 over the points $p \in P$. Let $\theta : N \to M$ be a natural projection. Clearly, for every point $x \in M \backslash P$, $\theta^{-1}(x)$ consists of two distinct points $y_1 \cup y_2 \in N$. If now $(v(x), -v(x))$ is a bivector associated to x, then one carries $v(x)$ to y_1 and $-v(x)$ to y_2, so that in the projection to M, they yield again the initial bivector. A vector field $v : N \to TN$, obtained on the surface N, has $2(l+k)$ singularities and the corresponding flow $\pi^t : N \times I\!R \to N$ fits the conditions of Definition 0.1.2. Thus, $\mathcal{F} = (N, \pi^t, \theta)$ is the required foliation.

Let $\mathcal{F} = (N, \pi^t, \theta)$ be a foliation on a surface N. In the projection to $M = N/\theta$, one obtains a (non orientable) foliation $F = \mathcal{F}(\pi^t)/\theta$. For all points $x \in M \backslash \text{ Sing } \mathcal{F}$, there is a pair $(v(x), -v(x))$ of unit vectors tangent to the leaves of foliation F. Consider a continuous non negative function $f : M \to I\!R$ such that $f(x) = 0$ if and only if $x \in \text{Sing } F$. The set of pairs $(f(x)v(x), -f(x)v(x))$ yields us the required line element field $w : M \to BM$ on the manifold M. □

0.2.4 Curvature Lines

Denote by S a 2-dimensional compact orientable manifold and let E^3 be the 3-dimensional space endowed with the usual Euclidean metric. Suppose that $I(S, \mathrm{E}^3)$ is the space of immersions of S into E^3. Every $i \in I$ gives rise to a special pair of the orthogonal non orientable foliations $\mathcal{F}_1$ and $\mathcal{F}_2$. These are formed by the curves tangent to the directions of principal curvatures, $k_1(x)$ and $k_2(x)$, at each point $x \in S$. (The principal curvatures are the maximal and the minimal normal curvatures associated to the point x.)

At the points where $k_1(x) = k_2(x)$ the principal directions are undefined, and such points are called *umbilic*. Clearly, $\text{Sing } \mathcal{F}_1 = \text{Sing } \mathcal{F}_2 = \{x \in S | k_1(x) = k_2(x)\}$. Apart from the umbilical points, $\mathcal{F}_1 \perp \mathcal{F}_2$, so that $\mathcal{F}_1$ is easily recovered from $\mathcal{F}_2$, and vice versa. Further we fix one of the two foliations on S and denote it by $\mathcal{U}$: $\mathcal{U} = \mathcal{F}_1$ or $\mathcal{U} = \mathcal{F}_2$. In a local chart

$U \subset \mathbb{R}^2$, $\mathcal{U}$ is tangent to a 'direction element' defined by the differential 2-form

$$\Omega = A(u,v)du^2 + 2B(u,v)dudv + C(u,v)dv^2.$$

The real-valued functions $A, B, C : \mathbb{R}^2 \to \mathbb{R}$ are uniquely defined by the coefficients of the first and the second fundamental forms of the surface. Dupin's Theorem [275] implies that at every $x \in S, A(u,v) + C(u,v) \equiv 0$. The umbilical points are precisely the points where

$$A(0,0) = B(0,0) = C(0,0) = 0.$$

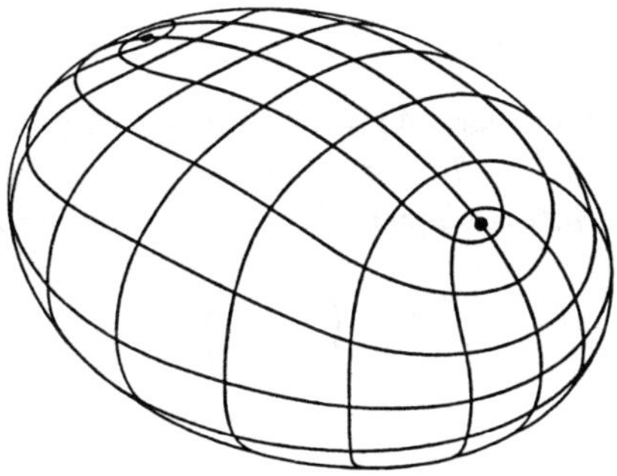

Fig. 0.1. Lines of the principal curvature.

Example 0.2.4 (Hilbert-Cohn Vossen) *Let $S^2 \hookrightarrow \mathrm{E}^3$ be an embedding of the topological sphere S^2 into the Euclidean space E^3 given by the equation*

$$\frac{x^2}{a^2} + \frac{y^2}{b^2} + \frac{z^2}{c^2} = 1, \quad a, b, c \in \mathbb{R}.$$

Consider the following three surfaces, which can be regarded as the result of continuous deformation of S^2 in E^3:

Surface	*Parameters*	*Number of umbilics*	*Type of umbilics*
Ideal sphere	$a = b = c$	∞	*non-isolated*
Geoid	$a = b \neq c$	*2*	*nodal points*
Ellipsoid	$a \neq b \neq c$	*4*	*thorns*

(The curvature lines of the ellipsoid are shown in Fig. 1.)

0.2.5 A–Diffeomorphisms

Let $f : M \to M$ be a diffeomorphism acting on a compact manifold M. The diffeomorphism $f \in \text{Diff}\ (M)$ is called *Anosov (A–) diffeomorphism* if the tangent bundle TM can be decomposed into the (continuous) Whitney sum of two subbundles E^s and E^u,

$$TM = E^s \oplus E^u,$$

such that for a Riemannian metric $|| \bullet ||$ on TM, one can find reals $C > 0$ and $\lambda > 1$, such that, for all positive $n \in \mathbb{Z}$,

$$||Tf^n v|| \leq C\lambda^{-n}||v||, \quad v \in E^s$$

$$||Tf^n u|| \geq C\lambda^n||u||, \quad u \in E^u.$$

Proposition 0.2.4 (Anosov-Arnold-Thom) *Consider $\mathbb{R}^2$ as a universal covering of the torus $T^2 \simeq \mathbb{R}^2/\mathbb{Z}^2$. Suppose that $A : \mathbb{R}^2 \to \mathbb{R}^2$ is a linear mapping given by an integer hyperbolic unimodular matrix, that is,*

$$\mathrm{A} = \begin{pmatrix} a & b \\ c & d \end{pmatrix},$$

where:

(i) a, b, c and d are integer numbers;
(ii) $ad - bc = \pm 1$;
(iii) the eigenvalues of A *does not lie at the unit circle in* $\mathbb{C}$.

Then $A : \mathbb{R}^2 \to \mathbb{R}^2$ generates Anosov diffeomorphism $f : T^2 \to T^2$ on the torus.

Proof. By the condition (i), the mapping $x \mapsto Ax$ preserves the lattice $\mathbb{Z} \times \mathbb{Z}$, so the mapping $f : T^2 \to T^2$ is defined correctly. Condition (ii) guarantees A is invertible and A^{-1} is given by an integer matrix. This implies that f is a diffeomorphism. And, finally, condition (iii) yields the splitting $E^s \oplus E^u$, so that all requirements for an Anosov diffeomorphism are satisfied for the Euclidean metric on T^2. □

Theorem 0.2.1 *The periodic points of an A–diffeomorphism $f : T^2 \to T^2$ are everywhere dense in T^2. Moreover, the eigenspaces E^s and E^u are orthogonal and have an irrational slope on T^2.*

Proof. Let $A : \mathbb{R}^2 \to \mathbb{R}^2$ be a linear mapping satisfying conditions (i)-(iii). We confine ourselves to the case det A $= 1$. The point $x \in \mathbb{R}^2$ is periodic with the period n if and only if $A^n x - x \in \mathbb{Z}^2$ (in other words, x is a *fixed* point of the linear mapping A^n). The set of all n-periodic points is denoted by

$$X_n = (A^n - Id)^{-1}(\mathbb{Z}^2).$$

The set of all periodic points in $\mathbb{R}^2$ is covered by the subgroup $X_\infty = \cup_{n=1}^{\infty} X_n$ of the additive abelian group $\mathbb{R}^2$. All closed subgroups of the abelian group $\mathbb{R}^2$ are well known. They can be either: (i) discrete, or (ii) a discrete family of parallel lines, or trivially, (iii) all of $\mathbb{R}^2$.

The following proposition rules out the possibility (i).

Lemma 0.2.1 *The subgroup X_∞ contains a convergent (fundamental) sequence.*

Proof. Notice first that if $\lambda > 1$ is the eigenvalue of A, then $1/\lambda < 1$ is another eigenvalue of A (this follows from the condition det A = 1). Denote by y_λ and $y_{\frac{1}{\lambda}}$ the eigenvectors which correspond to λ and $1/\lambda$. It is easy to see that $\mathrm{A}^n y_\lambda = \lambda^n y_\lambda$ and $\mathrm{A}^n y_{\frac{1}{\lambda}} = \frac{1}{\lambda^n} y_{\frac{1}{\lambda}}$, so that the eigenvalues of the linear operator $A^n - Id$ are $\lambda^n - 1$ and $(1-\lambda^n)\lambda^{-n}$. The eigenvalues of the inverse operator $(A^n - Id)^{-1}$ are given by

$$\mu_n = \frac{1}{\lambda^n - 1}, \qquad \nu_n = \frac{\lambda^n}{1-\lambda^n}.$$

Suppose now that (z_1, z_2) is a standard basis of the lattice $\mathbb{Z}\times\mathbb{Z}$. For $i = 1,2$ it is possible to decompose $z_i = \alpha_i y_\lambda + \beta_i y_{\frac{1}{\lambda}}$. The lattice X_n is generated by the vectors

$$(A^n - Id)^{-1} z_i = \mu_n \alpha_i y_\lambda + \nu_n \beta_i y_{\frac{1}{\lambda}}.$$

It remains to notice that $\lim_{n\to\infty} \mu_n = 0$ and $\lim_{n\to\infty} \nu_n = -1$. So $\lim_{n\to\infty}(A^n - Id)^{-1} z_i = -\beta_i y_{\frac{1}{\lambda}} \in X_\infty$, and the lemma follows. □

It remains to consider the cases (ii) and (iii). The lattice X_∞ contains an A-invariant subgroup of $\mathbb{R}^2$. This is an eigenspace of the linear operator A, and case (iii) is excluded. Hence, the A-invariant subgroup coincides with one of the eigenspaces E^s or E^u of A.

Let us show that $E^s(E^u)$ has the irrational slope α in $\mathbb{R}^2$. If, to the contrary, $\alpha \in \mathbb{Q}$, then one obtains a circle. By the property (iii) of Anosov diffeomorphisms, A must uniformly contract on $E^s \simeq S^1$ and expand on $E^u \simeq S^1$. Both cases are impossible for S^1. Therefore E^s and E^u are of irrational slope in $\mathbb{R}^2$. This corresponds to the everywhere dense leaf $E^s(E^u)$ winding on the torus and consisting of the periodic points of f. Thus $\overline{X}_\infty = T^2$ and the set Per f is dense in T^2. □

Let $f : M \to M$ be a homeomorphism of a compact surface M. Then f is called *pseudo-Anosov* if the following conditions are satisfied:

(i) there exists a pair of f-invariant foliations $\mathcal{F}^u$ and $\mathcal{F}^s$ with the common non empty set of singularities $\sigma_1, ..., \sigma_r$ which are k-prong saddles, $k \geq 3$;
(ii) $\mathcal{F}^u$ and $\mathcal{F}^s$ are transversal everywhere except at the points $\sigma_1, ..., \sigma_r$;
(iii) $\mathcal{F}^u$ and $\mathcal{F}^s$ are *measured* foliations with the transversal measures μ_u and μ_s;

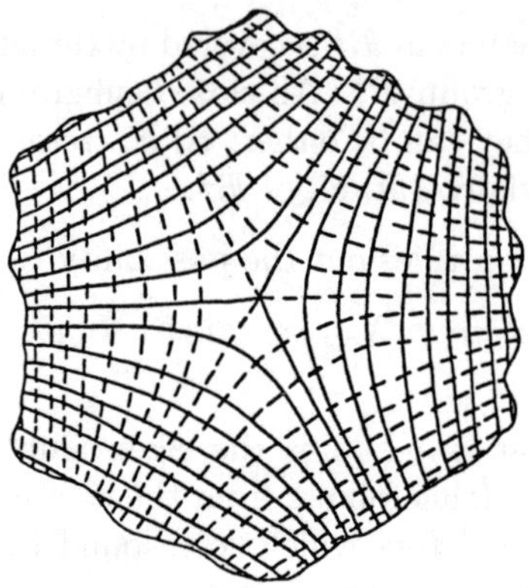

Fig. 0.2. Foliations $\mathcal{F}^u$ and $\mathcal{F}^v$ near a tripod singularity.

(iv) there exists a real number $\lambda > 1$, such that, for every arc $\alpha \subset M$, it is true that $\mu_u(f(\alpha)) = \lambda^{-1}\mu_u(\alpha)$ and $\mu_s(f(\alpha)) = \lambda\mu_s(\alpha)$;
(v) f is a diffeomorphism everywhere except at the points $\sigma_1, ..., \sigma_r$.

A typical configuration of leaves of foliations $\mathcal{F}^u$ and $\mathcal{F}^s$ in the vicinity of a k-prong saddle σ is shown in Fig. 2 (case $k = 3$).

0.3 Constructions

0.3.1 Suspension

Let $P = \{p_1, ..., p_n\}$ be a finite set of points, lying on the unit circle S^1. This yields a partition $v = \{v_1, ..., v_n\}$ of S^1 into a disjoint union $\cup_{i=1}^n v_i = S^1 \backslash P$ of open intervals. Let π be a permutation of the index set $N = \{1, ..., n\}$ and let $\varepsilon = (\varepsilon_1, ..., \varepsilon_n)$ be a vector with the coordinates $\varepsilon_i = \pm 1, i \in N$.

Definition 0.3.1 *The interval exchange transformation is a mapping $d(v, \pi, \varepsilon) : S^1 \to S^1$ which is continuous on every v_i, rearranges v_i according to the permutation π, and preserves or reverses orientation on v_i along with the sign of ε_i. If $\varepsilon_i = 1$ for all $i \in N$, the interval exchange transformation is called oriented. Otherwise, the interval exchange transformation is said to have flips.*

Let $d : S^1 \to S^1$ be an interval exchange transformation. By Susp d one understands a foliation $\mathcal{F}$ on a surface M such that d is the induced mapping on its global cross-section S^1. Suspensions over the interval exchange transformations are not unique. But they can be made unique if by fixing the set Sing (Susp d). Below we show the possible constructions for the various types of interval exchange transformations.

(i) The oriented interval exchange transformations. Let $d(v, \pi, \varepsilon)$ be an oriented interval exchange transformation acting on $S^1 \backslash P = \cup_{i=1}^{n} v_i$ by the formula $v_i \mapsto v_{\pi(i)}$. Take an irrational foliation on T^2 induced by a flow ϕ^t, and delete points $p \in P$ on S^1 obtaining partition v given by d. Since every permutation π can be produced by a finite number of elementary transpositions $i \leftrightarrow \pi(i)$, it is sufficient to show how to modify ϕ^t to obtain these transpositions. This can be done as follows.

Denote by $\phi^t(v_i)$ and $\phi^t(v_{\pi(i)})$ the stripes which are trajectories of intervals v_i and $v_{\pi(i)}$ under the action of the flow ϕ^t. We modify ϕ^t inside these stripes, bringing them to a pair of Cherry cells. Next one blows up stable and unstable nodes inside the Cherry cells, gluing up a handle linking two holes. The modified flow ϕ_1^t has two extra 4-saddles and acts on a surface of genus 2. Clearly, ϕ_1^t permutes v_i and $v_{\pi(i)}$, leaving other intervals untouched. Proceeding similarly in the other cases, one obtains a flow ϕ_k^t on an orientable surface of genus k. The transitive flow $\psi^t = \phi_k^t$ with $2k$ saddle points is called *a suspension* over the oriented interval exchange transformation $d(v, \pi, \varepsilon) : S^1 \to S^1$.

(ii) The interval exchange transformation with flips. Let $d : S^1 \to S^1$ be an interval exchange transformation with at least one flip. This means, that for some $i \in N$ the mapping $v_i \mapsto v_{\pi(i)}$ reverses orientation of v_i. It was shown previously how to exchange $i \leftrightarrow \pi(i)$ keeping their orientations. It remains to modify the oriented case to take care of the non-oriented one. We do it for a pair of nearby intervals (this can always be done as follows from item (i)). Suppose v_i and v_{i+1} are such intervals and let $\psi^t(v_i \cup v_{i+1})$ be the corresponding stripe, filled with the orbits of ψ^t, as constructed in (i). Clearly, the mapping $v_i \mapsto v_{i+1}$ reverses the orientation of v_i, leaving the other intervals untouched. Proceeding similarly, it is possible to realize all reversions in d, at the cost of appearance of extra (non orientable) singularities of *thorn* (1-saddle) and *tripod* (3-saddles) types. The resulting transitive non orientable foliation $\mathcal{F}$ is called a *suspension* over the interval exchange transformation with flips.

Example 0.3.1 *Let $d : S^1 \to S^1$ be an oriented interval exchange transformation of unit circle S^1, decomposed into $v_1 \cup v_2 \cup v_3 = S^1 \backslash \{0, 1/3, 2/3\}$, and let π act by the formula $v_1 \mapsto v_1, v_2 \mapsto v_3, v_3 \mapsto v_2$. Suppose also that Sing (Susp d) consists of a 6-separatrix saddle. In this case the realization manifold for Susp d is a pretzel $T^2 \# T^2$.*

0.3.2 Measured Foliations

Measured foliations are suspensions over the interval exchange transformations which preserve the ergodic measure on the intervals and such that Sing Susp consists of the p-saddles, $p \geq 3$. This measure may be not unique (strictly ergodic), but it is known that the number of such measures is finite (provided the number of intervals is finite). Below measured foliations are defined via the closed 1-forms, which is more elegant and is due to Thurston [279], [100], Hubbard and Masur [147].

Definition 0.3.2 (Hubbard-Masur-Thurston) *Let M be a compact C^∞ surface of genus $g > 1$ without boundary. A measured foliation $\mathcal{F}$ on M with singularities of order $k_1, ..., k_n$ at points $x_1, ..., x_n$ is given by an open cover U_i of $M \backslash \{x_1, ..., x_n\}$ and non-vanishing C^∞ real valued closed 1-form ϕ_i on each U_i, such that*

(i) $\phi_i = \pm\phi_j$ *on* $U_i \cap U_j$;
(ii) at each x_i there is a local chart $(u, v) : V \to \mathbb{R}^2$ such that for $z = u + iv$, $\phi_i = Im\ (z^{k_i/2}dz)$ on $V \cap U_i$, for some branch of $z^{k_i/2}$ in $U_i \cap V$.

The set of pairs (U_i, ϕ_i) is called an atlas for $\mathcal{F}$.

As it follows from the definition, apart from the singular points, measured foliations look like a non singular volume-preserving flows. In singularities, the substitution $z \mapsto re^{i\psi}$ brings ϕ_i, mentioned in (ii), to the form

$$\phi_i = r^{\frac{k_i}{2}} [\sin(\frac{k_i}{2} + 1)\psi dr + r\cos(\frac{k_i}{2} + 1)\psi d\psi].$$

It can be readily established that ϕ_i are closed differential 1-forms, that is, $d\phi_i = 0$ for all $k_i \geq 1$. To see what kind of singularities are generated by the above formula, let us consider a vector field v_i, given by the system of differential equations

$$\frac{dr}{dt} = -r\cos(\frac{k_i}{2} + 1)\psi, \qquad \frac{d\psi}{dt} = \sin(\frac{k_i}{2} + 1)\psi.$$

Clearly, v_i is tangent to a foliation given by the equation $\phi_i = 0$. Our primary interest is to study the behavior of trajectories of v_i in a narrow stripe $\Pi = \{(r, \psi) | -\varepsilon \leq r \leq \varepsilon, 0 \leq \psi \leq 2\pi\}$. There are exactly $k_i + 2$ equilibria $p_n \in \Pi$, which have the coordinates $(0, \frac{2\pi n}{k_i+2})$, where $n \in \mathbb{N}$ varies from 0 to $k_i + 2$. The linerization of the vector field v_i at these points yields

$$A(p_n) = \begin{pmatrix} (-1)^{n+1} & 0 \\ 0 & (-1)^n(\frac{k_i}{2} + 1) \end{pmatrix}.$$

Therefore all of the p_n are saddle points. One maps the half-strip $r \geq 0$ to the neighbourhood of the singular point x_i. Generally, a singular point x_i of the order k_i is a $(k_i + 2)$-prong saddle of a measured foliation $\mathcal{F}$.

Let M be a compact surface and let $\mathcal{F}$ be a measured foliation on M. By a *measure* μ on $\mathcal{F}$ one understands a line element $||\phi||$ associated to the point $x \in M$ induced in each $U_i \ni x$ by $||\phi_i(x)||$. It measures a 'transversal length' of $\mathcal{F}$, since μ vanishes in direction tangent to the leaves of $\mathcal{F}$.

Take a global cross-section S^1 of the measured foliation $\mathcal{F}$. $\mathcal{F}$ induces an interval exchange transformation $T : S^1 \to S^1$ on the unit (with respect to the measure μ) circle S^1. One can always suppose that T is *oriented*, passing otherwise to the measured foliation $\widetilde{\mathcal{F}}$ on a ramified double covering surface $\widetilde{M}$. Let $n \geq 2$ be a positive integer and let $\alpha = (\alpha_1, ..., \alpha_n)$ be a probability (i.e. $\alpha_1 + ... + \alpha_n = 1$) vector with positive components α_i. One sets

$$\beta_0 = 0, \qquad \beta_i = \sum_{j=1}^{i} \alpha_j, \qquad v_i = [\beta_{i-1}, \beta_i) \subset S^1.$$

If π is a permutation associated with the interval exchange transformation $T(v, \pi, \varepsilon)$, then one defines a probability vector

$$\alpha^\pi = (\alpha_{\pi^{-1}(1)}, \alpha_{\pi^{-1}(2)}, ..., \alpha_{\pi^{-1}(n)}),$$

so that the corresponding β_i^π and $v_i^\pi, 1 \leq i \leq n$, can be constructed. The transformation T, defined by the measured foliation $\mathcal{F}$, acts on S^1 by piecewise order-preserving isometries of intervals v_i:

$$Tx = x - \beta_{i-1} + \beta^\pi_{\pi(i)-1}$$

for every $x \in S^1$ and every $1 \leq i \leq n$.

There is a special class of interval exchange transformations defined by measured foliations. Let π be a permutation associated with $T : S^1 \to S^1$. Say that π is *irreducible* if, for each $1 \leq j \leq n-1$ $\pi(1, 2, ..., j) \neq (1, 2, ..., j)$ holds. An irreducible interval exchange transformation T is called *irrational* if the only rational relation between numbers $\alpha_1, ..., \alpha_n$ is given by the equality $\alpha_1 + ... + \alpha_n = 1$. Clearly, measured foliations define irrational transformations of the circle.

The measure on $\mathcal{F}$ is not strictly (or uniquely) ergodic. The following statement due to M. Keane [163] gives an estimate of the number of invariant measures.

Theorem 0.3.1 (Keane). *Let $T : S^1 \to S^1$ be an irrational interval exchange transformation of the unit circle. Then there are at most finitely many ergodic invariant probability measures on S^1.*

Proof. Suppose that there are n intervals. Denote by $D^\infty = \cup_{i=1}^{n-1} O(\beta_i) \cup \{1\}$ the union of orbits $O(\beta_i) = \{T^k(\beta_i) | k \in \mathbb{Z}\}$ of the endpoints of the intervals v_i. It follows from the irrationality of T that D^∞ is dense in S^1. Moreover, each interval $I \subset S^1$ with the endpoints outside of D^∞ splits into the subintervals $I_1, ..., I_l$ with $l \leq n+2$ in such a way that the intervals

$$I_1, T(I_1), ..., T^{k_j - 1}(I_1), ..., I_l, T(I_l), ..., T^{k_j-1}(I_l)$$

are pairwise disjoint and cover S^1. Consider a σ-field $B(I)$ on S^1 with the atoms $\mathcal{J} = \{T^k(I_j) | 1 \leq j \leq l, 0 \leq k \leq k_j\}$ and let μ be a T-invariant measure on S^1.

Our main purpose now is to show that if $A_1, ..., A_m$ is a T-invariant partition [1] of S^1, then m is finite. Indeed, set $B_i = \{J \in \mathcal{J} | \mu(A_i \cap J) \geq (1-\varepsilon)\mu(J)\}$ for $1 \leq i \leq m$ and $B_0 = S^1 \setminus \cup_{i=1}^m B_i$. Then $B_0, B_1, ..., B_m$ is a $B(I)$-measurable partition of S^1. Clearly, $\mu(A_i \triangle B_i) < \varepsilon$ and $\mu(B_0) < m\varepsilon$ for I chosen small enough. If $T^k(I_j) \subseteq B_i$, then by the T-invariance of A_i one estimates

$$\mu(A_i \cap T^s I_j) = \mu(A_i \cap T^k I_j) > (1-\varepsilon)\mu(T^k I_j) = (1-\varepsilon)\mu(T^s I_j)$$

for any $0 \leq s \leq k_j$, so that $T^s(I_j) \subseteq B_i$. Because B_i is non-empty and contains at least one $T^s(I_j)$, it follows that $m \leq l \leq n+2$, so that m is *finite*.

By the assumption, $T(A_i) = A_i$ for $1 \leq i \leq m$ and $\cup_{i=1}^m A_i = S^1$, so that

$$\mu_i(A_j) = \begin{cases} 1, & \text{if } i = j \\ 0, & \text{otherwise} \end{cases}$$

are *ergodic components* of T. Setting $\mu = \frac{1}{m}(\mu_1 + ... + \mu_m)$ yields us a T-invariant measure on S^1. Thus the total number of the invariant measures is finite. □

The theorem above says that the number of distinct ergodic measures on foliations with a finite set of singularities is at most finite. More exact estimates obtained by W. Veech [286] and E. A. Sataev [259]. The proof of the following statement is omitted (it can be found in [259]).

Theorem 0.3.2 (Sataev). *Let n and k be a pair of natural numbers such that $n \geq k$ and let M be a compact orientable surface. Then there exists a C^∞ orientable measured foliation $\mathcal{F}$ on M which has exactly k invariant ergodic measures and $2n-2$ singular points, all of which are the 4-separatrix saddles.*

As it was conjectured by M. Keane and proved by H. Masur [200] and W. Veech [287], 'almost all' (in the sense of natural measure in the space of foliations) measured foliations have a unique (strictly ergodic) invariant measure, which is a multiple of their Lebesgue measure.

0.3.3 Affine Foliations

Let $\alpha = (\alpha_1, ..., \alpha_n)$ be a probability vector, let $v_i = [\beta_{i-1}, \beta_i)$ be a semiinterval in S^1 with the endpoints in β_i defined as in the previous subsection, and let π be a permutation on the set $N = \{1, 2, ..., n\}$. Take α^π, β^π and v_i^π as earlier, and consider the piecewise linear mapping

[1] That is $T(A_i) = A_i$ for each $0 \leq i \leq m$ and $\cup_{i=1}^m A_i = S^1$

$$T_a x = a_i x - \beta_{i-1} + \beta^{\pi}_{\pi(i)-1},$$

where $a_i > 0$ are positive reals corresponding to each $v_i \subset S^1$.

Definition 0.3.3 *Let $T_a : S^1 \to S^1$ be a piecewise linear mapping defined on the unit circle S^1. By affine foliation $\mathcal{F}_a$ on a manifold M, one understands a suspension $\mathcal{F}_a = Susp\ T_a$ over the piecewise linear mapping T_a.*

It is clear that an affine foliation $\mathcal{F}_a$ is measured if and only if $a_i \equiv 1$ for all $i \in N$.

0.3.4 Labyrinths

Let us first define certain foliations of the half disc

$$D^+ = \{(x, y) | x^2 + y^2 \leq 1, y \geq 0\}.$$

Denote by $\mathcal{F}_0$ a foliation of D^+ topologically equivalent to the foliation by half circles $x^2 + y^2 = C$, with $\mathcal{F}_0$ orthogonal to the x-axis except at the singularity $x = y = 0$ (the vertical segments $\{-1/8 \leq y \leq 0, -1 \leq x \leq 1\}$ extend $\mathcal{F}_0$ so that $x = y = 0$ is a *thorn*). Consider a foliation box in which the leaves of $\mathcal{F}_0$ are vertical and whose bottom is on the x-axis. If L is the leaf of $\mathcal{F}_0$ that goes to the thorn, then this modification of $\mathcal{F}_0$ is called a *spreading* of the leaf L. The new foliation has two more singularities, one thorn and one tripod.

$\mathcal{F}$ is called a *standard foliation* of D^+ if it is obtained from $\mathcal{F}_0$ by the consecutive spreading of a finite number of leaves. A *labyrinth* $\mathcal{L}$ on a horizontal segment J is a finite collection of half discs, each with a standard foliation. Certain of these discs are placed on the top of J, going from one extremity of J to the other, with no two of these half discs touching. The remaining half discs are placed on the bottom of J and also reach from one extremity to the other. If a labyrinth $\mathcal{L}$ on a segment J lying in the disc D is given, one extends it to the foliation on D as follows. One sets the tripod singularities on all half discs of $\mathcal{L}$, except two: the extreme half discs at the bottom of $\mathcal{L}$.

Suppose now that $\alpha_1, ..., \alpha_p, \beta_1, ..., \beta_q$ is a set of positive real numbers satisfying $\sum_{i=1}^{p} \alpha_i = \sum_{i=1}^{q} \beta_i = 1$ and such that

$$m_1\alpha_1 + ... + m_p\alpha_p + n_1\beta_1 + ... + n_q\beta_q \in \mathbb{Z}$$

if and only if $m_1 = ... = m_p = n_1 = ... = n_q = 0$ for any $m_i, n_i \in \mathbb{Z}$. Place p half discs on the top of the unit interval J, $||J|| = 1$, and q half discs on the bottom of J, so that their diameters coincide with $\alpha_1, ..., \alpha_p$ and $\beta_1, ..., \beta_q$. One obtains a foliation $\mathcal{F}_{\alpha\beta}$ in $D_{\alpha_1} \cup ... \cup D_{\alpha_p} \cup D_{\beta_1} \cup ... \cup D_{\beta_q}$ by concentric half circles in the interior of half discs $D_{\alpha_1}, ..., D_{\alpha_p}, D_{\beta_1}, ..., D_{\beta_q}$. Suppose that $\mathcal{L}_{\alpha\beta}$ is a labyrinth in a unit disc D obtained by a finite number of spreadings, openings, and attaching thorns in $\mathcal{F}_{\alpha\beta}$. A (measured) foliation topologically equivalent to $\mathcal{L}_{\alpha\beta}$ is called a *Rosenberg labyrinth* in the disc D.

Proposition 0.3.1 *Let $\mathcal{L}_{\alpha\beta}$ be a Rosenberg labyrinth in the disc D and let α and β be rationally independent. Denote by Sep Sing $\mathcal{L}_{\alpha\beta}$ those leaves of $\mathcal{L}_{\alpha\beta}$ which are separatrices of the singularities in the labyrinth. Then each semi-leaf of $\mathcal{L}_{\alpha\beta}\setminus$ Sep Sing $\mathcal{L}_{\alpha\beta}$ which enters through the boundary ∂D is recurrent in the interior of D.*

Proof. Since $\alpha_1, ..., \alpha_p, \beta_1, ..., \beta_q$ are rationally independent, every leaf of the foliation $\mathcal{F}_{\alpha\beta}$ is dense in $D_{\alpha_1} \cup ... \cup D_{\alpha_p} \cup D_{\beta_1} \cup ... \cup D_{\beta_q}$. Let $I_{\alpha_2}, ..., I_{\alpha_{p-1}}$, $I_{\beta_2}, ..., I_{\beta_{q-1}}$ be finite set of disjoint intervals, attached to the points $\alpha_2, ..., \alpha_{p-1}, \beta_2, ..., \beta_{q-1}$ where the openings occur. Clearly, every leaf l passing through x which belongs to anyone of the intervals is recurrent in $D_{\alpha_1} \cup ... \cup D_{\alpha_p} \cup D_{\beta_1} \cup ... \cup D_{\beta_q}$. Since there exists a one-to-one correspondence between l and the leaves $\mathcal{L}_{\alpha\beta}\setminus$ Sep Sing $\mathcal{L}_{\alpha\beta}$ which cut the boundary ∂D, the proposition follows. □

Fig. 0.3. The simplest labyrinth.

Example 0.3.2 (Rosenberg) *Let $D = D_1 \cup D_2 \cup D_3$ be a plane region composed of three adjacent half discs D_1, D_2 and D_3, with the diameters $||D_1|| = \alpha, ||D_2|| = 1$ and $||D_3|| = 1 - \alpha$. Suppose that each of D_i is foliated by the concentric half circles, and let $\mathcal{F}_\alpha$ be the resulting foliation. Define Sing $\mathcal{F}_\alpha = \{\tau_i | 1 \leq i \leq 3\}$ to be the centers τ_i of the half discs D_i. Clearly, the return mappings T_i induced by the leaves of $\mathcal{F}_\alpha$ are as follows*

$$T_1(x) = \alpha - x, \quad T_2 = 1 - x, \quad T_3 = \alpha + 1 - x,$$

where $x \in [0, 1]$. If α is irrational, $\alpha \in \mathbb{R}\setminus\mathbb{Q}$, then each leaf $\mathcal{F}_\alpha\setminus$ Sing $\mathcal{F}_\alpha$ is dense in the interior of D; to see this, consider a composition $T_1 \circ T_2 \circ T_3 \circ T_2$ which is an irrational shift $x \mapsto x + 2\alpha - 1$ in $[0, 1]$ with the identified endpoints.

Now take an irrational foliation $\mathcal{F}_\alpha$ and open D along the leaf starting at α. The opening is thined sufficiently fast to obtain a continuous labyrinth $\mathcal{L}$, and one extends it (adding a tripod singularity at the boundary ∂D) to a standard labyrinth $\mathcal{L}$ in the disc, shown in Fig. 3.

0.3.5 Gluing Together

This procedure allows to obtain new foliations from old ones. Roughly, it describes how to obtain a foliation on the connected sum $N_1 \# N_2$ of two manifolds N_1 and N_2 provided they are endowed with smooth foliations $\mathcal{F}_1$ and $\mathcal{F}_2$ which behave 'well' at their boundaries ∂N_1 and ∂N_2. (The resulting foliation on $N_1 \# N_2$ is not necessarily smooth, but can always be chosen to be such.)

Let $\mathcal{F}_1$ and $\mathcal{F}_2$ be a pair of foliations given on the surfaces N_1 and N_2, respectively. Denote by $h : \partial N_1^* \to \partial N_2^*$ a homeomorphism between the boundary components $\partial N_1^* \subseteq \partial N_1$ and $\partial N_2^* \subseteq \partial N_2$ of the boundaries ∂N_1 and ∂N_2, respectively. Suppose that $\mathcal{F}_1$ and $\mathcal{F}_2$ are either both transversal or both tangent [2] to the boundary components ∂N_1^* and ∂N_2^* where the homeomorphism h is defined.

Definition 0.3.4 *By the sum of the foliations $\mathcal{F}_1$ and $\mathcal{F}_2$ one understands a foliation $\mathcal{F}_1 \# \mathcal{F}_2$ given on the connected sum $N_1 \# N_2$ of manifolds N_1 and N_2 such that:*

- *if $\mathcal{F}_1, \mathcal{F}_2$ are transversal to $\partial N_1^*, \partial N_2^*$, then a semi-leaf through $x \in \partial N_1^*$ is prolonged to N_2 by a leaf through $h(x) \in \partial N_2^*$;*
- *if $\mathcal{F}_1, \mathcal{F}_2$ are tangent to $\partial N_1^*, \partial N_2^*$, then leaves lying on its boundaries are identified.*

Example 0.3.3 *Let A_1 and A_2 be two annuli foliated by the Reeb component having a different 'sense of rotation'. Let $h : S_1^1 \to S_2^1$ be an orientation preserving homeomorphism which maps the 'smaller' boundary component of A_1 to the 'bigger' boundary component of A_2. Clearly, if h is chosen to be orientation reversing, one will obtain the topologically different foliation on $S^1 \times I$.*

Bibliographic Notes. Numerous examples of foliations on the surfaces are gathered in the monographs of Hector & Hirsch [141] and Godbillon [119]. The foliations given by closed differential 1-forms were studied by Moussu [207], Tischler [281], Novikov [231], Imanishi [150] and others. The flows on

[2] This means that the boundary component consists of the whole leaves of foliation $\mathcal{F}$ different from the set Sing $\mathcal{F}$.

higher genus surfaces defined by closed 1-forms of maximal rank were considered by Katok [160] and Zorich [314]. The bivector fields were used by Gutierrez [124] and Kadyrov [153] as an instrument for the investigation of non orientable foliations in the plane regions. The interval exchange mappings are nicely exposed in [162]. The suspensions with the $2k$-separatrix saddles, $k \geq 2$, over the oriented interval exchange transformations have been considered by Levitt [183]. The suspensions over the interval exchange transformations with the flips were studied by Gutierrez [128].

Part I

General Theory

1. Local Theory

1.1 Introduction

Typical singularities of non orientable foliations can be classified in a similar way as typical singularities of flows. For flows "typical" means that the Jacobian of the vector field does not vanish at the singular point. Such an approach due to M. Morse and S. Smale gives satisfactory results in the case of foliations as well.

Let M be a compact surface and let $\mathcal{F} = (M, \pi, \theta)$ be a C^r foliation on M. The aim of the qualitative theory is to give classification of all C^r foliations modulo topological equivalence. The space $\mathcal{F}^r(M)$ of all C^r foliations is too big to be classified. However, if one rules out "non typical" foliations, the problem of classification can be solved. "Typical" in our case is synonymous with "the majority of $\mathcal{F}$", or "almost all $\mathcal{F}$". Intuitively, "typical" must mean "everywhere dense" in $\mathcal{F}^r(M)$, the space of all C^r foliations on M endowed with the uniform C^r topology. But this is not enough – consider the rational set $\mathbb{Q} \subset \mathbb{R}$ which is dense in $\mathbb{R}$, but is not open. A more accurate notion is the following.

Definition 1.1.1 *By a massive subset $B \subset X$ of the topological space X one understands a set which is a countable intersection of open and everywhere dense subsets of X. The property is called C^r-typical (or generic) if it holds for a massive subset $B \subset X$ in the space $X = \mathcal{F}^r(M)$ of all C^r foliations on M.*

In 1937 A. A. Andronov and L. S. Pontryagin [5] introduced the "rough systems" which today are called structurally stable systems. In the context of foliations, this property may be formulated as follows.

Definition 1.1.2 *The foliation $\mathcal{F} \in \mathcal{F}^r(M)$ is called structurally stable if there exists a neighborhood U of $\mathcal{F}$ in the space $\mathcal{F}^r(M)$ such that each foliation $\mathcal{G} \in U$ is topologically conjugate to $\mathcal{F}$.*

It turns out that structurally stable foliations on the compact and orientable surfaces are C^r–typical. This fact is not at all evident and we reserve two later chapters for its proof.

1.2 Symmetry

Let $\mathcal{F} = (M, \pi^t, \theta)$ be a non orientable C^r foliation on the compact surface M. This means that the set W is non empty. Let $p \in W$ be a fixed point of the involution θ. Take a small neighborhood U of p which does not overlap other fixed points of θ. Since a backward orbit of every point $x \in U$ maps under θ to the forward orbit of x, we have $\pi^t(x) = \pi^{-t}(\theta x)$, where $\theta : U \to U$ is the restriction of θ to the neighborhood U. If one takes the t–derivative of both sides, then $v(x) = v(\theta x)$, where $v \in T_x U$ is a vector field in U tangent to π^t. There exists a chart (u, v) in U where $\theta : U \to U$ acts by the formula $(u, v) \mapsto (-u, -v)$. (Such a lemma has been proved in [67].) In particular, the vector field

$$v(x) = P(u,v)\frac{\partial}{\partial u} + Q(u,v)\frac{\partial}{\partial v}, \quad x \in U$$

must obey the following symmetry with regard to the action of group $\mathbb{Z}_2$ generated by the involution θ:

$$P(u,v) = P(-u,-v), \quad Q(u,v) = Q(-u,-v), \tag{1.1}$$

Equations (1.1) imply that the monomials of odd degree in the Taylor expansion of v must vanish. This corresponds to the vector field given by the system of differential equations

$$\frac{du}{dt} = P_2(u,v) + \phi(u,v), \qquad \frac{dv}{dt} = Q_2(u,v) + \psi(u,v), \tag{1.2}$$

where P_2 and Q_2 are homogeneous quadratic polynomials and $\phi(u,v) = O(u^2+v^2), \psi(u,v) = O(u^2+v^2)$ are higher order monomials of even degree.

1.3 Normal Forms

1.3.1 Typical Normal Forms

It will be helpful to simplify the "linear part" of (1.2):

$$\frac{du}{dt} = a^1_{11}u^2 + 2a^1_{12}uv + a^1_{22}v^2, \quad \frac{dv}{dt} = a^2_{11}u^2 + 2a^2_{12}uv + a^2_{22}v^2. \tag{1.3}$$

For the sake of brevity we let

$$\eta = -27(a^1_{22})^2(a^2_{11})^2 - 4a^1_{22}(a^1_{11} - 2a^2_{12})^3 + 18a^1_{22}a^2_{11}(a^1_{11} - 2a^2_{12}) \times$$

$$\times (a^2_{22} - 2a^1_{12}) - 4a^2_{11}(a^2_{22} - 2a^1_{12})^3 + (a^1_{11} - 2a^2_{12})^2(a^2_{22} - 2a^1_{12})^2,$$

$$\mu = (a^1_{11}a^2_{22} - a^1_{22}a^2_{12})^2 - 4(a^1_{11}a^2_{12} - a^1_{12}a^2_{11})(a^1_{12}a^2_{22} - a^1_{22}a^2_{12}).$$

Denote by Γ a space of vector fields of the form (1.2) and let $\Gamma_0 \subset \Gamma$ be subspace in Γ satisfying the conditions $\eta \neq 0, \mu \neq 0$. Clearly Γ_0 is open and dense in the space Γ endowed with the $C^r (r \geq 2)$ topology.

Lemma 1.3.1 *The quadratic system (1.3) of the class Γ_0 can be brought by a linear non degenerate transformations $q \in GL_2(\mathbb{R})$ to one of two normal forms:*

$$\frac{du}{dt} = \alpha u^2 + (\beta - 1)uv, \quad \frac{dv}{dt} = (\alpha - 1)uv + \beta v^2, \tag{1.4}$$

$$\alpha\beta(\alpha + \beta - 1) \neq 0, \quad \alpha, \beta \in \mathbb{R}.$$

$$\frac{du}{dt} = \alpha u^2 + \beta uv - v^2, \quad \frac{dv}{dt} = (\alpha + 1)uv + \beta v^2, \tag{1.5}$$

$$\alpha[\beta^2 + (\alpha + 1)^2] \neq 0, \quad \alpha, \beta \in \mathbb{R};$$

Proof. The condition $\mu \neq 0$ means that the quadratic polynomials to the right of (1.3) are relatively prime. In other words, the system

$$\frac{du}{dv} = \frac{a_{11}^1 u^2 + 2a_{12}^1 uv + a_{22}^1 v^2}{a_{11}^2 u^2 + 2a_{12}^2 uv + a_{22}^2 v^2}$$

is irreducible. Let us consider a differential 1-form

$$\omega = (a_{11}^2 u^2 + 2a_{12}^2 uv + a_{22}^2 v^2)du - (a_{11}^1 u^2 + 2a_{12}^1 uv + a_{22}^1 v^2)dv =$$

$$= a_{11}^2 u^2 du + (2a_{12}^2 uvdu - a_{11}^1 u^2 dv) + (a_{22}^2 v^2 du - 2a_{12}^1 uvdv) - a_{22}^1 v^2 dv,$$

which vanishes on the vectors defined by (1.3). The operator d in the above formula is linear on linear transformations $q \in GL_2(\mathbb{R})$, so one can replace ω by the cubic form

$$a_{11}^2 u^3 + (2a_{12}^2 - a_{11}^1)u^2 v + (a_{22}^2 - 2a_{12}^1)uv^2 - a_{22}^1 v^3. \tag{1.6}$$

The form (1.6) is called a *comitant form* [264] of system (1.3). Its discriminant coincides with η, whose positive or negative values define a pair of normal forms

$$uv(v - u), \qquad v(u^2 + v^2).$$

These forms are "normal forms" of (1.6) by the action of the linear group $GL_2(\mathbb{R})$. They correspond to the distribution of roots in cubics: $(3, 0)$ – three reals, or $(1, 2)$ – one real and two complex conjugates. Comparing with (1.6), we have:

$$a_{11}^2 = a_{22}^1 = 0, \quad a_{22}^2 - 2a_{12}^1 = 1, \quad 2a_{11}^2 - a_{11}^1 = -1,$$

$$a_{11}^2 = a_{22}^2 - 2a_{12}^1 = 0, \quad 2a_{12}^2 - a_{11}^1 = 1, \quad a_{22}^1 = -1.$$

To obtain the normal forms of the lemma it remains to denote a_{11}^1 by α and a_{22}^2 by β, and substitute in (1.3). □

1.3.2 Degenerate Normal Forms

Let us consider the expansion (1.2) as a formal infinite sum of monomials of even degrees,

$$\frac{dx}{dt} = P_2(x,y) + \sum_{m=2}^{\infty}\sum_{l=0}^{2m} A_l^{2m} x^l y^{2m-l},$$
$$\frac{dy}{dt} = Q_2(x,y) + \sum_{m=2}^{\infty}\sum_{l=0}^{2m} B_l^{2m} x^l y^{2m-l}. \tag{1.7}$$

Lemma 1.3.2 *The quadratic system is reduced by a linear non degenerate transformation $q \in GL_2(I\!R)$ to one of the following normal forms,*

(i) in the typical (non degenerate) case to:

$$v_0^{(1)} = [\alpha x^2 + (\beta - 1)xy]\frac{\partial}{\partial x} + [(\alpha - 1)xy + \beta y^2]\frac{\partial}{\partial y},$$
$$\mu = \alpha\beta(\alpha + \beta - 1) \neq 0;$$
$$v_0^{(2)} = [\alpha x^2 + \beta xy - y^2]\frac{\partial}{\partial x} + [(\alpha + 1)xy + \beta y^2]\frac{\partial}{\partial y},$$
$$\mu = \alpha[\beta^2 + (\alpha + 1)^2] \neq 0$$

(ii) in the degenerate case of codimension 1 to:

$$v_1 = [\alpha x^2 + xy]\frac{\partial}{\partial x} + [(\alpha - 1)xy + y^2]\frac{\partial}{\partial y}, \quad \mu = \alpha \neq 0$$

(iii) in the degenerate case of codimension 2 to:

$$v_2^{\pm} = xy\frac{\partial}{\partial x} + [\pm x^2 + y^2]\frac{\partial}{\partial y}, \qquad \mu = \pm 1 \neq 0.$$

Proof. The normal forms $v_0^{(1)}$ and $v_0^{(2)}$ have been obtained earlier. The normal forms v_1 and $v_2^{\pm}$ correspond to one double and one simple, or one triple root of the cubic form (1.6). In the respective cases (1.6) can be brought to one of the two forms

$$u^2 v, \qquad u^3.$$

Comparing these forms with the cubic (1.6), one comes to the following series of conditions

$$a_{11}^2 = a_{22}^2 - 2a_{12}^1 = a_{22}^1 = 0, 2a_{12}^2 - a_{11}^1 = 1;$$

$$a_{11}^2 = 1, 2a_{12}^2 - a_{11}^1 = a_{22}^2 - 2a_{12}^1 = a_{22}^1 = 0.$$

They correspond to the following normal forms

$$\frac{dx}{dy} = \frac{\alpha x^2 + \beta xy}{(\alpha - 1)xy + \beta y^2}, \quad \frac{dx}{dy} = \frac{\alpha x^2 + \beta xy}{x^2 + \alpha xy + \beta y^2},$$

where α and β denote the coefficients a_{11}^1 and a_{22}^2. To obtain normal forms v_1 and $v_2^\pm$ it remains to substitute $y \mapsto \beta^{-1}y$ in the first case and $x\sqrt{|\beta|} \mapsto x, \alpha x + \beta y \mapsto y$ in the second case (sgn β is responsible for $\pm$ in $v_2^\pm$). □

In what follows, the quadratic monomials in system (1.7) are supposed to be reduced to one of the normal forms $v_0^{(1)}, v_0^{(2)}, v_1$ or $v_2^\pm$. The statement below shows how much (1.7) may be simplified by the formal transformation of the coordinates and time.

Theorem 1.3.1 *The formal vector field (1.7) of degeneracy of codimension ≤ 2 can be reduced to one of the following formal orbital normal forms (the lower index shows the codimension in degenerate cases),*

I. Non resonant normal forms:

$$w_0^{(1)} = [\alpha x^2 + (\beta - 1)xy]\frac{\partial}{\partial x} + [(\alpha - 1)xy + \beta y^2 + \sum_{m=2}^{\infty} C_m y^{2m}]\frac{\partial}{\partial y};$$

$$\alpha(\alpha - 1)[2\alpha(1 - n) - 1][2\beta(1 - k) - 1][|(2l - 3)\alpha + 1 - 2l| + |\beta - 1|] \neq 0$$
$$(\ \forall n, k, l \geq 2);$$

$$w_0^{(2)} = [\alpha x^2 + \beta xy - y^2 + \sum_{m=2}^{\infty} C_m y^{2m}]\frac{\partial}{\partial x} + [(\alpha + 1)xy + \beta y^2]\frac{\partial}{\partial y}$$
$$(\ \alpha[2\alpha(n - 1) - 1][|\alpha + 1| + |\beta|] \neq 0 \quad (\forall n \geq 2));$$

$$w_1^* = [\alpha x^2 + xy + \sum_{m=2}^{\infty} C_m x y^{2m-1}]\frac{\partial}{\partial x} + [(\alpha - 1)xy + y^2]\frac{\partial}{\partial y}$$
$$(\ \alpha[2\alpha(1 - n) - 1] \neq 0 \quad (\forall n \geq 2));$$

$$w_2^\pm = (xy + \sum_{m=2}^{\infty} C_m x^{2m-2} y^2)\frac{\partial}{\partial x} + [\pm x^2 + y^2]\frac{\partial}{\partial y}.$$

II. Resonant normal forms of codimension 1:

$$\widetilde{w}_1^{(1)} = w_0^{(1)} + \begin{cases} D x^{2n} \frac{\partial}{\partial y}, & \text{if } \alpha = \frac{1}{2(1-n)} \\ \sum_{m=2}^{\infty} D_m x^{2m} \frac{\partial}{\partial x}, & \text{if } \alpha = 0 \\ \sum_{m=2}^{\infty} C_m x y^{2m-1} \frac{\partial}{\partial y}, & \text{if } \alpha = 1 \\ D y^{2k} \frac{\partial}{\partial x}, & \text{if } \beta = \frac{1}{2(1-k)}; \end{cases}$$

$$\widetilde{w}_1^{(2)} = w_0^{(2)} + \begin{cases} D x^{2n} \frac{\partial}{\partial y}, & \text{if } \alpha = \frac{1}{2(n-1)} \\ \sum_{m=2}^{\infty} C_m x^{2m} \frac{\partial}{\partial x}, & \text{if } \alpha = 0. \end{cases}$$

III. Resonant normal forms of codimension 2:

$$\widetilde{w}_2^{(1)} = w_0^{(1)} + \begin{cases} Cy^{2k}\frac{\partial}{\partial x} + Dx^{2n}\frac{\partial}{\partial y}, & \text{if } \alpha = \frac{1}{2(1-n)}, \beta = \frac{1}{2(1-k)} \\ (\sum_{m=2}^{\infty} C_m x^{2m} + Dy^{2k})\frac{\partial}{\partial x}, & \text{if } \alpha = 0, \beta = \frac{1}{2(1-k)} \\ Cxy^{2l-1}\frac{\partial}{\partial x}, & \text{if } \alpha = \frac{2l-1}{2l-3}, \beta = 1; \end{cases}$$

$$\widetilde{w}_2^{(2)} = w_0^{(2)} + [\sum_{m=2}^{\infty}(C_m xy^{2m-1} + D_m y^{2m})]\frac{\partial}{\partial y}, \text{ if } \alpha + 1 = \beta = 0;$$

$$\widetilde{w}_2^{*} = w_1^{*} + \begin{cases} Cx^{2n}\frac{\partial}{\partial y}, & \text{if } \alpha = \frac{1}{2(1-n)} \\ \sum_{m=2}^{\infty} D_m x^{2m}\frac{\partial}{\partial x}, & \text{if } \alpha = 0. \end{cases}$$

Proof. Let us consider vector field (1.7) with 2-jet, reduced to the form $v_0^{(1)}$. We apply the substitution $(x,y) \mapsto (u,v)$, acting by the formula

$$u = x + \sum_{m=2}^{\infty}\sum_{l=0}^{2m-1} a_l x^l y^{2m-l-1}, \quad v = y + \sum_{m=2}^{\infty}\sum_{l=0}^{2m-1} b_l x^l y^{2m-l-1},$$

and we rescale (1.7) multiplying the right-hand sides by function

$$\mu(x,y) = 1 + \sum_{m=2}^{\infty}\sum_{l=0}^{2m-2} \mu_l x^l y^{2m-l-2}.$$

Direct computations yield the following homology equations:

$$\begin{aligned} A_l &= (3\alpha - 2m\alpha + 2m - l)a_{l-1} + (\beta - 1)b_{l-1} + \\ &+ (l - 1 - 2\beta m + 2\beta)a_l - \alpha\mu_{l-2} + (1-\beta)\mu_{l-1}, \\ B_l &= (2\alpha - 2m\alpha + 2m - l - 1)b_{l-1} + (\alpha - 1)a_l + \\ &+ (3\beta - 2\beta m + l)b_l + (1-\alpha)\mu_{l-1} - \beta\mu_l, \end{aligned} \tag{1.8}$$

where $l = 0, 1, ..., 2m$ (we suppose here $a_{2m} = b_{2m} = \mu_{2m} = \mu_{2m-1} = 0$). If the system (1.8) admits a solution for all $l = 0, 1, ..., 2m, m \geq 2$, then the vector field (1.7) is equivalent to $v_0^{(1)}$. Let us evaluate the rank of the fundamental matrix M of the linear system (1.8) of dimension $(6m-1) \times (4m+2)$. (Rank of M coincides with the number of A_l and B_l in (1.7), which can be deleted by an appropriate choice of transformation parameters a_l, b_l and μ_l).

It will be shown that Rank $M \leq 4m + 1$. This means that, even in the typical case, a normal form admits at least one functional module. We set $\mu_2 = \mu_3 = ... = \mu_{2m-2} = 0$. Next one considers a square matrix $\overline{M}$ of dimension $(4m+2)$, which is composed as follows. We fix the order $A_0, B_0, ..., A_i, B_i, ..., A_{2m}, B_{2m}$, for the lines of (1.8) and the order $a_0, b_0, a_1, b_1, a_2, \mu_0, \mu_1, b_2, a_3, b_3, ..., a_{2m-1}, b_{2m-1}$ for the unknown variables $a_0, ..., a_{2m-1}, b_0, ..., b_{2m-1}, \mu_0, \mu_1$. The resulting matrix $\overline{M}$ has a diagonal block structure, $\overline{M} = \mathrm{diag}(S, T)$. Block T is an upper diagonal square matrix of dimension $4m-5$, which corresponds to the lines $B_3, A_4, B_4, ..., A_{2m}, B_{2m}$ and the columns $b_2, a_3, b_3, ..., a_{2m-1}, b_{2m-1}$. The diagonal of matrix T consists of the elements

$$
\begin{aligned}
&3\alpha - 2m\alpha + 2m - l \qquad (l = 3, ..., 2m), \\
&2\alpha - 2m\alpha + 2m - k - 1 \quad (k = 4, ..., 2m).
\end{aligned} \tag{1.9}
$$

Block S is given by

$$
\begin{pmatrix}
\begin{matrix}2\beta-1\\-2\beta m\end{matrix} & & & & & & \\
\alpha-1 & 3\beta-2\beta m & & & & -\beta & \\
\begin{matrix}3\alpha+2m-\\2m\alpha-1\end{matrix} & \beta-1 & 2\beta-2\beta m & & & 1-\beta & \\
 & \begin{matrix}2\alpha+2m-\\2m\alpha-2\end{matrix} & \alpha-1 & \begin{matrix}3\beta+1\\-2\beta m\end{matrix} & & 1-\alpha & \\
 & & \begin{matrix}3\alpha+2m-\\2m\alpha-2\end{matrix} & \beta-1 & \begin{matrix}2\beta+1\\-2m\beta\end{matrix} & -\alpha & 1-\beta \\
 & & & \begin{matrix}2\alpha+2m-\\2m\alpha-3\end{matrix} & \alpha-1 & & 1-\alpha \\
 & & & & \begin{matrix}3\alpha+2m-\\2m\alpha-3\end{matrix} & & -\alpha
\end{pmatrix}.
$$

(i) First let us consider the general (non resonant) case in which all diagonal elements (1.9) are non zero, so that Rank $T = 4m-5$. This means that monomials $A_i x^i y^{2m-i} (i = 4, ..., 2m)$ and $B_i x^i y^{2m-i} (i = 3, ..., 2m)$ can be killed by an appropriate choice of parameters $a_3, ..., a_{2m-1}, b_2, ..., b_{2m-1}$. After elementary transformations in the matrix S (i.e., taking various linear combinations of rows and columns in S) one obtains the matrix

$$
\widetilde{S} = diag(1, R, 1, 1, 1, 1), \qquad R = \begin{pmatrix} 1 & 1 \\ 1 & 1 \end{pmatrix}.
$$

Clearly, Rank $S = 6$ which implies that one of the monomials $B_0 y^{2m}$ or $A_1 x y^{2m-1}$ in the expansion (1.7) cannot be deleted (for brevity, it is assumed to be $B_0 y^{2m}$). Therefore, one comes to the typical normal form $w_0^{(1)}$.

(ii) Let us consider the resonant cases of codimension 1 and 2. These correspond to the rational values of α. Suppose first that one of the following relations holds

$$
\alpha = \frac{2m-l}{2m-3}, \qquad \alpha = \frac{2m-k-1}{2m-2},
$$

These correspond to the resonant monomials $A_l x^l y^{2m-l}$ and $B_k x^k y^{2m-k}$ in (1.7). In the case $l \neq 2m$ and $k \neq 2m$ the corresponding monomials are killed by parameters $\mu_i \quad (i = 2, ..., 2m-2)$. If otherwise $l = 2m$ and $\alpha \neq 0$, then $A_{2m} x^{2m}$ also can be killed by an appropriate choice of the number μ_{2m-2}.

If both $l = 2m$ and $\alpha = 0$, or $k = 2m$ and $\alpha = \frac{1}{2(1-n)}$, then the monomials $A_{2m} x^{2m} (m \geq 2)$ and, correspondingly, $B_{2n} x^{2n} (n \geq 2)$ cannot be removed in

(1.7). Also β-resonance is possible. If $\beta = \frac{1}{2(1-k)}$, then one obtains the resonant monomial $A_0 y^{2k}$. A degenerate case also occurs when $\alpha = 1$: for every $m \geq 2$ $B_1 x y^{2m-1}$ cannot be removed.

Degenerate cases of codimension 2 can be obtained as combinations of the above considered cases (taking into account that relations $\alpha = \frac{2l-1}{2l-3}, \beta = 1$ yield a new resonant monomial $A_1 x y^{2l-1}$). Thus, one arrives at normal forms $w_0^{(1)}, \widetilde{w}_1^{(1)}$ and $\widetilde{w}_2^{(1)}$.

A similar argument leads to the other normal forms. All of them can be deduced from the homology equations given below:

for the normal form $w_0^{(2)}$:

$$\begin{aligned} A_l &= (3\alpha - 2m\alpha - 2m + l)a_{l-1} + \beta b_{l-1} + 2\beta(1-m)a_l - \\ &\quad - 2b_l + (l+1)a_{l+1} - \alpha\mu_{l-2} - \beta\mu_{l-1} + \mu_l, \\ B_l &= (2\alpha - 2m\alpha - 2m + l + 1)b_{l-1} + (\alpha+1)a_l + \beta(3-2m)b_l + \\ &\quad + (l+1)b_{l+1} - (\alpha+1)\mu_{l-1} - \beta\mu_l; \end{aligned}$$

for the normal form w_1^* :

$$\begin{aligned} A_l &= (3\alpha - 2m\alpha + 2m - l)a_{l-1} + b_{l-1} + 2(1-m)a_l - \alpha\mu_{l-2} - \mu_{l-1}, \\ B_l &= (2\alpha - 2m\alpha + 2m - l - 1)b_{l-1} + (\alpha-1)a_l + (3-2m)b_l + \\ &\quad + (1-\alpha)\mu_{l-1} - \mu_l; \end{aligned}$$

for the normal form $w_2^{\pm}$:

$$\begin{aligned} A_l &= \mp(2m - l + 1)a_{l-2} + b_{l-1} + 2(1-m)a_l - \mu_{l-1}, \\ B_l &= \mp(2m - l + 1)b_{l-2} \pm 2a_{l-1} + (3-2m)b_l \mp \mu_{l-2} - \mu_l \end{aligned}$$

The theorem is proved. □

Note that the normal forms are not uniquely defined, and in some cases admit further simplifications. One can show that in the resonant cases only a finite number of unremovable monomials are possible. This correlates with the general results of Bogdanov [52].

1.4 Structurally Stable Singularities

1.4.1 Blowing–up Method

The idea of the method is to decompose a singularity with a certain degree of complexity into simpler singularities. The resolving process is inductive and, in general, infinite, but if one imposes restrictions on the vector field (such as the Lojasiewicz inequality and existence of a characteristic orbit) then the process is finite. In this case the topological structure near the singularity is determined by its finite jet, and this jet is called (*topologically,* or C^0) *sufficient.* Clearly, this is yet another way to say that the singularity is structurally stable with regard to higher order perturbations.

Definition 1.4.1 *The singular point $X(0) = 0$ of the vector field X is called elementary if at least one eigenvalue of $X(0)$ does not vanish.*

All elementary singularities of vector fields in the plane are exhausted by nodes, saddles, foci or centres (two non zero eigenvalues) and saddle-nodes (one zero eigenvalue). The idea behind the following process is to decompose the non elementary singularities into a finite collection of elementary ones lying on the projective line $\mathbb{R}P^1$.

Blowing-up process. Consider a natural mapping $f : \dot{\mathbb{R}}^2 \to \mathbb{R}P^1$ of the pinched plane $\dot{\mathbb{R}}^2 \simeq \mathbb{R}^2 \backslash 0$ to the projective line $\mathbb{R}P^1$. Each point $x \in \mathbb{R}^2 \backslash 0$ corresponds to a straight line, connecting 0 with x. Denote by $\mathcal{M}$ the subset $\{(x, f(x)) | x \in \dot{\mathbb{R}}^2\}$ of $\dot{\mathbb{R}}^2 \times \mathbb{R}P^1$. The closure $\overline{\mathcal{M}} = \{(x, f(x)) | x \in \mathbb{R}^2\}$ of $\mathcal{M}$ in $\mathbb{R}^2 \times \mathbb{R}P^1$ is a Möbius band. The natural projection $\pi : \mathbb{R}^2 \times \mathbb{R}P^1 \to \mathbb{R}^2$ brings $\overline{\mathcal{M}}$ to $\mathbb{R}^2$, so that $\pi^{-1}(0)$ is a projective line $L \simeq \mathbb{R}P^1$ which is called a *pasted projective line.* The mapping $\pi : \overline{\mathcal{M}} \backslash L \to \dot{\mathbb{R}}^2$ is a diffeomorphism.

Lemma 1.4.1 *To every smooth (not ∞-flat in 0) vector field X which is not ∞-flat at 0 there corresponds a smooth field of directions, α, given everywhere in a vicinity of L on $\overline{\mathcal{M}}$, except a finite number of points of L which are called singular. The projection $\pi : \overline{\mathcal{M}} \backslash L \to \dot{\mathbb{R}}^2$ brings foliation $\mathcal{F}(\alpha)$, generated by the field α, to the foliation $\mathcal{F}(X)$ generated by X. In the vicinity of each singular point of L, the field α is given by a smooth vector field $\widetilde{X}$ which is not ∞-flat at the singularity.*

The projective transformation π acts by formulas

$$(x, y) \mapsto (x, \eta), \qquad \eta = y/x,$$

$$(x, y) \mapsto (\theta, y), \qquad \theta = x/y,$$

where (x, y) is a local chart which covers the vicinity of 0, and $(x, \eta), (\theta, y)$ is a pair of charts which cover the projective line L. Each step of the blowing-up process decomposes $X(0)$ into simpler singularities lying on L provided that $X(0)$ is 'good'.

Definition 1.4.2 *The vector field X satisfies the Lojasiewicz inequality at x_0 if*

$$|X(x)| \geq C|x - x_0|^p$$

as $x \to x_0$ for some positive reals C and p.

(The set of germs which obey the Lojasiewicz inequality is non empty and rich. In particular, such a set includes analytic germs.) The following decomposition theorem has been proved in [92].

Theorem 1.4.1 (Takens–Dumortier) *Suppose that X is a C^∞ vector field which satisfies the Lojasiewicz inequality at the singular point $X(0) = 0$. Then there exists a finite sequence of blowing-up procedures leading to a vector field $\widetilde{X}$ given in the vicinity of the projective line L such that Sing $\widetilde{X} \subseteq L$. Each $x \in$ Sing $\widetilde{X}$ is one of the following:*

(1) an isolated elementary singularity given in a standard chart (u,v) by
(i) $H_n^\pm = \pm u\partial/\partial u \pm v\partial/\partial v, \quad v \geq 0$ (hyperbolic node);
(ii) $H_s^\pm = \pm u\partial/\partial u \mp v\partial/\partial v, \quad v \geq 0$ (hyperbolic saddle);
(iii) $P^\pm = \pm v\partial/\partial v \pm u^2\partial/\partial u, \quad v \geq 0$ (saddle-node of order 2);
(2) an isolated 2-saddle, given by

$$C^\pm = \pm u\partial/\partial u \mp v\partial/\partial v, \quad u \geq 0, v \geq 0$$

(3) an element of one of the finite number of closed singular intervals

$$I^\pm = \pm v\partial/\partial v, \quad 0 \leq u \leq 1, \quad v \geq 0.$$

Furthermore, the position of Sing $\widetilde{X}$ at L and their properties (1)-(3) depends on a finite jet of X at 0.

Consider triples $(\widetilde{X}, L, W)$, where $\widetilde{X}$ and L are as above, and W is a 'word' on the alphabet $\{H_n^\pm, H_s^\pm, P^\pm, C^\pm, I^\pm\}$, which corresponds to the one-by-one position of singularities at the projective line L. Equivalence relation $E : W_1 \sim W_2$ is a cyclic permutation of symbols in words. A *'singularity scheme'* is the equivalence class of words for a relation E. The following statement, proved in [92], is a sufficient condition of C^0-equivalence.

Theorem 1.4.2 (Takens–Dumortier) *If $(\widetilde{X}, L_1, W_1)$ and $(\widetilde{Y}, L_2, W_2)$ are two triples defined as above, and if they have the same 'singularity scheme', $W_1 \sim W_2$, different from the word $C_1^\pm, ..., C_n^\pm$, then there exists a homeomorphism h from the neighbourhood U of the singularity $X(0)$ of the vector field X to the neighbourhood V of the singularity $Y(0)$ of the vector field Y which conjugates the orbits of X and Y near 0, preserving their orientation.*

The claim that the 'singularity scheme' W is not $C_1^\pm, ..., C_n^\pm$ can be replaced by the assumption that $X(0)$ has a characteristic orbit. [1]

For the proofs of Theorems 1.4.1 and 1.4.2 the reader is referred to [92]. □

1.4.2 Fundamental Lemma

Lemma 1.4.2 *Let H, P and E denote hyperbolic, parabolic and elliptic sectors, respectively, in a given singularity scheme. The topological phase portraits of vector fields $\xi \in \Gamma_0$ in the neighbourhood of the origin are of one of the following five types:*

[1] By a characteristic orbit one understands a semi-orbit $O^\pm(x) = \{X^t(x) | t \in [0, \pm\infty)\}$ which enters $X(0)$ with a definite tangent direction as $t \to \pm\infty$.

(i) HPHP, (ii) HHHHHH, (iii) PEPE, (iv) HH and (v) EE.

Moreover, they are precisely the vector fields which are C^0-structurally stable in the neighborhood of the origin with regard to the perturbations from Γ.

Proof. We begin with the system

$$\begin{aligned} \frac{dx}{dt} &= \alpha x^2 + \beta xy - y^2 + \phi(x,y), \\ \frac{dy}{dt} &= (\alpha+1)xy + \beta y^2 + \psi(x,y), \end{aligned} \tag{1.10}$$

$$\mu = \alpha[\beta^2 + (\alpha+1)^2] \neq 0, \quad \alpha, \beta \in \mathbb{R}.$$

After the first step of blowing-up, the system (1.10) in the local chart (x,η) where $\eta = y/x$, assumes the form

$$\begin{aligned} \frac{dx}{dt} &= \alpha x + \beta x\eta - x\eta^2 + \frac{1}{x}\phi(x,x\eta), \\ \frac{d\eta}{dt} &= \eta + \eta^3 - \frac{\eta}{x^2}\phi(x,x\eta) + \frac{1}{x^2}\psi(x,x\eta), \end{aligned} \tag{1.11}$$

and in the local chart (θ, y), where $\theta = x/y$, it can be written as:

$$\begin{aligned} \frac{d\theta}{dt} &= -1 - \theta^2 - \frac{\theta}{y^2}\psi(\theta y, y) + \frac{1}{y^2}\phi(\theta y, y), \\ \frac{dy}{dt} &= \beta y + (\alpha+1)\theta y + \frac{1}{y}\psi(\theta y, y). \end{aligned} \tag{1.12}$$

At the axis $x = 0$ the system (1.11) has a unique singular point $(0,0)$, which is elementary (a node when $\alpha > 0$ and a saddle when $\alpha < 0$). At the axis $y = 0$ system (1.12) has no singular points and therefore no trajectory of system (1.10) tends to the origin in the direction $x = 0$. Thus, after the first step of the blowing-up process, we obtain hyperbolic singular points. It follows readily from (1.11) and (1.12) that the perturbation terms ϕ and ψ do not change the scheme of elementary points of the singularity. This means that the 2-jet

$$\frac{dx}{dt} = \alpha x^2 + \beta xy - y^2, \qquad \frac{dy}{dt} = (\alpha+1)xy + \beta y^2$$

of the germ (1.10) is C^0-sufficient. This corresponds to the case (iv) $(\alpha < 0)$ and case (v) $(\alpha > 0)$ of the lemma. Note that the conjugating homeomorphism h satisfies the equality $h(-x,-y) = -h(x,y)$. Thus, in the above cases the germs Γ_0 are structurally stable in the class Γ.

Blowing-up the singularity at the origin of the system

$$\frac{dx}{dt} = \alpha x^2 + (\beta - 1)xy + \phi(x, y),$$
$$\frac{dy}{dt} = (\alpha - 1)xy + \beta y^2 + \psi(x, y), \tag{1.13}$$

$$\mu = \alpha\beta(\alpha + \beta - 1) \neq 0, \quad \alpha, \beta \in I\!R,$$

we rewrite it in the local charts (x, η) and (θ, y) as follows:

$$\begin{aligned} \frac{dx}{dt} &= \alpha x + (\beta - 1)x\eta + \frac{1}{x}\phi(x, x\eta), \\ \frac{d\eta}{dt} &= -\eta + \eta^2 - \frac{\eta}{x^2}\phi(x, x\eta) + \frac{1}{x^2}\psi(x, x\eta); \end{aligned} \tag{1.14}$$

$$\begin{aligned} \frac{d\theta}{dt} &= -\theta + \theta^2 - \frac{\theta}{y^2}\psi(\theta y, y) + \frac{1}{y^2}\phi(\theta y, y), \\ \frac{dy}{dt} &= \beta y + (\alpha - 1)\theta y + \frac{1}{y}\psi(\theta y, y). \end{aligned} \tag{1.15}$$

The system (1.14) has two singular points at the axis $x = 0 : M_1(0, 0)$ and $M_2(0, 1)$. At M_1 the eigenvalues of the system are $\lambda_1 = \alpha, \lambda_2 = -1$ and, at $M_2, \lambda_1 = \alpha + \beta - 1, \lambda_2 = 1$. By the condition on the system (1.13), the points M_1 and M_2 are elementary.

The system (1.15) has a unique singular point $M_3(0, 0)$ which lies on the axis $y = 0$. M_3 is an elementary singularity since at this point $\lambda_1 = -1, \lambda_2 = \beta \neq 0$. Thus, the scheme of elementary points of the singularity of the system (1.13) does not depend on terms of order higher than 2. As before, this means that the 2-jet

$$\frac{dx}{dt} = \alpha x^2 + (\beta - 1)xy, \qquad \frac{dy}{dt} = (\alpha - 1)xy + \beta y^2.$$

of the germ (1.13) is C^0-sufficient, and defines completely the topology of orbits near the origin. Here again the conjugating homeomorphism h commutes with the involution θ given by $h(-x, -y) = -h(x, y)$. Phase portraits correspond to the cases (i)-(iii) of the lemma. Thus, in the above cases the germs Γ_0 are structurally stable in the class Γ. This completes the proof of the Fundamental Lemma. □

Remark 1.4.1 *The phase portraits (iii) and (v) of the Lemma are topologically (but not C^1) equivalent.*

1.4.3 Classification

Suppose that $p \in Sing\ \mathcal{F}$ is a non orientable singularity of the foliation $\mathcal{F} = (M, \pi^t, \theta)$ on the surface M/θ. Denote by H, P and E respectively hyperbolic, parabolic and elliptic sectors of p. We adopt the terminology presented in the table below and Fig. 1.1.

Sectors	Fig.	Name	Index	Form	Conditions
HHH	(i)	tripod	$-1/2$	(1.13)	$\alpha\beta(\alpha+\beta-1)<0$, $(\alpha-1)(\beta-1)(\alpha+\beta)>0$
HP	(ii)	sun-set	$+1/2$	(1.13)	$\alpha\beta(\alpha+\beta-1)>0$
H	(iii)	thorn	$+1/2$	(1.10)	$\alpha<0$
EP	(iv)	apple	$+3/2$	(1.13)	$\alpha\beta(\alpha+\beta-1)<0$, $(\alpha-1)(\beta-1)(\alpha+\beta)<0$
E	(v)		$+3/2$	(1.10)	$\alpha>0$

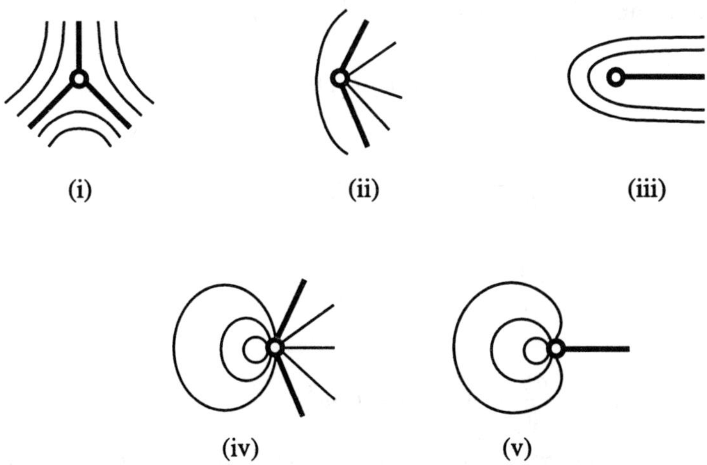

Fig. 1.1. Structurally stable non orientable singularities.

Theorem 1.4.3 *Let $\mathcal{F}^r(M)$ be a topological space of C^r smooth foliations on a surface M. The list of typical non orientable singularities of foliations $\mathcal{F} \in \mathcal{F}^r(M)$ is exhausted by those of Fig. 1.1. Moreover, they, and only they, are (locally) structurally stable.*

Proof. This follows from the Fundamental Lemma. □

Remark 1.4.2 *The topological structure in the vicinity of the points (iv) and (v) are equivalent. (This motivates the use of the same name, 'apple', in both cases.)*

1.5 Bifurcations

To finish the local theory, we shall look at non typical singularities of degeneracy degree ≤ 2. Surprisingly, no new phase portraits different from those of the Fundamental Lemma emerge after such bifurcations.

Theorem 1.5.1 *The phase portraits of vector fields having normal forms* $\widetilde{w}_1^{(1)}$ *(*$\widetilde{w}_1^{(2)}$*) are topologically conjugate to cases (i)-(iii) ((iv) and (v)) of the Fundamental Lemma. In the case of the normals form* w_1^* *and* $\widetilde{w}_2^*$*, one gets phase portraits topologically conjugate to cases (iii) and (i) of the Fundamental Lemma. And, finally, in the case of the normal forms* w_2^- *and* w_2^+ *one comes to cases (iv) and (v) of the Fundamental Lemma.*

Proof. Let us consider 4-jet of the normal form $\widetilde{w}_1^{(1)}$ in the case $\alpha = 0$:

$$\frac{dx}{dt} = (\beta - 1)xy + D_2x^4, \quad \frac{dy}{dt} = -xy + \beta y_2 + C_2y^4. \tag{1.16}$$

Unfolding the singularity $(0,0)$ of system (1.16) we rewrite it in the local charts (x, η) and (θ, y), where $\eta = yx^{-1}, \theta = xy^{-1}$:

$$\begin{aligned} \frac{dx}{dt} &= (\beta - 1)x\eta + D_2x^3, \\ \frac{d\eta}{dt} &= -\eta + \eta^2 - D_2x^2\eta + C_2x^2\eta^4; \end{aligned} \tag{1.17}$$

$$\begin{aligned} \frac{d\theta}{dt} &= -\theta + \theta^2 + D_2\theta^4y^2 - C_2\theta y^2, \\ \frac{dy}{dt} &= \beta y - \theta y + C_2y^3. \end{aligned} \tag{1.18}$$

The vector field (1.17) has two equilibria $M_1(0,0)$ and $M_2(0,1)$, which lie on the axis $x = 0$. M_1 is an elementary (but not hyperbolic) point. Its topological type depends on the coefficient $D_2 \neq 0$. It is known (see, for instance, [3]) that M_1 is a topological node when $D_2 < 0$, and a topological saddle when $D_2 > 0$. The point M_2 is a hyperbolic equilibrium with the eigenvalues $\lambda_1 = \beta - 1 \neq 0$ and $\lambda_2 = 1$. Thus, if $D_2 \neq 0$ then 4-jet of (1.7) is topologically sufficient. (Otherwise $D_2 = 0$ and the 4-jet is unsufficient. However it can be easily shown that in this case (of codimension 2) the topology of M_1 is defined by the coefficient $D_3 \neq 0$ of the 6-jet of $\widetilde{w}_1^{(1)}$.)

The origin of the system (1.18) is a hyperbolic point with eigenvalues $\lambda_1 = -1$ and $\lambda_2 = \beta \neq 0$. It means that after the first step of the blowing-up procedure all rest points obtained are elementary, and the singularity $(0,0)$ of (1.21) is resolved.

The singularity $(0,0)$ will be called *weak* if not all rest points obtained after the first step of the blowing-up are hyperbolic (although all of them can be elementary). In this sense system (1.16) obeying conditions $\beta(\beta-1)<0, D_2>0$, has a weak singularity in $(0,0)$, which is topologically equivalent to (ii) (weak (ii)-singularity); if $\beta(\beta-1)D_2>0$, it has weak (i)-singularity; if $\beta(\beta-1)>0, D_2<0$, it has weak (iii)-singularity.

The resonances of the normal form $\widetilde{w}_1^{(1)}$, which correspond to $\alpha=\frac{1}{2(1-n)}$, $\alpha=1$ and $\beta=\frac{1}{2(1-k)}$, do not influence the scheme of elementary points related with the singularity at $(0,0)$.

Now let us consider 4-jet of the normal form $\widetilde{w}_1^{(1)}$ under the conditions $\alpha=\beta=0$ (it can be checked that other double resonances are not *materialized*; see [33] for the terminology):

$$\frac{dx}{dt}=-xy+D_2x^4,\quad \frac{dy}{dt}=-xy+C_2y^4. \tag{1.19}$$

In the local charts (x,η) and (θ,y) the system (1.19) takes the form

$$\frac{dx}{dt}=-x\eta+D_2x^3,\quad \frac{d\eta}{dt}=-\eta+\eta^2-D_2x^2\eta+C_2x^2\eta^4; \tag{1.20}$$

$$\frac{d\theta}{dt}=-\theta+\theta^2+D_2\theta^4y^2-C_2\theta y^2,\quad \frac{dy}{dt}=-\theta y+C_2y^3. \tag{1.21}$$

As earlier, there are two equilibria M_1 and M_2, which lie at the axis $x=0$ of the system (1.20). M_1 is an elementary point (not hyperbolic) and is a topological node when $D_2<0$, and a topological saddle when $D_2>0$. The point M_2 is a hyperbolic saddle. The origin of the system (1.21) is an elementary point with one eigenvalue equal to zero. This is a node if $C_2<0$ and a saddle if $C_2>0$.

Thus, the singularity at the origin of the system (1.19) is unfolded, and it is a weak (ii)-singularity when $D_2>0, C_2>0$; a weak (i)-singularity when $D_2C_2<0$; and a weak (iii)-singularity when $D_2<0, C_2<0$.

The other cases are treated likewise. □

Bibliographic Notes. The local theory first appeared in [62]. The normal forms and the bifurcations have been considered in [31] and [65]. R. Bogdanov [52] evaluated the maximal number of functional modules in the formal expansions of vector fields in the plane. The normal forms for pairs of vector fields was a topic of the doctoral thesis of M.–F. Michel [203]. The recent paper of L. V. Detchenya and A. P. Sadovskii [90] is in the same vein.

2. Morse–Smale Foliations

2.1 Rough Foliations

Of course, rough means structurally stable. Introduced by Andronov and Pontryagin, this property unifies foliations whose "phase portrait" is not affected by the small "irregularities" in the topological space $\mathcal{F}^r(M)$. Fortunately, rough foliations are typical on compact orientable surfaces. If the surface is non orientable or non compact, then rough foliations are no longer typical.

2.1.1 Main Theorem

Let $\mathcal{F}^r(M)$ be the space of all C^r smooth foliations $\mathcal{F} = (M, \pi^t, \theta)$ on a compact surface M. $\mathcal{F}^r(M)$ is endowed with the C^r uniform (Whitney) topology and the equivalence relation in $\mathcal{F}^r(M)$ is topological conjugacy.

Lemma 2.1.1 *There exists an open and dense subset $\mathcal{F}_0^r$ of $\mathcal{F}^r(M)$ such that every vector field in $\mathcal{F}_0^r$ has only finitely many singular points.*

Proof. Denote by V and W the set of orientable and non orientable singularities of foliation $\mathcal{F} \in \mathcal{F}^r(M)$. To obtain the lemma, one should apply the standard argument of the transversality theory in the neighbourhoods $U(q)$ of points $q \in V$ [234], and $\mathbb{Z}_2$-equivariant transversality theory, due to Field [105], in the neighbourhoods $U(p)$ of points $p \in W$. □

Let $p \in W$ be a singular point of the foliation $\mathcal{F} \in \mathcal{F}^r(M)$ and let $U(p)$ be a neighbourhood of p. A component of the intersection of a leaf of the foliation $\mathcal{F}$ with the neighborhood $U(p)$ of a (singular) point p is called a *local leaf.* We say that a leaf l *tends* to the critical elements $p_i (i = 1, 2)$ if $\bar{l} \backslash l = p_1 \cup p_2$ (the case $p_1 = p_2$ is admitted). Finally, a *local separatrix* for a singular point p is a local leaf l^0 which tends to p and at least on one side of l^0 near p there are local leaves which do not tend to the point p. The leaf l that contains the local separatrix l^0 of the point p is called a *(global) separatrix* for p. The following statement establishes the structural stability of a certain class of foliations on orientable surfaces.

Theorem 2.1.1 (Andronov-Pontryagin-Peixoto) *Let M be an orientable surface. In the space $\mathcal{F}^r(M)$ of all C^r-smooth foliations on M, there exists an open and dense subspace $\mathcal{F}_0^r$ which consists of structurally stable foliations. These foliations are called Morse-Smale, and admit the following geometrical description:*

(i) the foliations from $\mathcal{F}_0^r$ have only finitely many orientable singularities and closed orbits, all of them being hyperbolic (i.e., saddles, nodes or foci, or else hyperbolic cycles);
(ii) the set of non orientable singularities is also finite and is exhausted by the five structurally stable singularities (i)-(v), mentioned in the Fundamental Lemma;
(iii) every leaf of $\mathcal{F} \in \mathcal{F}_0^r$ distinct from critical elements of items (i) and (ii) tends to a critical element;
(iv) there are no separatrix connections between critical elements (i.e., there are no leaves tending to singular points p and q which are separatrices for both p and q; the case $p = q$ is not excluded).

2.1.2 Structural Stability.

Let $\mathcal{F}_0^r = (M, \pi^t, \theta)$ be a Morse-Smale foliation on a compact orientable surface M. Denote by $\xi : M \to TM$ the corresponding vector field on M obeying $\mathbb{Z}_2$-symmetry. By $A = \{x_1, \cdots, x_k\}$ we denote the set of orientable critical elements (i.e., singular points and closed orbits) of the vector field $\xi \in \mathcal{F}_0^r(M)$. Let $B = \{y_1, \cdots, y_m\}$ be the set of non orientable singularities.

In what follows, we are concerned with a manifold with boundary M^* obtained from the manifold M by the polar blowing-up of all singular points of the set B. The natural blowing-down map $p : M^* \to M$ is a diffeomorphism everywhere except at the points lying on the pasted circles $\{S_i(B)\}_{i=1}^m$ in M. The union $\cup_{i=1}^m S_i(B)$ of all such circles represents the boundary ∂M^* of the manifold M^*. It consists of whole trajectories of the system $\xi \in \mathcal{F}_0^r(M^*)$.

Remark 2.1.1 *Instead of the manifold M^* with the boundary ∂M^* one could consider the manifold M^{**}, which is produced from two copies of the manifold M^* by pasting them together along the circles $S_i(B)$. The manifold M^{**} is boundaryless and the corresponding vector fields $\xi \in \mathcal{F}_0^r(M^{**})$ on the manifold M^{**} have only hyperbolic critical elements.*

We shall define now the stable and unstable manifolds of the elements which constitute the sets A and B. Since the points $x \in A$ are hyperbolic critical elements, their stable $W^s(x)$ and unstable $W^u(x)$ manifolds are defined as usual. For $y \in B$, the point y of the manifold M is decomposed by the blowing-up process into hyperbolic equilibrium states: saddles and nodes.

The nodal points which are situated on ∂M^*, will be called *semi-nodes.* The set of whole trajectories that enter the seminode as $t \to +\infty (t \to -\infty)$ is called the two-dimensional stable (unstable) manifold of the seminode.

By a saddle element of the given non-trivial singular point we mean the (possibly disconnected) subset of the polar blowing-up circle that consists of all saddle points and, possibly, of whole trajectories, if any, which join saddle points. One easily sees that in the case under consideration, there exist four types of saddle elements.

Each trajectory that does not lie at the polar blowing-up circle and tends to one of the saddle points of the given saddle element as $t \to +\infty (t \to -\infty)$ is called the stable (respectively, unstable) separatrix of the saddle element. By the stable (unstable) manifold of the saddle element we mean the union of all its stable (unstable) separatrices.

Thus, on each polar blowing-up circle, we distinguish a finite number of semi-nodes and at most one saddle element. In what follows by *nontrivial critical elements* we mean the semi-nodes and the saddle elements. Following our comments above, we discern eight types of critical elements, namely:

- nodes,
- semi-nodes,
- closed orbits,
- saddles and saddle elements of four kinds:

(i) a circle containing two saddle points and two arcs connecting them;
(ii) a circle containing six saddle points and six arcs connecting them;
(iii) a pair of isolated saddle points;
(iv) two arcs of the circle with ends at the saddle points.

Definition 2.1.1 *For the given vector field $\xi \in \mathcal{F}_0^r(M^*)$ we define the phase diagram D as a set of its critical elements (trivial as well as non-trivial) with the following partial order: for $\alpha_1, \alpha_2 \in D$, we write $\alpha_1 \le \alpha_2$ if $W^u(\alpha_1) \cap W^s(\alpha_2) \neq \emptyset$ (in other words, there exists a trajectory which goes from α_1 to α_2).*

Clearly, by condition (iv) of the theorem and by the fact that no one pair of sinks (sources) can be joined by a trajectory of the field $\xi \in \mathcal{F}_0^r(M^*)$, the order $\le$ is partial (i.e., not defined for some pairs of critical elements). Bearing this in mind, the phase diagram of the field $\xi \in \mathcal{F}_0^r(M^*)$, $dim M^* = 2$, breaks into three levels: (i) unstable elements (nodes, semi-nodes and closed orbits); (ii) saddles and saddle elements; (iii) stable elements (nodes, semi-nodes and closed orbits). Thus, the ordering is not defined for elements within every level.

Definition 2.1.2 *Let $\xi, \tilde{\xi} \in \mathcal{F}_0^r(M^*)$ and $D, \tilde{D}$ be their phase diagrams. We shall say that D is isomorphic to $\tilde{D}$ if there exists a one-to-one correspondence $h : D \to \tilde{D}$ such that:*

(i) $\alpha \in D$ is a critical element if and only if $h(\alpha) \in \tilde{D}$ is a critical element of the same kind;
(ii) if $\alpha_1, \alpha_2 \in D$, then $\alpha_1 \le \alpha_2$ if and only if $h(\alpha_1) \le h(\alpha_2)$.

Our purpose now is to show that, after small perturbations of the field $\xi \in \mathcal{F}_0^r(M^*)$, one obtains fields from the same class with isomorphic phase diagrams. The main technical tool we use is the notion of filtration.

Definition 2.1.3 *Let $\xi \in \mathcal{F}_0^r(M^*)$. By a filtration associated with ξ we mean a chain $M_0 \subset M_1 \subset \cdots \subset M_k = M^*$ of compact submanifolds M_i of the manifold M^* such that:*

(i) for each i, the boundary ∂M_i can be represented as a disjoint union of smooth curves, each lying entirely in M^* or in ∂M^*;
(ii) the vector field ξ is transversal to $\partial M_i \backslash \partial M^*$ and $\xi_t(M_i) \subset int M_i, t > 0$;
(iii) the maximal ξ_t-invariant subset of $M_{i+1} \backslash M_i$ consists of a unique trivial or non-trivial critical element a_{i+1}. In other words, $\cap_{t\in R}\xi_t(M_{i+1}\backslash int M_i) = \alpha_{i+1}$.

Lemma 2.1.2 *Assume $\xi \in \mathcal{F}_0^r(M^*)$. There exists a filtration associated with ξ.*

Proof. Let $\alpha_1, \alpha_2, \cdots, \alpha_j$ be the sinks (i.e. stable nodes, semi-nodes and closed orbits) of the field ξ. Choose disjoint neighbourhoods $V_1, V_2, \cdots, V_j$ of these points with boundaries transversal to the field ξ everywhere except at the points of ∂M^*. Assume $M_1 = V_1, M_2 = M_1 \cup V_2, \cdots, M_j = M_{j-1} \cup V_j$.

Furthermore, let $\alpha_{j+1}, \cdots, \alpha_s$ be the set of all saddle elements and saddles of the field ξ. We consider first the stable separatrices of the saddle element α_{j+1} (i.e., the components of the manifold $W^s(\alpha_{j+1})\backslash\alpha_{j+1}$) and we construct in the vicinity U of the critical element α_{j+1} some sections S_m which are transversal to the m-th separatrix of the saddle element α_{j+1}. If U is taken small enough, then the trajectories which pass through the end-points of the section S_m, cross ∂M transversely. Utilizing flow tubes of the field ξ, one can choose, near the arcs of trajectories, some transversal curves $c_n (n = 2m)$ which connect the end-points of S_m with ∂M_j. Assume now V_{j+1} is a domain bounded by the curves S_m, c_n and by a segment of ∂M_j. Let $M_{j+1} = M_j \cup V_{j+1}$. It is easily verified that $\cap_{t\in R}\xi_t(V_{j+1}) = \alpha_{j+1}$ and the M_{j+1} meets the restrictions imposed on the filtration. By arguing likewise for each saddle element and each saddle, we arrive at a chain of manifolds $\emptyset = M_0 \subset \cdots \subset M_s$.

Finally, let $\alpha_{s+1}, \cdots, \alpha_k$ be the sources (i.e., unstable nodes, semi-nodes and closed orbits) of the field ξ and let $V_{s+1}, \cdots, V_k$ be their disjoint neighbourhood with the boundaries transversal to ξ everywhere except for the points of ∂M^*. Define $M_{s+1} = M\backslash(int V_{s+2} \cup \cdots \cup int V_k), M_{s+2} = M\backslash(int V_{s+3} \cup \cdots \cup int V_k), \cdots, M_k = M^*$. One easily sees that $\emptyset = M_0 \subset M_1 \subset \cdots \subset M_k = M^*$ is the required filtration.

The following proposition establishes that the set $\mathcal{F}_0^r(M^*)$ is open in the space $\mathcal{F}^r(M^*)$. This an implication of the fact that the submanifolds $\partial_0 M_j = \partial M_j \backslash \partial M^*$ are transversal to the vector field $\xi \in \mathcal{F}_0^r(M^*)$.

Lemma 2.1.3 *Assume $\xi \in \mathcal{F}_0^r(M^*)$. Then there exists such a neighbourhood U of the field ξ that whenever $\zeta \in U$ then $\zeta \in \mathcal{F}_0^r(M^*)$ and the phase diagram of the field ζ is isomorphic to that of ξ.*

We proceed now to the proof of structural stability of the vector fields from $\mathcal{F}_0^r(M^*)$. To state this more precisely, we have to establish a homeomorphism h, close to the identity, which brings the trajectories of a perturbed field $\zeta \in U$ into the trajectories of the unperturbed one.

Part 1. We suppose first that the vector field ξ has no closed orbits. Consider a sink α of the field ξ and the corresponding sink $\alpha(\zeta) = h(\alpha)$, where $\zeta \in U$ and h is the isomorphism of the phase diagrams D and $\tilde{D}$ established earlier.

Assume V is a disc which contains the sink α, belongs to the stable manifold of the sink α and has a boundary ∂V transversal to all $\zeta \in U$. Moreover, let $\alpha(\zeta) \in V \subset W^s(\alpha)$. Assume $\alpha_1, \alpha_2, \cdots$ are the saddles and saddle elements of the field ξ such that $\alpha_1 \leq \alpha$ and let $p_1, p_2, \ldots$ be the points where the unstable separatrices $W^u(\alpha_j)$ cross ∂V. Likewise, construct the points $p_1(\zeta), p_2(\zeta), \ldots$ for the field ξ.

We remind the reader that the vector fields ξ and ζ satisfy $\mathbb{Z}_2$-symmetry with respect to the smooth involution θ defined on the manifold M^*. The following statement is valid.

Remark 2.1.2 *Assume that $\xi \in \mathcal{F}_0^r(M^*)$. The involution θ maps the sinks of the field ξ into the sources of the field ξ, and vice versa. The involution θ maps each stable separatrix of a saddle s (a saddle element σ) into an unstable separatrix of the saddle $\theta(s)$ (the saddle element σ).*

We consider further the source $\theta(\alpha)$ and assume $\beta_1, \beta_2, \ldots$ is a set of saddle and saddle elements of the field ξ such that $\beta_i \geq \theta(\alpha)$. By $q_1, q_2, \cdots$ we denote those points where the stable separatrices $W^s(\beta_i)$ of the elements β_i intersect the boundary $\theta(\partial V)$ of the source $\theta(\alpha)$. Note in passing that $q_i = \theta(p_i)$.

Now we shall define a topological equivalence h between the vector fields ξ and ζ. To this end, we shall first establish a homeomorphism between the sets $\partial V \cup q_i$ constructed for the fields ξ and ζ. Assume $h(\alpha) = \alpha(\zeta), h(\alpha_i) = \alpha_i, h(p_i) = p_i(\zeta), h(q_i) = \theta h \theta^{-1}(q_i)$. Extend the homeomorphism h between the sets $\partial V \backslash p_i$ and $\partial V \backslash p_i(\zeta)$ in an arbitrary continuous way, demanding however that our extension be uniformly continuous regarding the whole family of the vector fields $\zeta \in U$.

We deal in the same way with all other sinks of the field ξ (recall that there are finitely many sinks).

Thus, the homeomorphism h is defined on the sets $\partial V \cup \{q_i\}$. Every point of the manifold $M^* \backslash \partial M^*$ (except equilibria) reaches the boundary ∂V of some sink or passes through one of the points q_i. We see at once how to extend the homeomorphism h to $M^*\backslash\partial M^*$ via the equation $h\xi_t z = \zeta_t hz$, where $z \in \partial V \cup \{q_i\}$. [1]

One easily sees that h is invertible (the mapping h^{-1} can be constructed in the same manner, replacing ξ by ζ).

It remains to prove that h is continuous. Our proof is much the same as in [234]. The continuity of h is evident at sinks, sources and in the neighbourhoods of saddles and saddle elements. We shall prove the continuity of h at the points of stable separatrices of the saddle elements. Let us consider a polar blowing-up of the non orientable singularity of "thorn" type (other cases are similar).

Let $x \in W^s(T)$, where T is a saddle element of thorn type. Consider a sequence $hx_n \to hx$. Notice that by the continuity of the involution θ it is enough to prove that h is continuous at the point $\theta x \in W^u(T)$.

Assume S and $\tilde{S}$ are segments transversal to ξ which pass through the points q and $\tilde{q}$ of the polar blowing-up circle. Furthermore, let I_i be an arc of the boundary ∂V_i of the sink α such that $p_i \in I_i$. Applying the flow ξ_t to $S, \tilde{S}$ and I_i one obtains tubular families for $W^u(m)$ and $W^s(m)$, respectively. It is easily verified that the mapping h carries the leaves of tubular families of the flow ξ_t onto the leaves of the tubular families.

By the λ-lemma [234], the leaves $\xi_t(S), \xi_t(\tilde{S})$ and $\xi_t(I_i)$ approach (as $t \to +\infty$ and $t \to -\infty$, respectively) in the C^1-topology the manifolds $W^s(m)$ and $W^u(m)$. Taking projections π_s and π_u of the points $\theta(x_n)$ onto $W^s(m)$ and $W^u(m)$ yields the continuity of h.

Part 2. When the vector field $\xi \in \mathcal{F}_0^r(M^*)$ has closed orbits, the construction of a conjugacy is more complicated. However, because the closed hyperbolic orbits are trivial critical elements, our proof in this case is similar to [234]. Thus, we have proved that the field $\xi \in \mathcal{F}_0^r(M^*)$ is structurally stable.

Remark 2.1.3 *The above homeomorphism h commutes with the involution θ.*

Indeed, by our comments above, the homeomorphism h was established first between the sets $\partial V \cup \{q_i\}$, on which h commutes with θ. Since the flows ξ_t and ζ_t satisfy $\mathbb{Z}_2$-symmetry with regard to the involution θ, one obtains immediately that the homeomorphism h commutes with the involution θ, i.e., $h \circ \theta = \theta \circ h$.

2.1.3 Density

The proof that the set of the vector fields $\mathcal{F}_0^r(M^*, B, \theta)$ with a fixed collection of non-trivial singularities B and the involution θ is dense in the space

[1] In fact, the homeomorphism h need not be a 'conjugacy' between two flows (i.e., time-respecting). We are interested only in topological equivalence between ξ_t and ζ_t.

$\mathcal{F}^r(M^*, B, \theta)$ runs in much the same way as for the Morse-Smale vector fields on two-dimensional oriented manifolds (see, e.g. [234]). In what follows, we freely use the terminology of [234].

Assume that M^* is an oriented compact 2-manifold. If $\xi \in \mathcal{F}^r(M^*)$ has singular points (orientable, or non orientable), all hyperbolic, and there exists a non-trivial recurrent trajectory, then ξ may be approximated by a field ζ which has one more saddle connection than ξ. Fix a point p lying on the non-trivial recurrent trajectory and a neighborhood O_p so small that $O_p \cap \theta(O_p) = \emptyset$. The required perturbation $\Delta\xi$ will be chosen to vanish outside of $O_p \cup \theta(O_p)$. In fact, define $\Delta\xi$ on O_p just as in [234] and extend it to $\theta(O_p)$ by means of the involution θ. One can similarly delete all non-trivial recurrent trajectories $\gamma_i, \theta(\gamma_i)$, at the expense of the appearance of extra saddle connections $s_i, \theta(s_i)$.

Thus, a field ξ which has quasi-minimal sets can be approximated by a field ζ which has only trivial recurrent trajectories and a finite number of saddle connections s_i.

Now it remains to show how the saddle connections s_i which do not lie at the boundary ∂M^* can be destroyed in a finite number of steps by the small perturbations of the field ζ.

Consider a saddle connection s. By arguing as above, we construct a perturbation localized in a small neighborhood O_p of a point $p \in S$ and then extend it to $\theta(O_p)$. The rest of the proof is exactly the same as in [234].

Thus, the proof that $\mathcal{F}_0^r(M^*, B, \theta)$ is dense in the space $\mathcal{F}^r(M^*, B, \theta)$ is complete and the Andronov-Pontryagin-Peixoto theorem follows. □

2.2 Classification of Morse-Smale Foliations

Andronov-Pontryagin-Peixoto theorem characterizes structurally stable foliations on compact orientable two-dimensional manifolds. Two important open questions are:

(i) how to discern between two topologically inequivalent foliations in this class;

(ii) how to describe the variety of topological classes. Both (i) and (ii) will lead us to the construction of a discrete invariant of Morse-Smale foliations called a *Peixoto graph.*

2.2.1 Rotation Systems

Both Morse-Smale flows and Morse-Smale foliations given on a surface define a special kind of graphs embedded into the surface. Problems related with the embeddings of graphs belong to the subject of topological graph theory; see [121] for an introduction.

Definition 2.2.1 *An (oriented) graph X is defined to be a set of vertices V_X and a set of edges E_X such that an (ordered) pair of endpoints of every edge $e \in E_X$ lies in V_X. In other words, X is the set $E_X \cup V_X$ endowed with an 'incidence structure' $I_X = \{V_X(e)|e \in E_X\}$. For brevity, an oriented graph is called an orgraph.*

We omit subscripts for the vertex and the edge sets. If both E and V are finite, X is said to be finite. Every finite graph X can be geometrically represented as a drawing (dot–and–line structure), where dots correspond to the vertices $v \in V$ and lines (segments) correspond to the edges $e \in E$ such that the incidence structure I is respected. Although drawings are easily visualized, they should be treated carefully since a drawing is not a graph but only a representation of its incidence structure.

Let X and X' be two finite graphs. A graph map $f : X \to X'$ consists of a vertex function $V_X \to V_{X'}$ and an edge function $E_X \to E_{X'}$ such that the incidence structure is preserved (this means that for every $e \in E$, the vertex function f_V and the edge function f_E are such that the endpoints a, b of e are sent to the endpoints p, q of $f_E(e) : p = f_V(a), q = f_V(b)$). For orgraphs we demand also that f preserves the orientation.

Definition 2.2.2 *A graph map $f : X \to X'$ between two graphs X and X' is called an isomorphism if both its vertex function f_V and edge function f_E are one-to-one and onto (surjective). Two graphs X and X' are called isomorphic if there exists an isomorphism $f : X \to X'$.*

Let $u, v \in V$ be vertices of a graph X. *A walk* w on X from u to v of a length n means an alternating sequence of vertices and directed edges such that an initial vertex $v_0 = u$ and a final vertex $v_n = v$ for $i = 1, ..., n$. If $u \neq v$ a walk is called open, otherwise it is said to be closed. An open walk is called *a path* if all its vertices are distinct. A path on X such that its initial vertex coincides with the final vertex is called *a cycle* on X.

The central concern of topological graph theory is the problem of embedding of a graph into a surface. One classical problem is that of embedding of a graph into the plane (or 'planarity' of a graph). It is known, e.g., that the complete graph K_5 is not planar, that is, a drawing of the graph in the plane (or sphere) always has self-intersections. However, it is easy to see how to avoid intersections on higher genus surfaces: it is sufficient to glue on an extra handle and to draw one edge over 'the bridge', and the other under it. Each graph can be embedded into a surface of a sufficiently high genus.

Definition 2.2.3 *An embedding $i : X \hookrightarrow M$ of a graph X into a surface M is a 1-1 continuous map of X, taken as a topological space, into the topological space M. Two embeddings i_1 and i_2 of X in a surface M are equivalent if there exists a homeomorphism $h : M \to M$ such that $h \circ i_1 = i_2$ (in other words, h maps the image $i_1(X)$ to the image $i_2(X)$).*

If one takes an embedding $i : X \hookrightarrow M$ of a connected graph X in M, then the set $M \backslash i(X)$ is the union of open regions V_m. Clearly, gluing on handles to each V_m, it is possible to obtain embeddings of X into surfaces of arbitrarily high genus. An embedding $i : X \hookrightarrow M$ is called *2-cell (or cellular)* if all open regions V_m are homeomorphic to an open disc. Throughout this section only cellular embeddings are considered.

Definition 2.2.4 *Let X be a finite graph and let $v \in V(X)$ be its vertex. A rotation of a vertex v is an oriented cyclic order (defined up to cyclic permutations) of all edges incident to v. A rotation system R (or, simply, a rotation) of a graph X is a union of all local rotations over all vertices of X.*

Rotations give rise to a certain system of faces ($\equiv$ cycles) on X. The following so-called Face Tracing Algorithm allows one to determine all faces of a graph X endowed with rotation R.

Face Tracing Algorithm, see [121]. Let X be a finite graph and R be its rotation system. Take an arbitrary vertex $v_1 \in V(X)$ and an edge a_{v_1} incident to v_1. Let v_2 be a vertex of X connected with v_1 by the edge a_{v_2} and let b_{v_2} be an edge of the vertex v_2 which lies *to the right* in the cyclic order from a_{v_1}. Moving along the edge b_{v_2} to a vertex v_3, we shall define an edge c_{v_3} which lies *to the right* of b_{v_2}. Proceeding inductively, we stop the process at an edge z_{v_n} if two forthcoming edges will be again a_{v_1} and b_{v_2}. Hereby a cycle $a_{v_1}, b_{v_2}, ..., z_{v_n}$ of a length n, which defines a face F_1 on X, will be traced. To trace the next face F_2 one should start with an edge which lies *to the right* of any edge of the face F_1 and such that a corner between them did not occur in F_1 – and apply the above construction. All faces $F_1, F_2, ..., F_m$ on X will be traced when no unused corners remain.

The fundamental theorem, proved independently by Lothar Heffter [143] and Jack Edmonds [94], shows how rotations and embeddings of X are related.

Theorem 2.2.1 (Heffter-Edmonds) *Let X be a finite graph endowed with a rotation system R. Then there exists a 2-cell embedding of X into an orientable surface M such that one of the two rotations induced by this embedding coincides with R. Moreover, two embeddings are equivalent if and only if they have equivalent rotation systems.*

Remark 2.2.1 *Each embedding $i : X \hookrightarrow M$ induces a pair of rotation systems R and R^*, where R^* is the mirror image of R (i.e., can be obtained from R by reversing the cyclic order of all local rotations in R). The corresponding embeddings $i(X)$ and $i^*(X)$ are conjugate by a homeomorphism $h : M \to M$ which is not homotopic to id_M.*

Proof of theorem. We shall only give a sketch of the proof. It will be a geometric one – for other (more extent) expositions of the question, we recommend the monographs of Gross-Tucker [121] and Ringel [255].

Let X be a finite graph endowed with a rotation system R. As was mentioned earlier, according to the Face Tracing Algorithm, the rotation R

gives rise to a certain (finite) system of faces $F_1, ..., F_m$ on X, which we denote here

$$\begin{aligned} F_1 &= a_1 b_1 ... z_1 \\ F_2 &= a_2 b_2 ... z_2 \\ &\vdots \\ F_m &= a_m b_m ... z_m \end{aligned}$$

where $\{a_i ... z_i\}_{i=1}^m$ is the set of edges of X. Each face in the above system is a disc domain bounded by the respective edge sequence and every $e_k \in E_X$ occurs exactly twice in the list: $e_k = F_i \cap F_j, i \neq j$.

We shall prove the theorem by an explicit construction of a fundamental domain of a manifold M. It will be a polygon (namely, a 4g-gon) Δ, filled with the system of faces $\Delta = F_1 \cup ... \cup F_m$, and such that the boundary components of $\partial\Delta$ are identified in a natural way. Next we shall obtain $M \simeq \Delta/\Gamma$ gluing Δ by the action of the absolutely discontinuous group Γ generated by the system $F_1, \ldots, F_m$.

Take a face $\Delta_1 = F_1$ and place it arbitrarily in the plane, denoting all the boundary edges $a_1, b_1, ..., z_1$. Each edge e_k of this sequence will occur once again in a row F_k (case $k = 1$ is not excluded), so that F_1 has a common edge with every face of the sequence $F_{a_1}, F_{b_1}, ..., F_{z_1}$. We glue these faces to the polygon Δ_1 along its boundary $\partial\Delta_1$. Taking the edges, lying in the boundary $\partial\Delta_2$ of the polygon $\Delta_2 = F_1 \cup F_{a_1} \cup F_{b_1} \cup ... \cup F_{z_1}$, we use the list of faces to find all adjacent faces of Δ_2 and glue them correspondingly together to obtain a polygon Δ_3. Proceeding inductively one stops when all faces will be used. As a result, we come to a polygon $\Delta_p = \Delta$. The boundary $\partial\Delta$ of the polygon consists of the edges which we denote *abcba*... and since each edge on $\partial\Delta$ has its counterpart also in $\partial\Delta$, all the sides of the polygon Δ are pairwise identified. We call a vertex point p of the graph X an inner point if $p \in \text{int } \Delta$. Otherwise, it is called a corner point. Clearly, if a corner point p has a pair of its incident edges identified, then it can be made an inner point. In what follows, we suppose that for the polygon Δ all such identifications are made.

Now Δ is a $2n$-gon (a polygon with $2n$ sides) and there are no adjacent sides in $\partial\Delta$ which are identified. Let us consider the following cases:

(i) If $n = 2g$, then Δ is a $4g$-gon. In this case Δ is a fundamental domain of a surface $M \simeq \Delta/\Gamma$, where Γ is an absolutely discontinuous group generated by the identifications on $\partial\Delta$. Clearly, we have a 2-cell embedding $X \hookrightarrow M$, where M is an orientable surface of genus g;

(ii) If $n = 2g - 1$, then Δ is a $(4g-2)$-gon. To arrive at the 'standard' case (i) one must add 2 extra edges to $\partial\Delta$. This can be done as follows.

Take an edge $e_0 \in E_X$ incident to a corner point $p \in \partial\Delta$ and split p to p_1 and p_2, moving them aside. The edge e_0 will also be split in $e_0^{(1)}$ and

$e_0^{(2)}$. On the boundary $\partial\Delta$ there will appear an extra edge p_1p_2. Similarly, we add an extra edge q_1q_2, where $q_1 = \Gamma(p_1), q_2 = \Gamma(p_2)$. Now one arrives at the case (i), identifying p_1p_2 with q_1q_2. So far, we obtain an embedding $\overline{X} \hookrightarrow M$, where $\overline{X}$ has 2 extra triangle faces in M. One squeezes these faces, bringing the point p_1 to the point p_2 and q_1 to q_2. Now we shall obtain a 2-cell embedding $X \hookrightarrow M$, where M is an orientable surface of genus g. □

2.2.2 Equivalence Criterion

Let $\mathcal{F} = (M, \pi^t, \theta)$ be a Morse-Smale foliation on a compact surface M. Denote by $P \cup Q \cup C$ the set of critical elements of the flow π^t consisting of m non orientable singularities $p \in P$, $2n$ orientable singularities $q \in Q$ and $2r$ cycles $c \in C$.

Lemma 2.2.1 *Suppose that C is non empty. The triple (M, π^t, θ) can be replaced by a triple $(N, \varepsilon^t, \theta)$, defined on a surface N, which is produced from M by cutting along each of $2r$ cycles and by the consequent squeezing of the banks of each cut into a point. The flow ε^t has $4r$ nodal singular points more than π^t, and has no cycles. The manifold N is, in general, disconnected.*

Lemma 2.2.2 *Let $(N, \varepsilon^t, \theta)$ be the foliation of Lemma 2.2.1. Suppose that P is non empty and contains singularities of sun-set and apple types. In the neighborhood of these points, there exists a homotopy $h_\tau : N \times [0,1] \to N, h_\tau \circ \theta = \theta \circ h_\tau, \tau \in [0,1]$, which maps ε^t to a flow χ^t, as shown in Fig. 2.1 (where (b) is the homotopy deformation of a sun-set (a), and (d) is the homotopy deformation of an apple (c)). Every operation of this kind adds 2*

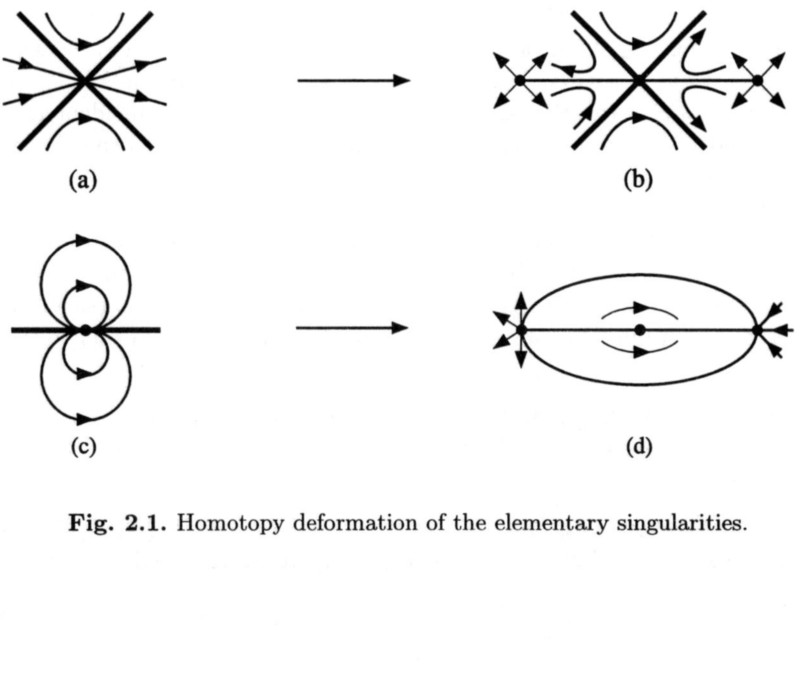

Fig. 2.1. Homotopy deformation of the elementary singularities.

extra nodes to the set Q, and after all possible operations, χ^t will have only tripods and thorns in the set P.

The proof of both statements is straightforward, and is left to the reader. Thus, a chain of invertible transformations $(M, \pi^t, \theta) \to (N, \varepsilon^t, \theta) \to (N, \chi^t, \theta)$ is defined.

Definition 2.2.5 *By a scheme of a Morse-Smale foliation $\mathcal{F} = (M, \pi^t, \theta)$ one understands a tripartite orgraph X with vertex set $V_X = Sing\chi^t$ and edge set $E_X = Sep\chi^t$. An upper level, which we denote by α, and a lower level, which we denote by ω, are filled, correspondingly, with sources and sinks of the flow χ^t; a middle level, denoted by σ, is filled with 2,4 and 6- separatrix saddles. An oriented incidence structure I_X is established as follows: for every separatrix, there corresponds an edge which joins the saddle on σ-level with a source (sink) on $\alpha(\omega)$-level, so that the direction of the edge coincides with the direction of the flow on the corresponding separatrix. Moreover,*

- *(i) one marks disjoint pairs of sources (sinks) on X which have been obtained from unstable (stable) hyperbolic cycles of π^t as in Lemma 2.2.1;*
- *(ii) one marks subgraphs in X, each having 2 edges and 3 vertices, which appear as a result of the homotopy h_τ; see Lemma 2.2.2;*
- *(iii) if a connectivity component $N_0 \subseteq N$ of a triple (N, χ^t, θ) has no saddles, then the corresponding subgraph $X_0 \subseteq X$ is a pair source – sink, joined by an edge; in this particular case $N_0 = S^2$ and the restriction of χ^t to N_0 is a polar flow.*

An example of two inequivalent Morse–Smale foliations with isomorphic orgraphs can be found in [66].Let $(M, \pi^t, \theta) \to (N, \varepsilon^t, \theta) \to (N, \chi^t, \theta)$ be a Morse-Smale foliation. Fix a clockwise (counter-clockwise) direction at each singular point $p \in N$ of flow χ^t. Fix also a clockwise (counter-clockwise) direction at the corresponding vertices $v \in V_X$ of the orgraph X. The flow χ^t induces *a local rotation* at v. Cyclic order on the edges incident to v is established in the same order as they lie in N with respect to the clockwise (counter-clockwise) direction, fixed at $p \in N$.

Let $R(R^*)$ be a clockwise (counter-clockwise) rotation system induced by χ^t at vertices of X. A scheme X of a foliation $\mathcal{F}$, together with a rotation system $R(R^*)$, is denoted by $X^R(X^{R^*})$ and is called a *Peixoto orgraph* of the foliation $\mathcal{F}$. An isomorphism $\cong$ on X^R is defined by isomorphism of underlying orgraphs X, taken together with the equivalence of their rotation systems.

Theorem 2.2.2 (Equivalence) *The Morse-Smale foliations $\mathcal{F}_1$ and $\mathcal{F}_2$ are topologically conjugate if and only if their Peixoto orgraphs are isomorphic: $X^{R_1} \cong X^{R_2}$.*

Proof. Necessary condition. Let $(M, \pi^t, \theta) \to (N, \varepsilon^t, \theta) \to (N, \chi^t, \theta)$ be the chain introduced earlier. It is evident that if two Morse-Smale foliations $\mathcal{F}_1$

and $\mathcal{F}_2$ are conjugate by a homeomorphism h, then $X^{R_1} \cong X^{R_2}$. Indeed, consider the restriction of $h : N \to N$ to the set $Sep\chi_1^t \mapsto Sep\chi_2^t$ which establishes isomorphism $X_1 \cong X_2$ of the respective orgraphs. The equivalence of the rotation systems $R_2 \sim R_1$ or $R_2 \sim R_1^*$ follows from the fact that h preserves 'the corner structure' of the embeddings $i : X \hookrightarrow N$.

Sufficient condition. Suppose that $X^{R_1} \cong X^{R_2}$. One must build a homeomorphism $h : N \to N$ which conjugates flows χ_1^t and χ_2^t and commutes with the involution θ. As a first step, one should construct h which establishes the equivalence of the two embeddings $i_1, i_2 : X \hookrightarrow N$ defined by equivalent rotation systems $R_1 \sim R_2$. For this, we use the notion of a dual graph, X^*, associated with the embedding $i : X \hookrightarrow N$.

Definition 2.2.6 *Denote by $i : X \hookrightarrow N$ a 2-cell embedding of an orgraph X into a surface N. Let $N\backslash i(X) = F_1 \cup ... \cup F_m$ be a union of open disc regions in N. A dual graph X^* associated to i is a graph with the vertex set $V_{X^*} = \{F_1, ..., F_m\}$. An edge $e^* \in E_{X^*}$ between F_i and F_j should be drawn (case $i = j$ is not excluded) if and only if there is an edge $e \in E_X$ between F_i and F_j, i.e., $e \subseteq \overline{F}_i \cap \overline{F}_j$.*

Lemma 2.2.3 *Let $i_1, i_2 : X \hookrightarrow N$ be two embeddings of X with the corresponding rotation systems R_1 and R_2. If $R_2 \sim R_1$ or $R_2 \sim R_1^*$, then dual graphs X_1^* and X_2^* defined by i_1 and i_2 are isomorphic.*

Proof of lemma. According to the face tracing algorithm, the rotations R_1 and R_2 generate on X the following systems of cycles $F_1^{(i)}, ..., F_m^{(i)}$, i=1,2, which are given by the table

$$\begin{array}{ll}
F_1^{(1)} = a_1^{(1)} b_1^{(1)} ... z_1^{(1)} & \qquad F_1^{(2)} = a_1^{(2)} b_1^{(2)} ... z_1^{(2)} \\
F_2^{(1)} = a_2^{(1)} b_2^{(1)} ... z_2^{(1)} & \qquad F_2^{(2)} = a_2^{(2)} b_2^{(2)} ... z_2^{(2)} \\
\vdots & \qquad \vdots \\
F_m^{(1)} = a_m^{(1)} b_m^{(1)} ... z_m^{(1)} & \qquad F_m^{(2)} = a_m^{(2)} b_m^{(2)} ... z_m^{(2)}
\end{array}$$

Since $R_2 \sim R_1(R_1^*)$, the two systems of cycles $F_j^{(1)}$ and $F_j^{(2)}$ (which are the vertices of X_1^* and X_2^*) will be the same, so that one easily obtains a vertex function $f_V : F_i^{(1)} \to F_j^{(2)}$ on dual graphs.

Now take a cycle $F_k^{(1)} = a_k^{(1)} b_k^{(1)} ... e_k^{(1)} ... z_k^{(1)}$ and the corresponding cycle $F_{k'}^{(2)} = f_V(F_k^{(1)})$, $F_{k'}^{(2)} = a_{k'}^{(2)} b_{k'}^{(2)} ... e_{k'}^{(2)} ... z_{k'}^{(2)}$, $1 \leq k, k' \leq m$. By the definition of a rotation system, the edge $e_k^{(1)}$ will be met exactly once in a row $F_l^{(1)}$ (the case $k = l$ is not excluded). In this case one draws an edge $\overline{e}_k^{(1)}$ (between $F_k^{(1)}$ and $F_l^{(1)}$) which is an edge of the dual graph X_1^*. Proceeding similarly with $F_{k'}^{(2)}$ and $F_{l'}^{(2)}$, one obtains an edge function $f_E : \overline{e}_k^{(1)} \to \overline{e}_{k'}^{(2)}$ between the dual graphs X_1^* and X_2^*. Since $f = f_V \cup f_E$, an isomorphism f is constructed and the lemma is proved. □

Let $N_0 = N \backslash O(V_X)$ be a surface with a boundary obtained from N by removing small open neighbourhoods $O(V_X)$ of vertices of X in the given cellular embedding $X \hookrightarrow N$. Let us consider a union of disc regions $D_1 \cup \ldots \cup D_m = N_0$ such that $\overline{D}_i \cap \overline{D}_j \in E_X (i \neq j)$.

Lemma 2.2.4 *The dual graph X^* can be obtained by a homotopy of N_0 in which the D_i are transformed into vertices of X^* and $\overline{D}_i \cap \overline{D}_j$ go to the edges of X^*.*

Proof. One easily constructs a homotopy $h_t : N_0 \times [0,1] \to N_0, t \in [0,1]$ such that $h_0 = id_{N_0}, h_1(N_0) = X^*$. □

To finish the proof of the theorem, one must invert the homotopy h_t so that the homeomorphism $h_0 : X_1^* \to X_2^*$, defined on dual graphs, deforms into a homeomorphism $h_1 : N_0 \to N_0$.

Let $D_i^{(1)}$ and $D_i^{(2)}$ $(1 \leq i \leq m)$ be disc regions as mentioned in Lemma 2.2.4, such that $h_0(D_i^{(1)}) = D_i^{(2)}$, where h_0 is the vertex function $V_{X_1^*} \mapsto V_{X_2^*}$ defined in Lemma 2.2.3. The boundary ∂D_i of D_i consists of segments $e_k \cup b_k$, where $e_k = \overline{D}_i \cap \overline{D}_j$ and $b_k \subset \partial N_0$.

First define a homeomorphism h on $\partial D_i = B \cup E$, where $B = \cup b_k, E = \cup e_k$. It may very well happen that on some e_k, h is already defined (since e_k touches other regions of N_0). In this case we fix $h|_{e_k}$ and extend it to the other points of ∂D_i in an arbitrary but continuous way. It is a trivial exercise to prolong h to the interior of D_i (e.g., it can be done using a level set of some function in D_i).

Thus, we obtain a homeomorphism $h_1 : N_0 \to N_0$ which is a homotopy deformation of a graph isomorphism h_0. Choose now $h = h_1$ and extend it to the whole N by 'squeezing' the holes $O(V_X)$ to the vertex points of X. The homeomorphism h is the required one, and conjugates the two embeddings $i_1, i_2 : X \hookrightarrow N$ of the graph X. The second part of the theorem is proved.

Now we can establish a topological conjugacy between the Morse-Smale foliations (N, χ_1^t, θ) and (N, χ_2^t, θ). Consider the embeddings $i_1(X)$ and $i_2(X)$ of the orgraph X corresponding to χ_1^t and χ_2^t. From the above arguments it follows that there exists a homeomorphism $h : N \to N$ mapping $i_1(X)$ to $i_2(X)$. This homeomorphism conjugates foliations on N defined by flows χ_1^t and χ_2^t. Since X is an orgraph, h preserves orientation on leaves of the foliations, giving an orbital conjugacy of flows: $\chi_2^t = h^{-1} \circ \chi_1^t \circ h$. Note also that, by $\mathbb{Z}_2$-symmetry of X (defined by graph involution $\theta : X \to X$), h commutes with θ, that is, $h \circ \theta = \theta \circ h$. □

2.2.3 Realization of the Graphs

One may wish to know when an arbitrary orgraph with a rotation system is the Peixoto orgraph of concrete Morse-Smale foliation. To study this question let us start with some definitions.

An *abstract scheme* is a tripartite orgraph $\mathcal{X}$ (disconnected, in general) with 2,4 and 6- valent vertices σ_2, σ_4 and σ_6, located in the middle level which we denote by σ. Every σ-vertex has an equal number of in-going and out-going edges. Every σ-vertex is connected arbitrarily with vertices of upper (α) and lower (ω) levels with the help of in-going and out-going edges, respectively. Furthermore:

(i) in the set of α and ω-vertices, there is a distinguished set (possibly empty) of disjoint pairs of vertices;
(ii) if $\mathcal{X}$ has σ_2 or σ_6 vertices, then there is an involution $\theta : \mathcal{X} \to \mathcal{X}$ which fixes σ_2 and σ_6 and moves σ_4 vertices (if any) to each other; vertices of α-level maps under θ to vertices of ω-level, and vice versa;
(iii) for some vertices $s \in \sigma_2$ and/or $s \in \sigma_6$ 2-edge subgraphs a-s-b are marked, where $a \in \alpha, b \in \beta$; these subgraphs are invariant under the involution θ;
(iv) if a connected component $\mathcal{X}_0 \subseteq \mathcal{X}$ has no vertices of σ-level, then it consists of unique α and ω-vertices joined by a unique edge.

By a *standard* molecule and *Andronov* molecule one denotes a connected orsubgraph of an orgraph $\mathcal{X}$, pictured in Fig. 2.2 (a) and (b). They correspond to the elementary cells of Morse-Smale foliations.

An abstract scheme $\mathcal{X}$ is called an *abstract* Peixoto orgraph $\mathcal{X}^{\mathcal{R}}$ if there exists a rotation system $\mathcal{R}$ such that every cycle traced on $\mathcal{X}$ by $\mathcal{R}$ is isomorphic either to a standard or to an Andronov molecule.

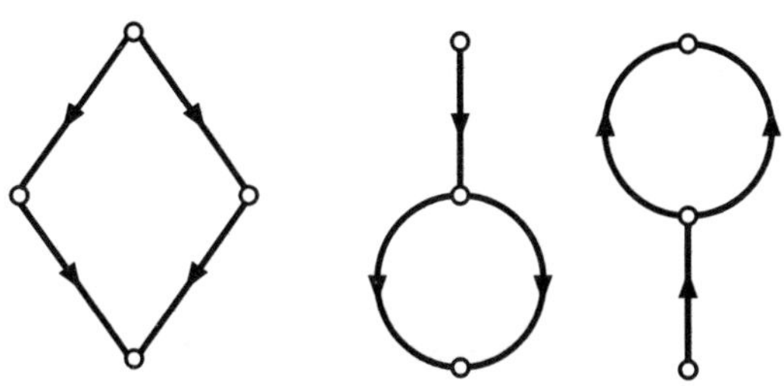

(a) (b)

Fig. 2.2. The standard and the Andronov molecules.

Theorem 2.2.3 (Realization) *Let X^R be a finite orgraph endowed with rotation system R. For the realization of X^R as a Peixoto orgraph of a Morse-Smale foliation, it is necessary and sufficient for X^R to be isomorphic to an abstract Peixoto orgraph $\mathcal{X}^{\mathcal{R}}$.*

Proof. Necessary condition. Let $\mathcal{F} = (M, \pi^t, \theta)$ be a Morse-Smale foliation on a surface M. Then it is a straightforward exercise to check that the Peixoto orgraph X^R satisfies all the requirements of the theorem.

Sufficient condition. Let $X = X_1 \cup ... \cup X_n$ be an orgraph consisting of n connected components. (We remind the reader that $n > 1$ if and only if the flow π^t has cycles contractible to a point.) The realization theorem will be proved for each $X_i (1 \le i \le n)$ independently, then we shall form $N = N_1 \# ... \# N_n$ by gluing N_i, where N_i is a realization manifold for X_i.

(i) Suppose that for some X_i condition (iv) is satisfied. Then $M_i = S^2$ and χ^t is a polar flow on the sphere.
(ii) Suppose that X_i is an orgraph with σ-level vertices, endowed with the rotation system R_i. By the Heffter-Edmonds theorem, there exists a 2-cell embedding of X_i into an orientable surface N_i. By definition, all cycles of R_i are either standard or Andronov molecules (a), (b), as shown in the figure above. Thus, the elementary cells of the embedding $X_i \hookrightarrow N_i$ will only be those shown in (c) and (d) of the figure. Since X_i is an orgraph, the directions on all separatrices are fixed and it remains to foliate the interiors of the cells in a standard way, as sketched in (c) and (d). By virtue of (ii), the flow χ^t will be involutive and the obtained triple (N_i, χ_i^t, θ) represents a Morse-Smale foliation on N_i.
(iii) If all the X_i have been realized as Morse-Smale foliations (N_i, χ_i^t, θ) on surfaces M_i, then according to condition (i) one must make a 'blow-up' of those pairs of sinks (sources) which were punctured, and glue together the corresponding banks of cuts. According to (iii), one makes necessary homotopy operations along the subgraphs marked, obtaining triples $(N_i, \varepsilon_i^t, \theta)$. And, finally, one comes to a Morse-Smale foliation $\mathcal{F} = (M, \pi^t, \theta)$, where $\pi^t = \varepsilon_1^t \# ... \# \varepsilon_n^t$ is a flow with hyperbolic cycles on the manifold $M = N_1 \# ... \# N_n$.

Since all the cases have been considered, the theorem is proved. □

2.2.4 Example

The example below illustrates Morse-Smale foliations, their Peixoto orgraphs and their associated rotation systems. One can readily construct own examples.

Example 2.2.1 *Let $\mathcal{F} = (T^2, \pi^t, \theta)$ be a Morse-Smale foliation with Sing $\mathcal{F}$={2 thorns, tripod, apple}. The projection to the surface $T^2/\theta = S^2$ is shown in Fig. 2.3 (a). The scheme X of the foliation $\mathcal{F}$ is given by Fig. 2.3 (b). The rotation system R is represented by the table:*

$v \in V_X$	σ_6	$\sigma_2^{(1)}$	$\sigma_2^{(2)}$	$\sigma_2^{(3)}$	α	ω
local rotations	$a_2b_3a_3b_4a_4b_2$	a_1b_1	a_5b_5	a_6b_6	$a_1a_2a_5a_3a_6a_4$	$b_4b_5b_2b_6b_3b_1$

(Exercise: find all 'faces' of X induced by R and check that its number fits the Euler equality $(\#V) - (\#E) + (\#F) = \chi(M)$, where V, E, F denote vertices, edges and faces of X and $\chi(M)$ is the Euler characteristic of the surface M).

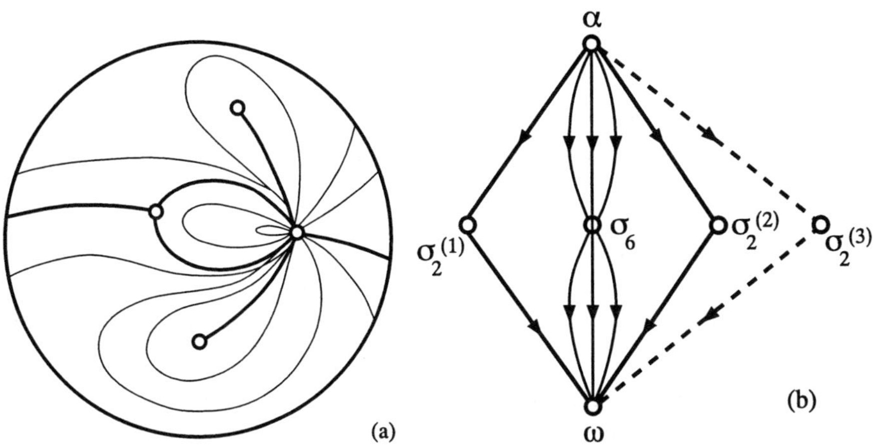

Fig. 2.3. Sphere with 4 non orientable singularities.

2.3 Gradient–like Foliations

Yet another way to classify Morse-Smale foliations is via their Lyapunov functions. Lyapunov functions work well even for foliations with separatrix connections, which are called *gradient-like* foliations. Further, we deal with the orientable foliations induced by the flows on M.

2.3.1 Lyapunov Function

For the gradient foliations on 2-dimensional manifolds one can introduce a real valued (Lyapunov) function $f : M \to \mathbb{R}$ which decreases along the trajectories of the flow,

$$L_v f \leq 0.$$

The notion of a Lyapunov function can be extended also to the Morse-Smale class, as proposed by K. Meyer [202]. Suppose that $f : M \to \mathbb{R}$ is a C^∞ function on the manifold M. Let

$$Singf = \{x \in M | Df(x) = 0\}$$

be the set of critical points of f. Denote by $\Delta_m \subseteq$ Sing f the set of points in Sing f, where the Hessian

$$D^2 f(x) = \begin{pmatrix} \frac{\partial^2 f}{\partial x_1^2} & \frac{\partial^2 f}{\partial x_1 \partial x_2} \\ \frac{\partial^2 f}{\partial x_1 \partial x_2} & \frac{\partial^2 f}{\partial x_2^2} \end{pmatrix}$$

has nullity m. Let $x \in \text{Sing } f$ be a critical point of the class Δ_0. In the appropriate coordinates at x, f acts by the formula

$$(x_1, x_2) \mapsto f(0,0) + Q(x_1, x_2),$$

where Q is a non-singular quadratic form in x_1, x_2 whose index is the same as the index of the Hessian at x.

Definition 2.3.1 *A smooth function $f : M \to \mathbb{R}$ is called a ξ-function on M if*

(i) Sing $f = \Delta_0 \cup \Delta_1$;
(ii) Δ_0 consists of a finite number of critical points $x_1, \dots x_M$ of M;
(iii) Δ_1 is a disjoint union of a finite number of circles, $C_1 \cup \dots \cup C_N$ such that the index of f is constant on each C_k;

ξ-functions are closely related to the vector fields $v : M \to TM$ on M. Let v be a C^∞ vector field on M and let us denote by L_v the Lie derivative in the direction of the vector field v.

Definition 2.3.2 *A ξ-function on M is called a ξ-function for the vector field v (or, the Lyapunov function of v) if the following conditions are satisfied.*

(i) $L_v f(x) < 0$ for all $x \in M \setminus$ Sing f; that is, f is decreasing along the trajectories of v. In other words, the trajectories of v are transversal to the level lines $f^{-1}(\mathbb{R})$ of f;
(ii) Sing $v \cap \Delta_1 = \emptyset$. In other words, the equilibria of v are distinct from the Δ_1-critical points of f;
(iii) there is a real positive $K > 0$ such that in a small neighborhood $U(C_k)$ of each circle C_k,

$$-L_v f(x) \geq K \inf_{y \in C_k} d^2(x, y), \quad x \in U(C_k).$$

The set of Lyapunov functions is non-empty because of the following statement.

Theorem 2.3.1 (Meyer) *If v is a Morse-Smale foliation, then there exists a ξ-function for v.*

Proof. The idea is to construct a filtration for v (that is, an appropriate family of transversal curves which split M into a chain of submanifolds). Then the standard argument of the Lyapunov theory is applied; see [202]. □

Note that the same construction has been used to prove the structural stability of Morse-Smale foliations. It is not surprisingly therefore that a partial converse of the above theorem is true.

Theorem 2.3.2 (Meyer) *Let $\mathcal{F} \in \mathcal{F}^\infty(M)$ be a C^∞ foliation on M. If there exists an ξ-function for $\mathcal{F}$, then $\mathcal{F}$ satisfies conditions (i)-(iii) of the definition of Morse-Smale foliation. Moreover, v can be approximated in the C^r-topology by Morse-Smale foliations.*

As it follows from the above theorem, ξ-functions define vector fields of 'Morse-Smale type', whose only difference from usual Morse-Smale vector fields is that they admit saddle connections. However, if one sharpens the notion of ξ-functions, this difference can be eliminated.

Definition 2.3.3 *If f is an ξ-function for a vector field $v \in \mathcal{X}^\infty(M)$ such that:*

- *all the sources of v lie in the (level) set $f^{-1}(1)$;*
- *all saddle points of v lie in the set $f^{-1}(0)$;*
- *all the sinks of v lie in the set $f^{-1}(-1)$,*

then f is called a special ξ-function for v.

From the construction of the ξ-function carried out earlier, it follows that f can always be made a special ξ-function.

Corollary 2.3.1 *Let M be a compact two-dimensional manifold and $v \in \mathcal{X}^\infty(M)$. Then a necessary and sufficient condition for v to be Morse-Smale is the existence of a special ξ-function for v.*

Proof. Indeed, suppose to the contrary that there exists a saddle connection. Then f must decrease along this connection. This is in contradiction with the choice of $f^{-1}(0)$ as a level set of all saddles of v. □

The rest of this section is devoted to the question how exactly the Lyapunov functions represent a given Morse-Smale foliation $\mathcal{F}$. For this a notion of equivalence of two (special) ξ-functions is needed. We use a standard one. See

R. Thom, La stabilité topologique des application polynomiales, L'Enseignement Mathematiques 8 (1962), 24-33.

Definition 2.3.4 *Two functions f, g from M to $\mathbb{R}$ are said to be topologically equivalent if there exist homeomorphisms $h_1 : M \to M$ and $h_2 : \mathbb{R} \to \mathbb{R}$ such that the following diagram commutes:*

$$\begin{array}{ccc} M & \xrightarrow{\;f\;} & \mathbb{R} \\ {\scriptstyle h_1}\big\downarrow & & \big\downarrow{\scriptstyle h_2} \\ M & \xrightarrow{\;g\;} & \mathbb{R} \end{array}$$

In general, two ξ-functions for topologically equivalent vector fields are not topologically equivalent since the ξ-functions are defined quite arbitrarily on the singular points $x_1, \ldots, x_M$ and the closed orbits $C_1, \ldots, C_N$. However, if one imposes the equivalence of two ξ-functions on these elements then a kind of uniqueness theorem can be proven [202].

Theorem 2.3.3 (Meyer) *Let $\mathcal{F}_1$ and $\mathcal{F}_2$ be a pair of Morse-Smale foliation on M which are topologically equivalent under the homeomorphism h. Let f_1 and f_2 be ξ-functions for v and w, respectively. If f_1 and f_2 are topologically equivalent on the set of singular points and closed orbits of $\mathcal{F}_1$ and $\mathcal{F}_2$, then they are topologically equivalent.*

Clearly, the special ξ-function fits the assumptions of the above theorem, so that the following statement is true.

Corollary 2.3.2 *Let M be a compact two-dimensional manifold. Let $\mathcal{F}_1$ and $\mathcal{F}_2$ be a pair of C^∞ foliations on M with f_1 and f_2 as their special ξ-functions. Then if f_1 and f_2 are topologically equivalent, then $\mathcal{F}_1$ and $\mathcal{F}_2$ are topologically equivalent.*

Proof. See [202]. □

2.3.2 Lyapunov Graph

By collapsing the components of level sets of the Lyapunov functions of Morse-Smale foliation to distinct points one comes to a *Lyapunov graph*. Directed edges of the graph correspond to the 'order' between the level sets, and directions on the edges correspond to directions of 'growth' of the Lyapunov function between adjacent level sets. K. A. de Rezende and R. D Franzosa [87] studied the realization and equivalence problem for Lyapunov graphs. As far as the equivalence problem is concerned, they solved it in negative by proving that Lyapunov graphs are not complete invariants. (There exist inequivalent Morse-Smale foliations with isomorphic Lyapunov graphs.)

The situation with the realization problem is more hopeful. Rezende & Franzosa found necessary and sufficient conditions for a given Lyapunov graph to correspond to a certain Morse-Smale flow with Lyapunov function f. Below some details are given.

Suppose that $f : M \to I\!R$ is a Lyapunov function associated to a Morse-Smale flow f^t on M. If f has finitely many critical values then it determines a Lyapunov graph. A *Lyapunov graph* X is obtained from f by collapsing to distinct points each component of each level set of f. An edge is drawn between $v_1, v_2 \in X$ if and only if there is an orbit of f^t along which the Lyapunov function decreases from $f^{-1}(c_1)$ to $f^{-1}(c_2)$, where c_1, c_2 are the critical values of f corresponding to v_1, v_2. The edge $e \in E$ is oriented in the direction in which f is decreasing. An *abstract Lyapunov graph* is a finite, connected, oriented graph X which possesses no oriented cycles and the vertices of which are labeled with Λ's taking the values { source, sink, saddle, periodic orbit }. Herewith the 'orientation' is prescribed for the Λ's corresponding to saddles and periodic orbits depending on whether the latter are contained in a cylindric strip (the orientable case) or Möbius strip (the non-orientable case). The equivalence relation on the set of Lyapunov graphs is defined so as to retain the labeling of the underlying graphs.

Let X be a finite Lyapunov orgraph (oriented graph). If $v \in V$ is a vertex of X then by e_v^+ (e_v^-) one denotes the number of in-going (out-going) edges of v. The *cycle rank* of a connected graph X is the number $\beta = 1 - |V| + |E|$, where $|E|$ is the total number of edges and $|V|$ is the total number of vertices of X. The following realization theorem for abstract Lyapunov graphs is valid.

Theorem 2.3.4 *([87]) Let M be closed surface. An abstract Lyapunov graph X is associated to a Morse-Smale foliation $\mathcal{F}$ with a Lyapunov function f on M if and only if the following conditions are satisfied:*

(1) (local conditions) If the vertex v is labeled with Λ and
  (a) Λ is a source (sink) then the number of exiting (entering) edges e_v^- (e_v^+) is equal to one;
  (b) Λ is a saddle then $1 \le e_v^+ \le 2$, $1 \le e_v^- \le 2$ and $e_v^+ + e_v^- \le 3$; moreover, Λ is orientable if and only if $e_v^+ + e_v^- = 3$;
  (c) Λ is a repelling (attracting) periodic orbit then $e_v^- \le 2$ ($e_v^+ \le 2$); moreover, Λ is orientable if and only if $e_v^- = 2$ ($e_v^+ = 2$).

(2) (global conditions)
  (a) If M is orientable, the cycle rank β of X must be equal to the genus of M;
  (b) If M is non-orientable, twice the cycle rank β of X plus the number of non-orientable Λ's must be equal to the genus of M.

Proof. We give a sketch of the proof, referring the reader to [87]. The local conditions come out from the step-by-step calculation of Conley's homology index $CH_i(\Lambda)$ for each Λ entering the Morse-Smale foliation. This index can be evaluated from the long exact sequence for Λ. Note also that the range of i is $\{0, 1, 2\}$.

As far as the global conditions are concerned, they are also almost evident. Actually they correspond to the Euler formula for the triangulation of M. (Attention should be paid to the non-orientable case since non-orientable Λ's 'modifies' the Euler formula by adding a Möbius strip to M.) □

2.4 Connected Components of Morse–Smale Foliations

One can look at the space $\mathcal{F}_0 \subset \mathcal{F}^r(M)$ of Morse-Smale foliations as consisting of 'connected components'. The equivalence class associated to corresponding Morse-Smale vector field $v \in \mathcal{F}_0$ can occupy several connected components if one demands the conjugating homeomorphism $h : M \to M$ to be isotopic to the identity. This approach reveals strong connections with Nielsen's theory of automorphisms of compact surfaces.

One says that flows $f^t,\ g^t \in X_0$ are *isotopically equivalent* if there exists a continuous mapping $F : [0,1] \to X^r(M)$ such that $F(0) = f^t,\ F(1) = g^t$, and for all $t \in [0,1]$ F_t belongs to one and the same equivalence class of Morse-Smale foliations. Isotopically equivalent Morse-Smale foliations form a *connected component.* Clearly, there exist topologically equivalent Morse-Smale foliations lying in the different connected components. For example, two foliations with isomorphic Peixoto graphs and mirror symmetric rotation systems. In general, the larger the group $Aut\ M$ is, the more connected components equivalent Morse-Smale foliations may have.

In 1977 C. Gutierrez and W. de Melo [129] described connected components of the space of Morse–Smale flows on orientable compact surfaces. Although the results were formulated for flows, they are certainly true for Morse-Smale foliations.

A gradient-like flow is said to be *polar* if it has only one source and one sink. Let f^t be a polar Morse–Smale flow with a single source $\alpha(f^t)$. Two ω-separatrices of every saddle σ of f^t along with $\alpha(f^t) \cup \sigma$ form a closed simple curve in M that we call a *stable cycle* of f^t.

Theorem 2.4.1 *([129]) Let $f^t,\ g^t$ be two polar Morse–Smale flows. Then $f^t,\ g^t$ are isotopically equivalent if and only if there is a bijection between the stable cycles of f^t and the stable cycles of g^t such that corresponding cycles are homotopic on M.*

Proof. See [129]. □

Let f^t be a Morse–Smale flow. There exist transversal circles C_1, C_2 bounding disjoint disks D_1, D_2, respectively, such that D_1 (D_2) contains all the sources (sinks) of f^t. The *stable cycle* of a saddle $\sigma \in M \setminus (D_1 \cup D_2)$ is a simple closed curve which is the union of ω-separatrices of σ and an interval contained in $int\, D_1$. A point $p \in C$ is said to be of type 1 (type 2) with respect to C_1, C_2 if it belongs to the $\omega(\alpha)$-separatrix of a saddle contained in D_1 (D_2). A point $p \in C_1$ is of type 3 if it belongs to the ω-separatrix of a saddle in $M \setminus (D_1 \cup D_2)$.

Let $\tilde{\pi}(f^t)$ denote the set of all 6-collections $(C_1,\ C_2,\ F,\ S,\ T,\ \Gamma)$ where F, S, T are sets of points of type 1, 2, 3, with respect to C_1, C_2 and Γ is the set of stable cycles of f^t. We say that $(C_1,\ C_2,\ F,\ S,\ T,\ \Gamma)$ is *equivalent* to $(\bar{C}_1,\ \bar{C}_2,\ \bar{F},\ \bar{S},\ \bar{T},\ \bar{\Gamma})$ if there is an orientation preserving diffeomorphism $h : C_1 \to \bar{C}_1$ with the following properties:

- $h(F) = \bar{F},\ h(S) = \bar{S},\ h(T) = \bar{T}$;
- If $p,\ q \in C_1$ belong to ω-separatrices of the same saddle of f^t then $h(p),\ h(q)$ belong to the ω-separatrices of the same saddle;
- The stable cycles $h(p) \in T$ are homotopic.

Let $\pi(f^t)$ denote the set of equivalence classes of $\bar{\pi}(f^t)$. For a Morse–Smale flow f^t the set $\pi(f^t)$ is finite.

Theorem 2.4.2 *([129]) Let f^t, g^t be two gradient-like Morse–Smale flows. Then f^t, g^t are isotopically equivalent if and only if $\pi(f^t) = \pi(g^t)$.*

Proof. This consists in the construction of 'filtrations' for the respective flows; see [129]. □

Consider a Morse–Smale flow f^t with a non-empty set $P(t)$ of periodic trajectories. Let $M_1, \dots, M_l$ be the connected components of $M \setminus P(f^t)$. If we assume that each hole of M_i $(i = 1, \dots, l)$ is a fixed point then $f^t|_{M_i}$ is a Morse–Smale flow without periodic trajectories. We can define sets $\pi(f^t|_{M_i})$, $i = 1, \dots, l$ in the obvious way. Let $\pi(f^t) = (\pi(f^t|_{M_1}), \dots, \pi(f^t|_{M_l}))$. The equality $\pi(f^t) = \pi(\varphi^t)$ makes sense only when the set of periodic trajectories of $f^t \in \Sigma(M)$ and $\varphi^t \in \Sigma(M)$ coincide. We say that $\pi(f^t)$ is *equivalent* to $\pi(g^t)$ if there is a $\varphi^t \in \Sigma(M)$ isotopically equivalent to g^t such that $\pi(f^t) = \pi(\varphi^t)$.

Theorem 2.4.3 *([129]) Flows f^t, $g^t \in X_0$ are isotopically equivalent if and only if $\pi(f^t)$ is equivalent to $\pi(g^t)$.*

Proof. This is similar to the proof of Theorem 2.4.2. □

2.5 Degrees of Stability

Definition 2.5.1 (Andronov-Leontovich) *The foliation $\mathcal{F} \in \mathcal{F}^r \backslash \mathcal{F}_0^r$ is said to be first-order structurally stable if there exists a neighbourhood U of $\mathcal{F}$ in the space $\mathcal{F}^r$ such that each structurally unstable foliation $\mathcal{G} \in U$ is topologically conjugate to $\mathcal{F}$.*

(The first-order stability can be thought of as a relative stability in the space of unstable foliations. As was shown by Sotomayor [268], this set coincides with the set of transversal intersections of typical one-parameter families of foliations with the boundary of $\mathcal{F}_0^r$.)

Suppose that $\mathcal{F} = (M, \pi^t, \theta)$ is an orientable foliation on a compact orientable surface M. The following result is due to Andronov and Leontovich [2]; see also [4], [268], [269].

Theorem 2.5.1 (Andronov-Leontovich) *A C^r-foliation $\mathcal{F} \in \mathcal{F}^r$ is first-order structurally stable if and only if it belongs to the (non empty) set*

$$\mathcal{F}_1^r \stackrel{\text{def}}{=} A \cap (\cup_{i=1}^3 B_i),$$

where

- *A is a set of C^r-foliations every leaf of which is either a singularity, a closed leaf (cycle) or, a regular leaf tending to a singularity or cycle;*

- *B_1 is a set of C^r-foliations which have one singular point with zero eigenvalue (saddle-node of multiplicity 2) or fine focus of cyclicity 1, all other singular points and cycles are hyperbolic and there are no saddle connections;*
- *B_2 is a set of C^r-foliations with the unique non-hyperbolic double [2] cycle, all other cycles and singular points are hyperbolic and there are no saddle connections; moreover, there are no pairs of saddle separatrices which wind around the double cycle;*
- *B_3 is a set of C^r-foliations which have a unique saddle connection, and if it makes a saddle loop, then the saddle value $\lambda_1 + \lambda_2$ (where λ_i are the eigenvalues of the saddle) does not vanish; all singular points and cycles are hyperbolic; moreover, the separatrix of a saddle cannot wind around (from the inside) the separatrix loop.*

(It can be easily seen that conditions $B_1 - B_3$ are derived by violating one of the conditions of Theorem 2.1.1.) In the case of non orientable foliations, the above list must be extended by

- B_4 is a set of C^r-foiations which have only one weak non orientable singularity (see Chapter 1), all other non orientable singularities are typical and structurally stable, orientable singularities and cycles are hyperbolic, and there are no saddle connections;
- B_5 is a set of C^r-foliations which have orientable singularities, hyperbolic cycles and typical non orientable singularities; these foliations have no saddle connections between orientable (hyperbolic) saddles, but have one of the saddle connections, shown in Fig. 2.4 (i)-(xi);

moreover, the saddle values in cases (i) and (ii) and the contour value in case (iii) do not vanish. The separatrices of saddles cannot wind on the separatrix loop of cases (i) and (ii).

The bifurcations of foliations of class B_4 have been studied earlier. Our main purpose here will be the study of bifurcations in class B_5 (figures (i)-(xi)). Furthermore, as it can be easily seen, in cases (iv)-(xi) the break-up of separatrix connections (by rotation of the corresponding vector fields, e.g.) will lead to stabilization of the separatrices on (hyperbolic) cycles or singular points, cf [234]. Thus, these bifurcations yield no extra cycles or singular points of the foliation $\mathcal{F}$. We shall study cases (i)-(iii), which we call *tripod*, *sun-set* loop bifurcation and *thorn* bifurcation. Later we shall see that these kinds of bifurcations generate cycles of $\mathcal{F}$.

Let $\mathcal{F} \in \mathcal{F}^r(M)$ be a C^r-foliation with a singularity (orientable or not) which has a saddle sector. Let p_1 and p_2 be the points in $W^s(0)$ and $W^u(0)$, respectively. Furthermore, let S_1 and S_2 be a C^2 segments through p_1 and p_2, transversal to $\mathcal{F}$.

[2] That is, the return mapping $x \mapsto h(x), h(0) = 0$ has a double root at 0.

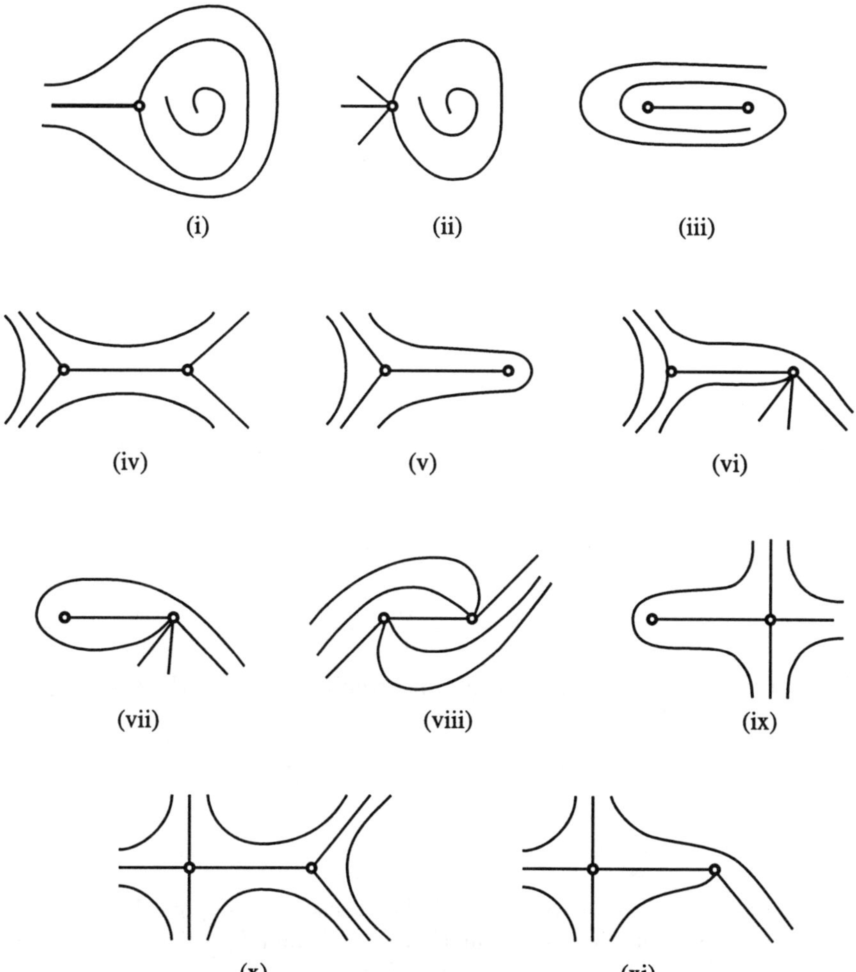

Fig. 2.4. Saddle connections.

Definition 2.5.2 *The Dulac (or transition) mapping near the saddle sector of a singularity is the mapping* $T : S_1 \to S_1$ *defined by the leaves of* $\mathcal{F}$ *between* $S_1 \backslash p_1$ *and* $S_2 \backslash p_2$.

Lemma 2.5.1 *Let* 0 *be a structurally stable non orientable singularity of tripod, sun-set or thorn types. Then the Dulac mappings* $T : S_1 \to S_1$ *in the vicinity of* 0 *are given by the following table:*

Name	Form	Conditions	Dulac function	Saddle value
tripod	$w_0^{(1)}$	$\alpha\beta(\alpha+\beta-1)<0$, $(\alpha-1)(\beta-1)\times$ $\times(\alpha+\beta)>0$	$y=\lvert x\rvert^{\alpha/\beta}+\ldots$	$\alpha-\beta\neq 0$
sun-set	$w_0^{(1)}$	$\alpha\beta(\alpha+\beta-1)>0$	$y=\lvert x\rvert^{\alpha/\beta}+\ldots$	$\alpha-\beta\neq 0$
thorn	$w_0^{(2)}$	$\alpha<0$	$y=e^{\pi\beta/\alpha}\lvert x\rvert+\ldots$	

Proof. The germs of structurally stable non orientable singularities are C^0-equivalent to either of their 2-jets:

$$w(u,v)=[\alpha u^2+(\beta-1)uv]\frac{\partial}{\partial u}+[(\alpha-1)uv+\beta v^2]\frac{\partial}{\partial v};$$

$$w(u,v)=[\alpha u^2+\beta uv-v^2]\frac{\partial}{\partial u}+[(\alpha+1)uv+\beta v^2]\frac{\partial}{\partial v}.$$

Both of the above normal forms are integrable and admit the first integrals given, respectively, by

$$|u|^\beta|v|^\alpha|u-v|^{1-\alpha-\beta}=C$$

$$\frac{\alpha+1}{2}\log|\frac{u^2}{v^2}+1|+\beta\arctan\frac{u}{v}+\log|v|=C.$$

It is classical that the germ $T(0)$ of the transition mapping $T:S_1\to S_2$ does not depend on the particular choice of S_1 and S_2. Thus, we consider $S_1=(-1,x), S_2=(y,1)$ and $S_1=(-1,x), S_2=(1,y)$ for the above two cases, respectively. The transition mapping $y=T(x)$ is implicitly given by

$$|x|^\alpha|1+x|^{1-\alpha-\beta}=|y|^\beta|1+y|^{1-\alpha-\beta},$$

$$\frac{\alpha+1}{2}\log|1+x^2|-\beta\arctan\frac{1}{x}-\alpha\log|x|=$$

$$=\frac{\alpha+1}{2}\log|1+y^2|+\beta\arctan\frac{1}{y}-\alpha\log|y|.$$

It readily follows from the above formulas that the main part of $T(0)$ for the normal form (1.9) and (1.11) is as stated in the lemma. □

Let $\mathcal{F}_0$ be a C^∞ foliation with a tripod (sun-set) at the point $\tau\in M$. We assume that there exists a regular leaf γ which tends along both ends to τ (see figures (i) and (ii)) and let $\Gamma = \gamma \cup \tau$. Γ is called a tripod (sun-set) loop or homoclinic saddle connection. Take a point $p \in \gamma$ with a transversal segment through p and let x be a parameter of this segment with $I = \{x | x \geq 0\}$ lying inside the loop Γ. Next let $\mathcal{F}_\mu = (M, \pi_\mu^t, \theta_\mu)$ be a one-parameter family of C^∞-foliations which are smooth in the parameter μ, and such that $\mu = 0$ corresponds to $\mathcal{F}_0$. Denote by $x_1 = \phi(\mu)$ and $x_2 = \psi(\mu)$ the points at which tripod's separatrices (the first and the second, respectively) hit the segment I for the first time. Clearly, ϕ and ψ are smooth, and $\phi(0) = \psi(0) = 0$. Suppose also that the family $\mathcal{F}_\mu$ is chosen so that

$$\frac{d}{d\mu}(\phi - \psi)(0) = 0.$$

Theorem 2.5.2 *Let $\mathcal{F}_\mu$ be the above mentioned family of foliations on a compact surface M, and suppose the saddle value of the tripod (sun-set) does not vanish for all values of μ. Then in an ε-vicinity $O_\varepsilon(\gamma)$ of the loop γ: (i) there are no cycles if $\mu \leq 0$; (ii) there is a unique hyperbolic cycle if $\mu > 0$.*

Proof. Let us first prove the existence and uniqueness of the cycle. Consider the return mapping $x \mapsto x + H(x, \mu)$, where $H(x, \mu)$ is the shift function for a set of points $X \subseteq I$. Clearly, $H = \Gamma_1 \circ \Gamma_2 - Id$, where Γ_1 is the Dulac function in the vicinity of the tripod (sun-set), and Γ_2 is a transition mapping along the regular part of the loop. We know that $\Gamma_1(x) = |x|^{\alpha(\mu)/\beta(\mu)} + \ldots$, while $\Gamma_2 = \sum_{m \geq 1} a_m(\mu) x^m$ is a regular function. The main part of the shift function is given by

$$H(x, \mu) = a_1 |x|^{\alpha(\mu)/\beta(\mu)} - x + \ldots$$

(Note that without loss of generality one can assume $\alpha(\mu) > \beta(\mu)$; otherwise, it is enough to invert the time.) Since we assumed $\alpha(\mu) \neq \beta(\mu)$, the partial derivative $\partial H / \partial x$ is defined, continuous and does not vanish at $x = 0$. On the other hand, $H(0, \mu) = \phi(\mu) - \psi(\mu)$ and we supposed $\partial H / \partial \mu = 0$ at $\mu = 0$. (Observe also that $H(0, 0) = 0$.) The equation

$$H(x, \mu) = 0$$

is the cycle equation (cycles correspond to the points x at which the shift function vanishes). By the Implicit Function Theorem, there exists a continuous invertible function $x = x(\mu)$, which satisfies $H(x, \mu) = 0$. This means that the cycle exists, is unique, depends continuously on μ and vanishes when $\mu \leq 0$ (since if $\mu < 0$ we are in the domain $x < 0$ where H is not defined).

Now we shall establish the hyperbolicity of the obtained cycle γ. Let $x \mapsto h(x)$ be the return mapping of the cycle γ, and let $p_\mu \in \gamma$ be the point of intersection with the transversal segment I. We shall check the following hyperbolicity criterion (see [61]):

$$|h'(0)| = \exp(\int_0^{T_\mu} div_\Gamma \mathcal{F}_\mu[\pi_\mu^t(p_\mu)]dt) \neq 1,$$

where T_μ is the period of γ. Splitting $T_\mu = T_{1\mu} + T_{2\mu}$, where $T_{1\mu}$ is the time of passage of the saddle sector and $T_{2\mu}$ is the time of passage near the regular part of loop, one can split the integral into the following parts:

$$\int_0^{T_\mu} div_\Gamma \mathcal{F}_\mu[\pi_\mu^t(p_\mu)]dt = \int_0^{T_{1\mu}} + \int_{T_{1\mu}}^{T_{1\mu}+T_{2\mu}}.$$

$T_{1\mu}$ grows infinitely as γ tends to the separatrix: $\lim_{x\to\phi(\mu)} T_{1\mu} = +\infty$ (it follows from the well known estimate $T_\mu \approx -\ln^{-1}|x-\phi(\mu)|$). Hence the first expression is bounded from above: $\int_0^{T_{1\mu}} \le T_{1\mu}(\beta-\alpha)/2 < 0$. The second one is uniformly bounded by the constant $-K$. Now the whole expression $|h'(0)|$ is $<< 1$, which satisfies the criterion $|h'(0)| \neq 1$. □

Let us consider the break-up of the thorn connection in to case (iii). This connection is covered by a separatrix contour $\Gamma = \gamma_1 \cup \gamma_2 \cup \nu_1 \cup \nu_2$ consisting of pair of heteroclinic saddle connections γ_1, γ_2 and pair of 2-saddles ν_1, ν_2. The involution θ sends γ_1 to γ_2 and fixes the points ν_1 and ν_2.

Definition 2.5.3 *Let $\mathcal{F}_0$ be a foliation with separatrix contour Γ. A contour value A is defined to be the divergence of the foliation $\mathcal{F}_0$ along the contour Γ,*

$$A = \int_\Gamma div\mathcal{F}_0.$$

Let $\mathcal{F}_\mu = (M, \pi_\mu^t, \theta_\mu)$ be a one-parameter family of C^∞-foliations which is continuous and smooth for $\mu \ge 0$ and with $\mu = 0$ corresponding to the foliation $\mathcal{F}_0$. Suppose also that the positive values of μ break-up the contour Γ.

Theorem 2.5.3 *Let $\mathcal{F}_\mu$ be the above family of foliations on a compact surface M, and suppose the contour value $A(\mu)$ does not vanish for all values of μ. Then in a one-sided ε-vicinity $O_\varepsilon^+(\Gamma) = \{\Gamma \times \varepsilon | \varepsilon \ge 0\}$ of the separatrix contour Γ: (i) there are no cycles if $\mu = 0$; (ii) there is a unique hyperbolic cycle if $\mu \ge 0$.*

Proof. Take a point $p \in \Gamma$ on the separatrix contour Γ and let $I = \{x | x \ge 0\}$ be a small transversal segment through p, directed inside Γ. Consider the return mapping $x \mapsto x + H(x,\mu)$ on I, where $H(x,\mu)$ is a shift function. $H(x,\mu)$ is decomposable into $H = \Gamma_1 \circ N_2 \circ \Gamma_2 \circ N_1 - Id$, where N_1 and N_2 are the Dulac functions in the vicinity of 2-saddles ν_1 and ν_2, while Γ_1 and Γ_2 are transition mappings along regular parts γ_1 and γ_2 of Γ. We know that $\Gamma_i(x) = \exp(\pi\beta_i(\mu)/\alpha_i(\mu))|x| + ...$, $i = 1,2$, while $N_1 = \sum_{m\ge1} a_m(\mu)x^m$ and $N_2 = \sum_{m\ge1} b_m(\mu)x^m$ are analytic for small x. The main part of the shift function is given by

$$\begin{aligned} H(x,\mu) &= [a_1(\mu)b_1(\mu)e^{\pi(\frac{\alpha_1(\mu)}{\beta_1(\mu)}+\frac{\alpha_2(\mu)}{\beta_2(\mu)})} - 1]|x| + ... \\ &= A(\mu)|x| + ... \end{aligned}$$

(Note that $A(\mu)$ coincides with the divergence of $\mathcal{F}_\mu$ along Γ.) Since we assumed $A(\mu) \neq 0$, the partial derivative $\partial H/\partial x$ is defined, continuous and does not vanish in $x = 0$. By the Implicit Function Theorem applied to the cycle equation $H(x,\mu) = 0$, there exists a continuous invertible function $x = x(\mu)$, which satisfies $H(x,\mu) = 0$. This implies that the cycle exists, is unique, depends continuously on μ and vanishes at $\mu = 0$. The hyperbolicity of the resulting cycle γ can be established with the help of the criterion mentioned above. Theorem is proved. □

Remark 2.5.1 *Since the above family $\mathcal{F}_\mu$ keeps $\mathbb{Z}_2$-symmetry of the corresponding vector field, a pair of hyperbolic cycles γ_1 and γ_2 appears: one in the one-sided tubular neighborhood $O_\varepsilon^+(\Gamma)$ and another in $O_\varepsilon^-(\Gamma) = \{\Gamma \times \varepsilon | \varepsilon \leq 0\}$. They are involutive: $\theta(\gamma_1) = \gamma_2$.*

One obtains the following theorem.

Theorem 2.5.4 *Let $\mathcal{F} \in \mathcal{F}^r$ be a (non orientable) foliation of class C^r. Then $\mathcal{F}$ is first-order structurally stable if and only if it belongs to the (non empty) set*

$$\mathcal{F}_1^r \stackrel{\text{def}}{=} A \cap (\cup_{i=1}^5 B_i).$$

Proof. This follows from the above theorems. (It only remains to notice that in class B_5 saddle separatrices cannot wind onto loops (i) and (ii); otherwise after loop bifurcation saddle connection must appear. This fact is established as in the orientable case.) □

Bibliographic Notes. The structural stability theorem for Morse-Smale flows on 2-manifolds has been proved by A. A. Andronov and L. S. Pontryagin [5] and M. M. Peixoto [236]. It was generalized to foliations in [67]. For special classes of foliations stability theorems have been considered by V. Guíñez [122] and I. Liousse [190]. Equivalence and realization problems for Morse-Smale flows have been studied by Meyer [202], Franks [110] and others using Lyapunov (energy) functions. M. M. Peixoto [236] introduced for this purpose certain 'distinguished graphs' endowed with combinatorial structure. This approach admits improvements; see [222]. Similar problems for foliations have been considered in [66]. The examples of Peixoto graphs are taken from [63]. First-order structural stability has been introduced by Andronov and Leontovich [2]. Sotomayor [268], [269] considered this class in connection with typical bifurcations of one-parameter families of vector fields on manifolds. For foliations the same problem is studied in the unpublished preprint [218]. Tripod and sun-set loop bifurcations of curvature lines have been taken up by Gutierrez and Sotomayor [133], [134].

3. Foliations Without Holonomy

The holonomy of a leaf is a mapping between the transversals to the leaf. If this mapping is the identity then the leaf is said to be "without" holonomy. When all leaves of a foliation $\mathcal{F}$ satisfy such a property then $\mathcal{F}$ is called a *foliation without holonomy.* Closed 1–forms define a class of orientable foliations without holonomy. Holomorphic quadratic differentials define a class of non orientable foliations without holonomy. The converse statements are true as well.

Foliations without holonomy constitute a residual set in the space $\mathcal{F}^r(M)$ which is a "complement" to the set of Morse–Smale foliations. Similar to the case of Morse–Smale foliations there are two types of "elementary cells" to which these foliations can be decomposed. Such cells are either "periodic" or "quasiminimal". In a contrast to the elementary cells of Morse–Smale foliations the structure of the elementary cells of foliations without holonomy is much more complicated.

3.1 Periodic Components

Orientable Case

We work with the foliations of the form $\mathcal{F} = (M, \pi^t, \theta)$ assuming that Sing $\mathcal{F} < \infty$. Point $x \in M$ is called a non-wandering point of the flow π^t if for any neighbourhood U of x it holds $U \cap \pi^t(U) \neq \emptyset$ for some moments of time $|t| > T$ where $T > 0$ can be chosen arbitrary big. A flow is called *non-wandering* if all points of M are non-wandering. It is not hard to see that $\mathcal{F}$ is without holonomy if and only if π^t is a non–wandering flow.

Theorem 3.1.1 *Let π^t be a non–wandering flow on an orientable, closed surface M and such that all recurrent trajectories of π^t are trivial. Then M can be represented as*

$$M = M_1 \bigcup \ldots \bigcup M_k,$$

where Int M_i are pairwise disjoint regions which obey the following conditions:

(i) the boundary ∂M_i is a union of separatrix cycles;

(ii) *M_i is either a sphere S^2 with equatorial flow (in this case $M = M_i = S^2$), or an oriented annulus or disc filled-up with the periodic trajectories, or a whole torus filled-up with the periodic trajectories (in the latter case $M = M_i = T^2$).*

Proof. Because every recurrent trajectory of π^t is trivial and π^t is non-wandering, it follows that M is partitioned by the separatrix cycles into discs and annuli. If M' is a union of the separatrix cycles, then $M \backslash M'$ consists of regions filled-up with the periodic trajectories. Taking into account the exceptional cases of sphere and torus, theorem follows. □

Non Orientable Case

Theorem 3.1.2 *Let M be a compact surface, orientable or non-orientable. If π^t is a non-wandering flow on M with trivial recurrent trajectories, then M can be represented as a union*

$$M = M_1 \bigcup \ldots \bigcup M_k,$$

of regions with the pairwise disjoint interiors and such that the boundary of ∂M_i's is a union of the separatrix cycles of π^t. An interior of M_i's is either of the following:

(i) *a disc region with the centre-type singularity;*
(ii) *an annulus filled-up with the periodic trajectories;*
(iii) *a Möbius strip filled-up with the periodic trajectories;*

Moreover, there are four exceptional cases: M_i is a sphere with an equatorial flow (in this case $M = S^2$); a projective plane P^2 with a center, filled-up with the periodic trajectories (in this case $M = P^2$); a torus T^2 filled-up with the periodic trajectories (in this case $M = T^2$); or, finally, a Klein bottle K^2 filled-up with the periodic trajectories (in this case $M = K^2$).

Proof. This is similar to the proof of Theorem 3.1.1. □

3.2 Quasiminimal Sets

The quasiminimal set may occupy a connected region in M or be of a Cantor type. In this section the structure of both types of quasiminimal sets is studied.

3.2.1 Structure of a Quasiminimal Set

Theorem 3.2.1 (Cherry-Maier) *Let $\mathcal{F} = (M, \pi^t, \theta)$ be a foliation with a non trivial recurrent leaf. Denote by $\mathcal{Q}(x)$ a quasiminimal set of a recurrent leaf through the point $x \in M$. Then $\mathcal{Q}(x)$ contains a continuum of non trivial recurrent leaves of $\mathcal{F}$. Moreover, every leaf of the continuum is dense in the $\mathcal{Q}(x)$.*

Proof. We consider the covering flow π^t to $\mathcal{F}$. Our proof breaks into three parts. First we construct a small tubular flow box around the recurrent semi-orbit through x, such that it looks (in a natural metric on M) as a 'snake eating its tail'. Next an infinite procedure of the binary coding (similar to the Cantor diagonal process) is applied to prove that the set of the recurrent semi-orbits, lying in $\mathcal{Q}(x)$, is uncountable. Finally, the density of the newly obtained exceptional semi-orbits will be proved.

(i) Denote by $\gamma(x)$ an ω-recurrent semi-orbit of the flow π^t passing through the point $x \in M$. Let Σ be a small segment, π^t-transversal at x. Since, by the recurrence, $\gamma(x) \subset \omega(x)$, the set $\Sigma \cap \gamma(x)$ consists of an infinite countable number of points (not necessarily dense in Σ). Suppose that $y \in \Sigma \cap \gamma(x)$ is an 'inner' point of $\Sigma \cap \gamma(x)$ (that is in a natural metric on Σ, $\inf \Sigma \cap \gamma(x) < y < \sup \Sigma \cap \gamma(x)$).

Fix a metric on Σ and take a small neighbourhood $O_\varepsilon(y)$ of the point y on Σ. $\varepsilon > 0$ can be chosen so small that no one orbit through $O_\varepsilon(y)$ goes to a singularity of the flow π^t. Let $\Pi = \pi^t(O_\varepsilon(y))$ be a flow box, formed by the orbits through $O_\varepsilon(y)$ and the transversal Σ. Denote by σ_0 and σ_1 segments, lying in $O_\varepsilon(y)$. They satisfy the following conditions:

(1) $\text{int } \sigma_i \cap \gamma(x) \neq \emptyset$ for $i = 0, 1$;
(2) $\text{int } \sigma_0 \cap \text{ int } \sigma_1 = \emptyset$;
(3) Π is a standard flow box.

Indeed, because $\sigma_i \subset O_\varepsilon(y) \subset \Sigma$, there exists countable many points at σ_i, which belong to the recurrent semi-orbit $\gamma(x)$. So the condition (1) follows. Since $\gamma(x)$ is not periodic, the sets σ_0 and σ_1 are separated on Σ by a point y. Thus the condition (2) holds. Finally, Π is a standard flow box, because ε has been chosen sufficiently small and it does not contain any orbits going to a singularity. The condition (3) follows.

(ii) Now we shall describe an infinite binary coding on Σ, of the newly obtained segments σ_i. Let us repeat all the arguments of the item (i), applied to the segment Σ, now having in mind the segments σ_i. What is obtained will be a pair of segments σ_{00}, σ_{01} for the segment σ_0, and a pair of segments σ_{10}, σ_{11} for the segment σ_1.

In general, let $\sigma_{i_1 i_2 \ldots i_\tau}$ be an encoded segment, with the subscripts taking the values 0 or 1. After the procedure described in the item (i), this will give us a pair of segments, $\sigma_{i_1 i_2 \ldots i_\tau i_{\tau+1}}$, where

$$i_{\tau+1} = \begin{cases} 0, \text{ for the 'left' segments,} \\ 1, \text{ for the 'right' segments.} \end{cases}$$

Proceeding by the induction, one will obtain a diagram of inclusions of Fig 3.1. (An arrow between an upper set and a lower set means that the lower

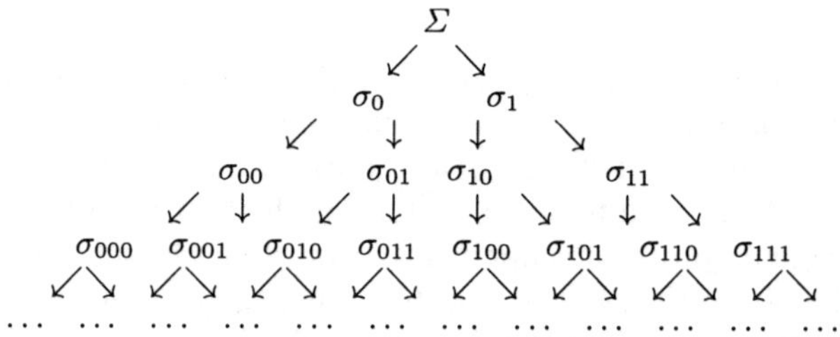

Fig. 3.1. The binary coding.

set is included into the upper set.) As a result, one obtains a family of closed segments, $\sigma_{i_1 i_2 \ldots i_\tau}$, $i_k \in \{0,1\}$, which obeys the following conditions.

(1) $\sigma_{i_1 i_2 \ldots i_{\tau+1}} \subset \sigma_{i_1 i_2 \ldots i_\tau}$;
(2) $\sigma_{i_1 i_2 \ldots i_\tau} \cap \sigma_{j_1 j_2 \ldots, j_\tau} = \emptyset$, if and only if $i_k \neq j_k$ for some $k \in \{1, \ldots, \tau\}$;
(3) int $\sigma_{i_1 i_2 \ldots i_\tau} \cap \gamma(x) \neq \emptyset$;
(4) in a fixed metric the length of the segments vanishes as τ grows, $\lim_{\tau\to\infty} |\sigma_{i_1 i_2 \ldots i_\tau}| = 0$.

Indeed, (1) and (2) follow from the encoding procedure and condition (2) of the item (i). By (1) of the item (i), the interior of every $\sigma_{i_1 i_2 \ldots i_\tau}$ must contain an infinite (countable) number of the common points with the recurrent semi-orbit $\gamma(x)$. So, (3) follows. Finally, the the length of the intervals, which enter a disjoint infinite family (of intervals) necessarily vanishes and condition (4) follows.

To every infinite sequence $i_1 i_2 \ldots i_k \ldots$ of the binary symbols $i_k \in \{0,1\}$, we put in correspondence a point

$$\xi_{i_1 i_2 \ldots i_k \ldots} = \cap_{k=1}^{\infty} \sigma_{i_1 i_2 \ldots i_k}.$$

(Note that by the Principle of Closed Segments, the above intersection is non-empty.) It follows from the property (4), that $\xi_{i_1 i_2 \ldots i_k \ldots}$ is a point. By the property (2) there exists uncountably many points of such kind, which form a continuum at Σ. Clearly, there exists a continuum of the semi-orbits $\pi^t(\xi_{i_1 i_2 \ldots i_k \ldots})$ passing through the segment Σ and distinct from $\gamma(x)$. Let us show that $\pi^t(\xi_{i_1 i_2 \ldots i_k \ldots})$ is recurrent.

Indeed, by the properties (3) and (4) for the ω-limit sets of points x and $\xi_{i_1 i_2 \ldots i_k \ldots}$, it holds $\pi^t(\xi_{i_1 i_2 \ldots i_k \ldots}) \subset \omega(x)$ and $\gamma(x) \subset \omega(\xi_{i_1 i_2 \ldots i_k \ldots})$. (Note that by the property (3) of the item (i), the limit set $\omega(\xi_{i_1 i_2 \dots i_k \dots})$ is different from a singular point.) The above inclusions imply that

$$\pi^t(\xi_{i_1 i_2 \dots i_k \dots}) \subset \omega(x) \subseteq \omega(\xi_{i_1 i_2 \dots i_k \dots}).$$

Thus, $\pi^t(\xi_{i_1 i_2 \dots i_k \dots}) \subset \omega(\xi_{i_1 i_2 \dots i_k \dots})$, and therefore the semi-orbit $\pi^t(\xi_{i_1 i_2 \dots i_k \dots})$ is exceptional.

(iii) The density of the semi-orbit $\pi^t(\xi_{i_1 i_2 \dots i_k \dots})$ in the $\mathcal{Q}$-set $\omega(x)$ follows from the inclusion $\omega(x) \subseteq \omega(\xi_{i_1 i_2 \dots i_k \dots})$. Finally, one obtains a continuum of the recurrent ω-orbits, lying in the quasiminimal set $\mathcal{Q}(x)$, all of which are dense therein.

Theorem is proved. □

Theorem 3.2.2 (Maier) *Let $\mathcal{F} = (M, \pi^t, \theta)$ be a foliation with a finite number of singularities on a compact surface M. Suppose that an open semi-orbit $l(x)$ of π^t through the point $x \in M$ lies in an α (or ω) limit-set of some other orbit, $l(y)$, of π^t. If in $\alpha(x)$ (or $\omega(x)$) there is a point, distinct from a singularity, then $l(x)$ is an open recurrent semi-orbit of $\mathcal{F}$.*

Proof. Suppose that $x_0 \in \omega(x)$ is a point which is not the singularity of $\mathcal{F}$. Let T be a transversal segment through x_0. There exists an infinite sequence of points $x_d \in l(x) \cap T$, such that $x_d \to x_0$. (Taking a subsequence in $\{x_d\}$ one can always assume that $\{x_d\}$ is monotone and belongs to one of the components $T \backslash x_0$.)

Suppose to the contrary, that $l(x)$ is not recurrent. Then there exists an interval $I \subset T$, such that $x_1 \in I$ and there are no other points of $l(x)$ at I. Roughly, to obtain a contradiction one must 'catch' $l(x)$ into a 'Bendixson sack', formed by a finite number of arcs of the semi-orbit $l(y)$. It can be done as follows.

First, let us show that there exists an infinite family of arcs $\{\lambda_m\}_{m=1}^{\infty}$ of the semi-orbit $l(y)$ such that for all m:

(i) the endpoints of λ_m (and only such points) lie at I;
(ii) λ_m intersects the set $T \backslash I$ only at the points of the segment $[x_d, x_{d+1}]$ for some $d = d(m) \in \{2, 3, \dots\}$.

Indeed, (i) is satisfied by the density of $l(y)$: $x_1 \subset \omega(y)$. Also because of the density of $l(y)$ and by the continuity of the orbits with respect to the choice of the initial point, the arcs λ_m can be chosen to hit $T \backslash I$, only 'between' two consequent points x_d and x_{d+1}. So the property (ii) holds.

Suppose that M' is a surface with a boundary, obtained from M by cutting along the segment I. (All notations for M' are kept the same.) Starting from the family $\{\lambda_m\}_{m=1}^{\infty}$, one can choose a finite number N of disjoint arcs (N depends on the genus of M, supposed to be finite), such that the arcs $\lambda_1, \dots, \lambda_N$ bound a disc region G on M'. The semi-orbit $l(x)$ enters G and remains inside G for all moments of time $t \geq t_0$. By the property (ii), region G is separated from the point x_0 and cannot belong to the ω-set of $l(x)$ what is a contradiction. □

Theorem 3.2.3 (Maier) *Let $\mathcal{F} = (M, \pi^t, \theta)$ be a foliation with a finite number of singularities on a compact surface M. Suppose that $l(x)$ and $l(y)$ are open recurrent semi-orbits of π^t. If $l(x) \subset \omega(y)$ then $l(y) \subset \omega(x)$ and vice versa.*

Proof. This is much the same as for Theorem 3.2.2. Suppose that $x_0 \in l(x)$ is an inner point of the semi-orbit $l(x)$. Denote by T a transversal segment through x_0. By the inclusion $l(x) \subset \omega(y)$ there is an infinite sequence $\{y_k\}_{k=1}^{\infty} = T \cap l(y)$ of the intersection points of $l(y)$ with T. (It can always be chosen monotone, lying to the one side of $T \backslash x_0$.)

Suppose to the contrary, that $l(y)$ does not belong to the ω-limit of $l(x)$. It implies that there exists an interval $I \subset T$, such that $y_1 \in I$ and $l(x) \cap I = \emptyset$. Since the semi-orbit $l(y)$ is recurrent, one can obtain an infinite sequence of arcs $\{\lambda_m\}_{m=1}^{\infty}$ of the semi-orbit $l(y)$ which obey the properties (i) and (ii) mentioned in the proof of Theorem 3.2.2. Arguing as before, one can choose a finite subsequence $\{\lambda_m\}_{m=1}^{N}$ of the above sequence, which yield a complete dissection of the manifold M'. Therefore the orbits of $l(y)$ through the segment I are 'caught' into a Bendixson sack separated from $l(x)$, and cannot approximate the semi-orbit $l(x)$. A contradiction. □

Examples below show that the claim Sing $\mathcal{F} < \infty$ of Theorems 3.2.2 and 3.2.3 cannot be dropped.

Example 3.2.1 (Aranson-Zhuzhoma) *Consider a disc region with a C^{∞} foliation of the following type. The semi-leaf $l(x)$ tends to the segment $[a, b]$ according with the formula $t \mapsto \sin(1/t)$ as $t \to +0$. The countable set of the tripods and thorns of the foliation $\mathcal{F}$ accumulate to the endpoints a and b of the segment $[a, b]$. The semi-leaf $l(y)$ approximates $l(x)$ 'winding around' closer and closer as $l(x)$ tends to $[a, b]$. Clearly, $l(x)$ lies in the limit set of $l(y)$ and the closure of $l(x)$ contains a point $\xi \in (a, b)$, which is not a singularity of $\mathcal{F}$. Nevertheless, $l(x)$ is not an open recurrent leaf of $\mathcal{F}$. Theorem 3.2.2 fails.*

Example 3.2.2 (Aranson-Zhuzhoma) *Here an example of a foliation $\mathcal{F}$ with Sing $\mathcal{F} = \infty$ on a pretzel $T^2 \# T^2$ is constructed, for which the Theorem 3.2.3 fails. For this let us take two samples of tori, T_1^2 and T_1^2, on which an irrational foliation $\mathcal{F}_1$ and the Denjoy foliation $\mathcal{F}_2$, respectively, are fixed. (By the Denjoy foliation on T^2 one understands a non-singular foliation with a minimal set, $\mathcal{Q}$, of the Cantor type.)*

Now one removes the disc regions, D_1 and D_2, from T_1^2 and T_2^2, so that $\mathcal{F}_i, i = 1, 2$ are transversal to ∂D_i everywhere, except the points A_i, C_i. (Disc D_2^2 is chosen such that it intersects with $\mathcal{Q}$ only at the points A_2 and C_2 of the *boundary.) Now one identifies T_1^2 and T_2^2 with a help of a homeomorphism $h : \partial D_1 \to \partial D_2$ acting on their (common) boundary D_i^2. h acts by the formula $A_1 \mapsto B_2, B_1 \mapsto C_2, C_1 \mapsto D_2, D_1 \mapsto A_2$, being a rotation through an irrational angle at other points of D_i. Herewith all the points $A_i - D_i, i = 1, 2$ are declared to be the singular points of $\mathcal{F}$.*

The newly obtained foliation $\mathcal{F} = \mathcal{F}_1 \#_h \mathcal{F}_2$ on the pretzel $T^2 \# T^2$ fits the following properties. Sing $\mathcal{F} = \infty$. There is an open recurrent leaf $l(x)$ (different from the separatrices of the foliation $\mathcal{F}_2$), such that the closure of $l(x)$ contains a leaf $l(y)$, where $l(y) \in \mathcal{Q}$ is a leaf in the minimal set $\mathcal{Q}$. (This happens since $l(x)$ comes arbitrary close to the point $A_2 = D_1$.) On the other hand, the leaf $l(y)$ is also an open recurrent leaf of the foliation $\mathcal{F}$, but its closure does not contain $l(x)$. Therefore Theorem 3.2.3 fails in this case.

3.2.2 Blowing–Down

If a foliation $\mathcal{F}$ with non trivial recurrent leaf is given, it is possible in many cases to simplify the structure of the respective quasiminimal sets by a 'blowing-down' process. The idea is similar to that of a 'squeezing' the of Cherry cell on the torus; as a result, one obtains an irrational foliation on the torus. Such a procedure can be generalized to the higher genus surfaces; the resulting foliation belongs to a 'highly transitive class' [114]. (By such one understands foliations with Sing $\mathcal{F}$ consisting of n-separatrix saddles and every other leaf being dense in M.) Conversely, let a transitive foliation, $\mathcal{T}$, be given. 'Blowing-up' some leaves of $\mathcal{T}$ one obtains a recurrent foliation $\mathcal{F}$. The relevant technique is exposed in below.

Blowing-Down Homotopy

Let $\mathcal{T} = (M, \psi^t, \theta)$ be a transitive foliation on M. (Further we deal with $\mathcal{T}$ in terms of its covering flow $\psi^t : M \times I\!R \to M$ and all properties of ψ^t are checked to be θ–invariant.)

Definition 3.2.1 (Gardiner) *An orbit cluster, $\mathcal{C}$, in $\mathcal{T}$ is either a non trivial recurrent orbit of ψ^t or else a saddle point in Sing ψ^t, taken together with its separatrices.*

Definition 3.2.2 (Gardiner) *Let $\mathcal{F} = (M, \pi^t, \theta)$ be a foliation on M with a non trivial recurrent leaf. A closed, surjective mapping $\tau : M \to M$ is called a blowing-down homotopy from the flow π^t to a transitive flow ψ^t on M if the following holds:*

(i) $\tau^{-1}(x)$ is compact and contractible for every $x \in M$;
(ii) $\tau^{-1}(\mathcal{C})$ is invariant under π^t for each orbit cluster $\mathcal{C}$ of ψ^t; moreover, τ preserves the direction of the flow π^t at every non singular point $x \in \tau^{-1}(\mathcal{C}) \cap \overline{\tau^{-1}(\mathcal{C})}$ provided $\tau(x)$ is non singular;

(iii) *if* $\tau = \mu \circ \nu$, *where* $\mu, \nu : M \to M$ *satisfy (i) and (ii), then* μ *is a topological equivalence;*
(iv) *the induced mapping* $\tau_* : \pi_1(M) \to \pi_1(M)$ *is an isomorphism.*

Proposition 3.2.1 *Let* $\tau : M \to M$ *be a blowing-down homotopy of the foliation* $\mathcal{F} = (M, \pi^t, \theta)$ *to a transitive foliation* $\mathcal{T} = (M, \psi^t, \theta)$. *Then the following is true:*

(1) *if* γ *is a recurrent orbit of* ψ^t, *then* $\tau^{-1}(\gamma)$ *contains at most 2 recurrent orbits, lying at the boundary of* $\tau^{-1}(\gamma)$; *if* $\tau^{-1}(\gamma)$ *consists of a single orbit, then this orbit is also recurrent;*
(2) *the image of a recurrent orbit of* π^t *is either a recurrent orbit of* ψ^t *or else a union of 3 components: an* α*-recurrent orbit, a singular point and an* ω*-recurrent orbit;*
(3) π^t *has at most one* $\mathcal{Q}$*-set, whose image under* τ *is the whole* M.

Proof. (1) Let $x \in M$ be a non-singular point of the flow ψ^t in an orbit cluster $\mathcal{C}$. Choose a small flow box D about x bounded by two transversal arcs, T_0 and T_1, and two arcs of orbits of the flow ψ^t. Denote by $\mathcal{C}' = \mathcal{C} \cap D$ a part of $\mathcal{C}$ which lie in D.

Let $\widetilde{D} = \tau^{-1}(D)$ be a disc region of the flow π^t containing $\tau^{-1}(x)$. Let $\widetilde{x}_1 \in \tau^{-1}(D)$ be a recurrent point of the flow π^t, and let $C_1 = O(\widetilde{x}_1) \cap \widetilde{D}$ be arc of the orbit through $\widetilde{x}_1$, lying in $\widetilde{D}$. If we suppose that $\tau^{-1}(x)$ contains an arc transversal to C_1 then by the invariance of $\tau^{-1}(x)$ with regard to the flow π^t, one can construct a non trivial closed transversal curve lying completely in $\tau^{-1}(\mathcal{C})$; this contradicts (iv). Therefore $\widetilde{x}_1$ must belong to $\tau^{-1}(\mathcal{C}) \cap \overline{\tau^{-1}(\mathcal{C})}$.

Now if $\widetilde{x}_2$ is a recurrent point of π^t lying in $\tau^{-1} \setminus C_1$, we denote by $C_2 = O(\widetilde{x}_2) \cap \widetilde{D}$ an arc of the orbit through $\widetilde{x}_2$, lying in $\widetilde{D}$. It is clear that $\tau^{-1}(T_0), \tau^{-1}(T_1), C_1$ and C_2 bound a disc region $\widetilde{D}'$ containing $\tau^{-1}(x)$. A part of any orbit of π^t, crossing $\widetilde{D}'$ more than once can be used as a boundary of a disc region containing either $\widetilde{x}_1$ or $\widetilde{x}_2$, which is a contradiction (since both $\widetilde{x}_1$ and $\widetilde{x}_2$ are supposed recurrent). Thus again $\widetilde{x}_1, \widetilde{x}_2$ must be in $\tau^{-1}(\mathcal{C}) \cap \overline{\tau^{-1}(\mathcal{C})}$. By similar arguments one proves the second part of (1).

(2) Let us show that the mapping $\tau : M \to M$ being restricted to a recurrent orbit, γ, of π^t is one-to-one. Indeed, if two points $p, q \in \gamma$ map to a single point of the flow ψ^t, then so must the whole arc $[p, q]$ (otherwise (ii) will fail). Thus one of the alternatives of item (2) holds true.

(3) Finally, if γ is a unique recurrent orbit of the flow π^t, then $\mathrm{cl}\,(\tau(\gamma)) = M$, and therefore $\tau(\,\mathrm{cl}\,\gamma) = M$. If π^t has more than one recurrent orbit closure ($\mathcal{Q}$-set), then each must map onto M. In this case one finds a non trivial simple closed curve C which does not intersect one of the $\mathcal{Q}$-sets. Therefore the continuous mapping τ sends $M \setminus C$ to M. This contradiction completes the proof of (3). □

Irreducible Foliations

Let $\mathcal{F} = (M, \pi^t, \theta)$ be a C^∞ foliation of the compact surface M. Suppose that $\mathcal{F}$ admits at least one $\mathcal{Q}$-set.

Definition 3.2.3 *Foliation $\mathcal{F}$ on M with a quasiminimal set $\mathcal{Q}(x)$ is called irreducible if every non trivial closed curve on M meets the $\mathcal{Q}$-set in at least one α or ω-recurrent point.*

Theorem 3.2.4 (Gardiner) *Let $\mathcal{F} = (M, \pi^t, \theta)$ be a recurrent foliation with finitely many singular points. Then $\mathcal{F}$ has a blowing-down homotopy if and only if $\mathcal{F}$ is irreducible.*

Proof. ($\Rightarrow$) The sufficient part is well evident. If $\mathcal{F}$ has a blowing-down homotopy $\tau : M \to M$ to a highly transitive foliation $\mathcal{T}$, then a transversal cycle C goes by τ to a transversal cycle $\tau(C)$. Since $\mathcal{T}$ is highly transitive, every recurrent leaf of $\mathcal{T}$ intersects $\tau(C)$ and hence this holds true for C and recurrent leaves of $\mathcal{F}$. That is $\mathcal{F}$ is irreducible.

($\Leftarrow$) The proof of the necessary part has two steps. First, a decomposition, $\mathcal{D}$, of the manifold M into compact contractible sets is given. The set $\mathcal{D}$ is homeomorphic to M by a natural homeomorphism $h : M \to \mathcal{D}$. Secondly, the mapping $\tau' : M \to \mathcal{D}$ is constructed, which being composed with h, yields the required blowing-down homotopy $\tau = h \circ \tau'$. (Roughly, τ' is a 'suppression' of the 'blind' intervals on the transversal C, which correspond to the points of C where non-recurrent orbits of the respective flow hit C. In other words, one obtains a highly transitive foliation which is a suspension over interval exchange transformations of the unit circle $S^1 \cong C$.) Consider first an auxiliary

Construction. Let $\mathcal{F} = (M, \pi^t, \theta)$ be a C^∞ foliation with at least one $\mathcal{Q}$-set on a compact surface M and such that Sing $\mathcal{F} < \infty$. Let C be a closed transversal curve for $\mathcal{F}$ meeting $\mathcal{Q}$ and no other quasiminimal set of $\mathcal{F}$. Let $\phi : U \to C$ be the first return mapping on C. Denote by $\mathcal{P}$ a partition of C whose elements are the closures in C of components of $C \backslash \mathcal{Q}$, and the points of $\mathcal{Q}$ not included in these closures. ($\mathcal{P}$ is indeed a partition since by the recurrence of $\mathcal{Q}$ no two components of $C \backslash \mathcal{Q}$ can have a common endpoint.)

Define two types of elements $A \in \mathcal{P}$ of a partition $\mathcal{P}$. For this let N be a neighbourhood of A in C, for which the first return mapping is defined, $N \backslash A \subset U$. Let $A = [x, y]$ be an arc of C (case $x = y$ is not excluded). Take a pair of monotonic sequences $x_n \to x$ and $y_n \to y$ lying in $N \backslash A$. It yields us the convergent sequences $\phi(x_n) \to \phi(x), \phi(y_n) \to \phi(y)$. We say that

- A is of type I, if either the arc $[\phi(x), \phi(y)] \in \mathcal{P}$ or $[\phi(y), \phi(x)] \in \mathcal{P}$;
- A is of type II, otherwise.

(Note that in the latter case $\phi(x)$ must be the endpoint of an element $A_{-1} \in \mathcal{P}$ and $\phi(y)$ must be an endpoint of an element $A_1 \in \mathcal{P}$.) Resume the procedure now for the elements A_{-1} and A_1, using the mapping $f^{-1} : f(U) \to C$. It is easy to see that A_{-1} and A_1 are of type II. Continuing the above process one comes to a sequence

$$A_{-n}, A_{-n+1}, \ldots, A_{-1}, A, A_1, \ldots, A_n.$$

The process terminates when $A_{-n} = A_n$, which must happen because each element A_i contains a point of intersection of C with a separatrix, the total number of which is finite. It is evident also that when $|i|$ is even the separatrix through the intersection point goes immediately to a sink, while when $|i|$ is odd the separatrix comes out immediately from a source. The above sequence $\{A_i\}_{i=-n}^{i=n}$ of the elements of the partition $\mathcal{P}$ is called a *saddle set* in $\mathcal{P}$, and its *index* is defined to be

$$Ind\{A_i\}_{i=-n}^{i=n} = 1 - n.$$

(One can think of a saddle set as a set of points in C where the separatrices of a saddle-type singularity hit the first and the last time the closed transversal C. The index of the saddle set coincides with the Euler-Poincaré index of the above singularity.)

We say that the elements A and A' of a partition $\mathcal{P}$ are *related* if there is a sequence $A_0 = A, A_1, \ldots, A_n = A'$ such that for all $0 \leq i < n$ either A_i and A_{i+1} belong to the same saddle set, or A_i is of type I and $A_{i+1} = \mathrm{cl}\, A_i$. By a *family* $\mathcal{F}(A)$ of A one understands a collection of all elements in $\mathcal{P}$, which are related to A. Finally, by a *core* $\mathcal{F}^*(A)$ of the family $\mathcal{F}(A)$ in $\mathcal{P}$ one means the set of all elements of $\mathcal{F}(A)$, which either (i) belong to a saddle set in $\mathcal{F}(A)$ or (ii) belong to a sequence $(A_0, A_1, \ldots, A_n)$ in $\mathcal{F}(A)$ such that A_0 and A_n are in a saddle set and for $0 \leq i < n$ either A_i and A_{i+1} are in the same saddle set, or else A_i is of type I and $A_{i+1} = \mathrm{cl}\, A_i$. (Note that every saddle set of A must belong to $\mathcal{F}^*(A)$.) The *index* of $\mathcal{F}(A)$ is defined to be the sum of the indices of its saddle sets. This remark completes a description of the auxiliary construction.

Let now $\mathcal{F} = (M, \pi^t, \theta)$ be irreducible with a recurrent orbit closure $\mathcal{Q}$. Let C be a transversal closed curve for $\mathcal{F}$, and let $\phi : U \to C$ be the first return mapping of $\mathcal{F}$ on C. Fix an orientation on C, and suppose that $\mathcal{P}$ is the partition of C, described above. Consider an element $A = [x, y]$ of the partition $\mathcal{P}$.

(I) Let A be of type I. Denote by $\Lambda^-(x, \phi(x))$ and $\Lambda^+(y, \phi(y))$ the arcs of orbits of π^t, joining the points x and $\phi(x)$, y and $\phi(y)$. Assume also that the segments A and $\overline{A} = \phi(A)$ are disjoint on C. Consider a closed curve $\lambda = A \cup \Lambda^+ \cup \overline{A} \cup \Lambda^-$. The lemma below implies that λ is nullhomotopic, and bounds a disc region, which we further denote by $R(A)$.

Lemma 3.2.1 *Either $x = y$ is α or ω-recurrent and lies in U, or λ is homotopic to a closed curve meeting no non-trivially α or ω-recurrent points of π^t.*

Proof of the lemma. If $x = y$ and $x \in U$ then the conclusion is evident. Let $x \notin U$. It means that $\omega(x) = \sigma$, where σ is a saddle point of π^t. If, further x is α-recurrent, then we deform the arc $O^+(x) \cup \sigma$ to σ and lemma follows.

Assume now that $x \neq y$. If x is α or ω-recurrent and lies in U, we choose a point $x' \in \text{int } A$, such that $[x, x'] \subset U$. Denote by λ' a closed curve obtained from λ by replacing the arc $[x, x'] \cup [x, \phi(x)] \cup [\phi(x), \phi(x')]$ by the arc $[\phi(x), \phi(x')] \subset \overline{A}$. (Clearly λ' is homotopic to λ.)
Suppose now that $x \notin U$. Then $\omega(x)$ is a saddle point σ. Choose again a point $x' \in \text{int } A$, and let λ' be a closed curve obtained from λ by deforming the arc $[x, x'] \cup O^+(x) \cup \sigma$ to an arc from x' to σ disjoint from $O^+(x) \cup$ int $[x, x']$. (The above operation is again a homotopy.)
In the same manner one proceeds with the endpoint y and the segment $[\phi(x), \phi(y)] \subset \overline{A}$. Finally, one obtains a sequence of closed curves $\lambda \sim \lambda' \sim \lambda''$, which belong to the same homotopy class, and such that λ'' meet no non-trivially recurrent point of π^t. Lemma follows. □

Suppose to the contrary that λ is not contractible. Then $R(A)$ contains a simple closed curve, C', meeting C exactly once and, by Lemma 3.2.1, meeting no non-trivially α or ω-recurrent points. Since C' meet C in a unique point, C' must be non-trivial and $\mathcal{Q} \cup C' = \emptyset$. A contradiction with $\mathcal{F}$ chosen irreducible. This finishes the first step of the proof for the type I partitions.

(II) Suppose that A is of type II. In a similar way, take a closed curve, λ, bounding a disc region, $R(A)$, in M. Note that for the type II saddle sets of partition $\mathcal{P}$, the Euler-Poincaré index of the curve λ coincides with the index of the respective saddle set. (In a contrast to the case I, where index of λ was 0.) It implies that $R(A)$ must contain at least one singular point of the flow π^t. If A and A' belong to different saddle sets of $\mathcal{P}$, then $R(A)$ and $R(A')$ can only intersect on C (recall that there is a finite number of saddle sets in $\mathcal{P}$). Now if $\mathcal{F}(A)$ is a family in $\mathcal{P}$ and $\mathcal{F}^*(A)$ is its core, we put

$$R^*(\mathcal{F}(A)) = \cup_{A \in \mathcal{P}} R(A),$$

where $A \in \mathcal{F}(A)$ if A is of type II, and $\overline{A} \in \mathcal{F}^*(A)$ if A is of type I. There exists a simple closed curve, λ^*, consisting if arcs of orbits and arcs of C which bounds a disc on M containing $R^*(\mathcal{F}(A))$, with the Euler-Poincaré index equal to the index of the family $\mathcal{F}(A)$. Note finally, that every point of Sing π^t is in some $R(A)$. Therefore the sum of indices of the families of π^t is equal to the Euler characteristic of the surface M,

$$\sum_{\mathcal{F} \in \mathcal{P}} Ind \quad \mathcal{F}(A) = \chi(M).$$

Now we pass to the second part of the proof. For this it is necessary to construct a mapping $\tau' : M \to \mathcal{D}$ for the obtained partition $\mathcal{D}$. Roughly, one has to show how to 'deform' the two-dimensional 'cells' of partition $\mathcal{D}$ to one-dimensional 'cells' (or, simply, to closed arcs). Let us treat the 'cells' of types I and II separately.

(i) Let $A = [x, y]$ be an element of $\mathcal{P}$ of type I, such that A and $\overline{A}$ are not both contained in any core $\mathcal{F}^*(A)$ in $\mathcal{P}$ and such that $R(A)$ contains at least one singular point. It can be easily seen how to 'suppress' the given cells to a simple arc, which joins points x and y in C (case $x = y$ is not excluded). Herewith the region $R(A)$ goes by a natural mapping $\tau' : M \to M$ to a corresponding one-dimensional 'cell'.

(ii) Consider the region $R^* = R^*(\mathcal{F}(A))$ for a family $\mathcal{F}(A)$ of $\mathcal{P}$ with index $1 - n$. The image of R^* under mapping τ' must be a collection of $2n$ arcs emerging from a single point. This point is a saddle, σ, of the Euler-Poincaré index $1 - n$. A natural mapping $\tau' : M \to M$ sends the region R^* to a collection of $2n$ half-open simple arcs which tend to the saddle point σ. Resuming the above operation over all families in C, one comes to a highly transitive flow on M.

It remains to consider a composition $\tau = h \circ \tau'$ of the above obtained mapping τ' with the homeomorphism of decomposition h, established in the first part of the proof. It is a straightforward exercise to check that the above homotopy τ satisfies all the axioms (i)-(iv) of the blowing-down homotopy. Moreover, it can be checked immediately that τ commutes with the involution θ fixed in the triple $\mathcal{F} = (M, \pi^t, \theta)$, being defined correctly for the factor space $N = M/\theta$ with an 'underlying' foliation π^t/θ. This completes the proof of theorem. □

Corollary 3.2.1 *A foliation $\mathcal{F}$ on a closed two-manifold with Sing $\mathcal{F} < \infty$ is irreducible if and only if it has a unique $\mathcal{Q}$-set which meets every non-trivial simple closed curve in a non-trivial α or ω-recurrent point.*

Proof. This follows from the above considerations. It is sufficient to notice that any simple closed curve meeting no non-trivial α or ω-recurrent point must be non trivial. □

Corollary 3.2.2 *If $\tau : M \to M$ and $\tau' : M \to M$ are blowing-down homotopies for a foliation $\mathcal{F} = (M, \pi^t, \theta)$ to highly transitive foliation $\mathcal{T}$ and $\mathcal{T}'$, respectively, then $\mathcal{T}'$ is topologically equivalent to $\mathcal{T}$ via a homeomorphism $h : M \to M$ such that τ' and $h \circ \tau$ take any point of M to the same orbit cluster of $\mathcal{T}'$. In other words the following diagram commutes:*

$$\begin{array}{ccc} \mathcal{F} & \xrightarrow{Id} & \mathcal{F} \\ \tau \downarrow & & \downarrow \tau' \\ \mathcal{T} & \xrightarrow{h} & \mathcal{T}' \end{array}$$

Proof. By definition of the orbit cluster, there exists a recurrent orbit, γ, of $\mathcal{F}$, which maps under τ to a recurrent orbit $\tau(\gamma)$ of the foliation $\mathcal{T}$, and under τ' to a recurrent orbit $\tau'(\gamma)$ of the foliation $\mathcal{T}'$. Choose a point $x \in \gamma$ and construct a non-trivial transversal curve, C, for $\mathcal{F}$, passing through x, which is disjoint from the inverse images, τ^{-1} (Sing $\mathcal{T}$) and $(\tau')^{-1}$ (Sing $\mathcal{T}'$), of singular points of foliations $\mathcal{T}$ and $\mathcal{T}'$.

Let $y = \tau(x)$ and let $\mathcal{N}$ be an annulus neighbourhood of $\tau(C)$ containing no singular points of $\mathcal{T}$. Since C is transversal to $\mathcal{F}$, $\tau(C)$ must be transversal to $\mathcal{T}$, and one can thus find a transversal curve, C_y, in $\mathcal{N}$, passing through y. Let $f : U \to C_y$ be the first return mapping for $\mathcal{T}$ on C_y, and let $y_n = f^n(y)$ for $n \in \mathbb{N}$. Similarly, define $y', C_{y'}, f' : U' \to C_{y'}$ and y'_n for $\mathcal{T}'$. In addition, let us denote by $\phi : U \to C$ and by x_n the corresponding first return mapping and the fundamental sequence for the foliation $\mathcal{F}$. Note that the points $\{y_n\}, \{y'_n\}, \{x_n\}$ lay on $C_y, C_{y'}, C$ in one and the same cyclic order.

Now define a homeomorphism h on C_y by setting $h(y_n) = y'_n$. Since $\{y_n\}$ is dense in C_y and $\{y'_n\}$ is dense in $C_{y'}$, h can be extended (not uniquely!) to a homeomorphism $C_y \to C_{y'}$. To extend $h|_C$ to the whole M, two steps are needed.

First, h is defined on the singular points of $\mathcal{T}$, and then on the arcs of saddle separatrices running between these singular points and C_y. For this one needs to establish a one-one correspondence between these points and arcs of $\mathcal{T}$ and those of $\mathcal{T}'$. It is done as follows. Take a point $p \in C_y$, such that $O^+(p)$ tends directly to a saddle point of $\mathcal{T}$. Let $\{y_{n_k}\}$ and $\{y_{m_k}\}$ be sequences on C tending to p from either side. By the inverse homotopy τ^{-1}, 'lifted' sequences $\{x_{n_k}\}$ and $\{x_{m_k}\}$ define an arc of $C\backslash$ cl $O(x)$. This allows (via the blowing-down homotopy τ') to define the respective points of $\mathcal{T}'$, defining h between the singular points of $\mathcal{T}$ and $\mathcal{T}'$.

Finally, h is defined on $M\backslash(C\cup$ Sing $\cup$ Sep) as follows. The complementary regions are rectangles of a parallel flow with h defined on the boundary of the rectangle. Herewith on the transversal 'ends' of rectangle one has $f(h(x)) = h(f(x))$. Take a parametrization of the rectangles by a standard flow-box $[0,1] \times [0,1]$. It is easy now to prolong h by the formula:

$$(u, v) \mapsto (u, (1-u)h((0,v)) + uh((1,v)).$$

(Clearly, the above extension agrees on the sides of the rectangles.) Since all regions of M are considered, corollary follows. $\square$

It is possible to extend Theorem 3.2.4 and Corollary 3.2.1 to foliations with a countable many singular points. For this it is necessary to describe correctly the sets $R(A)$ and the 'nearby' orbits. However if there is a continuum of singularities these results are no longer true.

Reducing Set of Curves

Let M be a closed two-manifold. If N is an open submanifold of M, let us denote by N^* a manifold obtained by compactifying each end of N with a point. If $\mathcal{F} = (M, \pi^t, \theta)$ is a foliation on M there is a foliation $\mathcal{F}|_N$, defined up to topological equivalence, whose leaves are leaves of $\mathcal{F}$ intersected with N. In turn, denote by $\mathcal{F}|_{N^*}$ a foliation on N^* obtained from $\mathcal{F}|_N$ by declaring each point of $N^* \backslash N$ a singular point of $\mathcal{F}|_{N^*}$. The foliation $\mathcal{F}|_N$ on an open manifold N is called *irreducible* if

- the boundary of N contains no non-trivial recurrent point of $\mathcal{F}$ and
- the foliation $\mathcal{F}|_{N^*}$ is irreducible.

Theorem 3.2.5 (Gardiner) *Let $\mathcal{F} = (M, \pi^t, \theta)$ be a C^∞ foliation with a finite number of singularities on a closed surface M. There exists a finite set $\mathfrak{C}$ of non-trivial closed curves on M such that*

(i) no curve of $\mathfrak{C}$ contains a non-trivially α or ω-recurrent point of $\mathcal{F}$, and
(ii) if $M_i, 1 \leq i \leq n$ are the components of $M \backslash \cup_{C \in \mathfrak{C}} C$ then for each i, either $\mathcal{F}|_{M_i}$ is irreducible, or M_i contains no non-trivially recurrent point of $\mathcal{F}$.

Proof. If $\mathcal{F}$ has no non-trivial recurrent points or is irreducible, then $\mathfrak{C} = \emptyset$. If $\mathcal{Q}$ is a recurrent orbit closure of $\mathcal{F}$, then there exist a non-trivial simple closed curve, C_1, on M meeting no non-trivial α or ω-recurrent point of $\mathcal{Q}$. Let N_1 be the component of $M \backslash C$ containing the recurrent points of $\mathcal{Q}$. Note that $\chi(N_1^*) > \chi(N)$. If $\mathcal{F}|_{N^*}$ is still not irreducible, we take another simple closed curve, C_2, such that $N_2 = M \backslash (C_1 \cup C_2)$ contains the recurrent points of the $\mathcal{Q}$-set. Here again, after the relevant compactification, $\chi(N_2^*) > \chi(N_1^*)$. The above procedure is resumed and it must terminate (due to the monotone growth of the Euler characteristic of compactified surfaces) for a finite k, such that $\mathcal{F}|_{N_k}$ is irreducible. Note that carrying out this procedure for every $\mathcal{Q}$-set of $\mathcal{F}$ will produce a finite collection of curves, satisfying (ii), but nor necessarily (i).

To prove item (i) one must apply a blowing-down homotopy $\tau : N_k^* \to N_k^*$ to the manifold N_k^* which takes $\mathcal{F}|_{N_k^*}$ to a highly transitive foliation $\mathcal{T}$ on N_k^*. Let γ be an orbit cluster of $\mathcal{T}$ such that $\tau^{-1}(\gamma)$ contains at least one point of $N_k^* \backslash N_k$, and assume that γ is a single recurrent orbit of $\mathcal{T}$. There exist points $p, q \in \gamma \cap \tau(N_k)$ such that $\alpha(x) = \mathcal{Q}$ for every $x \in \tau^{-1}(p)$ and $\omega(x) = \mathcal{Q}$ for every $x \in \tau^{-1}(q)$. Thus every singularity of $\tau^{-1}(\gamma)$ lies in $\tau^{-1}([p, q])$. Since no point of $\tau^{-1}(p)$ and $\tau^{-1}(q)$ can be a recurrent point of $\mathcal{F} \backslash \mathcal{Q}$, there is a closed curve in the boundary of $\tau^{-1}([p, q])$ which contains no singular point in $\tau^{-1}(\gamma)$. This curve can be deformed to a curve C^*, containing no recurrent points of $\mathcal{F}$. Now one obtains a closed curve, C, by adding one or two arcs of the family $\cup_{i=1}^k C_i$ lying in the boundary of $\mathcal{Q}$. The curve C must belong to $\mathfrak{C}$. Proceeding likewise for the orbit clusters of index $1-n$, one comes to the family $\mathfrak{C}$ of closed curves satisfying properties (i) and (ii). This completes the proof of the theorem. □

Corollary 3.2.3 *If $\mathfrak{C}$ and $\mathfrak{C}'$ are two reducing set of curves for $\mathcal{F}$, with $M_i, 1 \leq i \leq n$, the components of $M \setminus \cup_{C \in \mathfrak{C}} C$ and $M_i', 1 \leq i \leq n'$, the components of $M \setminus \cup_{C \in \mathfrak{C}'} C$, numbered so that each pair M_i and M_i' has a non-trivially recurrent orbit of $\mathcal{F}$ in common for $1 \leq i \leq m$, and that M_i and M_i' contain no non-trivially recurrent point of $\mathcal{F}$ for $i > m$, then, for each i with $1 \leq i \leq m$, there are blowing-down maps $\tau_i : M_i^* \to M_i^*$ to foliations $\mathcal{T}_i$ and $\mathcal{T}_i'$, respectively, and a topological equivalence $h_i : M_i^* \to M_i^*$ between $\mathcal{T}_i$ and $\mathcal{T}_i'$ such that for any non-trivially recurrent orbit γ of M_i^* it holds $\tau_i'(\gamma) = h_i \circ \tau_i(\gamma)$.*

Proof. This is similar to one of Corollary 3.2.2. □

Main Theorem

Theorem 3.2.6 (Gardiner) *Let $\mathcal{F} = (M, \pi^t, \theta)$ be a foliation with non trivial recurrent leaf, such that Sing $\mathcal{F} < \infty$. Then there exists a blowing-down homotopy $\tau : M \to M$ of M that brings $\mathcal{F}$ to a foliation $\mathfrak{T}$ which is highly transitive or possibly with a finite number of saddle-separatrix cycles splitting M into a certain number of pieces, M_i, in each of which $\mathcal{F}|_{M_i}$ is either highly transitive, or has no non-trivial recurrent leaves. Conversely, every foliation $\mathcal{F}$ with non trivial recurrent leaf and Sing $\mathcal{F} < \infty$, can be obtained from $\mathfrak{T}$ by a finite number of 'blowing-up' homotopies of certain leaves of $\mathfrak{T}$.*

Proof. Follows from Theorems 3.2.4 and 3.2.5. □

3.3 Decomposition

Let $f : S^1 \to S^1$ be the interval exchange transformation, that is an injective differentiable mapping defined everywhere except possibly a finite number of points. Let Dom f denote a domain of definition of f. Furhter we assume $|Df| = 1$ for all $x \in$ Dom f.

Definition 3.3.1 (Gutiérrez) *Suppose that $g : S^1 \to S^1$ is a continuous mapping from S^1 to S^1. g is said to be a covering mapping for f if it is:*

(i) injective;
(ii) Dom g is an open subset in S^1;

(iii) for some monotone continuous mapping $h : S^1 \to S^1$ of degree one $h(x) \in$ Dom f and $g \circ h = h \circ f$ for all $x \in$ Dom g.

Given f, g and h as above, we shall say that g covers f via h.

Let W_E be a quotient manifold obtained from the cylinder

$$Cyl = (S^1 \times [0,1]) \setminus \left(U_1 \bigcup U_2\right),$$
$$U_1 = [S^1 \setminus Dom f^{-1}] \times \{0\},$$
$$U_2 = [S^1 \setminus Dom f] \times \{1\},$$

by identifying point $(x, 1)$ with the point $(f(x), 0)$ for all $x \in$ Dom f. Let $\mathcal{F}$ be a continuous foliation on W_E. A pair $(\mathcal{F}, W_E)$ is said to be *a suspension of the pair* (g, f) if the following conditions are satisfied:

- $\mathcal{F}$ is transversal to $S^1 \times \{0\}$ and the set Sing $\mathcal{F}$ is either empty or compact;
- The forward Poincaré mapping $S^1 \to S^1$ induced by $\mathcal{F}$ is given by the formula $x \mapsto g(x)$.

Definition 3.3.2 (Gutiérrez) *Let $\phi^t : M \times I\!R \to M$ be a flow on a compact 2-dimensional manifold M. Let $\mathcal{Q}_1, \dots, \mathcal{Q}_m$ be the $\mathcal{Q}$-sets defined by recurrent orbits of ϕ^t and let $V_1, \dots, V_m$ be an open connected subsets of M. Every V_i is said to be a region of recurrence associated to $\mathcal{Q}_i$ if it obeys the following conditions:*

(i) There exists a circle $C_i \subset V_i$ transversal to ϕ^t and passing through $\mathcal{Q}_i$ such that the forward Poincaré mapping $g_i : C_i \to C_i$ induced by ϕ^t covers an interval exchange transformation $f_i : S^1 \to S^1$ which has every of its orbits dense in S^1 and which cannot be extended continuously to a bigger subset of S^1;

(ii) The pair $(\phi^t|_{V_i}, V_i)$ is topologically equivalent to a suspension (g, f). Moreover, the boundary ∂V_i of V_i contains only fixed points, regular orbits connecting these fixed points and finitely many transversal segments that connect fixed points. There are no arc of orbit of ϕ^t lying in V_i and joining two points of ∂V_i.

Theorem 3.3.1 (Gutiérrez) *Let $\mathcal{F} = (M, \pi^t, \theta)$ be a C^∞ recurrent foliation on a compact surface M such that Sing $\mathcal{F} < \infty$. Let $\mathcal{Q}_1, \dots, \mathcal{Q}_m$ be a finite number of quasiminimal sets of $\mathcal{F}$. Then there exist a finite sequence $V_1, \dots, V_m$ of open connected subsets of M such that:*

(i) If $i \neq j$ then $V_i \cap V_j = \emptyset$ and regions V_i contain all non-trivial recurrent leaves of $\mathcal{F}$ lying in $\mathcal{Q}_i$;
(ii) Each V_i is a region of recurrence associated to $\mathcal{Q}_i$;

(iii) *If V_i' is any other region of recurrence associated to $\mathcal{Q}_i$, then V_i and V_i' are homeomorphic. Moreover, when $\mathcal{F}$ has finitely many singular points and no other region of recurrence associated to $\mathcal{Q}_i$ contains fewer singular points than V_i (resp. V_i'), then the foliations $(\mathcal{F}|_{V_i}, V_i)$ and $(\mathcal{F}|_{V_i'}, V_i')$ are topologically equivalent;*

(iv) *Finally, the circle $C_i \subset V_i$ can be taken so that either $\mathcal{Q}_i \cap C_i = C_i$ or $\mathcal{Q}_i \cap C_i$ is a Cantor set.*

Proof. Roughly, the proof consists in constructing of $C_1, \ldots, C_m$ transversal closed curves corresponding to each quasiminimal set of $\mathcal{F}$. Then one takes a suspension flow Susp f_i defined for a certain interval exchange mapping f_i at each C_i. (This interval exchange transformation corresponds to a 'highly transitive' flow obtained from π^t by a blowing-down homotopy.) Susp f_i 'occupies' exactly an open region V_i, such that the boundary ∂V_i of V_i consists of separatrix connections and singular points of the foliation $\mathcal{F}$. We need the following auxiliary statement

Lemma 3.3.1 *Let U be an open subset of M such that there is no arc of trajectory of π^t contained in U and joining two points of ∂U. Suppose that γ is a non-trivial recurrent trajectory which does not lie at U, $\gamma \cap U = \emptyset$. Then there exists a circle C and an open set V containing C such that*

(i) *$U \cap V = \emptyset$ and either $\overline{\gamma} \cap C = C$ or, $\overline{\gamma} \cap C$ is a Cantor subset of C;*

(ii) *the restriction $\pi^t|_V$ is a region of recurrence associated to $\overline{\gamma}$. In other words, V obeys conditions (i) and (ii) of Definition 3.3.2.*

Proof of lemma. In the vicinity of γ it is easy to construct a closed curve $C \subset M \backslash U$ which is transversal to π^t and meets γ. Let $f : C \to C$ be the first return mapping induced by π^t. To prove the lemma one must show that the mapping

$$f_C : \mathcal{A}(C) \to \mathcal{A}(C)$$

is equivalent (up to a homotopy) to an interval exchange transformation. (Here $\mathcal{A}(C)$ is a partition of C by closed intervals $[a, b]$ and isolated points $a = b$ which correspond to the set $C \backslash \gamma$. Since $\mathcal{A}(C)$ is a partition it can be endowed with a quotient topology. In this topology an interval $[a, b]$ is identified with a point.) Then a suspension Susp $f_C : V \times I\!R \to V$ of the mapping f_C acting on an open region V is constructed and it is proved that V is the required region of recurrence.

(1) First let us notice that there exists such C that if $[a, b] \in \mathcal{A}(C)$, then $\gamma \cap [a, b]$ is either empty or $\{a\}$ or else $\{b\}$. Indeed, the set of points where (1) is violated corresponds to the case when $[a, b] \cap \gamma = \{a, b\}$. Therefore $b = f^N(a)$ for some positive integer N. But this contradicts the fact that γ is chosen to be non-periodic.

(2) $\mathcal{A}(C)$ is homeomorphic to C. In fact if $\overline{\gamma \cap C}$ contains a subinterval of C the statement is true. Otherwise when $\overline{\gamma \cap C}$ is a Cantor set, one takes

a Cantor function $\Delta : C \to C$ which is monotone, continuous, degree one mapping [135]. Clearly the quotient space C/Δ is homeomorphic to C. (One can think geometrically of this procedure as of blowing-down homotopy of 'squeezing' of the 'Cherry cells'.)

(3) The semi-orbit $\{f_C^n(x) | n \in \mathbb{N}\}$ is dense in $\mathcal{A}(C)$ for all $x \in \mathcal{A}(C)$. This fact follows from the recurrence of γ.

(4) f_C is topologically conjugate to an interval exchange transformation $f : S^1 \to S^1$ which has a dense orbit. Again, it is an implication of the recurrence of γ.

(5) Summing-up (1) - (4), we conclude that f_C is an interval exchange transformation having every orbit dense and $f : C \to C$ covers f_C via the quotient mapping $h : C \to \mathcal{A}(C)$. Note also that f_C cannot be extended continuously to a bigger subset of $\mathcal{A}(C)$. It follows easily now that $\overline{\gamma} \cap C = C$ or $\overline{\gamma} \cap C$ is a Cantor set of C. Item (i) of lemma is proved. □

Now we proceed to construct a suspension flow Susp f_C over an interval exchange transformation $f_C : S^1 \to S^1$. Let us introduce an equivalence relation $\sim$ on compact transversal segments Σ_1 and Σ_2. As usual let $f : \Sigma_1 \to \Sigma_2$ be the forward Poincaré mapping induced by the flow π^t. Suppose that $a, b \in \Sigma_1$. We say that $a \sim b$ if and only if there exists a closed subinterval $\overline{pq} \subset \Sigma_1$ containing $\{a, b\}$ and an open disc D_{pq} disjoint of $\Sigma_1 \cup \Sigma_2$ with the boundary

$$\overline{pq} \cup \overline{pf(p)} \cup \overline{qf(q)} \cup \overline{f(p)f(q)},$$

where $\overline{f(p)f(q)}$ denotes a closed subinterval of Σ_2 with the endpoints $f(p)$ and $f(q)$. The equivalence relation $\sim$ which is defined on an open subset of Σ_1 containing the domain of definition, Dom f, of f is called the *relation $\sim$ associated to f*. Any disc D_{pq} as above is called a *disc associated to the triple* $(p, q, \sim)$.

Let $\sim$ be the equivalence relation associated to $f : C \to C$. (The similar relation $\sim^*$ is associated with the mapping $f^{-1} : C \to C$.) The relation $\sim$ $(\sim^*)$ has finitely many equivalence classes $A_1, \dots, A_n$ $(A_1^*, \dots, A_n^*)$. For brevity further we assume $n \geq 2$. If now $p, q \in A_i \cap$ Dom f is a pair of distinct points then there exists a unique disc D_{pq} associated to the triple $(p, q, \sim)$. Let us set

$$V(A_i) = A_i \cup A_i^* \cup D_{pq},$$

where D_{pq} is a union of all open discs such that $p, q \in A_i \cap$ Dom (f). Then the following is true:

(6) For all $1 \leq i \leq n$ the set $V(A_i) \backslash (A_i \cup A_i^*)$ is an open disc whose boundary contains $A_i \cup A_i^*$. Moreover, $V(A_i)$ is disjoint from U: $V(A_i) \cap U = \emptyset$. Indeed, the first claim is evident. Let us prove that $V(A_i) \cap U = \emptyset$. For this it is enough to show that $\forall p, q \in$ Dom (f), $(\overline{D}_{pq} \backslash D_{pq}) \cap U = \emptyset$. Observe that $pf(p) \cup qf(q)$ is disjoint from U and there is no arc of trajectory of π^t,

contained in U and connecting two points of ∂U. Since $C \cap U = \emptyset$, this implies $(\overline{D}_{pq} \backslash D_{pq}) \cap U = \emptyset$ and proves the second claim of (6).

(7) Let a_i be the endpoint of A_i (respectively A_i^*). Then γ_a^+ (respectively γ_a^-) is contained in $\partial V(A_i) \backslash U$ and ω-limit set of $\gamma^+(a)$ (respectively α-limit set of γ_a^-) is a singular point. Moreover, the complement of these semi-orbits in $\partial V(A_i)$ contains only singular points and regular trajectories connecting these singular points.

Indeed, let a be an endpoint of A_i. The proof that $\gamma_a^+ \subset \partial V(A_i) \backslash U$ is similar to that of (6). Suppose that $\omega(\gamma_a^+)$ contains a regular point $x \in M$. Let Σ be a segment transversal to x and such that p lies in the interior of Σ. Since $C \backslash (\gamma_a^+ \backslash a) = \emptyset$, γ_a^+ meets Σ infinitely many times. But this is a contradiction with γ_a^+ chosen to lie at the boundary $\partial V(A_i)$. Therefore $\omega(\gamma_a^+)$ is a singular point.

Let $\widetilde{V}(A_i) = V(A_i) \cup \gamma_a^+ \cup \gamma_b^+ \cup \gamma_{a'}^- \cup \gamma_{b'}^-$ and let λ_i be a continuous simple curve embedded in $V(A_i) \backslash (A_i \cup A_i')$ which connects the singular points $\omega(\gamma_a^+)$ and $\omega(\gamma_b^+)$ where a and b (respectively a' and b') are the endpoints of A_i (respectively A_i'). Let us put

$$V = \cup_{i=1}^{n} \widetilde{V}(A_i).$$

We claim that V is the required region of recurrence for the quasiminimal set $\mathcal{Q}$. In fact, note that the set $V \backslash (\cup_{i=1}^{n} \lambda_i)$ is homeomorphic to $C \times (-1, 1)$. Since the mapping f covers f_C, and on account of the items (1)-(6) it can be easily derived that V is the region of recurrence mentioned in the lemma. These arguments complete the proof of lemma. □

Proof of theorem. In view of Lemma 3.3.1 is evident. To obtain the pairwise disjoint regions of recurrence $V_1, \ldots, V_m$ one must apply inductively Lemma 3.3.1 with regard to the simple closed curves $C_i \subset V_i$. (Note that these circles can be constructed for every $\mathcal{Q}$-set of the foliation $\mathcal{F}$ and such that $C_i \cap C_j = \emptyset$ if $i \neq j$.) Each region V_i corresponding to the quasiminimal set $\mathcal{Q}_i$ satisfies items (i) and (ii) of the theorem and there is a finite number of such regions. This proves items (i) and (ii) of the theorem.

Now let V_i' are another regions of recurrence associated to $\mathcal{F}$, and let C_i', f_i' and g_i' be similar objects to C_i, f_i and g_i, respectively. To prove (iii) observe that $g_i : \mathcal{A}(C_i) \to \mathcal{A}(C_i)$ and f_i are topologically conjugate. (This is also true for g_i' and f_i'.) It is easy to see now that any orbit of f_i defines a unique orbit of f_i' in such a way that:

(1) Each $x \in S^1 \backslash$ Dom f_i (resp. $y \in S^1 \backslash$ Dom f_i^{-1}) is in 1-1 correspondence with a unique $x' \in S^1 \backslash$ Dom f_i' (resp. $y' \in S^1 \backslash$ Dom $(f_i')^{-1}$);

(2) For $x \in S^1 \backslash$ Dom f and $y \in S^1 \backslash$ Dom f^{-1} there exists an interval $I \subset$ Dom f such that $x \in \overline{I}$ and $y \in \overline{f}(I)$ if and only if for some interval I' it holds $I' \subset$ Dom $f', x' \in I'$ and $y' \in \overline{f}'(I')$.

By the definition of suspension items (1) and (2) imply that V and V' are homeomorphic. (The rest of item (iii) of the theorem follows from the blowing-down theorem.) Finally, item (iv) is evident in view of construction of the mapping f_C.

Theorem is proved. □

3.4 Surgery

Surgery is a "transversal" decomposition of foliations. By such a decomposition we understand a collection of disjoint closed curves which intersect transversally each leaf of the foliation. Surgery allows to represent foliations as a suspension over the interval exchange transformations.

3.4.1 Surgery of Labyrinths

$\mathcal{F}$ is called *arational* if it has no compact leaves and Sing $\mathcal{F}$ consists of a finite number of thorns and tripods. The main result which we postulate in this paragraph is that surfaces with arational foliations can be decomposed into discs, annuli and pants (two-holed discs). In the discs and annuli $\mathcal{F}$ is a "standard extension" of the model labyrinths, while in pants $\mathcal{F}$ is regular of one of the two types given by Fig. 3.2. We begin with

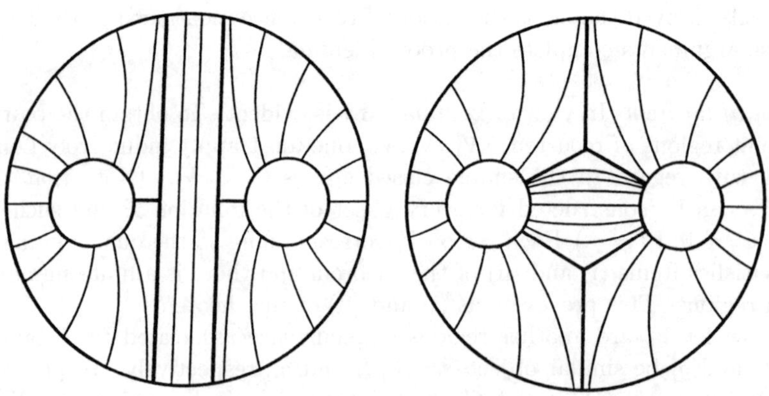

Fig. 3.2. The Rosenberg pants.

Lemma 3.4.1 (Rosenberg) *Let $\mathcal{F}$ be an arational foliation of D such that no leaf joins a point of ∂D to a thorn. Then there exists a labyrinth $\mathcal{L}$ in D, on a segment J, such that $\mathcal{F}$ is the standard extension of D of the labyrinth. (If no leaf enters ∂D and then leaves ∂D, then no leaf joins a point of ∂D to a thorn.)*

Proof. First let us prove an *extension lemma* which states that the holonomy liftings of an arc of a leaf are *closed* including a leaf passing through a singularity. (In other words, the 'transition mapping' near the singular point of thorn and tripod type can be extend to the 'ends' of the transversal segments.) More precisely, the following proposition is valid.

Proposition 3.4.1 *Let $\mathcal{F}$ be an arational foliation of a compact surface N. Let $l \in F$ be a leaf of $\mathcal{F}$ and $I = [x_0, x_1], J = [y_0, y_1]$ be two transversal arcs with $x_0, y_0 \in l$. Denote by ϕ a local holonomy map with $\phi(x_0) = y_0$ and let us assume that ϕ can be continuously extended from $[x_0, x_1)$ to $[y_0, y_1)$ so that $\phi(x_1) = y_1$. (The case when the leaf through x_1 passes through a singularity is not excluded.)*

Proof of proposition. Let l_1 be a half-leaf through x_1 chosen to follow ϕ continuously, i.e. l_1 is determined by the liftings l_t of the arc $l_0 \subset l$ between x_0 and y_0, parametrized so that l_t begins at $x_t \in I, 0 \le t < 1$. In case l_1 goes to a singularity then we go through the singularity and go out along the separatrix which follows the liftings l_t (if it is a thorn then we go back along the same leaf).

If $l_1 \cap \partial N \neq \emptyset$ then by the standard holonomy lemma, the leaves a_t containing l_t for t near 1, also intersect ∂N and for some real C near 1 the set

$$R = (\bigcup_{t \ge C} a_t) \cup l_1$$

is an embedded rectangle in N. Clearly, J goes from the edge of a_C of this rectangle to the edge l_1 and ϕ extends continuously to x_1.

Let now l_1 be non-compact. This implies that l_1 intersects some simple closed curve C infinitely often. Let us choose C transversal to $I \cup J$ and so that C intersects $I \cup J$ in a finite number of points. The number of points in $l_t \cap C$ is constant near each $t < 1$, for which the ends of l_t do not belong to C. Therefore, there exists a number K such that l_t meets C at most K times and K is independent of $t, 0 \le t < 1$. Take a point $p \in l_1$ so that the segment $a \subset l_1$ lying between x_1 and p intersects C $K+1$ times. Let in addition $p \notin C$.

Denote by H a small transversal arc at p, disjoint from C. For t near 1, l_t does not intersect H. Therefore $y_1 \in a$ and ϕ extends continuously to $[y_0, y_1]$. These arguments prove the proposition. □

Denote now by J an arc embedded in N which passes through all the thorns of $\mathcal{F}$ and is transversal to each thorn separatrix. We state that

Proposition 3.4.2 *J can be chosen to be transversal to $\mathcal{F}$.*

Proof of proposition. First let us construct a small isotopy of J which brings J in general position with respect to $\mathcal{F}$, so that there are only a finite number of points of J where J is tangent to $\mathcal{F}$ and no two such points lie on the same leaf. Indeed, let p be a contact point. In the vicinity of p $\mathcal{F}$ looks like the level curves $y = x^2$ with J a segment on the x-axis containing $p = 0$. Suppose $x_0 \in J, x_0 < 0, |x_0| < \varepsilon$ and let l_0 be an arc of the leaf through x_0 lying in a half-plane $y > 0$. We can lift l_0 for x near x_0 to the leaf through x so that both extremities of the lifted arc l_x are on J. We extend the lifting of l_0 with both ends on J, as far as possible. Let us consider now what possibilities can occur. (The lifted arcs are always in Int N, otherwise they must be tangent to ∂N, what is impossible.) Three cases are possible.

(i) l_0 can be lifted until one of the ends of the lifting goes to a thorn. Denote by l_1 this lifted arc joining a thorn x_1 to $y_1 \in J$. Let y be a point slightly to the right of y_1 and u a point slightly to the left of x_1 (in case x_1 is the end of J we do not need u). Let us change J by removing the arc on J between y and u adding a arc J_1 transversal to $\mathcal{F}$ between y and u. Clearly J_1 be in a neighbourhood of l_1. The newly obtained J must have one contact point less and still it passes through all the thorns. Let J_1 be an arc in this neighbourhood from y to x_1 transversal to $\mathcal{F}$ and tangent to the 'vertical' at y. Clearly one can extend J_1 from x_1 to u transversally to $\mathcal{F}$. (For example, it can be done by removing the arc J between y and u and adding the extended J_1, see [256] for details.)

(ii) One can lift l_0 until it comes to a singularity described by Proposition 3.4.1. Let this singularity be a tripod with the endpoints x_1 and y_1 on J. Let u be a point of J lying slightly to the right of y_1. We replace the arc on J between u and v by an arc J_1 transversal to $\mathcal{F}$ and close to l_1.

Let now the singularity, where the lifting stops, be of a thorn type. Denote by B the arc of J between x_1 and y_1 containing p. Let $z_0 \in J \backslash B$ be close to y_1 and $z_1 \in J \backslash B$ be close to x_1. Remove now the arc on J between z_0 and z_1 containing p and replace it by J_1. It is not hard to see that the obtained J has one fewer contact point and passes through all the thorns of the foliation $\mathcal{F}$.

(iii) Let, finally, the leaf l_1 through x_1 be tangent to J at $y_1 = \phi(x_1)$. (The liftings between l_0 and l_1 may have tangencies y with J but $y \neq \phi(x)$, i.e. the first point in the ϕ-holonomy image where $\mathcal{F}$ becomes tangent to J.) Choose a rectangular neighbourhood R of l_1 foliated by $\mathcal{F}$ horizontally and let J_1 be an arc transversal to $\mathcal{F}$ in R going from the upper left-hand corner to the lower right-hand corner. The new segment J_1, which replaces J, is constructed in the obvious way.

In cases (i)-(iii) we have constructed J which is transversal to $\mathcal{F}$, but not necessary embedded in D. However simple arguments, which we omit here, show that J can be chosen without self-intersections, see [256] for the relevant details. Proposition is proved. □

Proposition 3.4.3 *Let J be the arc mentioned in Proposition 3.4.2 and let l be a half-leaf of the foliation $\mathcal{F}$. If l does not go to a tripod then l goes either* *to J or to ∂D. If l goes to a tripod then one of the other two separatrices of the tripod goes to J. If l comes from ∂D and goes to a tripod then two other separatrices of the tripod go to J.*

Proof of proposition. Suppose that l does not go neither to a tripod nor to the ∂D. If l does not intersect J then a simple closed curve C can be constructed which is transversal to $\mathcal{F}$ and met l infinitely often. C is disjoint from J. (It is not hard to see that such C really exists. The rest of the argument can be found in [256].)

Curve C bounds a disc $B \subset D$ and $J \subset B$ because of the Euler characteristic: each thorn has the Euler-Poincaré index $+1/2$ while such index for a tripod is $-1/2$, and therefore to obtain after summing-up $\chi(B) = +1$, B must contain thorns. Now J is disjoint from ∂B and J passes through the thorns. It follows that $J \subset B$.

Let A be an annulus in D bounded by ∂B and ∂D. Since ∂A is transversal to $\mathcal{F}$ and there are no thorns in A, there can be no tripods there (since $\chi(A) = 0$). Hence A is foliated without singularities. Since there are no compact leaves in A the foliation $\mathcal{F}$ is the trivial interval fibration. Because l intersects C at least twice, there is a point of intersection where l enters A. Hence l goes to ∂D, which is a contradiction proving the first claim of the proposition.

Now suppose that l goes to a tripod and let l_1, l_2 be the other separatrices of the tripod. Assume to the contrary, that l_1 and l_2 do not go to J. Then by the first assertion of proposition 4.2.3, they both go to ∂D. $l_1 \cup l_2$ separates D into two discs D_1, D_2 and each disc must contain a thorn for Euler characteristic reasons. But J joins all the thorns and therefore J must intersect $l_1 \cup l_2$, what is a contradiction. Similar arguments prove the last claim of proposition. □

Proof of Lemma 3.4.1. Let $x_1, \ldots, x_n$ be the points of ∂D whose leaves go to the thorns $y_1, \ldots, y_n$ before intersecting J. Such points exist by Proposition 3.4.3 and our hypothesis that no leaf goes from ∂D to a thorn. The other two separatrices of every y_i go to J. For brevity, let us assume J to be horizontal. For each y_i the two separatrices leaving y_i intersect J for the first time coming from the top or from the bottom. We shall see that there are two possibilities. Either there exist two y_i, say y_1 and y_2, such that the separatrices of y_1 and y_2 intersect J from the top and bottom. For $y_3, \ldots, y_n$ the separatrices intersect J from the same side. The other possibility is that there is one thorn, say y_1, whose two separatrices intersect J from the same side and the two separatrices of $y_2, \ldots, y_n$ intersect J for the first time from the same side. For every y_i there are three possibilities for the rest two separatrices:

(1) The two separatrices of y_i intersect J on the same side without 'engulfing' J.

(2) They intersect J on the same side and 'engulf' J.
(3) They intersect J on opposite sides.

Let us show first that not every y_i can be of type (1). Suppose to the contrary that y_i is a tripod whose separatrix l_i is the farthest to the left and hits J from above. For $x \in \partial D$ near x_i l_x intersects J at $\phi(x)$, to the left of the intersection $\phi(x_i)$ of l_i and J. By the extension lemma (Proposition 3.4.1), as x moves along ∂D away from x_i, $\phi(x)$ moves along J to the left of $\phi(x_i)$ and the holonomy liftings stop when l_x goes to a singularity. Because y_i was the farthest to the left, there exists some $x \in \partial D$ such that l_x goes to the left-hand thorn on J. This is a contradiction.

Thus, there exists at least one tripod of type (2) or (3). Suppose that y_1 is of type (2) and two separatrices of y_1 intersect J from above. Then each other y_i is of type (1) and their separatrices meet J from above. Let us take the separatrix of each y_i coming from ∂D and not intersecting J. If y_1 is of type (3), then the two separatrices of y_1 'engulf' one of the ends of J. Arguing as above we see that there must exist another tripod, say y_2, which also has the type (3) and whose separatrices 'engulf' the other end of J. Therefore, $y_3, \ldots, y_n$ are of type (1).

Next J is extended on both sides, until it reaches the separatrices of y_1 and y_2, if we are in the first case, or, until it reaches the separatrices of y_1, if we are in the second case. (It is not hard to proceed so that J remains transversal to $\mathcal{F}$ and passes through all the thorns of $\mathcal{F}$.) For the remainder of the proof let us assume that we are in the first case; the proof of the other case is similar.

We denote by J the extension we obtained by going to the separatrices of y_1 and y_2 transversally. Let y_3, y_4 be the tripods of $\mathcal{F}$ whose separatrices come from the points x_3, x_4 on ∂D, which bound a segment on ∂D none of whose leaves go to a tripod. Since each leaf starting in the interior of the segment $[x_3, x_4]$ of ∂D must go to $R \cap J$, there are no singularities in the interior of R and the interior of R is foliated by $\mathcal{F}$ in a trivial way (as a product). Let D_3, D_4 be the discs bounded by the separatrices of y_3, y_4 that go to J and the segments on J which they determine. Similarly, each of $y_5, \ldots, y_n$ determines a disc $D_5, \ldots, D_n$. However y_1 and y_2 each determine two discs, $D_1, \widetilde{D}_1, D_2, \widetilde{D}_2$: one on the top of J and the other on the bottom. We claim that

$$\mathcal{L} = D_1 \cup \widetilde{D}_1 \cup D_2 \cup \widetilde{D}_2 \cup D_3 \cup \ldots \cup D_n,$$

is a labyrinth. Indeed, consider a D_j. If there are no tripods in the interior of D_j, then it is clearly foliated by concentric half-circles (up to a diffeomorphism). If there is a tripod in $\text{Int}\ D_j$, denote it by z, then its three separatrices must go to $D_j \cap J$, and now one can apply an inductive arguments to the two discs in D_j determined by these three separatrices. Thus $\mathcal{L}$ is a labyrinth. This argument completes the proof of Lemma 3.4.1. □

Corollary 3.4.1 *Let $\mathcal{F}$ be an arational foliation of D^2. Then $\mathcal{F}$ is obtained by a finite number of consecutive spreadings of either:*

(a) the trivial arational foliation of D^2 with exactly two thorns or,
(b) a standard arational foliation of D^2 on a labyrinth $\mathcal{L}$ with no leaf joining ∂D^2 to a thorn. The spreadings are done on leaves going to ∂D^2.

Proof. Denote by $x_1, \ldots, x_n$ the points of ∂D^2 whose leaves go to the thorns $z_1, \ldots, z_n$. Applying Proposition 3.4.1 to the points near x_1 on ∂D^2 one obtains that $\mathcal{F}$ is either a trivial foliation of D^2 with exactly two thorns, or there is a tripod, say y_1, such that $\mathcal{F}$ is obtained by spreading the leaf through y_1. Let C_1 be the transversal curve constructed starting from ∂D^2 and the arc J_1. Clearly, C_1 bounds a disc D_1 and $\mathcal{F}$ on D_2 is an arational foliation with $n-1$ points on ∂D_1 going to the thorns. Let us apply the same construction to x_2. We shall see that for some i, D_i is trivially foliated (here we have the case (a)), or, alternatively, the disc D_n has an arational foliation F_n and no leaf joins ∂D_n to a thorn. Applying Lemma 3.4.1 we obtain the case (b) of the corollary. □

Lemma 3.4.2 (Rosenberg) *Let $\mathcal{F}$ be an arational foliation of an annulus $A = S^1 \times I$. $\mathcal{F}$ can be obtained by a finite number of consecutive spreadings of either:*

(a) the trivial product foliation $B \times I$, or

(b) a standard foliation of A on a labyrinth $\mathcal{L}$.

Proof. This is similar to those of Lemma 3.4.1, see [256] for the details. □

Now we are in position to formulate and prove the following result due to H. Rosenberg.

Theorem 3.4.1 (Rosenberg) *Let N be a compact orientable surface possibly with a non-empty boundary, and let $\mathcal{F}$ be an arational foliation on N. There is a decomposition of N into discs $D_1, \ldots, D_k$, annuli $A_1, \ldots, A_l$ and pants $P_1, \ldots, P_m$, such that $\mathcal{F}$ is arational in each disc and annulus and in each pair of pants $\mathcal{F}$ is topologically conjugate to one of the two foliations given by Fig. 3.2. The discs, annuli and pants are either pairwise disjoint, or coincide along their common boundary components.*

Proof. Roughly, the proof consists of finding the discs D_i and annuli A_j of the foliation $\mathcal{F}$, such that

$$N' = \left(\bigcup_{i=1}^{k} D_i\right) \cup \left(\bigcup_{j=1}^{l} A_j\right)$$

contains all the thorns of the foliation $\mathcal{F}$. Then the complement, $N \backslash N'$, is considered and it is shown that it has a decomposition into pants as desired.

Let J be a closed curve in N which is transversal to $\mathcal{F}$ and which passes through all the thorns of $\mathcal{F}$ (such curve always exists due to Proposition 3.4.2). To obtain simple closed curves (i.e curves without self-intersections) one must cut and paste J to $J_1, \ldots, J_n$ in the evident way. In case $N \neq S^2$ J can be chosen non-trivial, and henceforth such will be certain J_i. Further we discard those J_i, which have no thorns.

Fix J_1, and let $z_1, \ldots, z_k$ be the thorns on J_1 whose separatrices are locally on the same side, S_1, of J_1 (denote the opposite side by S_2). For each z_i the leaves of $\mathcal{F}$ near z_i in S_2 are small arcs joining two points of J_1. Let us lift these arcs applying the extension lemma, Proposition 3.4.1 (note that the liftings remain disjoint from the other J_i's by the transversality). Either the liftings stop at a tripod, or they stop at another thorn z_2 and J_1 bounds a disc D_0 foliated trivially by the liftings. If two distinct z's yield the same tripod, then we continue the liftings by lifting the two separatrices of this tripod which, together with an arc on J_1, bound a half-disc containing the third separatrix. These liftings either stop at another tripod, or generate a disc bounded by J_1. Assuming that J_1 does not bound the disc D_0 with $S_2 \subset D_0$, the set of all these liftings for all z_i form the half-discs $D_1, \ldots, D_l$ and the boundary of each D_j has one tripod. Next, a simple closed curve C_1, which is transversal to $\mathcal{F}$, is constructed. It is done in the usual way.

Consider the remaining thorns of J_1. If there are none of them, let C_2 be a simple closed curve transversal to $\mathcal{F}$ and $C_2 \subset S_1$ (it is enough to take C_2 isotopic to J_1, C_2 close to J_1). Now either $C_1 \cup C_2$ bounds an annulus, A_1, on which $\mathcal{F}$ is arational and $J_1 \subset A_1$, or C_2 bounds a disc D_1, [1] such that $\mathcal{F}$ is arational on D_1 and $J_1 \subset D_1$.

If there are remaining thorns on J_1, we use the same construction as for $z_1, \ldots, z_n$. If J_1 does not bound a disc D_0 with $S_1 \subset D_0$, then one obtains the half-discs $E_1, \ldots, E_r$ from the liftings the boundary of which contains one tripod. Finally we come to a simple closed curve C_2 transversal to $\mathcal{F}$, just in the same manner as for C_1. Hence $C_1 \cup C_2$ bounds an annulus A_1, or C_2 bounds a disc D_1, what is required.

When J_1 is null homotopic on both sides (that is $N = S^2$), then one directly constructs a simple closed curve C_1 transversal to $\mathcal{F}$, disjoint from the singularities (it can be done following J_1 and choosing the appropriate side of each thorn). Then C_1 separates S^2 into discs D_1, D_2 on which $\mathcal{F}$ is arational. This is again what required.

Finally, we repeat this analysis for every J_i and we obtain the discs $D_1, \ldots, D_k$ and annuli $A_1, \ldots, A_l$ whose union contains all the thorns. Moreover, they are pairwise disjoint and $\mathcal{F}$ is arational therein. Therefore, to finish the proof of theorem one needs the following

Proposition 3.4.4 *Let N be a compact orientable surface and $\mathcal{F}$ be an arational foliation on it without thorns. Then there is a decomposition of N*

[1] $D_1 = D_0 +$ tubular neighbourhood of J_1.

into the pants $P_1, \ldots, P_m$, intersecting at most along their common boundary components, and $\mathcal{F}$ is standard shown in Fig. 3.2 on every P_i.

Proof. Let y_1 be a singularity of $\mathcal{F}$ and l_1, l_2, l_3 be the separatrices of y_1. The leaf l_1 either goes to a boundary component of N or accumulates somewhere in N. In any case one can find a simple closed curve C_1 in N transversal to $\mathcal{F}$ and intersecting l_1 first time in a point x_1. In the same way we can find a simple closed curve C_2 transversal to $\mathcal{F}$ and intersecting l_2 for the first time at x_2. After necessary cutting and pasting $C_1 \cup C_2$ we can assume either $C_1 \cap C_2 = \emptyset$ or $C_1 = C_2$. Let us study both possibilities step-by-step.

(i) Let $C_1 \cap C_2 = \emptyset$. Denote by t_1 a point of C_1 close to x_1 with the leaf intersecting C_2 in a point t_2 close to x_2. The leaf $l(t_1)$ stays close to $l_1 \cup l_2$ between x_1 and x_2. Let $l[t_1]$ be the arc of $l(t_1)$ between t_1 and t_2. We use the extension lemma to lift $l[t_1]$ to the leaves joining C_1 to C_2 on the side of $l[t_1]$ other than $l_1 \cup l_2$. The liftings stop at a point $s_1 \in C_1$ whose leaf goes to a tripod, denote it by y_2, and one of the other separatrices of y_2 which goes to a point $s_2 \in C_2$.

Suppose that $y_1 \neq y_2$ and let $u_1, v_1 \in C_1$ be the points in the complement of the arc $[s_1, t_1]$ of C_1, u_1 close to s_1, v_1 close to t_1. Let u_2, v_2 be the points of C_2, defined similarly. Then, in the obvious way, we construct a simple closed curve C transversal to $\mathcal{F}$, which goes from u_1 to v_1 along C_1, then from v_1 to v_2, staying close to $l_1 \cup l_2$, then from v_2 to u_2 along C_2, and, finally, from u_2 to u_1, staying close to the separatrices of y_2 that intersect at s_1, s_2. The contour $C \cup C_1 \cup C_2$ bounds a pair of pants P_1 containing exactly y_1, y_2 as singularities and $\mathcal{F}$ in P_1 is standard. Now consider the complement of P_1 in N and repeat the above arguments with the remaining singularities. If there are no such, then the complement is a union of annuli foliated as a trivial interval bundle over S^1.

It remains to show that when $y_1 = y_2$, still a pair of pants P_1 containing two singularities can be found, and such that $\mathcal{F}$ is standard in P_1. Let $u_1 \in C_1$ be as before. Now $l[u_1]$ intersects C_1 in a point v_1 close to x_1. The liftings of $l[u_1]$ will stop at a tripod, denote it by y_2, distinct from y_1. Let l_3, l_4 be the separatrices of y_2, that have been obtained from these liftings. Let they intersect C_1 for the first time in the points r_1, r_2, respectively. Suppose that C is a simple closed transversal curve obtained by following a point $\overline{r}_1 \in C_1$ near r_1 to a point $\overline{r}_2 \in C_1$ near r_2, then following $l_1 \cup l_2$. Then $C \cup C_1 \cup C_2$ bounds a pair of pants P_1 as desired. The case (i) when $C_1 \cap C_2 = \emptyset$ is done.

(ii) Consider the possibility $C_1 = C_2$. If the third separatrix l_3 of y_1 goes to a closed transversal curve C_3 disjoint from C_1, then we proceed as in the case (i). So we can assume that l_3 intersects C_1 and so do l_1 and l_2. (One can assume that the three separatrices intersect C_1 on the same side of C_1, since if it takes place from the opposite side, we could easily construct a closed transversal C disjoint from C_1 intersected by this separatrix.)

Now let us consider the liftings of l_0 going away from $l_1 \cup l_2$. The lifting will be continued until it reaches a tripod, say y_2, and the two separatrices l_3, l_4 of y_2 that go to C_1. The pants P_1 which contain y_1 and y_2 and in which $\mathcal{F}$ is standard, are bounded by C_1. This argument completes the proof of Proposition 3.4.4 and henceforth the theorem. □

3.4.2 Surgery of Measured Foliations

Surgery of the labyrinths can be extended to foliations with a finite number of compact leaves and a singularity set consisting of the k-prong saddles. We begin by considering the case $k = 4$ (Morse saddles) and then we pass to the case of arbitrary k.

Foliations With Morse Saddles

Let M be a compact orientable surface of genus g possibly with b boundary components. Let $\mathcal{F}$ be a C^∞ foliation on M with Sing $\mathcal{F}$ consisting of 4-separatrix saddles. We always assume that $\mathcal{F}$ is transversal to ∂M. Clearly, in this case $\mathcal{F}$ must have $2g + b - 2$ singular points on M. The following terminology is adopted.

A leaf $l \in \mathcal{F}$ is called an *interior leaf* if it does not meet the boundary ∂M of the manifold M. There are four types of behaviour possible for a separatrix joining the saddle point:

(i) Separatrix comes back to the same saddle point;
(ii) Separatrix goes to another saddle point;
(iii) Separatrix goes to the boundary ∂M
(iv) Separatrix is 'open' in the sense that it does not follow (i)-(iii) and define a non-closed subset in M.

If a separatrix $l \in \mathcal{F}$ verifies either (iii) or (iv) it is called a *free separatrix*. Foliation $\mathcal{F}$ is said to have a *connection* if there exists a separatrix of $\mathcal{F}$, which is not free. Two separatrices l, l' of the same saddle $s \in$ Sing $\mathcal{F}$ are called *adjacent* if they bound a saddle sector of s. Otherwise they are called *opposite* separatrices.

Foliation $\mathcal{F}$ with the above set of singularities is always locally orientable. But globally it is not the case (e.g. $\mathcal{F}$ may have a Reeb component). Further we rule out such cases and deal only with *orientable foliations.* on M.

By *a cycle of leaves*, Γ, one understands an image of $h(S^1)$ by a homeomorphism $h : S^1 \to M$ consisting of a finite number of saddle points of $\mathcal{F}$ joined by the separatrices of types (i) and (ii). Note that any cycle Γ inherits an orientation on each of its separatrix components.

Definition 3.4.1 (Levitt) *A cycle of leaves Γ is called totally antisymmetric if the orientation of Γ is one and the same on all its separatrix components. In other words, all the separatrices of Γ are adjacent in all saddle points of Γ.*

Definition 3.4.2 (Levitt) *By a nearly transversal cycle of leaves of a foliation $\mathcal{F}$ one understands an image, $\Gamma = \phi(S^1 \times \{0\})$, of mapping $\phi : S^1 \times [0,1] \to M$ such that:*

- *there are points $t_1, \ldots, t_p \in S^1$ for which the restriction of ϕ on $S^1 \times [0,1]$ with $(t_i, 0)$ deleted, is an immersion and points $\phi(t,u)$ are saddles of $\mathcal{F}$ if and only if $t = t_i, u = 0$;*
- *there exists a non-singular foliation on $S^1 \times [0,1]$ such that (i) $S^1 \times \{0\}$ is a leaf and its image by ϕ consists of saddles $\phi(t,u)$ and separatrices joining them, and (ii) image of every other leaf contains in some leaf of $\mathcal{F}$.*

(In other words, a nearly transversal cycle is a totally antisymmetric cycle of leaves Γ with all of its separatrices of saddle points, which do not belong to Γ, lying on the same side of Γ.)

Finally, by an *immersed nearly transversal cycle of leaves* one understands a nearly transversal cycle, for which the image $S^1 \to M$ may have a 'double' point coinciding with a saddle of Γ, provided there exists a perturbation bringing Γ to a nearly transversal cycle.

Now we are going to define a decomposition of $\mathcal{F}$ into pants, similar to those pictured in Fig. 3.2. Recall that if M is an arbitrary surface, by *decomposition in pants* of M one understands a finite family $\mathfrak{P}$ of disjoint closed curves such that cutting along $\mathfrak{P}$ dissects M into pants (3-connected disc regions). These decomposition in pants is said to be *determined by* $\mathfrak{P}$.

Theorem 3.4.2 (Levitt) *Let $\mathcal{F}$ be an orientable foliation on a surface M, different from torus and annulus, admitting no totally antisymmetric cycles of leaves. Then there exists a decomposition of M into the pants $\mathfrak{P}$ transversal to $\mathcal{F}$. Moreover, the following holds:*

- *If $\mathcal{F}$ has no compact interior leaves, the foliation $\mathcal{F}_0$ induced by $\mathcal{F}$ on each (pair of) pants determined by $\mathfrak{P}$ is topologically equivalent to those of Fig. 3.3. The surface M with the foliation $\mathcal{F}$ can be recovered by pasting together the pants $\mathfrak{P}$.*
- *If $\mathcal{F}$ has one or several compact interior leaves then one can choose $\mathfrak{P}$ to be disjoint from these leaves. Each pants determined by $\mathfrak{P}$ contains pants induced by $\mathcal{F}$ and conjugate to $\mathcal{F}_0$. Every $\mathcal{F}$ on M can be obtained by pasting together all pants $\mathfrak{P}$ along their boundary components, and also an annulus which contain all compact interior leaves of $\mathcal{F}$.*

Corollary 3.4.2 *If $\mathcal{F}$ is a foliation with Morse saddles with no saddle connections between them, then $\mathcal{F}$ is decomposable into a finite number of pants.*

Proof. Let $q = 2g + b - 2$ be the total number of saddles of the foliation $\mathcal{F}$. (Note that $\mathcal{F}$ has at least one saddle, since $M \neq T^2, S^1 \times I$.) It will be proved that one can find q disjoint pants $\mathfrak{p}_i, (1 \leq i \leq q)$ immersed in M such that:

- each $\mathfrak{p}_i$ contains exactly one saddle of $\mathcal{F}$;

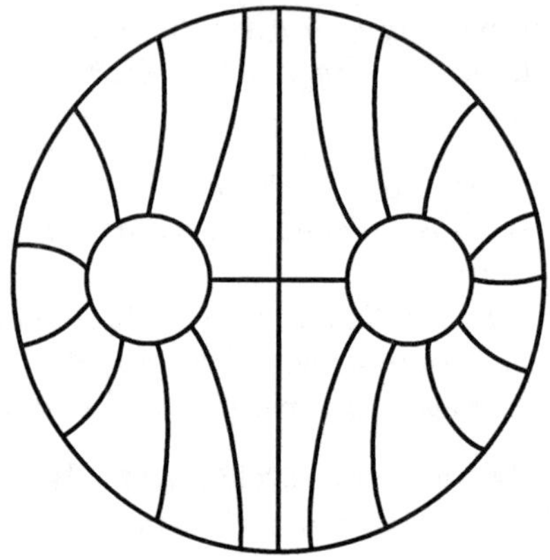

Fig. 3.3. The Levitt pants.

- the boundary components $\partial \mathfrak{p}_i$ are transversal to $\mathcal{F}$ and do not meet compact interior leaves of $\mathcal{F}$;
- the restriction $\mathcal{F}|_{\mathfrak{p}_i}$ of $\mathcal{F}$ is conjugate to $\mathcal{F}_0$.

Indeed, suppose that q curves $\mathfrak{p}_i$ exist (we shall construct such curves later). Denote by M_k a closure of $M \setminus \cup_{i=1}^{q} \mathfrak{p}_i$ and let $\mathcal{F}|_{M_k}$ be non-singular and transversal to ∂M_k. Then M_k is an annulus. Choosing one boundary component ∂M_k for each M_k which does not meet the boundary ∂M, one easily constructs a family of pants $\mathfrak{P}$. The only thing to be verified is that if pants do not contain compact interior leaves, then each foliation within pants is conjugate to $\mathcal{F}_0$. But this is trivial, noting that each pants contain one saddle point.

Now let us construct $\mathfrak{p}_i$. We begin with the behaviour of compact interior leaves of $\mathcal{F}$. Note that these leaves are non-trivial, and their number is finite.

Lemma 3.4.3 *Suppose that $\mathcal{F}$ is an orientable foliation on a surface M, which does not have immersed nearly transversal cycles of leaves. Let θ be an isotopy class of closed simple curves on M, containing at least two compact interior leaves of $\mathcal{F}$. There exist two compact interior leaves l_θ^1 and l_θ^2 in class θ such that all other compact interior leaves belonging to θ lie in an annulus bounded by l_θ^1 and l_θ^2.*

Proof. For all pairs of compact leaves in θ let us consider an interior of the annulus bounded by these leaves. Let U be the union of all such interiors. U is open and connected subset in M, which contains at least one compact leaf (provided that there are more than 2 compact leaves in θ).

The boundary of U consists of interior leaves of $\mathcal{F}$ and possibly of saddle points. If this boundary does not contain compact leaves, lemma follows. If there exists a leaf l in ∂U and which is not a separatrix of type (i) or (ii), then there exists an open region in M through which l passes infinitely many times. The same is true for a compact interval transversal to $\mathcal{F}$. But it is impossible since this interval must meet infinitely often all compact leaves in U (contradiction with compactness). It remains to consider a case when the boundary of U contains separatrices of types (i) or (ii). But the boundary which contains a saddle and one of these separatrices must contain also one of two adjacent separatrices, so that there exists an immersed nearly transversal cycle of leaves. But by the assumption this case is excluded. Lemma is proved. □

Let us prove the existence of $\mathfrak{p}_k$. Suppose that $p-1$ pants $\mathfrak{p}_k, (1 \leq k \leq p, 1 \leq p < q)$ are constructed, and let us show how to construct $\mathfrak{p}_p$. Denote by M' a closure of the component $M \setminus \cup_{k=1}^{p-1} \mathfrak{p}_k$, which contains at least one saddle. By hypothesis, $\mathcal{F}$ does not have totally antisymmetric cycles of leaves, what implies that $\mathcal{F}|_{M'}$ has two free opposite separatrices, l and l'. We are going to construct in the interior of M' two transversal curves, C and C', such that C (C') meets l (l'). Moreover, there exists a point x (x') in $C \cap l$ $(C \cap l')$ such that C (C') does not intersect l (l') between x (x') and s, where s is the saddle point. We demand also that C and C' meet no compact interior leaf of $\mathcal{F}$.

The existence of C and C' is clear if l and l' intersect $\partial M'$. Therefore let us suppose that l does not meet $\partial M'$. Let $y \in M$ be a point which is not a saddle point of $\mathcal{F}$ and in the vicinity of which l passes infinitely often. If a leaf through y is compact (necessarily interior), then l winds around this leaf and it is easy to construct C. If it is not compact, then slightly changing the argumentation [183] we come to the previous case. (The standard method of constructing transversal section uses 'foliation tubes'.)

If l' intersects C, one sets $C' = C$. If l' does not intersect C but meets $\partial M'$, one chooses C' to be a curve lying in the vicinity of $\partial M'$ to which l' tends. If l' does not intersect neither C nor $\partial M'$, one considers a surface obtained from M' by cutting along C, and then one applies the technique which allowed to construct C.

The theorem follows from a simple observation if we suppose $C = C'$. Then the orientability of $\mathcal{F}$ implies that l and l' reach C from the same side. The curves C_1 and C_2 on this figure can be drawn transversally and $C_1 \cup C_2 \cup C$ bounds the pants in M' which contain the saddle point s. These pants is the required one, which we denote by $\mathfrak{p}_p$.

If now $C \neq C'$ one draws a curve C_1 transversally to $\mathcal{F}$. The pants in this case are bounded by $C \cup C' \cup C_1$, and again they are the required one which we denote by $\mathfrak{p}_p$. (Note that $\mathcal{F}|_{\mathfrak{p}_p}$ is conjugate to $\mathcal{F}_0$.) This argument completes the proof of theorem. □

Remark 3.4.1 *It follows from Theorem 3.4.2 that the main obstacle for a complete decomposition of a foliation with a finite number of compact leaves* *into the Levitt's pants is the totally antisymmetric cycles which it may have. An example below shows that this claim is essential.*

Example 3.4.1 (Levitt) *Consider an irrational flow ϕ^t on torus T^2. Let us modify ϕ^t adding a loop of saddle in a vicinity of a point $x \in T^2$. We remove a disc region bounded by the loop, obtaining a torus with a hole. Now a second sample of torus with hole is taken, and one glues both tori along the holes not identifying singularities on its boundaries. One comes to a surface of genus $g = 2$ with a foliation $\mathcal{F}$ on it having a totally antisymmetric cycle of leaves. Foliation $\mathcal{F}$ admits no decomposition on pants and for it Theorem 3.4.2 fails.*

Foliations With k-Prong Saddles

Theorem 3.4.2 can be generalized to foliations $\mathcal{F}$ with saddles having an arbitrary even number of separatrices. Class $\mathcal{F}$ reduces to foliations with Morse saddles by a Whitehead homotopy operations over its saddle set. It will be proved that there exists a canonical decomposition of $\mathcal{F}$ into the pants, provided that there are no 'barriers' made of the separatrix cycles of $\mathcal{F}$.

As before, by M one understands a compact orientable surface (possibly with a boundary) different from S^2, D^2, T^2 and $S^1 \times I$. Foliation $\mathcal{F}$ on M is supposed orientable and transversal to ∂M if ∂M is non-empty. The set Sing $\mathcal{F}$ consists of even-separatrix saddles with at least 4 separatrices. (Further if $\mathcal{F}$ has only 4-separatrix saddles it will be called *simple*.)

Two foliations $\mathcal{F}$ and $\mathcal{G}$ are *equivalent by Whitehead* if one can obtain $\mathcal{G}$ from $\mathcal{F}$ by a finite number of Whitehead operations in the vicinity of saddles, which consist in blowing-up of a saddle with $k > 4$ separatrices to Morse saddles, and vice versa, see Fig. 3.4.

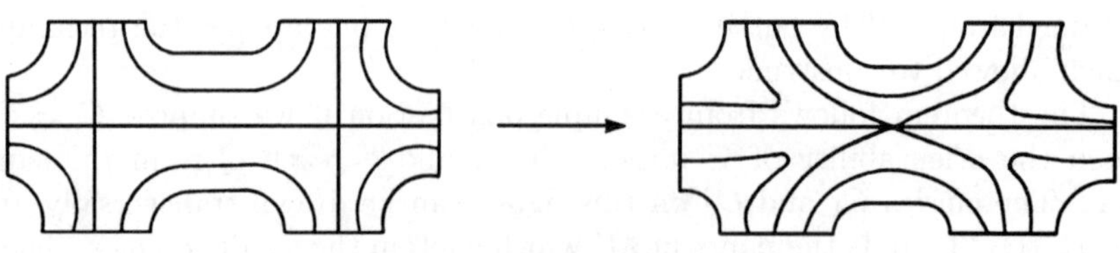

Fig. 3.4. The Whitehead homotopy

An important class of foliations $\mathcal{G}$ which we are going to consider further are *connected foliations* introduced by Novikov [231]. By such one understands $\mathcal{G}$ for which through any two points $x, y \in M$ one can draw a closed transversal curve γ. (The study of this class of foliations usually reduces to study of the interval exchange transformations on a transversal curve γ which intersects each 1-leaf of $\mathcal{G}$.) If such γ cannot be constructed, one can define an obstacle for this which is called *a barrier.* Roughly, a barrier is a leaf contour on which the transversal direction is 'one-sided': either directed 'within' contour, or 'outside' it. In our case it will be the following configurations of compact leaves.

Definition 3.4.3 (Levitt) *Let $\Gamma = \phi(S^1 \times \{0\})$ be a nearly transversal cycle of foliation $\mathcal{F}$ with leaves of $\mathcal{F}$ winding around it, or, possibly, lying around it as concentric circles. By a barrier $(\Gamma_1, \dots, \Gamma_k)$ of foliation $\mathcal{F}$ one understands a set of nearly transversal cycles or a set of homotopy classes of concentric closed leaves of $\mathcal{F}$.*

Theorem 3.4.3 (Levitt) *Let $\mathcal{F}$ be a foliation given on a compact (boundaryless) surface $M \neq S^2, T^2$, such that Sing $\mathcal{F}$ consists of saddle singularities. The following conditions are pairwise equivalent:*

(i) There exists a foliation $\mathcal{G}$ which can be obtained from $\mathcal{F}$ by Whitehead operations and admitting a canonical decomposition into pants, see Fig. 3.3;
(ii) There exists a foliation $\mathcal{G}$ equivalent to $\mathcal{F}$ which is a suspension over an interval exchange transformation;
(iii) There exists a foliation $\mathcal{G}$ equivalent to $\mathcal{F}$ such that two points $x, y \in M$ are not singular, then there exists a closed transversal curve γ through x and y. In other words, $\mathcal{G}$ is connected;
(iv) Foliation $\mathcal{F}$ has no barrier;
(v) There exists a foliation $\mathcal{G}$ equivalent to $\mathcal{F}$ and a closed differential 1-form ω which is transversal to $\mathcal{G}$;
(vi) There exists a foliation $\mathcal{G}$ equivalent to $\mathcal{F}$ and a simple closed curve γ which intersects all semi-leaves of $\mathcal{G}$;
(vii) There exists a foliation $\mathcal{G}$ equivalent to $\mathcal{F}$ such that all compact leaves and all nearly transversal cycles of $\mathcal{G}$ are crossed by a closed transversal curve.

Proof. We are going to prove successively, $(i) \Rightarrow (ii) \Rightarrow (vi) \Rightarrow (vii) \Rightarrow (i)$, $(ii) \Rightarrow (iii) \Rightarrow (vii)$ and $(i) \Rightarrow (v) \Rightarrow (iv) \Rightarrow (vii)$. (Clearly, this will close the equivalence chain of Theorem 4.3.2.) The only narrow places are $(vii) \Rightarrow (i)$ and $(iv) \Rightarrow (vii)$.

$(i) \Rightarrow (ii)$. If $\mathcal{G}$ admits a canonical decomposition into pants one can easily construct a closed curve C transversal to $\mathcal{G}$ and meeting all cycles of leaves of $\mathcal{G}$. Now, up to a Whitehead homotopy, foliation $\mathcal{F}$ is recovered upon an interval exchange transformation induced on C.

$(ii) \Rightarrow (vi)$. Follows from the definition of suspension.

$(vi) \Rightarrow (vii)$. Follows from the fact that every compact leaf is a semi-leaf and each closed curve intersecting all semi-leaves intersects also all nearly transversal cycles.

$(vii) \Rightarrow (i)$. This is an implication of item (a) of the following statement.

Lemma 3.4.4 *Let $\mathcal{F}$ be an orientable foliation of a surface M. There exists a foliation $\mathcal{G}$ equivalent to $\mathcal{F}$ and admitting a canonical decomposition into pants in either of the following two cases:*

(i) $\mathcal{F}$ has a semi-leaf which does not hit the boundary ∂M and all compact interior leaves of $\mathcal{F}$ as well as its nearly transversal cycles, are intersected by a closed transversal curve;
(ii) All semi-leaves of $\mathcal{F}$ meet ∂M and $\mathcal{F}$ does not have cycles of leaves.

Proof. Let $\mathcal{F}$ be an orientable foliation and let l be its free separatrix. Separatrix l is issued from a saddle s which we call *an extremity* of l and l is said to be positive if it goes out from s and negative if it comes to s. Two such separatrices are called *opposite* if they are free, of the same sign, and their extremities belong to the same connected component of a 1-complex $K(\mathcal{F})$ with edges formed by separatrices of types (i) and (ii).

Free separatrices are invariant with regard to the Whitehead homotopy, and it is easy to prove that two free separatrices l, l' having the same sign are opposite if and only if there exists a simple foliation equivalent to $\mathcal{F}$ for which l and l' are opposite. If there are no opposite separatrices, the behaviour of $\mathcal{F}$ is described by the following proposition.

Proposition 3.4.5 (Levitt) *Let $\mathcal{F}$ be a simple orientable foliation on M such that each semi-leaf of $\mathcal{F}$ hits the boundary ∂M and $\mathcal{F}$ has no opposite separatrices. Then:*

(i) ∂M has two components, one 'in-coming' and one 'out-going';
(ii) All saddles of $\mathcal{F}$ belong to a totally antisymmetric cycle of leaves;
(iii) All saddles of $\mathcal{F}$ has a property of a 'double passage'. [2]

Proof. This is elementary, see [183]. □

Before considering case (a) of Lemma 3.4.4, let us establish one additional auxiliary statement.

Proposition 3.4.6 (Levitt) *Let $\mathcal{F}$ and M be as before and $\mathcal{F}$ satisfies the hypothesis (a) of Lemma 3.4.4. Then there exists a finite family of disjoint transversal curves such that all semi-leaves of $\mathcal{F}$ which do not go to the boundary ∂M intersect one of curves of the family.*

Proof. Since $\mathcal{F}$ obeys the hypothesis (a) one can construct a finite family of disjoint transversal curves the union of which crosses all compact interior leaves of $\mathcal{F}$ as well as all nearly transversal cycles of $\mathcal{F}$. Suppressing if necessary some curves, one can suppose that there are no pairwise isotopic curves in the family. Let C be a family of maximal cardinality among those satisfying the above conditions. If a semi-leaf l does not hit neither ∂M nor C,

[2] That is for every neighbourhood of a saddle there exist a regular leaf which passes twice near this saddle.

it contains in its closure a non-compact regular leaf and there exists a closed transversal curve which hits l infinitely often and disjoint from $\mathcal{C}$. By choice of $\mathcal{C}$ this curve together with one of curves of $\mathcal{C}$ bounds an annulus. But this is a contradiction with the non-homotopy hypothesis. □

Consider now a set of pairs $(\mathcal{H}, \mathcal{C})$, where $\mathcal{H}$ is a foliation equivalent to $\mathcal{F}$ and $\mathcal{C} = (T_1, \ldots, T_n)$ is a finite family of disjoint transversal curves which intersect all semi-leaves of $\mathcal{H}$ which does not pass through ∂M. This set is non-empty and one can choose therein a pair $(\mathcal{H}_1, \mathcal{C}_1)$ for which the cardinality h of $\mathcal{C}_1$ is minimal. Cutting M along the curves $\mathcal{C}_1$ one can prove [182],[183] that $h = 1$. In other words, there exists one transversal closed curve which intersects all semi-leaves of $\mathcal{H}_1$. This completes the proof of case (a) of Lemma 3.4.4.

Similar arguments are applied to prove case (b) of the lemma. □.

(ii) $\Rightarrow$ *(iii)*. A suspension over the interval exchange transformation is always a connected foliation, since there exists a closed transversal curve passing through all semi-leaves and leaf connections of $\mathcal{F}$.

(iii) $\Rightarrow$ *(vii)* Is trivial.

(i) $\Rightarrow$ *(v)* If $\mathcal{G}$ is decomposable into pants, one easily constructs a closed 1-form ω transversal to $\mathcal{G}$ and such that the decomposition curves are the leaves of vector field $X : i_X\omega = 0$, and such that all regular leaves of X are compact.

(v) $\Rightarrow$ *(iv)* Let $\mathcal{G}$ and ω be as in (v). Let $(\gamma_1, \ldots, \gamma_k)$ be a family of closed leaves or nearly transversal cycles of $\mathcal{G}$. The integral

$$\gamma_i \mapsto \int_{\gamma_i} \omega$$

does not vanish and defines (de Rham) cohomology classes $H^1(M, \mathbb{Z})$ of ω. They correspond to the homology classes $[\gamma_i]$ of γ_i entering the family $(\gamma_1, \ldots, \gamma_k)$. The sum $[\gamma_1] + \ldots + [\gamma_k]$ is non-zero in $H_1(M, \mathbb{Z})$, and each $[\gamma_i]$ is invariant by the Whitehead homotopy.

Suppose to the contrary, that $\mathcal{G}$ has a barrier. One can find a submanifold $M_1 \subset M$ of manifold M with the boundaries consisting of compact leaves $\gamma_1, \ldots, \gamma_k$ and such that γ_i are 'entering' (resp. 'exiting') boundary for $\mathcal{G}$. Then there exists an orientation on M for which $[\gamma_1] + \ldots + [\gamma_k]$ is equal to zero in $H_1(M, \mathbb{Z})$. This is a contradiction.

Prove finally, *(iv)* $\Rightarrow$ *(vii)*. First let us notice that it is sufficient to prove that there exists a foliation $\mathcal{G}$ equivalent to $\mathcal{F}$ such that all compact leaves of $\mathcal{G}$ are intersected by a transversal closed curve. The absence of barrier allows us to reduce the study to a finite family $(\lambda_1, \ldots, \lambda_m)$ of simple closed leaves choosing them to lie in every isotopy classes of closed curves containing given compact leaf. We fix a transversal orientation on $\mathcal{F}$ and therefore such is fixed on $\lambda_i, (1 \le i \le m)$.

We start with a purely topological construction which uses λ_i and their orientation, but not the foliation itself. We are going to find a finite family of simple closed oriented curves (not necessarily disjoint) which cut λ_i transversally and coherent to the transversal orientation induced by $\mathcal{F}$. Moreover, each λ_i must intersect at least one curve of this family.

Consider the components $M \backslash (\lambda_1 \cup \ldots \cup \lambda_m)$ such that there is an oriented path joining λ_1 with this region and cutting all λ_i transversally and coherent to their transversal orientation. The closure of the union of such components is a connected submanifold of M with a boundary consisting of certain λ_i. By construction this submanifold can have only 'in-coming' boundaries. One can construct a closed oriented curve which cut transversally λ_i coherent to their orientation and also intersecting λ_1. If this curve does not meet all of λ_i, one constructs a second curve starting with λ_i which was not involved, obtaining finally the required family of curves.

For all $1 \le i \le m$ we denote by θ_i the total number of intersections of λ_i with the above constructed family. Further we 'forget' about these curves which serve only to determine $\theta_i \ge 1$ and possessing the following property: for each component M_1 of the surface obtained by cutting M along $\lambda_1 \cup \ldots \cup \lambda_m$ the sum of 'in-coming' points of the boundary is equal to the sum of 'out-going' points (this in fact is equivalent to absence of the barrier). Therefore the proof of theorem reduces to the proof of the following basic lemma.

Lemma 3.4.5 (Levitt) *Let M_1 be a component of a surface M obtained by cutting it along $\lambda_1 \cup \ldots \cup \lambda_m$ and let $\mathcal{F}_1$ be an induced foliation on M_1. There exists a foliation $\mathcal{G}_1$ equivalent to $\mathcal{F}_1$ and disjoint paths $\nu_1, \ldots, \nu_n$ transversal to $\mathcal{G}_1$ and such that:*

(i) Each ν_i connects an 'in-coming' component of ∂M_1 with an 'out-going' component of ∂M_1;
(ii) a component of ∂M_1 which is obtained by cutting of a leaf λ_i meets precisely θ_i of these paths.

(Clearly the above lemma implies the required result, since the foliations defined by lemma can be pasted together to give a foliation $\mathcal{G}$ equivalent to $\mathcal{F}$. The transversal curves of $\mathcal{G}$ are obtainable by connecting two-by-two paths $2\theta_i$ which are transversal to $\mathcal{G}$ and directed towards each λ_i.)

Proof of Lemma 3.4.5. First we are to show that there exist paths which satisfy the conditions of lemma being not transversal to $\mathcal{F}_1$, but being in a sense 'good'. (Further we specify what 'good' is: it roughly means that they are transversal to a foliation equivalent to $\mathcal{F}_1$. It will remain then to prove that by the Whitehead homotopy these paths can be made transversal simultaneously.)

Definition 3.4.4 *An oriented path η immersed in M_1 is said to be 'good' with respect to $\mathcal{F}_1$ if:*

- *η is piecewise smooth and the breakpoints of smoothness coincide with saddles of $\mathcal{F}_1$;*
- *η consists of smooth parts transversal to $\mathcal{F}_1$ and parts of separatrices joining either a saddle of $\mathcal{F}_1$ with one extremity of η, or separatrices joining two distinct saddles of $\mathcal{F}_1$;*
- *On each transversal part of η its orientation coincides with the transversal orientation of $\mathcal{F}_1$.*

Proposition 3.4.7 *Each 'in-coming' boundary of M_1 can be joined with each 'out-going' boundary by a 'good' path relatively $\mathcal{F}_1$.*

Proof. Notice that if γ is a compact leaf or a nearly transversal cycle of $\mathcal{F}_1$ then it is isotopic to a component of boundary ∂M_1. (This is a consequence of a choice of λ_i.) Let R be an 'in-coming' boundary of M_1. Let A be a set of points $x \in M_1$ such that there exists a 'good' path which joins R with x. Clearly, A contains R but does not contain another 'in-coming' boundary of M_1. We are going to prove that A contains all 'out-going' boundaries of M_1, and so the proposition will follow.

Denote by α a boundary of A. A leaf through α cannot pass twice in an open neighbourhood of a point of $\mathcal{F}_1$; α contains either compact leaves of $\mathcal{F}_1$, or separatrices of types (i) and (ii). Finally, α breaks M_1 into two regions: an interior, A, and exterior, denoted by B, which are disjoint.

The separatrices contained in α form a sub-complex K_α of 1-complex $K(\mathcal{F}_1)$ defined earlier. Note that a vertex of K_α belongs to at least two edges of K_α. First let us show that $K_\alpha \subset A$. Indeed, otherwise a connected component K_α^0 of K_α is not contained in A and this implies that K_α^0 has a nearly transversal cycle. But this cycle is isotopic to a component of boundary ∂M_1, which is (in view of absence of Reeb components) necessarily 'out-going'. It is a contradiction because K_α is in A. Therefore each separatrix issued from a saddle point of K_α is also contained in A.

Suppose now that that there exists an 'out-going' boundary S of M_1 not contained in A. We argue towards a contradiction, considering a connected component B_0 of B which contains S. Since $\mathcal{F}_1$ has no Reeb components, the boundary of B_0 cannot contain compact leaves. It contains however cycles of leaves, and because all separatrices of saddles of K_α are in A, it contains nearly transversal cycles of $\mathcal{F}_1$. But it is impossible, since such cycle is isotopic to a boundary component of ∂M_1, and $\mathcal{F}_1$ has no Reeb components. □

Now we can finish the proof of Lemma 3.4.5. From Proposition 3.4.7 it follows that there exist 'good' paths $\eta_1, \dots, \eta_n$ satisfying (i) and (ii) of the lemma.

They are not necessarily disjoint but slightly modifying them one can suppose that they are 'nearly disjoint', in a sense that there exists a nearby path of η_i which is disjoint from $\eta_j, i \neq j$.

By *complexity* $C(\eta_i)$ of a 'good' path η_i one understands a total number of saddles which it passes through. A complexity of a family of paths $(\eta_1, \ldots, \eta_n)$ is a sum

$$\sum_{i=1}^{n} C(\eta_i)$$

of complexities of its elements. We consider the tuples $(\mathcal{F}_1', \eta_1', \ldots, \eta_n')$, where $\mathcal{F}_1'$ is a foliation equivalent to $\mathcal{F}_1$ and η_i' are paths which are 'nearly disjoint' and satisfying conditions (i), (ii) of the lemma. Among these tuples one chooses an element $(\mathcal{G}_1, \nu_1, \ldots, \nu_n)$ of a *minimal complexity*, and further arguing towards a contradiction one obtains that it has a zero complexity (that is the transversal paths ν_i are disjoint).

First notice that ν_i are composed from pieces transversal to $\mathcal{G}_1$. If actually one of ν_i contains a separatrix connection it is possible to eliminate this connection by the Whitehead homotopy and diminish the complexity of given path. Take now a path ν_1. Without loss of generality one can assume $C(\nu_1)$ to be non-zero. Let s be a saddle point on ν_1. The separatrices emerging from s can be subdivided into two groups, each from one side of ν_1. By third item of Definition 3.4.4, each of these groups has an odd number of separatrices. Therefore one can replace $\mathcal{G}_1$ by a Whitehead equivalent foliation $\mathcal{G}_1'$. For the complexity of a 'good' path ν_1' one has

$$C(\nu_1') = C(\nu_1) - 1.$$

Therefore it is possible to diminish the complexity of each path not rising the complexity of the whole family. (Herewith all ν_i can be made transversal to $\mathcal{G}_1$.) This finishes the proof of Lemma 3.4.5 and Theorem 3.4.3. □

3.5 Number of Quasiminimal Sets

Spacious $\mathcal{Q}$-Sets

This important class is formed by those $\mathcal{Q}$-sets which cannot lie in a disc region $D \subset N$ of the surface N. Note that if $\mathcal{F}$ is a flow-induced foliation then all the $\mathcal{Q}$-sets of $\mathcal{F}$ (if any) must be spacious. (This is a consequence of the Poincaré-Bendixson theorem.) A labyrinth in the disc yields an example of the $\mathcal{Q}$-set which is not spacious.

Definition 3.5.1 *Let $\mathcal{Q}(x)$ be a quasiminimal set of a C^∞ foliation $\mathcal{F}$ on a surface N. $\mathcal{Q}(x)$ is called spacious if for each leaf $l \subset \mathcal{Q}(x)$ and each transversal segment γ intersecting $\mathcal{Q}(x)$ and no other $\mathcal{Q}$-set of $\mathcal{F}$, there are no arcs $l_0 \subset l$ such that $l_0 \cup \gamma$ bounds a disc region of N.*

(Clearly, if $\mathcal{Q}(x)$ contains a thorn or a sun-set singularity it cannot be spacious. In general, every $\mathcal{Q}$-set of the foliation $\mathcal{F}$ which contains an isolated singularity of index $+1/2$ cannot be spacious.)

Lemma 3.5.1 *Let γ be a closed transversal of a foliation $\mathcal{F}$ which intersects a quasiminimal set $\mathcal{Q}(x)$ of $\mathcal{F}$. Suppose also that γ bounds either a disc or is a boundary component of an annulus. In the latter case it is assumed that $\mathcal{Q}(x)$ does not intersect the other boundary component of the annulus. Then $\mathcal{Q}(x)$ cannot be spacious.*

Proof. In the disc or the annulus every arc with the endpoints at one and the same boundary component bounds a disc region. Therefore one obtains an arc $l_0 \subset l$ of the leaf l and a transversal segment γ such that $l_0 \cup \gamma$ bounds a disc. This means that $\mathcal{Q}(x)$ is not spacious. □

Theorem 3.5.1 *A foliation $\mathcal{F}$ cannot have spacious $\mathcal{Q}$-sets on the following surfaces:*

- *a sphere;*
- *a disc;*
- *an annulus;*
- *a disc with two holes (the pants);*
- *a projective plane;*
- *the Möbius strip;*
- *the Klein bottle.*

Proof. For the sphere and the disc the proof is evident since every simple closed curve on these manifolds is contractible. To prove the statement for the annulus, the pants and the Möbius strip it is sufficient to notice that every closed transversal curve which intersects a $\mathcal{Q}$-set on these manifolds, bounds either a disc or an annulus. Therefore by Lemma 3.5.1 the theorem follows.

Let us prove the theorem for a projective plane and for the Klein bottle. Suppose first that N is a projective plane. Let $\mathcal{Q}(x)$ be a quasiminimal set on N. Then there exists a closed transversal curve γ which intersects $\mathcal{Q}(x)$ and has a cylindric neighbourhood $\gamma \times [-\varepsilon, \varepsilon]$ in N. Since $N = \mathbb{R}P^2$ is a projective plane then γ bounds a disc region of N. By Lemma 3.5.1 $\mathcal{Q}(x)$ is not spacious.

Finally, let N be the Klein bottle. Suppose to the contrary that there exist a spacious $\mathcal{Q}$-set on N. Let γ be a closed transversal curve with a cylindric neighbourhood $\gamma \times [-\varepsilon, \varepsilon]$ which intersects the $\mathcal{Q}$-set. Two cases are possible.

(i) γ does not split N. In this case $N \backslash \gamma$ is an annulus A^2. The Klein bottle N can be restored upon A^2 by identifying its boundary components, γ_1 and γ_2, via an orientation reversing homeomorphism $\phi : \gamma_1 \to \gamma_2$. Let m_1, m_2, m_3 be three consequent intersection points of an open recurrent semi-leaf $l^+ \subset \mathcal{Q}$ with the curve γ. Since $\mathcal{Q}$ is spacious each arc of l^+ must have its endpoints lying on the different boundary components of A^2. Let $[m_1, m_2]$ be an arc of γ which contain the point m_3. Since ϕ reverses the orientation, these points are situated on the boundary of A^2. Therefore the semi-leaf l^+ meets γ only at the points of the arc $[m_2, m_3]$ which does not contain m_2. This is a contradiction with the recurrence of l^+.

(ii) γ splits N. Denote by $ab, bc \subset l^+$ the arcs of an open recurrent semi-leaf $l^+ \subset \mathcal{Q}(x)$ with the endpoints at γ. Since γ splits N, the arcs ab and bc lie at the different connected components of $N\backslash\gamma$. Both ab and bc intersect γ in a non-orientable way (that is a fixed positive direction at the arc does not coincide with a fixed normal orientation at γ). This implies that the arc $ac \subset l^+$ intersects γ in an orientable way.

The connected components N_1 and N_2 of $N\backslash\gamma$ are homeomorphic to a punctured projective plane (that is to $\mathbb{R}P^2$ with a hole). Because $\mathcal{Q}(x)$ is spacious, each of the arcs ab and bc must be one-sided. Therefore the arc ac is two-sided (has a cylindric neighbourhood). Moreover, since neither ab nor bc splits N_1 or N_2, the curve $ac \cup \sigma$ does not split N, where $\sigma \subset \gamma$ is a transversal segment at γ. It now remains to construct a closed transversal γ' near a two-sided arc ac, which does not split N and has a cylindric neighbourhood. Thus the case (ii) is reduced to the case (i). This argument completes the proof of the theorem. □

Corollary 3.5.1 *Let $N = T^2$ be a two-dimensional torus. Then every recurrent foliation $\mathcal{F}$ on N has at most one spacious $\mathcal{Q}$-set.*

Proof. Arguing towards a contradiction let us suppose that there exist two spacious $\mathcal{Q}$-sets at N. Then there is a closed transversal curve γ, which intersects one $\mathcal{Q}$-set and does not intersect the other. Since γ is non-contractible, $N\backslash\gamma$ is an annulus which contains a spacious $\mathcal{Q}$-set. This is a contradiction with Theorem 3.5.1. □

Theorem 3.5.2 (Aranson-Levitt-Zhuzhoma) *Let N be an orientable surface of genus g having n boundary components. Let $\mathcal{F}$ be a C^∞ foliation on N such that Sing $F < \infty$. Denote by S a spacious quasiminimal set of $\mathcal{F}$. Then the upper bound for the total number of spacious quasiminimal sets on N is given by the formula:*

$$|S| \leq g + \max\left\{[\frac{g+n}{2}] - 1; 0\right\},$$

where $[x]$ denotes the integer part of x. Moreover, the above estimate is sharp. In other words, there exists a foliation $\mathcal{F}$ on N for which the upper bound of the above estimate is attained.

Proof. The idea of the proof is to produce a dissection of N along closed transversal curves γ_i corresponding to each spacious quasiminimal set $\mathcal{Q}_i$. (Note that by Lemma 3.5.1, γ_i cannot be nullhomotopic.) Then one applies Theorem 3.5.1. The proof is inductive in g and n.

Step I. First let us prove the statement for the case $g = 0, n \geq 3$ (that is when N is a sphere with more than 2 holes; the cases $n = 0, 1, 2, 3$ follow from Theorem 3.5.1 where it was shown that $|\mathcal{S}| = 0$). Let us suppose that our estimate is true for $n = 0, \ldots, k$ where $k \geq 3$. Let us show that it is also valid for $n = k + 1$. (In fact we are to prove the inequality $|\mathcal{S}| \leq (n-2)/2$ which is equivalent to our statement under the assumption $g = 0$.)

Indeed, let $\mathcal{F}$ have at least one spacious $\mathcal{Q}$-set. We fix this $\mathcal{Q}$-set and denote it by $\mathcal{Q}^*$. There exists a closed transversal curve γ, which cuts $\mathcal{Q}^*$ and does not intersect other $\mathcal{Q}$-sets of $\mathcal{F}$. Since $g = 0$ γ splits N into two parts, N_1 and N_2, which are the spheres with n_1 and n_2 holes, respectively. Clearly, $n_1 + n_2 = n + 2$. Let $|\mathcal{S}_1|$ and $|\mathcal{S}_2|$ be a total number of the spacious $\mathcal{Q}$-sets at N_1 and N_2, respectively. Then

$$|\mathcal{S}_1| + |\mathcal{S}_2| = |\mathcal{S}| - 1.$$

By Lemma 3.5.1 neither $|\mathcal{S}_1|$ nor $|\mathcal{S}_2|$ will decrease, if we squeeze one hole which corresponds to the curve γ, both at N_1 and N_2. Therefore by the induction hypothesis it holds $|\mathcal{S}_1| \leq \frac{1}{2}(n_1 - 3)$ and $|\mathcal{S}_2| \leq \frac{1}{2}(n_2 - 3)$. Taking into account our previous estimates one obtains

$$|\mathcal{S}| = |\mathcal{S}_1| + |\mathcal{S}_2| + 1 \leq \frac{n_1 - 3}{2} + \frac{n_2 - 3}{2} + 1 = \frac{1}{2}(n-2).$$

Step II. Suppose that our statement is true for all $g = 0, \ldots, h$ and all $n = 0, \ldots, k$. (That is the inequality $|\mathcal{S}| \leq \frac{1}{2}(3g + n - 2)$ is valid. Notice that when $N \neq T^2$ this inequality is equivalent to the estimate of the theorem.) Let us show that it holds true also for the case $g = h, n = k + 1$. As before one takes a transversal closed curve γ which cuts a quasiminimal set $\mathcal{Q}^*$, and only it. Then N is dissected along γ. Two cases are possible.

(i) γ splits N into two connected components, N_1 and N_2, which are the surfaces of genus g_1 and g_2 with n_1 and n_2 holes, respectively. Clearly, $g_1 + g_2 = g$ and $n_1 + n_2 = n + 2$. As it was done before we squeeze one hole at N_1 and one hole at N_2, which corresponds to the curve γ. This operation does not change the rest number of the spacious $\mathcal{Q}$-sets both at N_1 and N_2. With slight modifications, one obtains the following estimate:

$$|\mathcal{S}| = |\mathcal{S}_1| + |\mathcal{S}_2| + 1 \leq \frac{3g_1 + n_1 - 3}{2} + \frac{3g_2 + n_2 - 3}{2} + 1 = \frac{1}{2}(3g + n - 2).$$

(ii) γ does not split N. Producing a cut along γ one comes to a surface N' of genus $g - 1$ with $n + 2$ holes. Clearly, N' has by one $\mathcal{Q}$-set less than N. By the induction hypothesis for N' it holds

$$|\mathcal{S}| - 1 \leq \frac{3(g-1) + (n+2) - 2}{2} = \frac{3g + n - 3}{2}.$$

It gives us the following estimate

$$|S| \le \frac{1}{2}(3g+n-1) = \frac{1}{2}[3g+(n+1)-2].$$

for the surface of genus g with $n+1$ holes. This completes the proof for the case $g=h, n=k+1$. (Similar induction can be applied to the case $g=h+1, n=k$.)

It is not hard to prove that the above estimates are sharp. The relevant examples of foliation $\mathcal{F}$ with Sing $\mathcal{F}$ consisting of the n-prong saddles are constructed in [182], [186]. Theorem is proven. □

Number of the $\mathcal{Q}$-Sets on Orientable Surfaces

Here the general case is considered when the $\mathcal{Q}$-sets of the foliation $\mathcal{F}$ are not necessarily spacious. Surface N is supposed orientable possibly with n boundary components. As usual, the set Sing $\mathcal{F}$ is assumed to be finite.

Definition 3.5.2 *By a thorn number,* $\mathfrak{t}(\mathcal{F})$, *of the foliation* $\mathcal{F}$ *one understands the following finite number*

$$\mathfrak{t}(\mathcal{F}) = \{|Sing \quad \mathcal{F}| : Ind \quad (Sing\ \mathcal{F}) \ = \ 1/2\}.$$

In other words, $\mathfrak{t}(\mathcal{F})$ *is a total number of the singular points of foliation* $\mathcal{F}$ *whose Euler-Poincaré index is equal to 1/2.*

Proposition 3.5.1 *Let* $\mathcal{Q}(x)$ *be a quasiminimal of foliation* $\mathcal{F}$. *Then* $\mathcal{Q}(x)$ *is spacious if and only if* $\mathfrak{t}(\mathcal{F})|_{\mathcal{Q}(x)} = 0$. *In other words there are no singular points at* $\mathcal{Q}(x)$ *having the Euler-Poincaré index equal to 1/2.*

Proof. Indeed, let $\mathcal{Q}(x)$ have a singular point p such that Ind $p = 1/2$. Then it is easy to construct an arc $l_0 \subset l$ of a leaf l of $\mathcal{Q}(x)$ and a transversal segment T such that $l_0 \cup T$ bounds a disc region containing p. (Note that the rotation of the 'field of line elements' along the above closed curve will be exactly 1/2.) Therefore, $\mathcal{Q}(x)$ cannot be spacious.

Conversely, suppose that $\mathcal{Q}(x)$ is spacious. Since there are no closed nullhomotopic curves at $\mathcal{Q}(x)$, such that the rotation of the 'field of line elements' along these curves will be 1/2, the same holds true for the Euler-Poincaré index of singular points at $\mathcal{Q}(x)$. □

Theorem 3.5.3 (Aranson-Levitt-Zhuzhoma) *Let* $\mathcal{F}$ *be a* C^∞ *foliation on an orientable surface* N *of genus* g *with* n *boundary components. Let the set Sing* $\mathcal{F}$ *be finite and let* $\mathfrak{t}(\mathcal{F})$ *be the thorn number of* $\mathcal{F}$. *Then for the total number* $|\mathcal{Q}|$ *of the quasiminimal sets of* $\mathcal{F}$ *the following estimate is valid:*

$$|\mathcal{Q}| \le g + \max\left\{[\frac{g+n+\mathfrak{t}(\mathcal{F})}{2}] - 1; 0\right\},$$

where $[x]$ *denotes the integer part of* x. *Moreover, the above estimate is sharp. In other words, there exists a foliation* $\mathcal{F}$ *on* N *for which the upper bound of the above estimate is attained.*

Proof. Below we shall give two independent proves of the above statement. First applies an estimate given by Theorem 3.5.2 for the spacious $\mathcal{Q}$-sets. The second one uses a 'covering representation' of $\mathcal{F}$, given by the triple $\mathcal{F} = (M, \pi^t, \theta)$.

(I) Let $\mathcal{F}$ be a foliation on the surface N mentioned in the theorem, and let $\mathfrak{t}(\mathcal{F}) > 0$ be its thorn number. Then there exist $\mathfrak{t}(\mathcal{F}) = t$ singular points $p_1, \ldots, p_t$ such that each point can be surrounded by a closed curve $l_0 \cup \gamma$ mentioned in Definition 3.5.1. One can deform the foliation $\mathcal{F}$ in the neighbourhoods of p_i so that

- the newly obtained foliation F' has the same number of quasiminimal sets, $|\mathcal{Q}'| = |\mathcal{Q}|$;
- F' is transversal [3] to the boundary of the set of discs $D_1, \ldots, D_t$ centered at the points $p_1, \ldots, p_t$.

Let us remove the discs $D_1, \ldots, D_t$. This yields us a compact orientable manifold with $n + \mathfrak{t}(\mathcal{F})$ holes and the foliation $\mathcal{F}'$ on it. Clearly, $\mathcal{F}'$ has only spacious $\mathcal{Q}$-sets. It remains now to substitute the obtained number into the estimate for the total number of spacious $\mathcal{Q}$-sets given by Theorem 3.5.2. □

(II) Let $\mathcal{F} = (M, \pi^t, \theta)$ be a covering representation of the foliation $\mathcal{F}$ on the surface N. (That is $N = M/\theta, F = \pi^t/\theta$.) Clearly, the $\mathcal{Q}$-sets of the flow $\pi^t : M \times I\!R \to M$ are always spacious. According to the Riemann-Hurwitz formula, adding each pair of non-orientable singularities to $\mathcal{F}$ means that the genus of M grows by one. Since the Euler-Poincaré index of the singularity at M, which covers the singularity of index 1/2, is zero, one can easily include it into a standard irrational flow on torus. This torus is glued into M as a handle thus giving a manifold $M\#T^2$ with an extra $\mathcal{Q}$-set on it. Therefore when the thorn number grows by two, the possible number of $\mathcal{Q}$-sets must grow by one. The estimate of the theorem follows. □

(The exactness of the above estimate can be verified directly. The relevant examples are considered in [182], [186].)

Number of the $\mathcal{Q}$-Sets on Non-Orientable Surfaces

In this section the manifold N is supposed to be non-orientable; see an Appendix to this section for a reference. First, we establish an estimate for the maximal number of the spacious $\mathcal{Q}$-sets on N. Further, this estimate is extended to the general case of an arbitrary $\mathcal{Q}$-sets. As in the orientable case, the thorn number $\mathfrak{t}(\mathcal{F})$ of the foliation $\mathcal{F}$ plays here an important role.

[3] To obtain such a transversal one should take the arc l_0 of the recurrent leaf l and construct a transversal, γ_0, in a 'flow box' formed by the nearby leaves. Then one takes a closed curve $\gamma_0 \cup \gamma$ as a transversal boundary for the foliation F'. Note that this boundary is not smooth.

Lemma 3.5.2 *Let N_1 be a non-orientable compact surface of genus 1 with two boundary components (i.e. the projective plane $\mathbb{R}P^2$ with two holes). Let N_2 be a non-orientable surface of genus 2 with one boundary component (i.e. the Klein bottle with a hole). Then neither N_1 nor N_2 admits a foliation $\mathcal{F}$, Sing $\mathcal{F} < \infty$, with spacious $\mathcal{Q}$-sets.*

Proof. (I) First let us establish that there are no spacious $\mathcal{Q}$-sets at N_1 (projective plane with two holes). Suppose to the contrary that $\mathcal{S}$ is a spacious $\mathcal{Q}$-set of a foliation $\mathcal{F}$. Take a closed transversal γ which intersects $\mathcal{S}$ and does not intersect other $\mathcal{Q}$-sets (if any). We can suppose that γ has a cylindric neighbourhood $\gamma_\varepsilon = \gamma \times [-\varepsilon, \varepsilon]$ (otherwise there exists a curve γ' close to γ which 'winds' twice around γ and has a cylindric neighbourhood, and we set $\gamma = \gamma'$). γ splits N_1 and according to Lemma 3.5.3 and Corollary 3.5.2 there are the following possibilities:

- γ splits N_1 into an annulus and an annulus with a cross-cap;
- γ splits N_1 into a Möbius strip (i.e N_1 with a hole) and a disc with two holes.

It is easy to see that in the first case when $N_1 = N_0^{(2)} \# N_1^{(2)}$ (here $N_0^{(2)}$ denotes a sphere with two holes and $N_1^{(2)}$ denotes a projective plane with two holes) $\mathcal{S}$ cannot be spacious.

In the second case when $N_1 = N_1^{(1)} \# N_0^{(3)}$ ($N_1^{(1)}$ denotes a projective plane with a hole and $N_0^{(3)}$ denotes a sphere with three holes), let us take a recurrent leaf $l \subset \mathcal{S}$ and consider consequently the arcs l_1, l_2, l_3, l_4 lying alternatively at $N_1^{(1)}$ and $N_0^{(3)}$: $l_1, l_3 \subset N_1^{(1)}, l_2, l_4 \subset N_0^{(3)}$. By Remark 3.5.1 and the spaciousness of $\mathcal{S}$, the arcs l_1 and l_3 split $N_1^{(1)}$ into two discs D_1 and D_2.

Since $\mathcal{S}$ is spacious every arc γ in $N_0^{(3)}$ splits it into two annuli: $N_0^{(3)} = N_0^{(2)} \# N_0^{(2)}$. Therefore l_2, l_4 together with γ bounds a disc region, D_{24}, at $N_0^{(3)}$. Fix an orientation at l so that its positive direction corresponds to the sequence l_1, l_2, l_3, l_4. If l_4 belongs to D_1 then the positive semi-leaf l^+ must remain in $D_1 \cup D_{24}$. This contradicts the assumption that γ is recurrent. If l_4 enters D_2 then the negative semi-leaf l^- must remain in $D_1 \cup D_{24}$. This also contradicts the assumption of the recurrence. The first part of Lemma 3.5.2 is proved.

(II) Now let us establish that the surface N_2 (the Klein bottle with a hole) admits no spacious $\mathcal{Q}$-sets. As before suppose to the contrary that $\mathcal{S}$ is spacious in N_2. Let γ be a closed transversal curve, $\gamma \cap \mathcal{S} \neq \emptyset$. Cutting along γ let us consider all the possible cases.

(i) Suppose that γ does not split N_2. γ has a cylindric neighbourhood in N_2 and therefore cutting along γ yields an annulus with a hole, $N_0^{(3)}$. (By Corollary 3.5.2 a non-orientable surface is excluded, because the genus of N_2

is 2.) The boundary components, K_1 and K_2, of $N_0^{(3)}$ corresponding to γ are identified via an orientation preserving homeomorphism. Take a recurrent leaf l and consider its consequent arcs in $N_0^{(3)}$. If there exists an arc of l with the endpoints at K_2 then it bounds a disc region with a hole. Let us consider a positive semi-leaf l^+ lying next to the point a. Since $\mathcal{S}$ is spacious the semi-leaf l^+ cannot intersect K_1 twice. Then it necessarily cuts K_2. The rest of the proof repeats the argument used in Theorem 3.5.1 for the Klein bottle.

(ii) Suppose that γ splits N_2. Then the following cases are possible:

- N_2 falls into an orientable surface which is either a disc $N_0^{(1)}$ or an annulus $N_0^{(2)}$, and a non-orientable surface;
- N_2 falls into a pair of non-orientable surfaces, $N_1^{(1)}$ (a Möbius strip) and $N_1^{(2)}$ (a projective plane with two holes).

In the first case neither $N_0^{(1)}$ nor $N_1^{(2)}$ admits spacious $\mathcal{Q}$-sets. Therefore lemma follows. Let us consider the second case. Take a recurrent leaf $l \subset \mathcal{S}$ and consider the consequent arcs $l_1, l_2, l_3, \ldots \subset l$ so that $l_{2k+1} \subset N_1^{(1)}$ and $l_{2k+2} \subset N_1^{(2)}$. Since $\mathcal{S}$ is spacious and on account of Remark 3.5.1, the arcs l_{2k+1} are one-sided. If in addition at least one of the arcs is one-sided then there exists a segment $\gamma_0 \subset \gamma$ such that $l_{2k+1} \cup l_{2k+2}$ is a two-sided arc with the endpoints at γ. The closed curve $\gamma_0 \cup l_{2k+1} \cup l_{2k+2}$ is approximated by a closed transversal curve $\widetilde{\gamma}$. Since both l_{2k+1} and l_{2k+2} are one-sided, $\widetilde{\gamma}$ does not split $N_2^{(1)}$ and one comes to the item (i).

Finally, let all arcs l_{2k+2} be two-sided. Then the arcs l_2 and l_4 bound a disc region on N_2. (Indeed, l_2 and l_4 split N_2 into two regions one of which has a hole. Hence the other is homeomorphic to a disc.) Further arguments are similar to the proof of absence of spacious $\mathcal{Q}$-sets at the surface $N_1^{(2)}$ (projective plane with two holes). Lemma is proved. □

Theorem 3.5.4 (Aranson-Zhuzhoma) *Let N_p be a compact non-orientable surface of genus $p > 0$ possibly with n boundary components. Let $\mathcal{F}$ be a foliation of N_p such that Sing $F < \infty$. Denote by $|\mathcal{S}|$ the total number of spacious $\mathcal{Q}$-sets of the foliation $\mathcal{F}$. Then the following estimate is valid:*

$$|\mathcal{S}| \leq \max\left\{ \{\frac{3p+2n-5}{4}\}; 0 \right\},$$

where $[x]$ denotes the integer part of x. Moreover, the above estimate is sharp. (In other words, there exists a foliation $\mathcal{F}$ on N_p for which the upper bound of the above inequality is attained.)

Proof. (I) For the proof let us use the induction in p and n. By Lemma 3.5.2 and Theorem 3.5.1 our statement holds true for $p = 1, n = 0, 1, 2$ and

$p = 2, n = 0, 1$. Suppose that it is valid for $p = 1, n = 0, \ldots, k$. Let us prove it for $p = 1, n = 0, \ldots, k, k+1$.

Let $\mathcal{F}$ have at least one spacious $\mathcal{Q}$-set $\mathcal{S}$. Then there exists a transversal closed curve γ which intersects $\mathcal{S}$ and no other $\mathcal{Q}$-set of $\mathcal{F}$. Let us cut the surface $N_1^{(k+1)}$ (a projective plane with $k+1$ holes) along γ. Since $\mathcal{S}$ is spacious γ cannot bound a disc. It can always be assumed that γ has a cylindric neighbourhood. Since the genus of $N_1^{(k+1)}$ is 1 this one splits into two surfaces, S_0 and S_1. According to Corollary 3.5.2 one of these surfaces, say S_0, is orientable of genus 0. The other one is non-orientable of genus 1.

Let S_0 and S_1 have $|\mathcal{S}_0|$ and $|\mathcal{S}_1|$ spacious $\mathcal{Q}$-sets, and n_0 and n_1 holes, respectively. (In other words, $N_1^{(k+1)} = S_0 \# S_1, S_0 = N_0^{(n_0)}, S_1 = N_1^{(n_1)}$.) Clearly, the following equalities hold:

$$|\mathcal{S}_0| + |\mathcal{S}_1| = |\mathcal{S}| - 1, \quad n_0 + n_1 = k + 3.$$

The holes on S_0 and S_1 formed by the curve γ is glued by discs. According to Lemma 3.5.1 the number of spacious $\mathcal{Q}$-sets will remain the same (otherwise both at S_0 and S_1 there exists an annulus with a unique boundary component intersecting a $\mathcal{Q}$-set; this is impossible). Then by Theorem 3.5.3 and the assumption of the induction one comes to the inequality

$$|\mathcal{S}| = |\mathcal{S}_0|+|\mathcal{S}_1|+1 \le \frac{1}{2}[(n_0-1)-2]+\frac{1}{4}[2(n_1-1)-2]+1 = \frac{1}{4}[2(n_0+n_1)-6] = \frac{1}{2}k.$$

This proves the theorem for the case $p = 1, n \ge 0$.

(II) Let us suppose now that the theorem is true for all $p = 1, \ldots, r$ and all $n \ge 0$. First we shall prove that it is also true for $p = r+1$ and $n = 0$. Again let us take a closed transversal γ which intersects only one spacious $\mathcal{Q}$-set, and then cut the surface N_{r+1}^0 along γ. Consider step by step all possible cases offered by Corollary 3.5.2.

(i) $N_p^{(0)}$ splits into an orientable surface S_1 of genus p_1 and a non-orientable surface S_2 of genus p_2. That is $N_p^{(0)} = S_1 \# S_2$ and $p = 2p_1 + p_2$. Both S_1 and S_2 have a hole which we glue by a disc, D, and we extend $\mathcal{F}$ to $S_1 \cup D$ and $S_2 \cup D$ so that D contains a unique singularity which is a node. As usual by $|\mathcal{S}_1|$ and $|\mathcal{S}_2|$ one understands the total number of spacious $\mathcal{Q}$-sets at S_1 and S_2, respectively. The following equality is evident

$$|\mathcal{S}| = |\mathcal{S}_1| + |\mathcal{S}_2| + 1.$$

By Theorem 3.5.3 $\mathcal{F}$ cannot have spacious $\mathcal{Q}$-sets in a disc, and therefore $p_1 \ge 1$. We consider cases $p_1 = 1$ and $p_1 > 1$ separately.

Step 1. Let $p_1 = 1$ so that $S_1 \# D_1 = T^2$ is a torus. Then every arc of a recurrent leaf in $\mathcal{S}$ lying in S_1 dissects S_1 to an annulus. Therefore in S_1 there are no spacious $\mathcal{Q}$-sets. Since $p_2 = p-2$ the induction hypothesis yields

$$|\mathcal{S}| = |\mathcal{S}_2| + 1 \le \frac{1}{4}[3(p-2) - 5] + 1 = \frac{1}{4}(3p-7) \le \frac{1}{4}(3p-5).$$

Step 2. Let $p_1 > 1$. Then by Theorem 3.5.2 $|\mathcal{S}| \le \frac{1}{2}[3p_1 - 2]$ and the following inequality is valid

$$|\mathcal{S}| = |\mathcal{S}_1|+|\mathcal{S}_2|+1 \le \frac{1}{2}(3p_1-2)+\frac{1}{4}(3p_2-5)+1 = \frac{1}{4}[3(2p_1+p_2)-5] = \frac{1}{4}(3p-5).$$

(ii) Let $N_p^{(0)}$ splits into two non-orientable surfaces S_{p_1} and S_{p_2} of genus p_1 and p_2, respectively. In other words, $N_p^{(0)} = S_{p_1}\#S_{p_2}$ and $p = p_1 + p_2$, where $p_1, p_2 \ge 1$. By the induction hypothesis one obtains:

$$|\mathcal{S}| \le \frac{1}{4}(3p_1 - 5) + \frac{1}{4}(3p_2 - 5) + 1 = \frac{1}{4}(3p - 6) < \frac{1}{4}(3p - 5).$$

Therefore also in this case the required estimate is valid.

(iii) Suppose that γ does not split $N_p^{(0)}$ and cutting along γ one comes to a non-orientable surface N' of genus $p' \ge 1$. One can always suppose that γ has a cylindric neighbourhood and therefore $p' \le p - 2$. According to Lemma 3.5.1 it is possible to 'squeeze' at least one hole at N' not changing the number of spacious $\mathcal{Q}$-sets of $\mathcal{F}$. It yields us

$$|\mathcal{S}| \le \frac{1}{4}(3m + 2 - 5) + 1 \le \frac{1}{4}[3(p - 2) + 1] = \frac{1}{4}(3p - 5),$$

what proves the theorem in the above case.

(iv) It remains to consider the case when cutting along γ one obtains an orientable surface N of genus $m \le 1/2(p - 1)$. Without loss of generality we can suppose that γ is two-sided so that the boundary ∂N consists of two connected components, K_1 and K_2.

(1) Let $l_0 \subset l$ be an arc of a recurrent leaf $l \subset \mathcal{S}$ with the endpoints at K_1 or K_2 (for brevity let it be K_1). Suppose that l_0 does not split N. Gluing-up a disc to K_2 one comes to an orientable surface of genus $m - 1$ with two holes with $|\mathcal{S}| - 1$ spacious $\mathcal{Q}$-sets on it. By Theorem 3.5.3 it yields

$$|\mathcal{S}| \le \frac{1}{2}[3(m - 1) + 2 - 2] + 1 \le \frac{1}{2}[3(\frac{p-1}{2}) - 1] = \frac{1}{4}(3p - 5).$$

(2) Suppose that there exist arcs $l_0^{(1)}$ and $l_0^{(2)}$ which split N. Cutting along them one obtains three orientable surfaces of genus m_1, m_2, m_3, respectively. Herewith two of them have one hole, and the third has two holes. It can be written as $N = N_{m_1}^{(1)}\#N_{m_2}^{(1)}\#N_{m_3}^{(2)}$. Note that the genus p of the initial surface $N_p^{(0)}$ must be even and $m \le 1/2(p - 2)$. It follows from the estimate given by Theorem 3.5.3 that

$$\begin{aligned}|\mathcal{S}| &\le \frac{1}{2}(3m_1 + 1 - 2) + \frac{1}{2}(3m_2 + 2 - 2) + \frac{1}{2}(3m_3 + 1 - 2) + 1 = \frac{3m}{2}\\ &\le \frac{1}{4}(3p - 6) < \frac{1}{4}(3p - 5).\end{aligned}$$

(3) Let $l_0 \subset l$ be an arc of a recurrent leaf $l \subset S$ with one endpoint at K_1 and the other at K_2 and no other points at the boundary ∂N. Denote by (ab) and (bc) two consequent l_0-arcs of the foliation $\mathcal{F}$. Let us show that they cannot split N.

Indeed, suppose the contrary. Fix an orientation at l such that the points b, c lie at the positive semi-leaf $l^+ \subset l$ issued from a. Let us represent N as an annulus with a certain number of glued handles, such that the boundary components of the annulus are identified via an orientation preserving homeomorphism $\phi : K_1 \to K_2$. For the points a and b at the boundary denote by (ab) an arc of $\mathcal{F}$ having the 'positive' direction. Suppose next that $\phi^{-1}(c) \in (a\phi^{-1}(b))$. Then necessarily $\phi(a) \in (bc)$. The positive semi-leaf $l^+(c)$ issued from c must intersect K_2 first time in the point d which belong to the segment (cb). Therefore $\phi^{-1}(d) \in (\phi^{-1}(c)\phi^{-1}(b)) \subset (a\phi^{-1}(b))$. Similar arguments are applied to the semi-leaf $l^+(d)$. It follows that the point $\phi(a) \in (bc)$ cannot lie in the limit set of $l^+(c)$ what contradicts the recurrence of l.

If $\phi^{-1}(c) \in (\phi^{-1}(b)a)$, then $\phi(a) \in (cb)$. It implies that $l^+(c)$ meets only the arcs $(bc), (\phi^{-1}(b)a)$ (a contradiction with l chosen recurrent). Thus, the arcs $(ab), (bc)$ do not split N. Then we cut N along these arcs. As a result one obtains an orientable surface of genus $m-1$ with two holes. According to Theorem 3.5.3 the following estimate is true:

$$|S| \leq \frac{1}{2}[3(m-1)+2-2]+1 = \frac{1}{2}(3m-1) \leq \frac{1}{2}[\frac{3}{2}(p-1)-1] = \frac{1}{4}(3p-5).$$

The above considered case finishes the proof of the theorem when $p = r + 1, n = 0$.

(III) Suppose now that the estimate of the theorem is true for $p = r+1$ and $n = 0, \dots, k$. Repeating with slight modifications the arguments of the part (II) of the proof, one can readily obtain the conclusion of the theorem for the case $p = r+1, n = k+1$. The rest is left to the reader.

The above obtained estimate is sharp. The relevant example is discussed detailedly in [28]. (There an example of a foliation $\mathcal{F}$, Sing $F < \infty$ on a non-orientable surface is constructed. For the spacious $\mathcal{Q}$-sets of $\mathcal{F}$ the upper bound of the inequality, mentioned in the theorem, is attained.) This remark completes the proof of the theorem. □

Theorem 3.5.5 (Aranson-Zhuzhoma) *Let $\mathcal{F}$ be a C^∞ foliation on a non-orientable surface N of genus $p > 0$ possibly with n boundary components. Let the set Sing $\mathcal{F}$ be finite and let $\mathfrak{t}(\mathcal{F})$ be a thorn number of $\mathcal{F}$. Then for the total number $|\mathcal{Q}|$ of the quasiminimal sets of $\mathcal{F}$ the following estimate is valid:*

$$|\mathcal{Q}| \leq \max\left\{[\frac{3p+2n+2\mathfrak{t}(\mathcal{F})-5}{4}];0\right\},$$

where $[x]$ denotes the integer part of x. Moreover, the above estimate is sharp. In other words, there exists a foliation $\mathcal{F}$ on N for which the upper bound of the above estimate is attained.

Proof. If $\mathfrak{t}(\mathcal{F}) = 0$ the estimate follows from Theorem 3.5.4. Let $\mathfrak{t}(\mathcal{F}) > 0$ and the foliation $\mathcal{F}$ has $\mathfrak{t}(\mathcal{F}) = t$ singular points $p_1, \dots, p_t$ such that each p_i can be surrounded by a curve $l_0 \cup \gamma$ of Definition 3.5.1. Now one must deform the foliation $\mathcal{F}$ in the neighbourhoods $D_1, \dots, D_t$ of small discs centered at $p_1, \dots, p_t$, respectively. This deformation is chosen such that the newly obtained foliation F' has the same number of the $\mathcal{Q}$-sets.

Remove the disc regions $D_1, \dots, D_t$ from the surface N obtaining a non-orientable surface N' of genus p with $n+t$ holes. The total number of $\mathcal{Q}$-sets for $\mathcal{F}'$ is $|\mathcal{Q}'|$ and coincides with the total number of such sets for $\mathcal{F}$, that is $|\mathcal{Q}'| = |\mathcal{Q}|$. But the foliation $\mathcal{F}'$ has only spacious $\mathcal{Q}$-sets. Applying the estimate for the total number of spacious $\mathcal{Q}$-sets given by Theorem 3.5.4 one obtains immediately the estimate of theorem. □

Appendix: Non-Orientable Surfaces

Here a brief overview of some basic facts about non-orientable compact surfaces is given. For more elaborate exposition the following monographs are recommended:

B. A. Dubrovin, A. T. Fomenko, A. T. & D. B. Fucks, D. B, Modern Geometry. Methods and Applications. Moskva. Nauka. 1986 (Russian).

D. B. Fucks & V. A. Rokhlin, Introduction to Topology. Geometric Chapters. Moskva. Nauka. 1977 (Russian).

As a model surface in the non-orientable case one takes a two-dimensional sphere with p holes each glued by a cross-cap. (By a cross-cap one understands a Möbius strip glued to the model surface by a diffeomorphism between the boundary of the Möbius strip and the boundary component of the model surface.) The positive number $p > 0$ is called a genus of the non-orientable surface N which is written in this case as $N = N_p$. N_p is allowed to have $n \geq 0$ holes not glued by cross-caps. In this case they say that N_p has n boundary components, or simply, holes. It follows from the above considerations that if $p = 1, n = 0$ the model surface is diffeomorphic to a projective plane $\mathbb{R}P^2$, if $p = n = 1$ to the Möbius strip, if $p = 2, n = 0$ to the Klein bottle. The Euler characteristic of the surface $N = N_p$ with n holes is given by

$$\chi(N) = 2 - p - n.$$

(Note that in the above classification of the model surfaces, there are no 'handles' glued into S^2. As we shall see later when $p > 0$ the sphere with g handles, p cross-caps and n holes is diffeomorphic to a sphere with $2g + p$ cross-caps and l holes.)

Lemma 3.5.3 *Let N_{p_1} and N_{p_2} be a pair of non-orientable surfaces with $l_1 > 0$ and $l_2 > 0$ holes, respectively. Let ϕ be a diffeomorphism between a boundary component of N_{p_1} and a boundary component of N_{p_2}. Finally, let $N = N_{p_1} \#_\phi N_{p_2}$ be a connected sum of N_{p_1} and N_{p_2} along the diffeomorphism ϕ. Then:*

- *N is diffeomorphic to a model surface $N_{p_1+p_2}$ with $l_1 + l_2 - 2$ holes;*
- *if in addition $p_1 = 0$, but $g > 0$ is a number of handles glued into a model sphere N_0, then N is diffeomorphic to a model surface N_{2g+p_2} with l_1+l_2-2 holes.*

Proof. The first statement is evident. Let us prove the second statement. Assume first that that $g = 1, p_2 = 1$ and $l_1 + l_2 = 3$. Then N is diffeomorphic to a Möbius strip with a handle. Since the Möbius strip is one-sided, it is diffeomorphic to the Möbius strip with a 'twisted' handle. This surface can be split into a sphere with two holes and a cross-cap, and a disc with a 'twisted' handle (by such we understand a handle which is glued from the opposite sides of the disc). Clearly what we have is a sphere with three cross-caps and a hole.

In the general case one applies the induction by g. If $g > 0$ then N can be glued from a sphere with $g - 1$ handles, $p_2 - 1$ cross-caps and $l_1 + l_2 - 1$ holes and a sphere with one handle, one cross-cap and one hole. As it was shown before, the second manifold is diffeomorphic to a sphere with three cross-caps and a hole. The lemma follows. □

Further we shall use the following easy

Corollary 3.5.2 *Let N_p be a closed non-orientable surface of genus $p > 0$. Let γ be a simple closed curve which does not bound a disc on N_p. Then dissecting N_p along γ one comes to the following cases:*

(i) if γ splits N_p, then N_p falls into two compact surfaces N_{p_1} and N_{p_2} of genus p_1 and p_2, respectively. N_{p_1} and N_{p_2} are either both non-orientable and in this case $p_1 + p_2 = p$, or one of them, say N_{p_1}, is orientable and the other is non-orientable. In the latter case it holds that $2p_1 + p_2 = p$;
(ii) if γ does not split N_p, then one obtains a compact surface $N_{p'}$ of genus p'. $N_{p'}$ is either non-orientable and $p' \le p - 1$, or $N_{p'}$ is orientable and $2p' \le p - 1$.

Remark 3.5.1 *Let N be a non-orientable surface and let γ be a simple closed curve which is a boundary of a submanifold $N_1 \subset N$ with a hole.* [4] *Let l be an arc lying in N_1 with the endpoints at the boundary of N_1. Then the following conditions are equivalent:*

(i) l is two-sided;

[4] That is a Möbius strip. Note that γ splits M.

(ii) for a segment $\gamma_0 \subset \gamma$ the closed curve $\gamma_0 \cup l$ bounds a disc region on N_1;
(iii) l splits N_1.

Proof. (i) $\Rightarrow$ (iii) follows from that the genus of N_1 is 1. The implication (ii) $\Rightarrow$ (i) is evident. Finally, the implication (iii) $\Rightarrow$ (ii) is proven in [16]. □

Remark 3.5.2 *Let l_1 and l_2 be a pair of simple arcs on a non-orientable surface N_1 with a hole. Suppose also that the endpoints of these arcs belong to the boundary of N_1. If l_1 and l_2 are one-sided, then they split N_1 into two disc regions.*

Proof. By Corollary 3.5.2, neither l_1 nor l_2 splits N_1. Clearly, dissecting N_1 along either of the arcs l_1 or l_2 one obtains a disc region. □

We conclude by the following fundamental statement.

Theorem 3.5.6 *Every connected compact non-orientable manifold N is diffeomorphic to a model surface N_p with n holes, where $p > 0, n \geq 0$.*

Proof. This can be readily derived from Lemma 3.5.3. The rest of the argument is left to the reader. □

3.6 Application: Smoothing Theorem

The decomposition theorems proved in this Chapter can be applied to classification of foliations. The so-called 'smoothing theorem' shows how the 'topological' and 'smooth' properties of foliations are linked.

Theorem 3.6.1 (Gutiérrez) *Let $\mathcal{F} = (M, \pi^t, \theta)$ be a continuous foliation on a compact C^∞ surface M. Then there exists a C^1 foliation $\widetilde{\mathcal{F}} = (M, \widetilde{\pi}^t, \theta)$ on M, which is topologically conjugate to $\mathcal{F}$. Furthermore, the following conditions are equivalent:*

(i) any minimal set of π^t is trivial;
(ii) π^t is topologically conjugate to a C^2 flow;
(iii) π^t is topologically conjugate to a C^∞ flow.

Proof will be split into the series of lemmas and related notions. We give an idea of the proof referring the reader to the original work [126] in the case when more details are needed.

Smoothable Flow Boxes

Let $\pi^t : M \times I\!R \to M$ be a C^0-flow as in the Smoothing Theorem. Let Sing π^t be the set of fixed points of π^t. Let us assume that all the minimal sets of π^t are trivial. To prove that π^t is topologically equivalent to a smooth flow we decompose the manifold $M\setminus$ Sing π^t into the submanifolds which are 'almost' flow-boxes of π^t. (For the latter the smooth 'models' are constructed.) These submanifolds are further called 'ϕ-sequences' and they are densely distributed in $M\setminus$ Sing π^t but may not cover it. Using the Structure Theorem (Theorem 3.3.1), one arranges these T-sequences so as to obtain a new differentiable structure on $M\setminus$ Sing π^t with respect to which the foliation induced by π^t is smooth. Thus far the theorem will follow.

The μ-Coordinates

Denote by σ a closed segment with the endpoints a and b. Consider a measure μ on the Borel algebra of σ. Such a measure is said to be *distinguished* if it is bounded on σ $0 < \mu(\sigma) < \infty$ and the mapping

$$h : \sigma \longrightarrow [0, \mu(\sigma)]$$

given by $h(x) = \mu([ax])$ is a homeomorphism (here $[ax] \subseteq \sigma$ is a subsegment with the endpoints a and x). (This mapping h will also be referred to as *μ-homeomorphism.*)

Let $\Sigma = \{\sigma_i\}_{i\in I}$ be a family of segments such that for each $i \in I$, σ_i is either a circle or a segment. The measure μ is called a *distinguished measure on the family Σ* if for all $i \in I$ and for all subsegments $\lambda \subseteq \sigma_i$ the restriction $\mu|_\lambda$ is a distinguished measure.

Denote by θ a π^t-flow box with the transversal edges A_1 and A_2. Suppose that $\phi : A_1 \to A_1$ is the transition (Dulac's) mapping induced by π^t between A_1 and A_2. Let us assume that A_1 and A_2 are oriented and ϕ preserves the orientation fixed. Suppose also that both A_1 and A_2 are supplied with the distinguished measure μ. Denote by h_1 and h_2 the respective μ-homeomorphisms on A_1 and A_2. The mapping ϕ of the flow-box θ is said to be *μ-smooth* (*μ-C^1 smooth*) if the μ-coordinate expression for ϕ:

$$\tilde{\phi} = h_2 \circ \phi \circ h_1^{-1}$$

is $C^r, r \geq 2$-smooth (respectively, C^1-smooth).

Lemma 3.6.1 (Gutiérrez) *Suppose that the set Sing π^t of the singular points of the flow π^t is non-empty and finite. Then the submanifold $M\setminus$ Sing π^t can be expressed as the union of flow boxes θ_i such that:*

(i) Every compact set of $M\setminus$ Sing π^t is contained in finitely many flow boxes θ_i;
(ii) If $i \neq j$ then $\theta_i \cap \theta_j = \partial\theta_i \cap \partial\theta_j$;
(iii) There exists a natural number $n_0 \in \mathbb{N}$ and a distinguished measure μ on the transversal edges of θ_i such that for all $i \in \mathbb{N}$, θ_i is $\mu - C^1$ and for all $i \geq n_0$, θ_i is μ-smooth;
(iv) If all minimal sets of π^t are trivial then $n_0 = 1$.

Proof. The arguments are lagerly based on the Structure Theorem. Let us keep the notation used there. Let Δ be the union of the arcs of trajectory $[pq]$ such that $p, q \in \Gamma$ where Γ is a global cross-section, and T is the first return mapping such that $T(p) = q$. Then either T is discontinuous at p or T^{-1} is discontinuous at q. Denote by $\mathcal{Y}$ the set of connected components of

$$M \backslash \left(\textit{Sing } \pi^t \bigcup \Delta \bigcup \Gamma \right).$$

The closure $\overline{Y}$ of $Y \in \mathcal{Y}$ may fail to be a flow box only because of the 'transversal edges' meet each other. In this case $\overline{Y}$ can be expressed as the union of two flow boxes, Y_1 and Y_2, such that $Y_1 \cap Y_2$ is a global cross-section for $\pi^t|_{\overline{Y}}$ and also each $Y_1, i = 1, 2$ shares with Y exactly one transversal edge. Denote by $\mathcal{C}$ the union of all transversal edges of Y_i over all $Y \in \mathcal{Y}$. From the above construction it follows that:

(1) $\mathcal{C} \supset \Gamma$.

For given $i \in \mathbb{N}$ there are only finitely many connected components of $M \backslash($ Sing $\pi^t \cup \Delta \cup \Gamma)$ which intersect $M \backslash M_i$. Therefore the family $\{Y_i : i = 1, 2, Y \in \mathcal{Y}\}$ of the closure of the connected components of $M \backslash($ Sing $\pi^t \cup \Delta \cup \mathcal{C})$ can be enumerated as $\theta_1, \theta_2, \ldots, \theta_j, \ldots$ so that for all $i \in \mathbb{N}$ there exists $k(i) \in \mathbb{N}$ for which $\theta_j \cap (M \backslash M_i) = \emptyset$ as $j \geq k(i)$. From here we have

(2) $M \backslash$ Sing $\pi^t = \bigcup_j \theta_j$.

It is clear that item (i) is true because any compact subset of $M \backslash$ Sing π^t is contained in some $M \backslash M_i$. Item (ii) follows from the construction of θ_j's.

Finally, let us show that the distinguished measure μ can be extended to $\mathcal{C}$. Denote by $\tilde{T} : \mathcal{C} \to \mathcal{C}$ the first return mapping induced by π^t. Fix $Y \in \mathcal{Y}$ and assume that the flow box θ_j crosses $Y_1 \cap Y_2$ from Y_1 to Y_2. Let us set

$$\mu|_{Y_1 \cap Y_2} = \mu \circ \tilde{T}^{-1}|_{Y_1 \cap Y_2}.$$

This yields us that items (iii) and (iv) of the lemma are verified [126]. Lemma is proved. □

The ϕ-Sequences

Let ϕ be a first return mapping which is either T or T^{-1}. A finite sequence $\Sigma = \{\Sigma_1, \Sigma_2, \ldots, \Sigma_n\}$ of open, pairwise disjoint segments of Γ is said to be a *ϕ-sequence* if any two consecutive terms Σ_i and Σ_{i+1} satisfy $\phi(\Sigma_i) = \Sigma_{i+1}$.

Let $\Sigma = \{\Sigma_1, \Sigma_2, \ldots, \Sigma_n\}$ be a ϕ-sequence with *phi* $\in \{T, T^{-1}\}$. Then:

- Σ is said to be *attracting* if $\phi^n(\Sigma_1) \subset \Sigma_1$ and $\phi^n|_{\Sigma_1}$ admits continuous extension to $\overline{\Sigma}_1$. Moreover, this extension has a unique fixed point which is an attractor situated at one of the endpoints of Σ_1;

- Σ is said to be *periodic* if $\phi(\Sigma_n) = \Sigma_1$ and each $x \in \Sigma_1$ is a fixed point of ϕ^n;
- Σ is said to be *wandering* if Σ_1 is π^t-wandering;
- When Σ is attracting the half-open segment $\Sigma_1 \backslash \phi(\Sigma_n)$ will be called a *fundamental domain* of Σ;
- Finally, the union of all terms of Σ will be denoted by *Span* Σ.

Lemma 3.6.2 (Gutiérrez) *Suppose that the conditions (and notation) of the Structure Theorem are valid. Let $\Omega_{l+1}, \Omega_{l+2}, \ldots, \Omega_{l+k}$ are the non-trivial minimal sets of the flow π^t. For the natural numbers $i, j : l+1 \geq i, j \geq l+k$ the following is true: If $V_i^{(1)}$ (respectively $V_i^{(-1)}$) denotes the union of the transversal segments of ∂ Rec C_i from where the orbits enter (respectively exit) the region Rec C_i and $\tilde{\mu}$ is a distinguished measure on $V_i^{(1)} \cup V_i^{(-1)}$ then there exists a distinguished measure μ on C_i which extends $\tilde{\mu}$ and such that in μ-coordinates the mapping*

$$T_i|_{C_i \backslash V_i^{(-1)}} : C_i \backslash V_i^{(-1)} \longrightarrow C_i$$

is C^1 and has the derivative equal to 1 at all the endpoints of the connected components $C_i \backslash V_i^{(-1)}$.

Proof. This follows from the Structure Theorem that:

(1) Both $V_i^{(1)}$ and $V_i^{(-1)}$ have the same number of closed connected components. Moreover, if $V_{i1}^{(\pm 1)}, V_{i2}^{(\pm 1)}, \ldots, V_{is_i}^{(\pm 1)}$ are those $V_i^{(\pm 1)}$, then for all $(n, j) \in \mathbb{N} \times \{1, 2, \ldots, s_i\}$, the mapping $T^{\pm n}$ is defined in $V_{ij}^{(\pm 1)}$.

(2) There exists $J \subset \mathbb{N}$ and a family of closed non-trivial subintervals $\{Z_j\}_{i \in J}$ lying in the set $C_i \backslash (V_i^{(1)} \cup V_i^{(-1)})$ such that for all $(n, j) \in \mathbb{Z} \times J$ the mapping T^n is defined in Z_j. Moreover, the family

$$\{T^n(Z_j),\ T^{\pm m}(V_{ik}^{\pm}) \mid n \in \mathbb{Z},\ j \in J,\ m \in \mathbb{N} \cup \{0\},\ 1 \leq k \leq s_i\}$$

consists of pairwise disjoint intervals whose union is a dense subset of C_i.

Suppose now that $M = T^2$ is a torus so that Rec $C_i = M$. In this case by the Denjoy argument [88] there exists a measure μ on C_i such that:
(3) $\mu(C_i) = 1$, $\mu(C_i \backslash \bigcup_{n \in \mathbb{Z}} T^n(\mathbb{Z}_1)) = 0$ and $T_i : C_i \to C_i$ is $\mu - C^1$.

(Note that even if $J \subset \mathbb{N}$ is arbitrary the same idea may be applied to construct a measure μ on C_i which extends $\tilde{\mu}$ and satisfies the lemma.) Lemma is proved. □

Proof of the Smoothing Theorem

It is obvious that (iii) implies (ii). Notice also that by the Denjoy-Schwartz Theorem [260] we have that (ii) implies (i). Thus far we are only to prove that (i) implies (iii). So our main assumption is that

- All minimal sets of the flow π^t are trivial.

Suppose first that π^t has no fixed points, Sing $\pi^t = \emptyset$. Then M is either the torus or the Klein bottle. (The latter case we mention only for the sake of generality.) If π^t has no non-trivial recurrent orbits then by Neumann's Smoothing Theorem [210] π^t is topologically conjugate to a smooth flow. If π^t has non-trivial recurrent orbits, then M is necessarily the torus by the Aranson-Markley result [16], [195]. This fact implies that π^t is topologically conjugate to an irrational flow which is evidently smooth. Therefore we are to assume that:

- The set Sing π^t of the fixed points is non-empty.

Suppose that every subinterval of real axis $I\!R$ is provided with a canonical orientation. Let $M\setminus$ Sing $\pi^t = \cup_{i=1}^{\infty}\theta_i$, where θ_i are the flow boxes of π^t which satisfy the conditions of Lemma 3.6.2. A closed arc σ of a trajectory will be called a *tangent elementary arc* if it is contained in the boundary of some θ_i and intersects the set formed by all the corners of the flow boxes $\theta_1, \theta_2, \ldots$ exactly at their endpoints.

The smooth submanifold $M\setminus$ Sing π^t of M without the differentiable structure will be denoted by $\mathcal{M}$. The idea of the further analysis is a construction of a new smooth structure on $\mathcal{M}$. (Below some technical details are given.) For a θ_i fixed denote by A_{i1} and A_{i2} the transversal edges of θ_i. For $j = 1, 2$ there defined two μ-homeomorphisms

$$g_{ijk} : A_{ij} \longrightarrow [0, \mu(A_{ij}], \qquad k = 1, 2$$

determined by two possible orientations of A_{ij}. For each g_{ijk} choose a surjective continuous flow box of π^t: $\alpha_{ijk} : [1, 2] \times [0, \mu(A_{ij})] \to \theta_i$, such that

(i) For all $t \in [0, \mu(A_{ij})], \alpha_{ijk}([1, 2] \times \{t\})$ is an arc of trajectory of $\phi|_{\theta_i}$;
(ii) $(\alpha_{ijk})^{-1}$ as restricted to each tangent elementary arc $\sigma = [a, b]$ of θ_i coincides with $h_{\sigma(i)}$;
(iii) $(\alpha_{ijk})^{-1}|_{A_{ij}}$ is the μ-homeomorphism g_{ijk} defined above.

(The functions α_{ijk} play the role of the required new 'coordinates' in the flow boxes θ_i, see [126] for details.) Let again σ be an elementary arc of an π^t-orbit. Suppose that it is contained in $\partial\theta_i \cap \partial\theta_l$. Let A_{i1} and A_{l1} be 'first' transversal edges of θ_i and θ_l respectively. Let $[\varepsilon_{i1}, \varepsilon_{i2}]$ and $[\varepsilon_{l1}, \varepsilon_{l2}]$ be a subintervals of $[1, 2]$. We set $\varepsilon = \min\{\mu(A_{i1}, \mu(A_{l1})\}$. Define $\beta(\sigma) : (\varepsilon_{l1}, \varepsilon_{l2}) \times (-\varepsilon, \varepsilon) \to \mathcal{M}$ as follows

$$\beta(\sigma)(s, t) = \begin{cases} \alpha_{l1k}(s, t) & \text{if } t \geq 0 \\ \alpha_{l1k}(h_{\sigma(i)} \circ h^{-1}_{\sigma(l)}(s - \varepsilon_{l1}) + \varepsilon_{i1},\ \mu(A_{i1} + t) & \text{if } t < 0, \end{cases}$$

where $h_{\sigma(i)} : \sigma \to [0, \varepsilon_{i2} - \varepsilon_{i1}]$ and $h_{\sigma(l)} : \sigma \to [0, \varepsilon_{l2} - \varepsilon_{l1}]$ are smooth mappings. (Note that $\beta(\sigma)$ is well defined and all possible changes of coordinates are smooth.)

It remains to construct coordinate systems around the 'corners' of the flow boxes $\theta_1, \theta_2, \ldots$. Let p be a corner of an arbitrary θ_i. There are three cases to be considered.

(i) The 'corner' set is $\{\theta_i, \theta_l, \theta_n, \theta_s\}$ and p is a corner of all of them.
(ii) The corner set is $\{\theta_i, \theta_l, \theta_n\}$ anp p is an endpoint of two consecutive tangent elementary arcs of θ_l being a corner of θ_n and θ_i.
(iii) The corner set is $\{\theta_i, \theta_l, \theta_n\}$ anp p is an endpoint of two consecutive transversal elementary arcs of θ_l being a corner of θ_n and θ_i.

'Smooth' coordinates in the cases (i)-(iii) are introduced likewise to the earlier construction, see [126] for details.

We claim that the foliation $\pi^t|_{\mathcal{M}}$ on $\mathcal{M}$ is smooth and topologically equivalent to the continuous foliation on $\pi^t|_{M\setminus\ Sing\ \pi^t}$ on $M\setminus$ *Sing* π^t. Indeed, the elements of the above partition $\mathcal{A}$ are also smooth flow boxes for $\pi^t|_{\mathcal{M}}$ and this implies that it is a smooth foliation on $\mathcal{M}$. Since each element of $\mathcal{A}$ is a continuous flow box of $\pi^t : M \times I\!R \to M$ the identity map Id: $M\setminus$ *Sing* $\pi^t \to \mathcal{M}$ is a homeomorphism which provides the required topological equivalence. This proves the claim.

Let $\mathcal{F}$ be the above constructed smooth orientable foliation on $M\setminus$ Sing π^t such that $f(\mathcal{F}) = \pi^t|_{\mathcal{M}}$, where $f : M\setminus$ *Sing* $\pi^t \to \mathcal{M}$ is a diffeomorphism between two manifolds. We are going to show that there exists a smooth vector field $X \in \mathcal{X}^\infty(M)$ with the singularity set Sing π^t and such that $X|_{M\setminus\ Sing\ \pi^t}$ and $\mathcal{F}$ have the same phase portrait.

Indeed, let $X \in \mathcal{X}^\infty(M\setminus$ *Sing* $\pi^t)$ be such that the foliation which it induces is $\mathcal{F}$. The extension of X to the singularity set Sing π^t may be done by continuity, see [126] for a technique. Thus far we have a flow ψ^t of the vector field X which is smooth and equivalent under the diffeomorphism f to the flow π^t. This argument completes the proof of Smoothing Theorem. □

Remark 3.6.1 *It may be readily seen that if π^t admits non-trivial (Cantor-type) minimal sets, then the manifold $\mathcal{M}$ can be provided only with C^1 manifold structure. (The flow boxes θ_i are certainly μ-continuous, but the new smooth structure which one can associate to them is only C^1.)*

Bibliographic Notes. Cherry-Maier theorem is one of the earlest results on the structure of quasiminimal sets. Cherry [78] proved that there is a continuum of recurrent orbits in the Cantor-type quasiminimal set. The density of such orbits in the $\mathcal{Q}$-set has been proved by Maier [194]. Theorems 3.2.2

and 3.2.3 are due to Maier [194] who proved them for the flows on compact surfaces of finite genus. Aranson and Zhuzhoma [28] extended Maier's theorems to the case of foliations on compact surfaces. The counterexamples to the Maier theorems in the case of foliations with an infinite number of singularities, are taken from [28]. The idea of 'suppressing' of the Cherry cells is the oldest method of the theory. With the help of blowing-up of selected leaves of an irrational foliation, one can produce a Denjoy foliation on the torus; see Denjoy [88]. In this way Cherry [78] introduced his famous example. Gardiner [114] considered a general problem for the surfaces of genus $g \geq 2$.

The way in which the $\mathcal{Q}$-sets can border each other was understood by Maier [193], [194]. The structure theorem was established (more or less independently) by Gardiner [114], Gutierrez [126], Levitt [183], [187], Neumann [210], Aranson and Zhuzhoma [28]. Our exposition is close to [126]. The decomposition of closed orientable surface of genus $g \geq 2$ into a connected sum of $2g - 2$ pants is known to be a classical result, see e.g. [100]. In the context of transitive flows on 2-manifolds it was studied by Levitt [182], [183], while for the foliations with the tripod and the thorn singularities by Rosenberg [256]. In the series of papers [182], [183], [187] G. Levitt studied the decomposition theorems for recurrent flows either admitting the saddle-saddle connections or not. Theory of 'barriers' for the global cross-sections is known to belong to S. Novikov [231]. In the context of foliations on surfaces the 'barriers' were studied by Levitt [185]. The decomposition theorems of Levitt are exposed closely to the original papers [182],[183].

The early estimates of the maximal number of $\mathcal{Q}$-sets on surfaces are due to Maier [194]. In particular he proved that for the flow-induced foliations its total number cannot exceed g for the orientable surfaces of genus g, and in the case N is a non-orientable surface of genus p this number is $[(p-2)/2]$. Levitt [182], [186] generalized such estimates for the recurrent foliations with k-prong saddles, $k \geq 1$. He also introduced such an important characteristic of foliation as thorn number. Independently Plykin [245] introduced a notion of spacious $\mathcal{Q}$-set which arise naturally in the study of one-dimensional basic sets of A-diffeomorphisms on the compact surface. The most complete and comprehensive investigation of this problem has been undertaken by Aranson and Zhuzhoma [28]. They obtained the estimates for the total number of the $\mathcal{Q}$-sets (spacious or not) both for the orientable and non-orientable surfaces. Our exposition is close to [28].

4. Invariants of Foliations

We do not distinguish between two foliations if they are topologically equivalent. Such an equivalence relation splits the space of foliations into the equivalence classes which we are going to study in this chapter. This objective requires the following tasks:

(i) Find a constructive invariant which takes the same values on topologically equivalent foliations.
(ii) To describe all topological invariants which are admissible, i.e., which may be realized in the chosen class of foliations.
(iii) Find a standard representative in each equivalence class, i.e., for a given admissible invariant to construct a flow whose invariant "coincides" with the admissible invariant.

An invariant is called *complete* if its respective values on two foliations are the same if and only if the foliations are topologically equivalent. The 'if' part only yields a *relative* invariant of foliations.

There are two basic types of invariants related to foliations: discrete, like graphs (Lyapunov, Peixoto, etc) and continuous, like the real numbers (rotation numbers, homotopy rotation class, etc). Of course, in the first case we have a countable set of equivalence classes, while in the second there is a continuum of such classes. A mixture of these two types of invariants is not excluded.

4.1 Torus

Let $\mathcal{F} = (M, \pi^t, \theta)$ be a foliation of the compact surface M. We suppose that:

- The underlying manifold is a torus, $M = T^2$;
- The foliation $\mathcal{F}$ is orientable;
- $\mathcal{F}$ has no trivially recurrent leaves.

4.1.1 Minimal Foliations

Let $\phi : S^1 \to S^1$ be an orientation preserving homeomorphism of the circle S^1. Every such homeomorphism may be written in the form

$$\phi(x) = f(x) \quad mod \quad 1, \qquad 0 \le x < 1,$$

where $f(x)$ is a continuous monotone function defined for all $x \in \mathbb{R}$ in the real line $\mathbb{R}$ which may be chosen to satisfy the following 'additivity' property:

$$f(x+1) = f(x) + 1.$$

(In other words, the function f is a 'lifting' of the homeomorphism ϕ in the sense that it can be factorized mod 1 to obtain a homeomorphism at the unit circle S^1.) We shall say that the *function* f *represents* the homeomorphism ϕ. Clearly f 'inherits' the smoothness of ϕ, being of the same differentiability class as ϕ. We shall say that ϕ is a diffeomorphism of class C^k if f is C^k-smooth on $\mathbb{R}$. Note that the 'representation' of ϕ by the function f is not unique, as every pair of functions f_1, f_2 for which

$$f_2(x) = f_1(x) + n, \qquad \forall n \in \mathbb{N}, \qquad \forall x \in \mathbb{R},$$

represents the same homeomorphism $\phi : S^1 \to S^1$. (Fortunately, this is the only 'degree of freedom' allowed for the functions f representing ϕ.)

If the function f represents the homeomorphism ϕ, then the iterates ϕ^n of ϕ for all $n \in \mathbb{Z}$ are 'represented' by the functions f^n defined by the n-time composition of the function f:

$$f^n = f \circ \ldots \circ f.$$

All of the functions f^n possess the same properties as f, that is, they are continuous, monotonic and satisfy the 'additivity' property.

Theorem 4.1.1 *For any orientation preserving homeomorphism $\phi : S^1 \to S^1$ of the circle S^1 and any function $f(x)$ representing ϕ, the limit*

$$\lim_{n\to\infty} \frac{f^n(x)}{n} = \alpha$$

exists and does not depend on the choice of the point $x \in \mathbb{R}$. Moreover, the real number α is rational if and only if the homeomorphism ϕ is periodic (i.e., $\phi^k(x) = x$ for some power $k \neq 0$ and some point $x \in S^1$).

Proof. Since the function $f(x)$ is defined up to an integer n we obviously need only consider one function $f(x)$ representing ϕ since if otherwise $f_2(x) = f_1(x) + k$ then

$$\lim_{n\to\infty} \frac{f_2^n(x)}{n} = \lim_{n\to\infty} \frac{f_1^n(x)}{n} + k,$$

so that α is simply 'shifted' by $k \in \mathbb{Z}$. First let us assume that the limit exists for at least one point $x_0 \in \mathbb{R}$. Let us choose an arbitrary point $x \in \mathbb{R}$. Now take an integer m such that

$$x_0 + m \le x \le x_0 + m + 1.$$

Since the function f^n is monotonic the above inequality implies that

$$f^n(x_0 + m) \le f^n(x) < f^n(x_0 + m + 1),$$

and by the 'additivity' of $f^n(x)$

$$f^n(x_0) + m \le f^n(x) < f^n(x_0) + m + 1.$$

Therefore,

$$\lim_{n\to\infty} \frac{f^n(x) - f^n(x_0)}{n} = 0,$$

and this means that if the number α exists for one point $x_0 \in \mathbb{R}$ then it exists for all $x \in \mathbb{R}$ and does not depend on x.

Now let us prove the existence of the limit for some point x_0.

(i) First consider the case when some non-zero power of ϕ, say ϕ^k, has a fixed point,

$$\phi^k(x_0) = x_0.$$

The point x_0 is a fixed point for the homeomorphism ϕ^{-k} as well, and let us suppose $k > 0$. Taking a representing function, we have $f^k(x_0) = x_0 + r, r \in \mathbb{Z}$. Hence for any $l = 0, \pm 1, \pm 2, \ldots$,

$$f^{lk}(x_0) = x_0 + lr.$$

Any integer n may be written as $n = lk + s, 0 \le s < k$. Then

$$f^n(x_0) = f^{(lk+s)}(x_0) = f^s(f^{lk}(x_0)) = f^s(x_0 + lr) = f^s(x_0) + lr.$$

Therefore,

$$\lim_{n\to\infty} \frac{f^n(x_0)}{n} = \lim_{n\to\infty} \left(\frac{f^s(x_0)}{n} + \frac{lr}{lk+s} \right) = \frac{r}{k},$$

and the limit α exists and is rational.

(ii) Suppose that no power of ϕ has a fixed point. Let $k > 0$ be a positive integer. Since the difference $f^k(x) - x$ cannot take integer values then for all $x \in \mathbb{R}$ we have

$$x + r < f^k(x) < x + r + 1,$$

for some integer r depending on k. Take a natural number $n \in \mathbb{N}$ and apply the above inequality successively to the points $x_0 = 0; f^k(x_0); f^{2k}(x_0); \ldots f^{(n-1)k}(x_0)$. We obtain a system of inequalities for $l = 0, 1, \ldots, n-1$:

$$f^{lk}(x_0) + r < f^{(l+1)k}(x_0) < f^{lk}(x_0) + r + 1.$$

Taking the sum of all the inequalities we get

$$nr < f^{nk}(x_0) < n(r+1).$$

Now put $n = 1$ so as to obtain the inequality

$$r < f^k(x_0) < r + 1.$$

From the above two inequalities one easily derives (by multiplying the last inequality by n and estimating the difference between the two inequalities) the following estimate

$$\left| \frac{f^{nk}(x_0)}{nk} - \frac{f^k(x_0)}{k} \right| < \frac{2}{k}.$$

Because n and k are arbitrary, one can interchange them in the above formula and summing up further the two inequalities,

$$\left| \frac{f^n(x_0)}{n} - \frac{f^k(x_0)}{k} \right| < \frac{2}{n} + \frac{2}{k}.$$

From the last inequality it follows that $f^n(x_0)/n$ is a Cauchy sequence and therefore the limit

$$\alpha = \lim_{n\to\infty} \frac{f^n(x_0)}{n}$$

exists for $x_0 = 0$ and hence for all $x \in \mathbb{R}$.

It remains to show that if α is rational then some non-zero power of ϕ has a fixed point. This may be done in two steps.

(i) Suppose that $\alpha = 0$. Assume to the contrary that there are no such points. Then $f(x) \neq x$ for any x and therefore we can assume $f(x) > x$ for all real x. In particular $f(0) > 0$ and hence in view of the monotonicity of f, $f^n(0) > f^{(n-1)}(0) > \ldots > 0$. Thus $\{f^n(0)\}$ is a monotonically increasing sequence. Moreover, $f^n(0) < 1$. (Otherwise a contradiction with the assumption $\alpha = 0$ can be easily derived; see e.g. [80].) We have a sequence $\{f^n(0)\}$ which is monotonic and bounded. Such a sequence must have a limit, which we denote by $x_0 = \lim_{n\to\infty} f^n(0)$. Then we immediately have

$$f(x_0) = \lim_{n\to\infty} f(f^n(0)) = \lim_{n\to\infty} f^{(n+1)}(0) = x_0,$$

which means that x_0 is a fixed point of the homeomorphism ϕ, a contradiction.

(ii) Now suppose that α is a rational number, $\alpha = r/k$. Take the function $g(x) = f^k(x) - r$ which represents the homeomorphism ϕ^k. It is not hard to compute that

$$\lim_{n\to\infty} \frac{g^n(x)}{n} = \lim_{n\to\infty} \frac{f^{kn}(x)}{n} - r = k \lim_{n\to\infty} \frac{f^{kn}(x)}{kn} - r = 0.$$

Therefore there exists a fixed point of the homeomorphism ϕ^k. □

Definition 4.1.1 *Let $\phi : S^1 \to S^1$ be an orientation preserving homeomorphism of the circle S^1 and let $f(x)$ be a representing function of the homeomorphism ϕ. The real number*

$$\alpha = \alpha(\phi) = \lim_{n\to\infty} \frac{f^n(x)}{n} \quad mod \quad 1, \qquad x \in \mathbb{R},$$

defined up to an integer k, is called the rotation number associated with the homeomorphism ϕ.

Note that $0 \le \alpha(\phi) < 1$. One may consider $\alpha(\phi)$ as function on the compact metric space X of all orientation-preserving homeomorphisms of the circle endowed with the metric

$$dist(\phi_1, \phi_2) = \sup_{x\in S^1} dist(\phi_1 x, \phi_2 x) + \sup_{x\in S^1} dist(\phi_1^{-1} x, \phi_2^{-1} x).$$

Theorem 4.1.2 *The function $\alpha(\phi)$ is continuous at every point $\phi_0 \in X$.*

Proof. Let $\varepsilon > 0$ be a positive real number and choose a natural number $k > 1/\varepsilon$ and an integer r such that

$$\frac{r}{k} < \alpha(\phi_0) < \frac{r+1}{k}.$$

Suppose that the function $f_0(x)$ represents the homeomorphism ϕ_0 and let

$$\alpha(\phi_0) = \lim_{n\to\infty} \frac{f_0^n(x)}{n}$$

be a rotation number of ϕ_0 defined via f_0. Let us show (in two steps) that for all $x \in \mathbb{R}$ we have

$$f_0^k(x) > x + r.$$

(i) Indeed, if the above inequality holds for some but not all x then one can find an $x_0 \in \mathbb{R}$ for which $f_0^k(x_0) = x_0 + r$. But then

$$\alpha(\phi_0) = \lim_{l\to\infty} \frac{f_0^{lk}(x_0)}{lk} = \lim_{l\to\infty} \frac{x_0 + lr}{lk} = \frac{r}{k},$$

which contradicts the choice of k and r.

(ii) If we suppose $f_0^k(x) < x + r$ for all x then we immediately get $\alpha(\phi_0) \le r/k$, which again contradicts our choice of the numbers k and r. Thus the estimate $f_0^k(x) > x + r$ is proved. (In a similar way we can prove that $f_0^k(x) < x + r + 1$ for all x.)

Since $f_0^k(x) - x$ is a continuous periodic function, it follows from the above that for some $\eta > 0$

$$r + \eta < f_0^k(x) - x < r + 1 - \eta.$$

Since ϕ^k varies continuously with respect to ϕ one can find a positive δ such that for any homeomorphism $\phi \in X$ satisfying the condition dist $(\phi, \phi_0) < \delta$ and for some function $f(x)$ the inequality

$$|f^k(x) - f_0^k(x)| < \eta, \qquad x \in \mathbb{R},$$

is valid. For such a homeomorphism ϕ we have

$$r < f^k(x) - x < r + 1.$$

Repeating all of the previous arguments with regard to the function $f(x)$ we obtain

$$\frac{r}{k} < \lim_{n\to\infty} \frac{f^n(x)}{n} < \frac{r+1}{k},$$

so that

$$\left| \lim_{n\to\infty} \frac{f^n(x)}{n} - \alpha(\phi_0) \right| < \frac{1}{k} < \varepsilon.$$

Since $\alpha(\phi) = \lim_{n\to\infty} f^n(x)/n$ (mod 1) we have $|\alpha(\phi) - \alpha(\phi_0)| < \varepsilon$ for every vanishing value of ε. □

We shall concentrate on the orientation-preserving homeomorphisms $\phi : S^1 \to S^1$ with the irrational rotation numbers. To do this let us introduce a class of homeomorphisms ϕ_α which are the rotations of S^1 through an angle α acting by the formula:

$$\phi_\alpha x = x + \alpha \quad mod \quad 1.$$

Theorem 4.1.3 *Let $\phi : S^1 \to S^1$ be an orientation-preserving homeomorphism of the unit circle S^1 with irrational rotation number α. Then there exists a continuous mapping $h : S^1 \to S^1$ which conjugates the homeomorphism ϕ to the homeomorphism ϕ_α, that is, $h \circ \phi(x) = \phi_\alpha \circ h(x)$ for all $x \in S^1$. Moreover, the mapping h is a homeomorphism if and only if ϕ is a transitive homeomorphism* [1] .

Proof. For an elegant proof of the above theorem we apply an argument coming from the ergodic theory of homeomorphisms on S^1; see [80] for an introduction. We prove the two parts of the theorem in succession.

(i) Suppose that μ is a normalized Borel measure on S^1 invariant with respect to ϕ. (Such a measure exists for any homeomorphism ϕ of the circle; see, e.g., Theorem 1 of §8, Ch.1 of [80].) Take a point $x_0 = 0$ on S^1 and put

$$h(x) = \mu([x_0, x]),$$

for any $x \in S^1$. Since ϕ has no periodic points the measure μ must be continuous; so is the mapping h. Therefore for any points $x_1, x_2, x_3 \in S^1$,

[1] That is, each orbit of ϕ is everywhere dense in S^1.

$$\mu([x_1, x_3]) = \mu([x_1, x_2]) + \mu([x_2, x_3]) \quad mod \quad 1.$$

For an arbitrary point $x \in S^1$ one can write:

$$\begin{aligned} h(\phi x) &= \mu([x_0, \phi x_0]) = \mu([x_0, \phi x_0]) + \mu([\phi x_0, \phi x]) \quad mod \quad 1 = \\ &= \mu([x_0, \phi x_0]) + \mu([x_0, x]) \quad mod \quad 1 = \\ &= h(x) + \beta \quad mod \quad 1, \end{aligned}$$

where $\beta = \mu([x_0, \phi x_0])$. Let us show that $\beta = \alpha$. For this we suppose that $f(x), x \in \mathbb{R}$, is the function representing ϕ and that $0 \leq f(x_0) < 1$. Let ν be an infinite measure on $\mathbb{R}$ which coincides with the measure μ on every closed interval $[k, k+1]$ where $k \in \mathbb{Z}$. It follows from the invariance of μ that

$$\nu([f^n(x_0), f^{n+1}(x_0)]) = \nu([x_0, f(x_0)]), \quad n = 0, \pm 1, \pm 2, \dots$$

Therefore we have

$$\begin{aligned} \beta &= \mu([x_0, \phi x_0]) = \nu([x_0, f(x_0)]) = \\ &= \frac{1}{n} \sum_{k=0}^{n-1} \nu([f^k(x_0), f^{k+1}(x_0)]) = \frac{1}{n} \nu([x_0, f^n(x_0)]). \end{aligned}$$

Suppose that $z_n < f^n(x_0) < z_n + 1$ where $z_n \in \mathbb{Z}$ is an integer. Then obviously it holds $z_n < \nu([x_0, f^n(x_0)]) < z_n + 1$. Therefore,

$$\begin{aligned} \alpha &= \lim_{n \to \infty} \frac{f^n(x_0)}{n} = \lim_{n \to \infty} \frac{z_n}{n} \quad mod \quad 1 = \\ &= \lim_{n \to \infty} \frac{\nu([x_0, f^n(x_0)])}{n} = \beta \quad mod \quad 1. \end{aligned}$$

Since $0 \leq \alpha, \beta < 1$ the first part of the theorem is proved.

(ii) Indeed, if h is a homeomorphism then the trajectory $\{\phi^n x\}$ of any point $x \in S^1$ is the inverse image with respect to h of the trajectory $\{\phi_\alpha^n(h(x))\}$, which is dense in S^1. Hence $\{\phi^n x\}$ is also dense and therefore ϕ is transitive.

Conversely, if ϕ is transitive, then by the Denjoy argument, there exists a Lebesgue measure μ which is positive on any closed interval $[x, x_0], x \neq x_0$. Therefore the function $h(x)$ introduced in item (i) is a strictly monotone function. That is, the mapping $h : S^1 \to S^1$ is a homeomorphism (one-to-one mapping of S^1). □

4.1.2 Foliations With a Cantor Minimal Set

Let S^1 be the unit circle. By a Cantor set in S^1 one understands a perfect nowhere dense subset of S. Any Cantor set $\mathfrak{K}$ can be represented as

$$\mathfrak{K} = S \backslash \bigcup_{i=1}^{\infty} (a_i, b_i),$$

where $[a_i, b_i]$ are disjoint segments of S^1. The accessible and inaccessible points of $\mathfrak{K}$ are denoted by $\mathfrak{A}$ and $\mathfrak{I}$, respectively. Clearly,

$$\mathfrak{I} = S^1 \backslash \bigcup_{i=1}^{\infty} [a_i, b_i], \qquad \mathfrak{A} = \mathfrak{K} \backslash \mathfrak{I}.$$

The equivalence relation $\sim$ on S^1 identifies the points of S^1 as follows: $x \sim y$ mod $\mathfrak{K}$ if and only if $x = y$ or $x, y \in [a_i, b_i]$ for some $i \in \mathbb{N}$. By the *Cantor function* $p : S^1 \to S^1$ one understands a continuous mapping of S^1 onto S^1 which takes the same value at all points of the Cantor set $\mathfrak{K}$.

Given two Cantor functions p_1, p_2 coming from two distinct Cantor sets $\mathfrak{K}_1, \mathfrak{K}_2$ when can we 'conjugate' p_1 with p_2 on S^1? The lemma below gives the sufficient conditions for such a 'conjugacy'.

Lemma 4.1.1 *Let $\mathfrak{K}_1$ and $\mathfrak{K}_2$ be two Cantor sets in S^1 and let p_1 and p_2 be the Cantor functions associated with $\mathfrak{K}_1$ and $\mathfrak{K}_2$, respectively. Let h be a continuous mapping of S^1 onto S^1 such that $h^{-1}(x)$ is finite for all $x \in S^1$ and such that*

$$h \circ p_1(\mathfrak{I}_1) \subseteq p_2(\mathfrak{I}_2).$$

Then there exists a continuous mapping g of S^1 onto S^1 such that $g(\mathfrak{K}_1) = g(\mathfrak{K}_2)$ and $p_2 \circ g = h \circ p_1$. Moreover, g is a homeomorphism if and only if h is a homeomorphism and instead of the above inclusion the equality $h \circ p_1(\mathfrak{I}_1) = p_2(\mathfrak{I}_2)$ holds.

Proof. Notice that $p_2^{-1} \circ h \circ p_1$ defines a continuous mapping from $\mathfrak{I}_1$ into $\mathfrak{I}_2$. We are going to show that this mapping is uniformly continuous relative to $\mathfrak{I}_1$ and $\mathfrak{I}_2$. For any $\varepsilon > 0$ there exists a finite set $F_2 \subseteq S^1 \backslash p_2(\mathfrak{I}_2) = p_2(\mathfrak{A}_2)$ with the following property: if z, w are in $\mathfrak{I}_2$ and if $p_2(a) \notin F_2$ when $a \in \mathfrak{A}_2 \cap (z, w)$, then $|z - w| < \varepsilon$. Let $F_1 = h^{-1}(F_2) \subseteq S \backslash p_1(\mathfrak{I}_1) = p_1(\mathfrak{A}_1)$. The set F_1 is finite. Choose $\delta > 0$ such that, whenever $z, z' \in \mathfrak{I}_1$ and $|z - z'| < \delta$, we have either $F_1 \cap (p_1(z), p_1(z') = \emptyset$ or $F_1 \cap (p_1(z'), p_1(z)) = \emptyset$. Let $w = h \circ p_1(z)$ and $w' = h \circ p_1(z')$. Assuming that $F_1 \cap (p_1(z), p_1(z')) = \emptyset$ we have by connectivity that $h^{-1}(x) \cap F_1 = \emptyset$ either for all x in (w, w') or for all x in (w', w). Thus either $F_2 \cap (w, w') = \emptyset$ or $F_2 \cap (w', w) = \emptyset$. It follows from our choice of F_2 that $|p_2^{-1} \circ h \circ p_1(z) - p_2^{-1} \circ h \circ p_1(z')| = |p_2^{-1}(w) - p_2^{-1}(w')| < \varepsilon$, which establishes uniform continuity.

Since the mapping $p_2^{-1} \circ h \circ p_1$ is uniformly continuous, it has a unique extension, g, to the closure Cl $(\mathfrak{I}_1) = \mathfrak{K}_1$. Clearly $p_2 \circ g = h \circ p_1$, and $g(\mathfrak{K}_1) = \mathfrak{K}_2$ because $h \circ p_1(\mathfrak{I}_1)$ is dense in S^1.

It remains to extend g to the whole S^1. Let $\{(a_i, b_i)\}_{i=1}^{\infty}$ be the complementary intervals of $\mathfrak{K}_1$. Either $g(a_i) = g(b_i)$ or $g(a_i), g(b_i)$ are the end points of the complementary interval (a, b) of $\mathfrak{K}_2$. In the latter case suppose that g_i is a homeomorphism of $[a_i, b_i]$ onto $[a, b]$ which extends g. Collecting all the extensions together we define g on S^1 by

$$g(z) = \begin{cases} g(z) & \text{if } z \in \mathfrak{K}_1, \\ g(a_i) & \text{if } z \in (a_i, b_i) \text{ and } g(a_i) = g(b_i), \\ g_i(z) & \text{if } z \in (a_i, b_i) \text{ and } g(a_i) \neq g(b_i) \end{cases}$$

The second part of the lemma follows from the fact that the equality holds if and only if $g(a_i) \neq g(b_i)$ for all i. □

Now we are in position to introduce certain invariants which will allow us to refine our classification of the homeomorphisms of the unit circle S^1. Notice first that considering S^1 as a circle $|z| = 1$ lying in the complex plane $\mathbb{C}$ endows S^1 with a group structure compatible with the natural topology of S^1. Corresponding to any point $a \in S^1$ there is a homeomorphism $\chi_a : S^1 \to S^1$ which acts by multiplication:

$$\chi_a(x) = ax, \qquad x \in S^1.$$

Following Markley [197], by a *generator* γ of S^1 we understand an element $\gamma \in S^1$ such that its orbit $O(\gamma) = \{\gamma^n | n \in \mathbb{Z}\}$ is dense in S^1.

Let $h : S^1 \to S^1$ be a homeomorphism of S^1 without periodic points and suppose that the dynamical system (S^1, h) has a limit set $\mathfrak{K}$ of Cantor type. If (a, b) is a complementary interval of $\mathfrak{K}$, then $h(a, b)$ is also a complementary interval of $\mathfrak{K}$ and $h^n(a, b) \neq (a, b)$ for all $n \neq 0$. Therefore $x \sim y \bmod \mathfrak{K}$ if and only if $h(x) \sim h(y) \bmod \mathfrak{K}$, and h induces a homeomorphism g of S^1 such that $p \circ h = g \circ p$, where p is the Cantor function on S^1 associated with $\mathfrak{K}$. In other words, p is a homomorphism of (S^1, h) onto (S^1, g). We can always assume that $g = \chi_\gamma$, where γ is a generator of S^1. If γ is chosen so that Im $\gamma > 0$ then γ is uniquely defined by the homeomorphism h, which we denote by $\alpha(h)$ and refer to as the *rotation number* of h. [2]

Along with the invariant $\alpha(h)$ let us introduce another feature of circle homeomorphisms. Let (S^1, h) be as earlier and denote by $T(h) = p(\mathfrak{J})$ where p is a Cantor function associated with h and $\mathfrak{J}$ is a set of unattainable points relatively to a Cantor set. (Note that if h is transitive then $T(h) = S^1$.) It is easy to see that the invariant $T(h)$ is uniquely defined by h up to rigid rotations of S^1.

Lemma 4.1.2 *Let $A \subseteq S^1$ be a subset of S^1 and let γ be the generator of S^1 with Im $g > 0$. If $S^1 \backslash A$ is countable and $\chi_\gamma(A) = A$, then there exists a homeomorphism g of S^1 without periodic points such that $\alpha(g) = \gamma$ and $T(g) = A$.*

[2] Note that Markley rotation number of h coincides with the rotation number of the homeomorphism h; the author is grateful to N. Bykov for indicating this fact to him.

Proof. If $A = S^1$ we set $g = \chi_\gamma$ and we are done. Let now $A \neq S^1$. Let $\mathfrak{K}$ be a Cantor set in S^1. There exists a Cantor function p on S^1 such that $p(\mathfrak{J}) = A$. We apply Lemma 4.1.1 with $\mathfrak{K}_1 = \mathfrak{K}_2 = \mathfrak{K}, p_1 = p_2 = p$ and $h = \chi_\gamma$ so as to obtain a homeomorphism $g : S^1 \to S^1$ such that $g(\mathfrak{K}) = \mathfrak{K}$ and $p \circ g = \chi_\gamma \circ p$. Clearly g is neither periodic nor transitive. If $\mathfrak{K}'$ is the Cantor set of (S^1, g), then $\mathfrak{K}' \subseteq \mathfrak{K}$ and $p(\mathfrak{K}') = S^1$. Thus $\mathfrak{J} \subseteq \mathfrak{K}'$ and $\mathfrak{K} = \mathfrak{K}'$. So we have $\alpha(g) = \gamma$ and $T(g) = A$. □

Lemma 4.1.3 *Let θ_1, θ_2 be a pair of homeomorphisms of the circle S^1 with no periodic points and with Cantor limit sets $\mathfrak{K}_1, \mathfrak{K}_2$. Let p_1, p_2 be the Cantor functions associated with $\mathfrak{K}_1, \mathfrak{K}_2$ respectively, and let h be a continuous mapping of S^1 onto S^1. If h^{-1} is finite for all $x \in S^1$ and the following conditions are satisfied:*

(i) $h \circ p_1(\mathfrak{J}_1) \subseteq p_2(\mathfrak{J}_2)$,

(ii) $p_2^{-1} \circ h \circ \theta_1(x) = \theta_2 \circ p_2^{-1} \circ h \circ p_1(x)$,

for all $x \in \mathfrak{J}_1$, then there exists a continuous mapping g of S^1 onto S^1 such that $g(\mathfrak{K}_1) = \mathfrak{K}_2, p_2 \circ g = h \circ p_1$ and $g \circ \theta_1 = \theta_2 \circ g$. Moreover, g is a homeomorphism if and only if equality holds in (i) and h is a homeomorphism.

Proof. (Note that the above lemma is a modification of Lemma 4.1.1.) Let $\mathfrak{K}_1, \mathfrak{K}_2, p_1, p_2$ and h be as in Lemma 4.1.3. Suppose that $\mathfrak{K}_i$ are minimal under θ_i for $i = 1, 2$. Assume that, for all $x \in \mathfrak{J}_1$,

$$p_2^{-1} \circ h \circ p_1 \circ \theta_1(x) = \theta_2 \circ p_2^{-1} \circ h \circ p_1(x).$$

Let g be an extension of $p_2^{-1} \circ h \circ P_1$ to $\mathfrak{K}_1$. Then $g \circ \theta_1(x) = \theta_2 \circ g(x)$ for all $x \in \mathfrak{K}_1$ since g is continuous and $\mathfrak{J}_1$ is dense in $\mathfrak{K}_1$. By a slight modification of our extension of g to S^1 we can obtain $g \circ \theta_1(x) = \theta_2 \circ g(x)$ for all $x \in S^1$. In particular one can find a collection of complementary intervals $\{(a_i, b_i)\}_{i=1}^m$ such that every complementary interval of $\mathfrak{K}_1$ has a unique representation of the form $\theta_1^n(a_i, b_1)$. If $g(a_i) \neq g(b_i)$, then we define $g(z)$ as earlier. Finally, if $z \in \theta_1^n(a_i, b_i)$ and $g(a_i) \neq g(b_i)$ we define $g(z)$ by the formula

$$g(z) = \theta_2^n \circ g_i \circ \theta_1^{-n}(z).$$

The lemma follows. □

If one brings together the conclusions of Lemmas 4.1.1 – 4.1.3, then we obtain the following beautiful classification theorem for homeomorphisms of the unit circle without periodic points.

Theorem 4.1.4 (Markley) *Let ϕ_1 and ϕ_2 be homeomorphisms of the unit circle S^1 without periodic points which define the dynamical systems (S^1, ϕ_1) and (S^1, ϕ_2) with the discrete time on S^1 . Then (S^1, ϕ_1) is topologically conjugate to (S^1, ϕ_2) if and only if $\alpha(\phi_1) = \alpha(\phi_2)$ and $T(\phi_1) \equiv T(\phi_2)$.*

Proof. This follows from Lemmas 4.1.1 – 4.1.3. □

4.1.3 Foliations With Cherry Cells

Example of T. Cherry [78] gives rise to a wider class of foliations, the so–called Cherry type foliations. A foliation $\mathcal{F}$ on T^2 is said to be *of Cherry type* if the following conditions are satisfied:

(i) $\mathcal{F}$ has one quasiminimal set $\Omega(\mathcal{F})$ with finite number of fixed points $O_1, \dots, O_k$ belonging to it;
(ii) Each of $O_1, \dots, O_k$ is a Morse saddle;
(iii) Let $L_1^i, L_2^i, L_3^i, L_4^i$ be the separatrices of O_i $(i = 1, \dots, k)$ labeled clockwise around O_i. Then there exists such a separatrix, say, L_3^i, such that $\Omega(\mathcal{F}) \cap [\omega(L_3^i) \cup \alpha(L_3^i)] = O_i$. For the other separatrices L_1^i, L_2^i, L_4^i we have $\Omega(\mathcal{F}) \subset \omega(\alpha)(L_j^i)$, $j = 1, 2, 4$.

Let us call the separatrix L_3^i of the saddle O_i *black*. Directly from the definition it follows that the set $\Omega(\mathcal{F})$ is nowhere dense. According to [194] $\Omega(\mathcal{F}) \setminus \{O_1, \dots, O_k\}$ is locally homeomorphic to the Cartesian product of the Cantor set and the segment. Since $O_i \in \Omega(\mathcal{F})$, the separatrices L_1^i, L_2^i or L_1^i, L_4^i are non-trivially recurrent semi-trajectories, one of them being the Bendixson extension of the other. It is clear that both separatrices (L_1^i, L_4^i or L_1^i, L_2^i) belong to $\Omega(\mathcal{F})$ and are dense therein.

The component of the set

$$D(\mathcal{F}) = T^2 \setminus [\Omega(\mathcal{F}) \cup_{i=1}^k (L_1^i \cup L_2^i \cup L_4^i)]$$

containing at least one black separatrix will be called a *black cell.* (Each Cherry-type flow has such components). The other components of $D(\mathcal{F})$ we call *grey cells.* (There exist Cherry-type flows without such components). A black cell is said to be positive (negative) if the black separatrix belonging to it is a $\omega(\alpha)$-separatrix of the saddle which belongs to $\Omega(\mathcal{F})$. We add to the conditions 1)–3) of the Cherry-type flows an extra condition (such foliations are denoted as $Y(T^2)$):

4. Each black cell of $\mathcal{F}$ contains exactly one fixed point, which is a topological node; the flow f $\mathcal{F}$ has no other fixed points except this node and the saddles $O_1, \dots, O_k$.

Let $\mathcal{F} \in Y(T^2)$ and let $\Omega(\mathcal{F})$ be a quasiminimal set of $\mathcal{F}$. The point $m \in \Omega(\mathcal{F})$ will be called a *boundary point accessible from the inside* if there exists an arc λ with the end point m such that $\lambda \setminus m \subset T^2 \setminus \Omega(\mathcal{F})$. (For instance all the saddles $O_1, \dots, O_k$ are boundary points accessible from the inside.) Let $\hat{\Omega}(\mathcal{F})$ denote the set of points of $\Omega(\mathcal{F})$ which are not boundary points accessible from the inside. This set consists of the non-trivially recurrent trajectories.

Recall that the foliation on torus is said to be linear irrational if the $\mathbb{R}^2$-covering flow for it can be defined by the system

$$\frac{dx}{dt} = 1, \quad \frac{dy}{dt} = \mu, \quad \mu \in I\!R \setminus \mathbb{Q}.$$

Let $\mathcal{F}_0(k)$ denote a linear irrational foliation on T^2 with $k \geq 1$ fake saddles added. (A fake saddle is a singularity with exactly two saddle sectors). Gardiner's theorem (see Chapter 3) implies the following result.

Lemma 4.1.4 *Let $\mathcal{F} \in Y(T^2)$ and let the quasiminimal set $\Omega(\mathcal{F})$ contain $k \geq 1$ saddle points $O_1, \dots O_k$ and let $\mathcal{F}_0(k)$ be a linear irrational flow with k fake saddles such that $[\mu](\mathcal{F}) = [\mu](\mathcal{F}_0(k))$, where $[\mu]$ is the rotation number of the respective foliation. Then there exists a continuous map (not necessarily a homeomorphism) $h : T^2 \to T^2$ with the following properties:*

(i) $h(O_1), \dots h(O_k)$ are fake saddles of the flow $\mathcal{F}_0(k)$;
(ii) The mapping h transforms a black positive (negative) cell into the $\omega(\alpha)$ separatrix of the fake saddle;
(iii) The mapping h sends each trajectory of $\hat{\Omega}(\mathcal{F})$ into a trajectory of the flow $\mathcal{F}_0(k)$, in an orientation-preserving manner. Moreover, $h|_{\hat{\Omega}(\mathcal{F})}$ is a homeomorphism;
(iv) Let Ω_0 be a grey cell. Then
a) If an accessible boundary $\partial\Omega_0$ of Ω_0 accessible from the inside contains no saddle points then $h(\Omega_0 \bigcup \partial\Omega_0)$ is a trajectory of the flow $\mathcal{F}_0(k)$;
b) If $\partial\Omega_0$ contains one saddle then $h(\Omega_0 \cup \partial\Omega_0)$ consists of the fake saddle and both of the separatrices of the flow $\mathcal{F}_0(k)$;
c) If $\partial\Omega_0$ contains two saddles then $h(\Omega_0 \cup \partial\Omega_0)$ consists of two fake saddles and their separatrices;
(v) The images of the black cells are mutually disjoint;
(vi) The mapping h is homotopic to the identity;
(vii) $h[\Omega(\mathcal{F})] = T^2$.

Let $\mathcal{F}$ be of Cherry-type and suppose the foliation $\mathcal{F}_0(k)$ satisfies Lemma 4.1.4. Let $\mathcal{F}_0$ denote a linear irrational foliation derived from $\mathcal{F}_0(k)$ by deleting all fake saddles. Let $\chi_0(\mathcal{F}, h) = \bigcup h(\Omega_0)$, where the union is taken over all grey and black cells of the flow $\mathcal{F}$. Let $\chi(\mathcal{F}, h)$ denote the set of the corresponding leaves of $\mathcal{F}_0$ (i.e., the trajectories which intersect $\chi_0(\mathcal{F}, h)$). We assign a code (i.e., a sequence of numbers from the set $\{-1, 0, +1\}$ and, possibly, letters) to each trajectory l from $\chi(\mathcal{F}, h)$ in the following way:

(i) We assign -1 (+1) if l contains the image of a negative (positive) cell, and 0 if l contains the image of a grey cell;
(ii) If we assign two numbers $(-1; 0 \text{ or } 0; -1)$ to the trajectory l then we also assign the letter "l" or "r" (standing for 'left' or 'right', respectively) depending on the side from which the grey cell adjoints the black cell;
(iii) If we assign three numbers $(-1; 0; +1)$ to l then we also assign two letters "lr" or "rl" in the same way.

We call the set $\chi(\mathcal{F},h)$ with its assigned codes a *blowing-down scheme of the foliation* $\mathcal{F}$ and we denote it by $\chi(\mathcal{F},h)$. (Note that the set of trajectories with the code ± 1 is finite. The set of trajectories with code 0 is at most countable.) Let χ be an at most countable set of leaves of $\mathcal{F}_0$. Suppose that to each leaf χ we assign a code $\{-1,0,+1,l,r,lr,rl\}$ and:

(a) the code of each leaf consists of at least one and at most three different numbers;
(b) the number of leaves with code including ± 1 is finite;
(c) if a leaf code includes $+1$ and -1 then it includes 0 as well;
(d) if the code contains two numbers then it contains one letter "l" or "r" as well;
(e) if the code contains three numbers then it contains at least one of the pairs "rl" or "lr".

The above set of leaves with their assigned codes will be called an *abstract permissible scheme.* (From the definition of a foliation of Cherry type it follows that a corresponding scheme is permissible for any blowing-down map.) Two abstract permissible schemes χ_1, χ_2 are said to be *commensurable* if there exists a diffeomorphism $F: T^2 \to T^2$ such that:

(a) $F(\chi_1) = \chi_2$;
(b) The covering mapping $\widetilde{F}: \mathbb{R}^2 \to \mathbb{R}^2$ for F is of the form $\bar{x} = ax+by$, $\bar{y} = cx + dy + \xi$; here $\begin{pmatrix} a & b \\ c & d \end{pmatrix}$ is an integer unimodular matrix, $\xi \in \mathbb{R}$;
(c) The codes of the leaves $l \in \chi_1$, $f(l) \in \chi_2$ are the same for all leaves $l \in \chi_1$;
(d) The corresponding linear irrational foliations are orbitally F-conjugate.

Theorem 4.1.5 (Aranson-T. Medvedev-Zhuzhoma) *Two foliations $\mathcal{F}_1, \mathcal{F}_2$ of Cherry type are topologically equivalent if and only if their schemes $\chi(\mathcal{F}_1,h_1), \chi(f_2^t,h_2)$ are commensurable. (Here h_1, h_2 are the blowing-down mappings for f_1^t, f_2^t, respectively.)*

Proof. This runs in the similar way to that for Theorem 4.1.4. □

In particular, from Theorem 4.1.5 it follows that the scheme of a foliation of Cherry type is independent (up to commensurability) from the blowing-down mapping. Let $\mathcal{F}^{-1}$ denote a foliation derived from $\mathcal{F}$ by time-inversion.

Corollary 4.1.1 *Two foliations $\mathcal{F}_1, \mathcal{F}_2$ of Cherry type are topologically equivalent if and only if the schemes of $\mathcal{F}_1, \mathcal{F}_2$ or $f_1^{-t}, \mathcal{F}_2$ are commensurable.*

Theorem 4.1.6 (Aranson-T. Medvedev-Zhuzhoma) *Let χ be an abstract permissible scheme. Then it is realizable as a C^∞ Cherry-type foliation $\mathcal{F}$ on T^2 with the scheme χ.*

Proof. The proof is similar to those of Lemma 4.1.1. □

Let χ be an abstract permissible scheme. Let $|\chi^+|$, $|\chi^-|$, $|\chi^0|$ denote the cardinality of the set of trajectories with codes including $+1$, -1 and 0, respectively. The following theorem shows that the numbers $|\chi^-|, |\chi^+|, |\chi^0|$ (if $|\chi^0|$ is finite) and the rotation number $[\mu](\mathcal{F})$ do not determine the Cherry type foliation.

Theorem 4.1.7 (Aranson-T. Medvedev-Zhuzhoma) *There exists a continuum of mutually topologically distinct C^∞ foliations of Cherry type with the same rotation number and the same $|\chi^-|, |\chi^+|, |\chi^0|$ for each of the following cases:*

(a) $|\chi^-| + |\chi^+| + |\chi^0| \geq 4$;
(b) $|\chi^+| = |\chi^0| = 0,\ |\chi^-| \geq 2$;
(c) $|\chi^-| = |\chi^0| = 0,\ |\chi^+| \geq 2$;
(d) $|\chi^0| = 0, |\chi^+| \geq 1, |\chi^-| \geq 1$.

Proof. The proof is the same as that for Lemma 4.1.1. □

4.1.4 Analytic Classification

Let $\phi : S^1 \to S^1$ be an analytic homeomorphism of the unit circle S^1 with irrational rotation number α. Let $K > 0$ and $\sigma > 0$ be positive real numbers. The irrational number α is said *to be of type* (K, σ) if, for every pair of integers $p, q \in \mathbb{Z}$,

$$\left|\alpha - \frac{p}{q}\right| \geq \frac{K}{|q|^{2+\sigma}}.$$

Theorem 4.1.8 (Arnold) *Let $\Pi_r, r > 0$ be a strip $|Im\ z| < r$ in the complex plane $\mathbb{C}$ and let $a(z)$ be a bounded holomorphic function in this strip with the norm $||a||_r = \sup_{z\in\Pi_r} |a(z)|$. Suppose that there exists $\varepsilon > 0$ which depends only on K, σ and r and suppose that $a(x)$ is a 2π-periodic analytic real function on the real axis $Im\ z = 0$ of the complex plane such that $||a||_r < \varepsilon$. Then the diffeomorphism*

$$\phi(x) = x + 2\pi\alpha + a(x)$$

with rotation number α such that α is of type (K, σ) is analytically conjugate to rotation through the angle $2\pi\alpha$.

Proof. Denote by A a rotation through the angle $2\pi\alpha$ and by H the required diffeomorphism which conjugates A to the diffeomorphism ϕ, that is

$$H \circ \phi = A \circ H.$$

Let us represent H in the form $H(x) = x + h(x)$, where $h(x + 2\pi) = h(x)$. One arrives to the following functional equation with regard to h:

$$h(x + 2\pi\alpha) - h(x) = a(x + h(x)).$$

(This equation is often called *a homology equation of h*.) The expansion of the functions $a(x)$ and $h(x)$ in formal Fourier series gives us

$$a(x) = \sum_{k=-\infty}^{k=+\infty} a_k e^{ikx}, \qquad h(x) = \sum_{k=-\infty}^{k=+\infty} h_k e^{ikx}.$$

The substitution of the formal Fourier series into the homology equation brings us, after the identification of the coefficients before e^{ikx}, to the equations which formally determine the h_k's:

$$h_k = \frac{a_k}{e^{2\pi ik\alpha} - 1}.$$

(Note that to resolve the above equations it is necessary that the nominator and denominator vanish simultaneously; e.g. if $a_0 \neq 0$ then there are no solutions for h_0. If the rotation number α is irrational then the above equations for the h_k's define a formal solution of the homology equation.)

It remains now to establish the convergence of the formal solution. The first step in this direction is the following

Lemma 4.1.5 *If $f(z)$ is a 2π-periodic function analytic in the strip Π_r and continuous on its closure and $||f||_r \leq M$ then the Fourier coefficients f_k of f decrease with k as*

$$|f_k| \leq Me^{-|k|r}.$$

Proof. This follows from the estimate of the contour integral for the f_k's:

$$f_k = \frac{1}{2\pi} \int_C f(z)e^{-ikz} dz.$$

Indeed, for $k > 0$ the contour C consists of part of the real line, two vertical lines and part of the line $z = x - ir$ so that it follows easily that

$$f_k = \frac{1}{2\pi} \int_0^{2\pi} f(x - ir)e^{-k(r+ix)} dx \leq Me^{-kr},$$

since along the vertical lines the integral vanishes. (The case $k < 0$ is treated similarly by switching C to $z = x + ir$.) □

Lemma 4.1.6 *If $|f_k| \leq M \exp(-|k|r)$ then the function*

$$f(x) = \sum_{k=-\infty}^{k=+\infty} f_k e^{ikx}$$

is analytic in the strip Π_r. Moreover, it satisfies the inequality $||f||_{r-\delta} \leq 4M/\delta$ for some positive $\delta < r$.

Proof. The first part is evident. To prove the second part we make the direct calculations:

$$\|f\|_{r-\delta} \leq \sum_{k=-\infty}^{k=+\infty} |f_k||e^{ikz} \leq M \sum_{k=-\infty}^{k=+\infty} e^{-|k|r} e^{|k|(r-\delta)} =$$

$$= M \sum_{k=-\infty}^{k=+\infty} e^{-|k|\delta} \leq \frac{2M}{1-e^{-\delta}} \leq \frac{4M}{\delta}.$$

The lemma is proved. □

To solve the homology equation one should divide the coefficients in the Fourier series by the numbers $\exp(2\pi ik\alpha) - 1$. If the rotation number is irrational these numbers do not vanish for $k \neq 0$. However some of them come very close to zero. In fact, every irrational α admits an approximation by rationals p/q with the 'error'

$$\left|\alpha - \frac{p}{q}\right| < \frac{1}{q^2}$$

where q is chosen arbitrarily large. For $k = q$ the denominator $\exp(2\pi ik\alpha) - 1$ can be arbitrarily small. Below we establish that 'almost all' irrational numbers are 'separated' from the rationals by a distance which is a power of k.

Lemma 4.1.7 *Let $\sigma > 0$ be a real number. Then for almost all* [3] *real numbers α there exists a constant $K = K(\alpha, \sigma)$ such that*

$$\left|\alpha - \frac{p}{q}\right| \geq \frac{K}{|q|^{2+\sigma}},$$

for all non-zero integers p and q.

Proof. Let us consider the numbers $\alpha \in [0, 1]$ for which the above inequality (with p, q, K fixed) fails. These numbers form an interval of length $\leq 2K/q^{(2+\sigma)}$. The union of such intervals for all p is of measure $\leq CK$, where $C = 2\sum 1/q^{(1+\sigma)} < \infty$. But the set of points for which K does not exist is covered by sets of arbitrarily small measure. Therefore this set has zero measure on the interval $[0, 1]$. □

Corollary 4.1.2 *For an irrational number α of type (K, σ) the small denominators admit the following estimate from below:*

$$|e^{2\pi ik\alpha} - 1| \geq \frac{K}{2|k|^{1+\sigma}}, \qquad |k| > 0.$$

[3] In the sense of Lebesgue measure on $\mathbb{R}$.

Proof. Indeed, the distance between $k\alpha$ and the nearest integer is greater than $K/|k|^{1+\sigma}$ and for all arcs of S^1 it is not shorter than the length of the smallest arc divided by π. □

We make an extra assumption regarding the function $a(x)$. Namely, we suppose that $a(x)$ is 2π-periodic and has 'mean value' [4] equal to zero.

Lemma 4.1.8 *For almost all values of α the homology equation admits an analytic 2π-periodic solution (real if $a(z)$ is real). Moreover, there exists a constant $\nu = \nu(K, \sigma) > 0$ such that if α is of type (K, σ) then for all $\delta > 0$ less than $r < 1/2$ the following estimate holds:*

$$||h||_{r-\delta} \leq \frac{||a||_r}{\delta^\nu}.$$

Proof. We do it in the four steps as follows. (i) Lemma 5.2.4 implies that $|a_k| \leq M \exp(-|k|r)$ if $||a||_r \leq M$. (ii) If α is of type (K, σ) then one has $|h_k| \leq 2MK^{-1} \exp(-|k|r)|k|^{1+\sigma}$. (iii)It is known that a function of the form $x^m \exp(-\omega x), x \geq 0$ has a maximum at the point $x = m/\omega$. But $x^m \exp(-\omega x) \leq C\omega^{-m}$, where $C = (m/e)^m$. Therefore, for all $\omega > 0$,

$$|k|^{1+\sigma} e^{-\omega|k|} \leq C\omega^{-m}, \qquad m = 1 + \sigma.$$

(iv) Thus $|h_k| \leq M \exp(-|k|(r - \omega)2CK^{-1}\omega^{-m}$. By Lemma 4.1.6 we have $||h||_{r-\delta} \leq DM$ where $D = 8C/K\omega^m(\delta - \omega)$. We now set $\omega = \delta/2$. Then $D \leq \delta^{-\nu}$, where ν is sufficiently large. Combining items (i)-(iv) one obtains the conclusion of the lemma. □

Now we have to solve the homology equation for the 'second term' $\tilde{a} = a - a_0$ which measures the deviation of the function $a(x)$ from its mean value a_0. Denote by h_0 solution of the 'mean' homology equation, which defines a mapping H_0 acting by the formula $z \mapsto z + h_0(z)$. Let us consider the mapping $A_1 = H_0^{-1} \circ A \circ H_0$. Let $a_1(z)$ be a function defined by the equation

$$A_1 z = z + 2\pi\alpha + a_1(z).$$

(In other words we introduce a new coordinate system $z' = H_0(z)$ on the unit circle and we write the mapping A in these coordinates. The mapping $z_1 \mapsto A_1 z_1$ differs from rotation through the angle $2\pi\alpha$ by a certain 'residue', $a_1(z)$.)

The second approximation is constructed in the similar manner by substituting A_1 for A. Then one obtains $h_1(z)$ and passes to the mapping $A_2 = H_1^{-1} \circ A_1 \circ H_1$, etc. Finally one comes to a mapping $\mathcal{H}_n = H_0 \circ H_1 \circ \ldots \circ H_{n-1}$. Clearly, $A_n = \mathcal{H}_n^{-1} \circ A \circ \mathcal{H}_n$. To prove the Arnold's Theorem it remains to show that the sequence $\mathcal{H}_n$ is convergent if α is of type (K, σ) and if $||a||_r$ is sufficiently small. The 'limit' element of the sequence, $\mathcal{H}$, conjugates the mapping $\lim A_n = \mathcal{H}^{-1} \circ A \circ \mathcal{H}$ to a rotation through the angle $2\pi\alpha$.

[4] This is given by the integral $\int_0^{2\pi} a(x)dx$.

Lemma 4.1.9 *There exist positive constants $\kappa, \lambda > 0$ which depend solely on K and σ such that, for all $\delta \in (0, r), r < 1/2$, the inequality $||a||_r \leq \delta^\kappa$ implies*

$$||a_1||_{r-\delta} \leq \frac{||a||_r^2}{\delta^\lambda}.$$

(In fact this lemma establishes that the 'residue' $a_1(z)$ after the first change of coordinates has infinitesimal order 2 with regard to the initial 'residue' $a(z)$. Therefore the successive approximations give a 'square error' for each step of the iteration. Therefore on the n-th step one obtains the error ε^{2^n}, where ε is the initial error.)

Proof. The proof of the above lemma consists of the following five steps:
(i) Suppose that Ω is a convex domain in $\mathbb{C}^n$ and $h : \Omega \to \mathbb{C}^n$ is a differentiable mapping such that $||Dh|| = \sup_{z\in\Omega} ||Dh(z)|| < 1$. Then the mapping H given by the formula $z \to h(z)$ is a diffeomorphism from Ω to $H(\Omega)$.

Indeed, the eigenvalues of $Dh(z)$ are different from zero, and therefore H is a local diffeomorphism. The mapping h is contractible because $|dh| < q < 1$ and Ω is convex. But the difference between the images of two arbitrary points of the mapping H is less than the distance between these points, that is, H is a bijection.

(ii) Let us show that if κ is big enough then the mapping A_1 is analytic in the strip $\Pi_{r-\delta}$. Indeed, let $||a||_r \leq M = \delta^\kappa$. Then $|a_0| \leq M, ||\tilde{a}||_r \leq 2M$. According to Lemma 4.1.8 we have $||h_0||_{r-\omega} \leq 2M\omega^{-\nu}$. But $||dh_0/dz||_{r-2\omega} \leq 2M\omega^{-(\nu+1)}$. Now set $\omega = \delta/8$. If κ is big enough then the previous inequalities give us:

$$||a||_r < \omega, \quad ||h_0||_{r-\omega} < \omega, \quad ||dh_0/dz||_{r-2\omega} < \omega.$$

But by item (i) H_1 is a diffeomorphism of the strip $\Pi_{r-2\omega}$ and its image contains the strip $\Pi_{r-3\omega}$. Now we have

$$H_0(\Pi_{r-\delta}) \subset \Pi_{r-\delta+\omega}, \qquad A \circ H_0(\Pi_{r-\delta}) \subset \Pi_{r-\delta+2\omega} \subset \Pi_{r-3\omega}.$$

The diffeomorphism H_0^{-1} is defined on $A \circ H_0(\Pi_{r-\delta})$. Therefore the mapping $A_1 = H_0^{-1} \circ A \circ H_0$ is analytic in $\Pi_{r-\delta}$.

(iii) Let us evaluate the 'residue' $a_1(z)$. The relevant commutative diagram gives us

$$z + 2\pi\alpha + a_1(z) + h_0(z + 2\pi\alpha + a_1(z)) = z + h_0(z) + 2\pi\alpha + a(z + h_0(z)).$$

On account of the homology equation we have

$$a_1(z) = [a(z + h_0(z)) - a(z)] - [h_0(z + 2\pi\alpha + a_1(z)) - h_0(z + 2\pi\alpha)] + a_0.$$

Applying the Mean Value Theorem and the Cauchy inequality for the expression in the first set of brackets we arrive at the following estimate using the results of item (ii):

$$||a(z + h_0(z)) - a(z)||_{r-\delta} \leq \frac{M}{\delta}||h_0||_{r-\delta} \leq \frac{M^2}{\delta^u},$$

where u is a constant depending only on ν (that is, only on K and σ). For the expression in the second set of brackets,

$$|| \bullet ||_{r-\delta} \leq 2M\omega^{-(\nu+1)}||a_1||_{r-\delta} \leq M\delta^{-u_1}||a_1||_{r-\delta}.$$

Therefore we obtain the following estimate of the 'residue' $a_1(z)$:

$$||a_1||_{r-\delta}(1 - M\delta^{-u_1}) \leq |a_0| + M^2\delta^{-u}.$$

(iv) Let us now estimate the quantity $|a_0|$, exploiting the fact that the rotation number of the mapping A in the 'coordinates' of A_1 is equal to a rotation through $2\pi\alpha$. It follows that $a_1(z)$ vanishes at a real point z_0. Substitite z_0 in the formula for $a_1(z)$. This gives us $a_0 = a(z_0) - a(z_0 + h_0(z_0))$ and hence $|a_0| \leq M^2\delta^{-u}$.

(v) Notice finally that the estimates of items (iii) and (iv) imply $||a_1||_{r-\delta} \leq 4M^2\delta^{-u}$. These arguments complete the proof of the Lemma. □

Finally, let us study the convergence of the formal system of approximations obtained above. We subdivide our analysis into four steps.

(i) Let us study the mapping A_n obtained at the n-th step of the iteration procedure, which is defined in the strip with the ray r_n decreasing at every step of the approximation: $r_0 = r, r_n = r_{n-1} - \delta_{n-1}$. We choose a decreasing sequence of reals δ_n such that $\delta_n = \delta_{n-1}^{3/2}, \delta_0 < 1/2$. For δ_0 small enough,

$$\sum_{n=0}^{\infty} < r/2.$$

(ii) Let us construct a sequence of numbers M_n by setting $M_n = \delta_n^N$, where the number N is defined uniquely by K and σ will be chosen later on. Notice that $M_n = M_{n-1}^{3/2}$.

(iii) Suppose that $||a||_r \leq M_0$. Let us show that $||a_n||_{r_n} \leq M_n$. Indeed, Lemma 4.1.9 implies that if $N > \kappa$ then

$$||a_1||_{r-1} \leq M_0^2\delta_0^{-1} = \delta_0^{2N-\lambda}.$$

Or, $\delta_0^{2N-\lambda} < \delta_1^N = \delta_0^{3N/2}$ if $N > 2\lambda$. Now choose N to be greater than 2λ and κ. This implies

$$||a_1||_{r_1} \leq \delta_1^N = M_1.$$

Similarly one passes from a_{n-1} to a_n.

(iv) Finally, let us prove the convergence of the product $\mathcal{H}_n = H_0 \circ \ldots \circ H_{n-1}$ in the strip $\Pi_{r/2}$. Indeed, the diffeomorphism H_0 is analytic in Π_{r_1} and satisfies the inequalities

$$||h_0||_{r_1} \leq \delta_0, \qquad ||dh_0/dz||_{r_1} \leq \delta_0.$$

In the same manner one obtains for the H_{n-1}'s:

$$||h_{n-1}||_{r_n} \leq \delta_{n-1}, \qquad ||dh_{n-1}/dz|| \leq \delta_{n-1}.$$

But $\mathcal{H}_n$ is analytic in the strip Π_{r_n} and has a derivative which is bounded from above and from below by the constants $C = \Pi_{k=-\infty}^{k=\infty}(1+\delta_k)$ and $c = \Pi_{k=-\infty}^{k=\infty}(1-\delta_k)$, respectively. It follows that $\mathcal{H}_n$ is a diffeomorphism of Π_{r_n} and the sequence $\mathcal{H}_n$ is convergent in the strip $\Pi_{r/2}$. Indeed,

$$||\mathcal{H}_n - \mathcal{H}_{n+1}||_{\frac{r}{2}} \leq C||h_n||_{\frac{r}{2}} \leq C\delta_n.$$

Let H be a limit of the sequence $\mathcal{H}_n$. Taking limits in the relation $A \circ \mathcal{H}_n = \mathcal{H}_n \circ A_n$ one obtains $A \circ \mathcal{H} = \mathcal{H} \circ R_\alpha$ where R_α is a rotation through the angle $2\pi\alpha$. The theorem is proved. □

4.2 Homotopy Rotation Class

4.2.1 Surfaces of Genus $g \geq 2$

Let $\mathcal{F} = (M, \pi^t, \theta)$ be a foliation of the compact orientable surface M of genus $g \geq 2$. The following assumptions are made:

- $\mathcal{F}$ is orientable;
- π^t has a non-trivial recurrent orbit whose closure is all of M (*transitivity*);
- The set Sing π^t is *finite* and consists of saddle points;
- There are no separatrix connections between the saddle points.

Following [21] such a class of flows is referred to as a *T-class*. Suppose that γ is a non-trivial recurrent orbit of the flow $\pi^t : M \times \mathbb{R} \to M$ of the class T. Let $\widetilde{\gamma} \in p^{-1}(\gamma)$ be a 'covering' curve of the family $\{\widetilde{\gamma}_\omega\}$ in the universal covering $\mathbb{H}$. (It can be easily seen that $\tilde{\gamma}$ is a connected one-dimensional continuum in $\mathbb{H}$ which can be 'parametrized' in a natural way.) The following statement is fundamental.

Theorem 4.2.1 (Aranson-Grines) *The α- (ω-) limit set of the curve $\widetilde{\gamma}$ consists of a unique irrational* [5] *point lying at the absolute $\partial\mathbb{H}$.*

Proof. The proof splits into two parts. First it will be shown that the α- (ω-) limit set of $\widetilde{\gamma}$ consists of a unique point $x^* \in \partial\mathbb{H}$ lying on the absolute. Next it will be proved that x^* is an irrational point.

[5] By this one means the complement of the set of *rational* points on the absolute. The rational points are those which are fixed under the (extended) action of the Fuchsian group on the absolute; see the *Appendix*.

Part I. Take a global cross-section X of the flow π^t which is a simple closed curve on M. It can be easily derived from the properties of the universal covering that the set $p^{-1}(X)$ consists of a countable collection $\{\widetilde{X}_\omega\}$ of continuous disjoint curves on $\mathbb{H}$ such that:

(i) For each $\widetilde{X} \in p^{-1}(X)$ there exists an element $g \in G$ of the Fuchsian group G which leaves $\widetilde{X}$ invariant. Moreover, if $g \in G$ takes a point $x \in \widetilde{X}$ to a point $g(x) \in \widetilde{X}$ then $\widetilde{X}$ is invariant with regard to g;
(ii) The curve $\widetilde{X}$ has two (and only two) limit points lying at the absolute and which are fixed by all elements of G leaving $\widetilde{X}$ invariant; and
(iii) Two different curves $\widetilde{X}$ and $\widetilde{X}'$ from the family $p^{-1}(X)$ have no common points at the absolute.

Clearly every curve $\widetilde{X} \in p^{-1}(X)$ is also a cross-section for the 'covering flow' $\widetilde{\pi}^t$ on the unit disc $\mathbb{H}$; see [21] for the details. Every orbit of $\widetilde{\pi}^t$ either has no common points with $\widetilde{X}$, or intersects $\widetilde{X}$ at a unique point.

Since the orbit γ intersects the cross-section X in a countable infinite number of points in M the same will hold true for the orbit $\widetilde{\gamma}$ and the curve $\widetilde{X}$ in $\mathbb{H}$. Denote by $\partial\widetilde{X}^+$ and $\partial\widetilde{X}^-$ the 'ends' of the curve $\widetilde{X}$. Denote by $g_i \in G$ an element of the Fuchsian group which takes the curve $\widetilde{X}_0$ of the family $p^{-1}(X)$ into the curve $\widetilde{X}_i$ of the same family:

$$g_i(\widetilde{X}) = \widetilde{X}_i, \qquad 0 \leq i \leq \infty.$$

Let us show that if $\widetilde{X}_i$ are those curves in the unit disc which intersect the curve $\widetilde{\gamma}$ then $\lim_{i\to\infty} \widetilde{X}_i$ consists of a unique point at the absolute. Indeed, the limit of the sequence of curves $\widetilde{X}_i$ (as i goes to ∞) is a connected subset $\Lambda \in \mathbb{H}$. Let us show that Λ belongs to the absolute.

Suppose the contrary. Then for each point $z \in \Lambda$ and every $\varepsilon > 0$ the neighbourhood $U_\varepsilon(z)$ intersects a countable number of congruent (by isometries $g_i \in G$) curves $\widetilde{X}_i$. This implies that the simple closed curve $X \subset M$ must intersect the set $p(U_\varepsilon)$ in a countable number of arcs, which is impossible.

Denote by x^*_+ and x^*_- those points at the absolute which are the limit points for the sequences $\{\partial\widetilde{X}_i^+\}$ and $\{\partial\widetilde{X}_i^-\}$ for $i = 0, 1, \ldots$ Then by connectivity Λ is an arc of the absolute with the endpoints x^*_+ and x^*_-. It remains to show that $x^*_+ = x^*_- = x^*$.

Again, suppose to the contrary that $x^*_+ \neq x^*_-$. Let us draw an arc α through the points $\widetilde{X}_0^-$ and $\widetilde{X}_0^+$ which is orthogonal to the absolute and such that Int $\alpha \subset \mathbb{H}$. The set $p(\alpha)$ is a simple closed curve in M. (A similar construction is possible for the rest of the points $\widetilde{X}_i^-$ and $\widetilde{X}_i^+$ for $i = 1, 2, \ldots$.) It follows that the limit of the arcs α_i is the arc Λ. On the other hand it is easy to see that each α_i must have as a limit an arc which is orthogonal to the absolute. This is a contradiction which proves that $x^*_+ = x^*_-$.

Part II. As before let $\widetilde{\gamma}$ be a curve of the family $p^{-1}(\gamma)$ where γ is a recurrent orbit of the flow $\pi^t \times \mathbb{R} \to M$. Suppose to the contrary that the ($\alpha-$ or $\omega-$) limit point x^* of $\widetilde{\gamma}$ is a rational point of the absolute. Then one may construct a transversal curve $\widetilde{X}$ intersecting $\widetilde{\gamma}$ and having x^* as its 'endpoint'. Moreover, the image $p(\widetilde{X})$ is a simple closed transversal curve on M. Let us give more details of the above construction.

Denote by $g \in G$ an element of the Fuchsian group for which the point $s_g = x^*$ is a stable fixed point [6] and denote by u_g an unstable fixed point of g. Note that it follows from Part I that for the family of transversal curves $\{\widetilde{X}\}$ and sufficiently large $i > N(\varepsilon)$ the point x^* is an inner point of the arc $(\partial\widetilde{X}_i^+, \partial\widetilde{X}_i^-)$ lying completely in the ε-neighbourhood of x^*.

Therefore there is a number $i_0 > 0$ such that the pair of points $\partial\widetilde{X}_{i_0}^+, \partial\widetilde{X}_{i_0}^-$ is separated on the absolute by the pair u_g, s_g. Thus the semi-orbit $\widetilde{\gamma}$ will hit a countable number of transversal curves $\widetilde{X}_k = g^k(\widetilde{X}_{i_0})$.

Denote by a_0 (a_1) the intersection point of the semi-orbit $\widetilde{\gamma}$ with the curve $\widetilde{X}_0$ ($\widetilde{X}_1$). Let us fix a point $g(a_0)$ lying on the curve $\widetilde{X}_1$. (Clearly $g(a_0) \neq a_1$, for otherwise γ is a closed curve.) Let us consider a curve

$$\tau = [a_0, a_1] \bigcup [a_1, g(a_0)],$$

which consists of the arc $[a_o, a_1]$ of the semi-orbit $\widetilde{\gamma}$ and the segment $[a_1, g(a_0)]$ of the transversal $\widetilde{X}_1$. Now take the g-invariant curve

$$\tilde{\tau} = \bigcup_{k=-\infty}^{\infty} g^k(\tau),$$

which is obtained from τ by isometries. (Note that $\tilde{\tau}$ has its endpoints in x^* and g_u.)

By the recurrence of γ there are no congruent points on the arc $[a_0, a_1]$. Let us show that the segment $[a_1, g(a_0)]$ also admits no congruent points. For this it is enough to show that the above segment contains no points congruent to the point $g(a_0)$. Suppose to the contrary that such a point exists, and denote it by ξ, so that $\xi = h(g(a_0))$ for some element $h \in G$. (Note that the 'ends' $\partial\widetilde{X}_1^-$ and $\partial\widetilde{X}_1^+$ of $\widetilde{X}_1$ are fixed under the action of h.) Clearly the element $f = hg$ of the Fuchsian group G moves the point a_0 to the point ξ. Consider the semi-orbit $\widetilde{\gamma}' = f(\widetilde{\gamma})$ passing through the point ξ. Clearly $\widetilde{\gamma}'$ must have its limit point in x^*. Since the elements $g, f \in G$ both have a fixed point in x^* the same must hold true for the element $h = fg^{-1}$, a contradiction which proves that there are no congruent points on the segment $[a_1, g(a_0)]$.

The following two possibilities may occur:

(i) The curve τ has no pairs of the congruent points except a_0 and $g(a_0)$;

[6] That is, $g^k(x) \to x^*$ as $k \to \infty$ for all 'nearby' points $x \in \mathbb{H}$.

(ii) There exists a pair of such points, $\xi \in [a_0, a_1]$ and $\zeta \in [a_1, g(a_0)]$.

In case (i) the required curve $\widetilde{X}$ (see the top of Part II) is constructed by an appropriate 'smoothing' of the curve $\tilde{\tau}$; see [21]. Let us consider the case (ii).

The arc $[a_0, a_1]$ can intersect only a finite number of curves of the family $p^{-1}(X)$ and therefore there exists only a finite number of congruent points. Thus one may suppose that on the transversal segment $[\xi, g(a_0)]$ there are no points congruent to the points of $[a_0, a_1]$.

Denote by h an element of the modular group for which $\zeta = h(\xi)$. As it was shown before, the semi-orbit $\widetilde{\gamma}' = h(\widetilde{\gamma})$ has its limit point in x^* and therefore x^* is a fixed point for the element h. Consider the curve

$$\sigma = [a_0, \xi] \bigcup h^{-1}([\zeta, g(a_0)]).$$

Evidently there are no pairs of congruent points on σ different from a_0 and $h^{-1}g(a_0)$. Clearly the point x^* on the absolute is a fixed point for the element $f = h^{-1}g$. In this case the required curve $\widetilde{X}$ is constructed as in item (i) from the curve

$$\tilde{\sigma} = \bigcup_{k=-\infty}^{\infty} f^k(\sigma).$$

Now recall that if γ is a non-trivial recurrent orbit and X is a global cross-section on M such that $\gamma \cap X$ is non-empty, then γ must hit X at least one extra time. Therefore on $\mathbb{H}$ the semi-orbit $\widetilde{\gamma}$ must intersect another curve, $\widetilde{X}'$, from the family $p^{-1}(X)$ at a unique point. The curves $\widetilde{X}$ and $\widetilde{X}'$ are disjoint and have no common points at the absolute. Therefore each of them splits $\mathbb{H}$ into two parts. But in this case the curve $\widetilde{\gamma}$ cannot have x^* as its $\alpha-$ ($\omega-$) limit point. This is a contradiction which proves the theorem. □

Definition 4.2.1 (Aranson-Grines) *Let γ be a semi-orbit of a flow $\pi^t : M \times \mathbb{R} \to M$ and let $\widetilde{\gamma}$ be a semi-orbit on $\mathbb{H}$ such that $p(\widetilde{\gamma}) = \gamma$, where p is a covering mapping. Let also x^* be a limit point of the semi-orbit $\widetilde{\gamma}$ lying at the absolute $\partial\mathbb{H}$. The set*

$$\mu(\gamma) = \bigcup_{g \in G} g(x^*)$$

is called a homotopy rotation class associated to the semi-orbit γ on M.

(It follows from the above definition that a homotopy rotation class $\mu(\gamma)$ of a semi-orbit γ is a countable set of either rational or irrational points at the absolute. Therefore the homotopy rotation class distinguishes between the closed and the non-trivially recurrent semi-orbits of the flow π^t. Note also that there exist flows on M such that, for each of their semi-orbit γ, $\mu(\gamma) = \emptyset$.)

Theorem 4.2.2 (Aranson-Grines) *Let $\pi^t : M \times \mathbb{R} \to M$ be a flow on a compact surface M and let γ be a semi-orbit of the flow π^t. The homotopy* *rotation class $\mu(\gamma)$ of γ is an invariant of the topological conjugacy class of π^t up to a homeomorphism h of the absolute which leaves invariant the set of rational and irrational points on it. Moreover, h is induced in a unique way by an automorphism of the Fuchsian group G.*

Proof. Suppose that $h : M \to M$ is a homeomorphism of M which maps the orbits of π^t to the orbits of the conjugate flow ϕ^t. According to a general result of Nielsen [214] there exists a homeomorphism $\widetilde{h} : \mathbb{H} \to \mathbb{H}$ of the unit disc which 'covers' h. The homeomorphism $\widetilde{h}$ brings the 'orbits' of covering flow $\widetilde{\pi}^t$ to the 'orbits' of the flow $\widetilde{\phi}^t$. Appealing to Nielsen's result again, there exists an induced automorphism $h_* \in \operatorname{Aut} G$ of the Fuchsian group such that

$$h_*(g) = \tilde{h} g \tilde{h}^{-1}, \qquad g \in G,$$

where $\tilde{h}$ and $\tilde{h}^{-1}$ are homeomorphisms of the absolute induced by h and such that they leave invariant the set of rational and irrational points on it. But this means that the homotopy rotation classes of the semi-orbits of flows π^t and ϕ^t coincide. □

4.2.2 Classification

Transitive Foliations

Theorem 4.2.3 (Aranson-Grines) *Let π_1^t and π_2^t be a pair of C^∞ T-flows on the compact surface M. Suppose also that neither π_1^t nor π_2^t have equilibrium points which are two-separatrix saddles. Then for π_1^t and π_2^t to be topologically conjugate it is necessary and sufficient that there exist:*

(i) a homeomorphism $h : \partial\mathbb{H} \to \partial\mathbb{H}$ of the absolute which is induced by an automorphism of h_* of the Fuchsian group G, and
(ii) two semi-orbits γ_1 and γ_2 of the flows π_1^t and π_2^t respectively such that $\mu(\gamma_2) = h(\mu(\gamma_1))$.

Proof. Necessary conditions. The proof follows from Theorem 4.2.2.

Sufficient conditions. The argument is divided into a series of lemmas.

Lemma 4.2.1 *Suppose that the $\pi_1^t, \pi_2^t \in T$ have the non-trivial recurrent semi-orbits γ_1 and γ_2, respectively, such that $\mu(\gamma_1) = \mu(\gamma_2)$. Then both π_1^t and π_2^t admit (non-trivial) global cross-sections X_1 and X_2 which are homotopy equivalent to each other.*

Proof. Denote by $\{\widetilde{\gamma}_1\}$ and $\{\widetilde{\gamma}_2\}$ the complete preimages $p^{-1}(\gamma_1)$ and $p^{-1}(\gamma_2)$ of the curves γ_1 and γ_2. Since $\mu(\gamma_1) = \mu(\gamma_2)$ then for each semi-orbit $\widetilde{\gamma}_1 \in \{\widetilde{\gamma}_1\}$ there exists a unique semi-orbit $\widetilde{\gamma}_2 \in \{\widetilde{\gamma}_2\}$ having the same limit point x^* at the absolute. Denote by $\widetilde{\gamma}_i'$ the orbit which contains the semi-orbit $\widetilde{\gamma}_i$ for $i = 1, 2$. Two cases are possible (we consider these successively):

(i) $\widetilde{\gamma}_i'$ is not an α-separatrix of a saddle point of the flow π_i^t;
(ii) $\widetilde{\gamma}_i'$ is such a separatrix.

(i) Since $\widetilde{\gamma}_i$ covers a non-trivial recurrent orbit on M its α-limt set will be a point $x^{**} \neq x^*$ at the absolute. Fix $i = 1$. There exists a simple closed π_1^t-transversal cycle X_1 on M such that $\gamma_1 \cap X_1$ is non-empty. As before let $\{\widetilde{X}_1\}$ be the complete preimage of X_1 on $\mathbb{H}$. It follows from above that the limit of curves from the family $\{\widetilde{X}_1\}$ which hit $\widetilde{\gamma}_1$ is the point x^*. Moreover, the 'ends' $\partial^{\pm}\widetilde{X}_1$ of the curve $\widetilde{X}_1 \in \{\widetilde{X}_1\}$ are separated on $\partial\mathbb{H}$ by the points x^* and x^{**}. Such relation is denoted by

$$(\partial^+\widetilde{X}_1, \partial^-\widetilde{X}_1 || x^*, x^{**}).$$

Denote by z the point of intersection of $\widetilde{X}_1$ and $\widetilde{\gamma}_1$. Let z' be a point of intersection of $\widetilde{X}_1$ and $\widetilde{\gamma}_2$ and such that after z' the semi-orbit $\widetilde{\gamma}_2$ never hits $\widetilde{X}_1$ in the direction of growing time. (We also assume that a small arc $\lambda \subset \widetilde{X}_1$ which contains z is π_2^t-transversal – this is always possible by modifying locally $\widetilde{X}_1$.)

Similarly, for $i = 2$ there exists a simple closed π_2^t-transversal cycle X_2 on M which hits the curve γ_2. Repeating all the above argument for the case $i = 2$ we obtain at the absolute

$$(\partial^+\widetilde{X}_1, \partial^-\widetilde{X}_1 || a, b),$$

where a and b are taken to be arbitrary points on the arcs $(\partial^-\widetilde{X}_1, \partial^+\widetilde{X}_1)$ and $(\partial^-\widetilde{X}_2, \partial^+\widetilde{X}_2)$, respectively.

Denote by $\lambda_0 \subseteq \lambda$ a closed arc for which z' is an inner point and such that all $\widetilde{\pi}_1^t$-orbits intersecting λ_0 reach the curve $\widetilde{X}_2^{(1)}$ as time increases and the curve $\widetilde{X}_2^{(2)}$ as time decreases. (By $\widetilde{X}_2^{(1)}, \widetilde{X}_2^{(2)} \in p^{-1}(X_2)$ one denotes two curves in the preimage of X_2.) There exists a point $g(z'), g \in G$ such that the arc $g(\lambda_0)$ intersects $\widetilde{\gamma}_2^+$ in z' and the curve

$$\sigma_2 = [z', g(z')] \bigcup \tau_2$$

consisting of an arc $[z', g(z')]$ of the orbit $\widetilde{\gamma}_2$ and the transversal τ_2 contains no congruent points. In a similar way one constructs the curve

$$\sigma_1 = [z, g(z)] \bigcup \tau_1$$

for the curve $\widetilde{\gamma}_1$; see [21] for the details. It can be readily seen now that σ_1 and σ_2 are the required curves which cover on M two simple closed transversal cycles X_1 and X_2 which are homotopic to each other.

Case (ii) when $\widetilde{\gamma}_i^+$ is a saddle separatrix is treated similarly. □

The proof of the following two lemmas is rather technical and we omit it, referring the interested reader to the original work [21].

Lemma 4.2.2 *Suppose that the T-flows π_1^t and π_2^t have no 2-separatrix saddles in the singularity sets. Let γ_1 and γ_2 be non-trivial recurrent semi-orbits of π_1^t and π_2^t, respectively, such that $\mu(\gamma_1) = \mu(\gamma_2)$. Then for each covering semi-orbit $\widetilde{\gamma}_1 \in p^{-1}(\gamma_1)$ having $x^* \in \partial\mathbb{H}$ as its ω-limit point there exists a semi-orbit $\widetilde{\gamma}_2 \in p^{-1}(\gamma_2)$ which also has x^* as its ω-limit point, and vice versa.*

Lemma 4.2.3 *Suppose that the assumptions of Lemma 4.2.2 hold. Let $\widetilde{\gamma}_1$ and $\widetilde{\gamma}_2$ any two semi-orbits of covering flows $\widetilde{\pi}_1^t$ and $\widetilde{\pi}_2^t$ which have the same ω-limit point $x^* \in \partial\mathbb{H}$. Then the following is true:*

(i) If $\widetilde{\gamma}_1$ is not a separatrix then $\widetilde{\gamma}_2$ is not a separatrix either. Moreover, $\widetilde{\gamma}_1$ and $\widetilde{\gamma}_2$ have the same α-limit point at the absolute;
(ii) If $\widetilde{\gamma}_1$ is an α-separatrix of a (saddle) singular point s_1 then also $\widetilde{\gamma}_2$ is an α-separatrix of some (saddle) singular point s_2. In this case for each $\omega-$ $(\alpha-)$ separatrix of s_2 there exists an $\omega-$ $(\alpha-)$ separatrix of s_1 which tends to the same $\alpha-$ $(\omega-)$ limit point at the absolute, and vice versa.

Consider the flows π_1^t and π_2^t under the premises of Lemma 4.2.2. By Lemma 4.2.1, π_1^t and π_2^t admit transversal cycles X_1 and X_2 lying in the same homotopy class. As before let $\{\widetilde{X}_1\}$ and $\{\widetilde{X}_2\}$ be the complete preimages of X_1 and X_2, and $\widetilde{X}_1$ and $\widetilde{X}_2$ the representatives of these two classes which have the common points x^* and x^{**} at the absolute.

Let us establish a one-to-one correspondence between the points of the curves $\widetilde{X}_1$ and $\widetilde{X}_2$. Suppose that $z_1 \in \widetilde{X}_1$ is a point and l_1 is an orbit of $\widetilde{\pi}_1^t$ through z. Since π_1^t is a T-flow at least one of the limit points of l_1 belongs to the absolute. According to Lemma 4.2.2, there exists a unique orbit l_2 of the flow $\widetilde{\pi}_2^t$ which has the same limit point at the absolute.

It can be easily seen that $\widetilde{X}_2 \cap l_2$ is non-empty and consists of a unique point z_2. Let us fix the mapping

$$\tau_{\widetilde{X}_1} : z_1 \longmapsto z_2.$$

(Clearly the mapping $\tau_{\widetilde{X}_1}$ is invertible with inverse $\tau_{\widetilde{X}_2}$ constructed similarly by replacing index 1 by 2. Thus $\tau_{\widetilde{X}_1}$ is a one-to-one mapping.)

Lemma 4.2.4 *The mapping $\tau_{\widetilde{X}_1} : \widetilde{X}_1 \to \widetilde{X}_2$ is a homeomorphism which satisfies the equation:*

$$g \circ \tau_{\widetilde{X}_1} = \tau_{g(\widetilde{X}_1)} \circ g, \qquad g \in G.$$

Proof. Let $z_1 \in \widetilde{X}_1$ be an arbitrary point and let $z_2 \in \tau_{\widetilde{X}_1}(z_1) \subset \widetilde{X}_2$. Let us show that $\tau_{\widetilde{X}_1}$ is continuous in z_1. Indeed, since π_2^t is transitive one finds in a ε-vicinity of z_2 a pair of points z_2' and z_2'' lying on a small π_2^t-transversal such that the orbits passing through these points are lifted to the orbits on $\mathbb{H}$ which hit the absolute at the points $x_{i'}^*$ and $x_{i''}^{**}$ for $i = 1, 2$. By Lemma 4.2.2

there exist orbits l_1 and l_2 of the flow $\widetilde{\pi}_1^t$ which have limit sets consisting of the above mentioned points. Since $(\partial^-\widetilde{X}_1, \partial^+\widetilde{X}_1 \| x_{i'}^*, x_{i''}^{**})$ then $\widetilde{X}_1 \cap l_i$ is non-empty for $i = 1, 2$. Denote the newly obtained points by z_1' and z_1'' and notice that the point z_1 lies 'between' the above points. Clearly, by choosing the $\varepsilon > 0$ sufficiently small one readily establishes the continuity of $\tau_{\widetilde{X}_1}$ at the point z_1. (In the same manner one may show that the inverse mapping $\tau_{\widetilde{X}_1}^{-1}$ is continuous.)

Now let us prove that for every $g \in G$ the 'commutation' equality holds. Indeed, let $z_1 \in \widetilde{X}_1$ be an arbitrary point and let $\tau_{\widetilde{X}_1}(z_1)$ be the corresponding point on $\widetilde{X}_2$. Denote by l_1 and l_2 the orbits of $\widetilde{\pi}_1^t$ and $\widetilde{\pi}_2^t$ through z_1 and $\tau_{\widetilde{X}_1}(z_1)$. Let x^* be their common limit point at the absolute.

By the action of the Fuchsian group the orbits l_1, l_2 tend to the orbits $g(l_1), g(l_2)$ for each $g \in G$ which have the common point $g(x^*)$ at the absolute. The points $z_1 = l_1 \cap \widetilde{X}_1$ and $\tau_{\widetilde{X}_1}(z_1) = l_2 \cap \widetilde{X}_2$ tend to the points $g(z_1) = g(l_1) \cap g(\widetilde{X}_1)$ and $g \circ \tau_{\widetilde{X}_1}(z_1) = g(l_2) \cap g(\widetilde{X}_2$.

On the other hand, by the transitivity of π_2^t there exists a unique orbit, $g(l_2)$, which has $g(x^*)$ as its limit point at the absolute. Therefore the formula $g \circ \tau_{\widetilde{X}_1} = \tau_{g(\widetilde{X}_1)} \circ g$ follows. □

In the sequel we need the following construction. Consider an arbitrary simple closed π_1^t-transversal curve Y on M and take an arbitrary preimage $\widetilde{Y}$ in the family $p^{-1}(Y)$. Then one can always find a point $z_0 \in \widetilde{Y}$ which is in common with a certain ω-separatrix l^1 of an equilibrium O^1 of the flow $\widetilde{\pi}_2^t$. This simple observation clarifies the importance of the lay-out of the separatrices in the unit disc $\mathbb{H}$. In general let $\{O^i | l_1^i, l_2^i, \ldots, l_{2n_1}^i\}$ be an enumeration of the $2n_i$ separatrices adjacent to the saddle O^i, which we call a *scheme of the equilibrium.* (Notice that all of the above separatrices have their 'traces' at the absolute.) The following lemma is evident and we omit its proof (see the details in [21]).

Lemma 4.2.5 *The separatrix $l_{2n_{q+1}}^{q+1}$ of an equilibrium O^{q+1} $(q = 1, \ldots, k)$ hits the transversal curve $\widetilde{Y}$ in a point w^q and this curve is the first transversal curve which it hits as time increases.*

Denote by Φ_q $(q = 1, 2, \ldots, k)$ the simply connected regions bounded jointly by the transversal arcs $[z_0^q, z_0^{q+1}], [z^q, w^q]$, the separatrices $l_1^q, l_2^q, l_1^{q+1}, l_{2n_{q+1}}^{q+1}$ and the equilibria O^q, O^{q+1}. (By Lemma 4.2.4 such regions exist.) Let $\overline{\Phi}_q$ be the closure of Φ_q. Now we set

$$\overline{D} = \bigcup_{q=1}^{k} \overline{\Phi}_q .$$

Lemma 4.2.6 *The open region Int $\overline{D}$ is a fundamental region of the Fuchsian group G.*

Proof. (I) First let us show that in the region Int $\overline{D}$ there are no congruent points. Suppose to the contrary that there are such points, say z_1 and z_2. Then the orbits l_{z_1} and l_{z_2} of the flow $\widetilde{\pi}^t$ passing through these points are congruent to each other. Thus they cover one and the same orbit γ of the flow π^t on the surface M which passes through the point $z = p(z_1) = p(z_2)$.

The orbits l_{z_1} and l_{z_2} intersect the transversal curve $\widetilde{X}$ at some points z_1' and z_2' so that the arcs (z_1, z_1') and (z_2, z_2') of the orbits l_{z_1} and l_{z_2} do not intersect any other transversal curve from the family $\{\widetilde{X}\}$. Therefore, for times $t < t_0$, the orbit γ on M intersects the π^t-transversal cycle X at a point z' which is an image of the points z_1', z_2'. Therefore the points z_1' and z_2' are congruent, which is a contradiction since by the construction there are no congruent points on the arc $[z_0^1, z_0^{k+1}]$ of the curve $\widetilde{X}$; see Lemma 4.2.5.

(II) Let us show that the boundary $\partial\overline{D}$ of the region Int $\overline{D}$ consists of a finite number of pairwise congruent simple arcs. First consider $2(k+1)$ separatrices which cross the boundary $\partial\overline{D}$. They have the following ordering:

$$l_1^1, l_2^1, l_{2s_2}^2, l_2^2, \ldots, l_{2s_{k+1}}^{k+1}, l_1^{k+1},$$

where l_1^1 and its congruent l_1^{k+1} are ω-separatrices, and all the rest are α-separatrices. (The congruence relation is denoted by $\sim$.) Since $l_1^1 \sim l_1^{k+1}$, we have $O^1 \sim O^{k+1}$.

Notice that the action of the Fuchsian group preserves the cyclic ordering of separatrices adjacent to the singular point, $l_{2s_r}^r \sim l_2^1$ and the arc (O^r, w^{r+1}) of the separatrix $l_{2s_r}^r$ belongs to $\partial\overline{D}$ and $l_{2s_r}^r$ is a unique separatrix congruent to l_2^1 with an arc in $\partial\overline{D}$.

Taking a separatrix $l_{2s_{k+1}}^{k+1}$ one may find a unique point $z_0^p, (p = 2, 3, \ldots, k)$ on the arc (z_0^1, z_0^{k+1}) from which an ω-separatrix l_1^p of the equilibrium O^p, issues so that $l_1^p \sim l_{2s_{k+1}-1}^{k+1}$. Therefore $l_2^p \sim l_{2s_{k+1}}^{k+1}$ and the arc (O^p, z^p) of the separatrix l_2^p lies in $\partial\overline{D}$.

Now let O^i be an arbitrary equilibrium which enters the boundary $\partial\overline{D}$ and is different from O^1 and O^{k+1}. Suppose also that $l_2^i, l_{2s_i}^i$ are its α-separatrices, the arcs $(O^i, z^i), (O^i, w^{i-1})$ of which belong to $\partial\overline{D}$. Repeating the argument with regard to the separatrices l_2^1 and l_{2s_i}, we conclude that for either of them there exists a unique congruent separatrix the arc of which belongs to $\partial\overline{D}$. □

Now we are in position to complete the proof of the main statement. According to Nielsen [214] every automorphism $h_* : G \to G$ of the Fuchsian group defines a homeomorphism $h : M \to M$ of the surface M which induces h_*. (In other words, $h_* : g \mapsto \tilde{h} \circ g \circ \tilde{h}^{-1}$ where $\tilde{h}$ is a 'lifting' of h to the universal covering $\mathbb{H}$.) By a slight change in the proof [214] the homeomorphism h can always be supposed a diffeomorphism. Again due to Nielsen [215], the automorphism h_* is uniquely extended to the homeomorphism $h^* : \partial\mathbb{H} \to \partial\mathbb{H}$ of the absolute.

Let us denote be π_3^t a flow on M which is conjugate by a diffeomorphism d to the flow π_1^t. Let $\widetilde{\pi}_3^t$ be the corresponding covering flow on $\mathbb{H}$. Since for the semi-orbits γ_1 and γ_2 of π_1^t and π_2^t we have $\mu(\gamma_2) = d^*(\mu(\gamma_1))$ then for the semi-orbits γ_2 and $\gamma_3 = d(\gamma_1)$ we obtain $\mu(\gamma_2) = \mu(\gamma_3)$.

For the proof of the theorem it is sufficient to construct an auxiliary homeomorphism $\phi : M \to M$ which sends the orbits of the flow π_3^t into the orbits of the flow π_2^t. The required homeomorphism h which sends the orbits of π_1^t into the orbits of π_2^t will be a composition $h = \phi \circ g$ of the homeomorphisms ϕ and d. By an abuse of notation we simply write π_3^t as π_1^t, so that for π_1^t and π_2^t the condition $\mu(\gamma_1) = \mu(\gamma_2)$ holds, and we shall search for a homeomorphism $\phi : M \to M$ which sends the orbits of π_1^t into the orbits of π_2^t.

Note that reversing the time one obtains topologically conjugate flows, so we may suppose that γ_1 and γ_2 are positive semi-orbits of π_1^t and π_2^t. Let X_1 and X_2 be transversal cycles of π_1^t and π_2^t which satisfy Lemma 4.2.1 and let $\widetilde{X}_1 \in p^{-1}(X_1)$ and $\widetilde{X}_2 \in p^{-1}(X_2)$ be the transversal curves on $\mathbb{H}$ which have the common points x^* and x^{**} at the absolute.

We retain the notation of Lemmas 4.2.5 and 4.2.6. Let Int $\overline{D}$ be a fundamental region of the Fuchsian group which is bounded by the arcs of orbits and transversal arcs of the flow π_1^t. According to Lemma 4.2.4 the homeomorphism $\tau_{\widetilde{X}_0} : \widetilde{X}_0 \to \widetilde{X}_1$ sends the points $z_0^1, \dots, z_0^{k+1}$ of $\widetilde{X}_0$ into the points $z_1^1, \dots, z_1^{k+1}$ of the curve $\widetilde{X}_1$. According to the same lemma the homeomorphism $\tau_{\widetilde{X}_0}$ satisfies the equality $g \circ \tau_{\widetilde{X}_0} = \tau_{g(\widetilde{X}_0)} \circ g$, $g \in G$ and therefore $z_0^1 \sim z_1^{k+1}$, so that there are no congruent points on the arc (z_1^1, z_1^{k+1}).

By the construction of the homeomorphism $\tau_{\widetilde{X}_0}$ the ω-separatrices $l_2^1, \dots, l_2^{k+1}$ of the equilibria $O_2^1, \dots, O_2^{k+1}$ of the flow $\widetilde{\pi}_2^t$ pass through the points $z_1^1, \dots, z_1^{k+1}$, respectively. It is not hard to prove (see [21] for details) that the arcs (z_2^i, O_2^i) of the separatrices l_2^i cannot intersect any of the $\widetilde{\pi}_2^t$-transversal curves $\widetilde{X}_2$.

Notice that by Lemma 4.2.3 the flows π_1^t and π_2^t have the same number of equilibria and separatrices, so that the arc (z_1^1, z_1^{k+1}) contains no points through which the ω-separatrices of $\widetilde{\pi}_2^t$ may pass. Therefore one may construct, by analogy with the region Int $\overline{D}_1$ corresponding to the flow $\widetilde{\pi}_1^t$, the region Int $\overline{D}_2$, which is bounded by the transversal curves $\widetilde{X}_2^1, \dots, \widetilde{X}_2^k$ from $\{\widetilde{X}_2\}$ and the separatrices $l_1^1, l_2^1, \dots, l_{2s_{k+1}}^{k+1}, l_1^{k+1}$ and the equilibria $O_2^1, O_2^2, \dots, O_2^{k+1}$ of the flow $\widetilde{\pi}_2^t$. (Notice that the ordering of separatrices in Int $\overline{D}_2$ is chosen to be the same as for Int $\overline{D}_1$.)

As before let z_2^i, w_2^i be the intersection points of the transversal curve $\widetilde{X}_2^i$ with the separatrices l_2^i and $l_{2s_{i+1}}^{i+1}$. It can be derived from the construction [21] that the endpoints $\partial^\pm \widetilde{X}_2^i$ coincide with the endpoints $\partial^\pm \widetilde{X}_1^i$. Moreover, the arc (z_2^i, w_2^i) is the image of the arc (z_1^i, w_1^i) under the action of the homeomorphism $\tau_{\widetilde{X}_1^i}$.

Denote by $l_1^1, l_2^1, \ldots, l_{2(k+1)}^1$ the separatrices of π_1^t the arcs of which cross the boundary of $\overline{D}_1$. Suppose that the subset $l_1^1, l_2^1, \ldots, l_{k+1}^1$ is formed by non-congruent separatrices. Each separatrix from this subset hits at the point z_i a transversal curve of $\widetilde{\pi}_1^t$ crossing the boundary of Int $\overline{D}_1$. Now let us define an arbitrary homeomorphism

$$\tau_i : l_i^1 \longrightarrow l_i^2$$

between the separatrices l_i^1 of the flow π_1^t and the similarly defined separatrices l_i^2 of the flow π_2^t. (The above homeomorphism τ_i is extended by continuity on the set of equilibria O_1^i.) For the rest of the separatrices $l_{k+2}^1, \ldots, l_{2(k+1)}^1$, τ_i is extended by the action of the Fuchsian group:

$$\tau_{k+1+i} = g_i \circ \tau_i \circ g_i^{-1},$$

where $g_i \in G$ is an element of the Fuchsian group sending l_i^1 to l_{k+1+i}^1. Thus one obtains a homeomorphism $\tau : \partial\overline{D}_1 \to \partial\overline{D}_2$.

Notice that the regions Int $\overline{D}_1$ and Int $\overline{D}_2$ are 'foliated' by the orbits of $\widetilde{\pi}_1^t$ and $\widetilde{\pi}_2^t$ which cut the boundaries of the above regions. But in this case it is always possible to extend τ to Int $\overline{D}_1$ by the respective flows. Finally one extends τ to all of $\mathbb{H}$ by the formula:

$$\tau(z) = g \circ \tau \circ g^{-1}(z), \qquad g \in G,$$

where g is chosen so that $g^{-1}(z) \in \overline{D}_1$. Now recall that Int $\overline{D}_1$ is a fundamental region and therefore τ commutes with the G-actions: $\tau \circ g = g \circ \tau$. This implies that the mapping $\phi : M \to M$ is a homeomorphism sending the orbits of the flow π_1^t to the orbits of the flow π_2^t such that $\phi \circ p = p \circ \phi$, where $p : \mathbb{H} \to M$ is the covering mapping. □

Foliations With Minimal Set of Cantor Type A minimal set is called *non-trivial (exceptional)* if it is not a fixed point, a closed trajectory, or the whole surface M. In accordance with the list of minimal sets, an exceptional minimal set is nowhere dense and consists of a continuum of non-trivially recurrent leaves, each being dense in the minimal set. An exceptional minimal set is locally homeomorphic to the product of the Cantor set and a segment. Two minimal sets N_1, N_2 of the foliations $\mathcal{F}_1, \mathcal{F}_2$, respectively, are *topologically equivalent* if there exists a homeomorphism $\varphi : M \to M$ such that $\varphi(N_1) = N_2$ and φ maps the trajectories of N_1 onto the trajectories of N_2.

Let N be an exceptional minimal set of the foliation $\mathcal{F}$ on the surface M. A pair of trajectories $l_1, l_2 \subset N$ is called *special* if there exists a simply connected component Ω of $M \setminus N$ such that $\partial\Omega = l_1 \cup l_2$ (here $\partial\Omega$ is an accessible boundary of Ω).

Theorem 4.2.4 *([19],[20]) Let M be a closed orientable surface of genus $g \geq 2$. For $i = 1, 2$ let N_i be an exceptional minimal set of the foliation $\mathcal{F}_i$ without special pairs of trajectories. Then the minimal sets N_1, N_2 are topologically equivalent if and only if there exist semi-trajectories $l_1 \subset N_1, l_2 \subset N_2$ with the same homotopy rotation class. In particular, N_1, N_2 are topologically equivalent via a homeomorphism homotopic to the identity if and only if there are two semi-trajectories $l_1 \subset N_1$, $l_2 \subset N_2$ with the same homotopy rotation class.*

Proof. In view of Nielsen's results [213],[214]-[216], it is enough to prove the second part of the theorem. As before, one constructs two homotopic closed transversals C_1, C_2 to $\mathcal{F}_1$, $\mathcal{F}_2$, respectively. (C_1 and C_2 are not homotopic to zero.) Using C_1, C_2 one constructs canonical neighbourhoods $\Omega(N_1)$, $\Omega(N_2)$ of the sets N_1, N_2, respectively. First, a homeomorphism $\Omega(N_1) \to \Omega(N_2)$ is constructed. Next it is prolonged to the whole surface M. □

4.2.3 Properties of the Homotopy Rotation Class

All non-trivially recurrent semi-trajectories on the torus have the same asymptotic direction. This is not the case for orientable closed surfaces M of genus $g \geq 2$.

Theorem 4.2.5 *([18]) Suppose that foliation $\mathcal{F}$ (not necessarily transitive) on M has a non-trivially recurrent semi-leaf and let $\widetilde{\mathcal{F}}$ be the covering flow on $\mathbb{H}$. Then any point $\sigma \in \partial\mathbb{H}$ is the limit set for at most two semi-leaves of $\widetilde{\mathcal{F}}$ which cover non-trivially recurrent semi-leaves. Moreover, if $\mathcal{F}$ is transitive then any point of the absolute is the limit set for at most one semi-leaf of $\widetilde{\mathcal{F}}$.*

Proof. Suppose the contrary. Then there exist at least three semi-leaves $\widetilde{l}_1, \widetilde{l}_2, \widetilde{l}_3$ such that $\omega(\widetilde{l}_1) = \omega(\widetilde{l}_2) = \omega(\widetilde{l}_3) = \sigma \in \partial\mathbb{H}$. Suppose that $\widetilde{l}_2$ lies 'between' $\widetilde{l}_1$ and $\widetilde{l}_3$. Since the projection $\pi(\widetilde{l}_2)$ is non-trivially recurrent, there exists a non-trivially recurrent trajectory $\widetilde{l}$ congruent to $\widetilde{l}_2$ such that $\omega(\widetilde{l}) = \sigma$. Then σ is a rational point of the absolute, which is a contradiction. □

Cherry's theorem [77] says that a foliation $\mathcal{F}$ on M has a continuum of non-trivially recurrent leaves provided $\mathcal{F}$ has at least one non-trivially recurrent semi-leaf. Therefore there is a continuum of points $\sigma_\alpha \in \partial\mathbb{H}$ belonging to limit sets of trajectories of the covering foliation $\widetilde{\mathcal{F}}$. This is in contrast to the case of the torus. However, the set of points σ_α has Lebesgue measure zero [17].

A point $\sigma \in \partial\mathbb{H}$ is called *attainable* if there exists a foliation $\mathcal{F}$ on M and a covering foliation $\widetilde{\mathcal{F}}$ on $\mathbb{H}$ such that σ belongs to the limit set (α– or ω–) of a semi-trajectory of $\widetilde{\mathcal{F}}$. The set of all attainable points will be denoted by $A(G)$. In other words

$$A(G) = \bigcup(\lim(\widetilde{\mathcal{F}}) \cap \mathbb{H}),$$

where the union is taken over all covering foliations $\widetilde{\mathcal{F}}$.

Theorem 4.2.6 *([17]) Let $M = \mathbb{H}/G$ be a closed orientable surface of genus $g \geq 2$. Then the set $A(G)$ has zero Lebesgue measure on the absolute.*

Proof. Following G. Hedlund [142] we call a geodesic $\widetilde{L} \subset \mathbb{H}$ *transitive* if for any open intervals $U_1, U_2 \subset \partial\mathbb{H}$ there exists an element $g \in \Gamma$ such that one endpoint of $g(\widetilde{L})$ lies in U_1 and the other in U_2. Let $Tr(\Gamma)$ denote the set of endpoints of all transitive geodesics. G. Hedlund proved that the set $Tr(\Gamma) \subset \partial\mathbb{H}$ has full measure on the absolute. The theorem follows from the inclusion $Tr(\Gamma) \subset \partial\mathbb{H} \setminus A(\Gamma)$. Note that this inclusion is proper; see [17]. □

4.3 Non Orientable Surfaces

Denote by N_p a closed non orientable surface of genus $p \geq 1$. As has been noted, transitive orientable foliations exist only on the surfaces N_p, $p \geq 3$. The surface N_3 (a torus with a cross-cap) plays a fundamental role in the topological classification of such foliations. There exists an invariant on N_3 similar to the Poincaré rotation number, but there are no irrational foliations. The surfaces N_p, $p \geq 4$, however, admit irrational foliations. Therefore, the cases $p = 3$ and $p \geq 4$ are treated separately. Almost all the results of this section are obtained in [25].

4.3.1 Torus With the Cross-Cap

Lemma 4.3.1 (A. Katok) *Let $\mathcal{F}$ be a transitive foliation without fake saddles on N_3. Then $\mathcal{F}$ has exactly one fixed point, which is a saddle with four separatrices. Moreover, two separatrices of this saddle make a loop with a neighbourhood homeomorphic to the Möbius strip.*

Proof. By transitivity, $\mathcal{F}$ only has fixed points of saddle type. The index argument implies that $\mathcal{F}$ has a unique saddle of index -1. Again by transitivity, there exists a simple two-sided transversal C which is not contractable. Moreover, the return mapping $P : C \to C$ is defined everywhere except at points of intersection with ω-separatrices. Cutting along C we obtain a non orientable surface N of genus 1 with two boundary components C_1, C_2. Suppose that the separatrices do not form a loop. Then, by the non-orientability of N, one of the segments $(a, b), (b, a)$ must have a flip. Since P has no periodic points, it must have a wandering interval. This contradicts the transitivity of $\mathcal{F}$. □

The surface N_3 is produced from the torus T^2 by the amalgamation of the Möbius strip. Therefore there exist a covering N_∞ for N_3 which is the Euclidean plane with a countable number of the cross-caps which are congruent to each other by deck transformations. Lemma 4.3.1 allows us to assign to a transitive flow $\mathcal{F}$ on N_3 a number *rot* $(\mathcal{F})$ which is similar to the Poincaré rotation number. Notice that *rot* $(\mathcal{F})$ is defined with the help of a *point at the absolute*. By transitivity, *rot* $(\mathcal{F}) \in \mathbb{R}\backslash\mathbb{Q}$. The following two theorems solve the classification problem for the transitive foliations on N_3.

Theorem 4.3.1 (Aranson-Telnykh-Zhuzhoma) *Two transitive foliations without fake saddles on N_3 are topologically equivalent if and only if their rotation numbers are modular equivalent (i.e., are equivalent under transformation by an integer unimodular matrix).*

Theorem 4.3.2 (Aranson-Telnykh-Zhuzhoma) *For every $\alpha \in \mathbb{R}\backslash\mathbb{Q}$ there exists a transitive foliation $\mathcal{F}$ without fake saddles on N_3 such that rot $(\mathcal{F}) = \alpha$.*

Proof. The proof of both statements proceeds in the same way as for the case of the torus. □

4.3.2 Surfaces of Genus $p \geq 4$

As in the case of orientable surfaces, one can introduce the homotopy rotation class for transitive foliations on non orientable surfaces.

Theorem 4.3.3 (Aranson-Telnykh-Zhuzhoma) *Two transitive foliations on the surface $N_p, p \geq 4$ are topologically equivalent if and only if there exists a pair of recurrent semi-leaves whose homotopy rotation class is the same.*

Proof. This is similar to the proof given for transitive foliations on orientable surfaces. However, the construction of cross-sections is more elaborate since the group of deck transformations is larger than that for the orientable case. □

4.4 Discrete Invariants

4.4.1 Regular Foliations on the Sphere

A foliation is said to be regular if all of the recurrent leaves are trivial. In the case of foliations given by polynomial vector fields in the plane, E. A. Leontovich and A. G. Maier [180] singled out a family of trajectories (the so–called singular trajectories) which determine the topological structure of such foliations. We call these polynomial foliations on S^2.

Let $\mathcal{F}$ be a polynomial foliation on S^2 and let $l(x)$ be a leaf through $x \in S^2$. We denote by $N_\epsilon(A)$ the ϵ-neigbourhood of $A \subset S^2$. The leaf $l(x_0)$ is said to be *orbitally unstable* if there exists $\varepsilon_0 > 0$ such that for every $\delta > 0$ there is a point $x \in O_\delta(x_0)$ such that the trajectory $l(x)$ abandons the ε_0-neigbourhood of $l(x_0)$.

A polynomial foliation with a finite number of fixed points can only contain the following orbital unstable trajectories: fixed points, separatrices and limit cycles (one-sided at least). Such trajectories separate the regions filled by the trajectories with "similar" behaviour. They therefore play a key rôle in the topological classifications of foliations on the sphere. Denote by $B(\mathcal{F})$ a collection of orbital unstable trajectories of the foliation $\mathcal{F}$.

Theorem 4.4.1 *Suppose that v_1, v_2 are polynomial vector fields with a finite number of fixed points on the sphere. Then the corresponding foliations $\mathcal{F}_1$, $\mathcal{F}_2$ are topologically equivalent if and only if there exists a homeomorphism $\varphi : S^2 \to S^2$ such that:*

(i) $\varphi(B(\mathcal{F}_1)) = B(\mathcal{F}_2)$.
(ii) Each trajectory in $B(\mathcal{F}_1)$ is sent by φ to a trajectory in $B(\mathcal{F}_2)$.
(iii) φ sends one trajectory from each component of $S^2 \backslash B(\mathcal{F}_1)$ into a trajectory from $S^2 \backslash B(\mathcal{F}_2)$.

Proof. This is left to the reader. □

The family $B(\mathcal{F})$ (together with a set of leaves containing one leaf from each component of $S^2 \backslash B(\mathcal{F})$), is a complete topological invariant for orientable foliations on S^2. Note that the assumption of polynomiality in Theorem 4.4.1 is not essential: it was made to ensure the finiteness of family $B(\mathcal{F})$.

4.4.2 Orbit Complex

In 1976 D. Neumann and T. O'Brien [212] introduced a complete topological invariant for the regular orientable foliations on surfaces, called an *orbit complex.*

Orientable Case Let *Sing* $(\mathcal{F})$ denote the union of all singular trajectories of the foliation $\mathcal{F}$. Let R be a component of the set $M \backslash Sing\ (\mathcal{F})$. As was shown earlier, $\mathcal{F}|_R$ is topologically equivalent to one of the following types:

(i) A parallel strip foliation $\mathcal{F}_{St}$;
(ii) A parallel annular foliation $\mathcal{F}_{An}$;
(iii) A parallel spiral foliation $\mathcal{F}_{Sp}$;
(iv) A linear rational foliation $\mathcal{F}_{Rat}$.

Let us introduce a step-by-step ordering of the trajectories in the described above elementary regions, which will yield us the required invariant.

(1) Suppose that $\mathcal{F}|_R$ is orbitally topologically equivalent to $\mathcal{F}_{St}$. Every leaf $l \subset R$ divides R into the so-called *half canonical regions* R^+, R^-. Let $\delta R^{\pm}$ denote the accessible boundary of $R^{\pm}$ taken without the trajectory l. The one-dimensional leaves of δR^+ (or δR^-) can be ordered as follows. Let $l_1, l_2 \subset \delta R^+$ (or δR^-) be one-dimensional trajectories, $l_1 \neq l_2$. We write $l_1 < l_2$ if, for any two given disjoint transversal segments $\Sigma_1, \Sigma_2 \subset R^+ \cup \delta R^+$ any trajectory of R^+ intersects Σ_2 before Σ_1.

(2) Suppose that $\mathcal{F}|_R$ is topologically equivalent to $\mathcal{F}_{An}$. One can argue as in item (1) provided we first distinguish a one-dimensional trajectory of $\delta R^{\pm}$ as initial in the above ordering.

(3) Suppose $\mathcal{F}|_R$ is topologically equivalent to $\mathcal{F}_{Sp}$. Then $\mathcal{F}_{Sp}$ has a closed transversal C which divides R into the half regions R^+, R^-. All one-dimensional leaves of $\delta R^{\pm}$ (without the curve C) can be ordered as in item (1) provided we choose some leaf of $\delta R^{\pm}$ as initial.

Again let $\mathcal{F}$ be an orientable regular foliation on the orientable surface M. Let $M/\mathcal{F}$ be the quotient space (with the usual quotient topology) obtained by collapsing the leaves of $\mathcal{F}$ to points. To obtain a complete topological invariant we must add extra structure to the space $M/\mathcal{F}$. The invariant is denoted by $K(\mathcal{F})$. Let $p : M \to M/\mathcal{F}$ denote a natural projection in the quotient space. The image of each component R of $M \setminus Sing\ \mathcal{F}$ is declared to be a 1-cell of $K(\mathcal{F})$. The latter is an open arc if $\mathcal{F}|_R = \mathcal{F}_{St}$, $\mathcal{F}_{An}$, and a simple closed curve if $\mathcal{F}|_R = \mathcal{F}_{Sp}, \mathcal{F}_{Rat}$. Let it call an *open* and a *closed* 1-cell, respectively. The image of each singular leaf is declared to be a 0-cell of $K(\mathcal{F})$.

For $i = 0; 1$ each i-cell r is called a *line, circle* or *point i-cell* according to the topological type (open leaf, periodic leaf or fixed point) of the image $p^{-1}(x)$ of a point $x \in r$. (It is easy to see that this definition is independent of the choice of points $x \in r$.) The order defined on one-dimensional singular leaves in the accessible boundary of each canonical half-region $R^{\pm}$'s induces an order on the line and the circle 0-cells. We extend this order to the point 0-cells as follows.

Suppose that $p(l_0)$ is a point 0-cell and $p(l)$ is a line 0-cell. Let $p(l_0) < p(l)$ if $l_0 \in \alpha(l)$ and $p(l_0) > p(l)$ if $l_0 \in \omega(l)$. Otherwise $p(l), p(l_0)$ are not related. Suppose $\mathcal{F}|_R$ be topologically equivalent to $\mathcal{F}_{Sp}$. Then $p(R)$ is a closed 1-cell. All trajectories of R have the same ω-limit set $\omega(R)$ and the same α-limit set $\alpha(R)$. For any trajectories $l_1 \subset \omega(R) \cap \delta R$, $l_2 \subset \alpha(R) \cap \delta R$ we specify $p(l_1) < p(R) < p(l_2)$. A closed 1-cell $p(R)$ of $K(\mathcal{F})$ corresponding to the parallel spiral flow $\mathcal{F}|_R$ is said to be orientable or non-orientable.

By an *orbit complex* $K(\mathcal{F})$ of $\mathcal{F}$ one understands a factor-space $M/\mathcal{F}$ endowed with a cell structure, a fibre-type structure, an order structure, (and on some closed cells), the orientability type. Two orbit complexes $K(\mathcal{F}_1), K(\mathcal{F}_2)$ are *isomorphic* if there exists a homeomorphism $h : K(\mathcal{F}_1) \to K(\mathcal{F}_2)$ which preserves all of the above structures and which preserves the orientability of closed 1-cells. A one-dimensional singular trajectory l is called a *limit singular leaf* if $l \subset cl(Sing\ (\mathcal{F}) \setminus l)$.

Theorem 4.4.2 (Neumann-O'Brien) *Suppose $\mathcal{F}_1, \mathcal{F}_2$ are two regular foliations on an orientable surface M with isolated fixed points and no limit singular leaves. Then $\mathcal{F}_1, \mathcal{F}_2$ are topologically equivalent if and only if $K(\mathcal{F}_1), K(\mathcal{F}_2)$ are isomorphic.*

Proof. This follows from the construction of the orbit complex; see [212] for the details. □

Theorem 4.4.3 (Neumann-O'Brien) *Suppose that $\mathcal{F}_1, \mathcal{F}_2$ are two regular foliations on an orientable surface M having finitely many one-dimensional singular leaves. Then $\mathcal{F}_1, \mathcal{F}_2$ are topologically equivalent if and only if $K(\mathcal{F}_1), K(\mathcal{F}_2)$ are isomorphic.*

Proof. This is a modification of the proof of Theorem 4.4.2. □

Non-Orientable Case We now assume that M is a non-orientable surface and $\mathcal{F}$ an orientable foliation on M. Rather then directly define the structure of the space $M/\mathcal{F}$, D. Neumann and O'Brien take a simpler way by reducing the definition to the orientable case. Let $\widetilde{M}$ be an orientable 2-fold covering manifold of M and let $q : \widetilde{M} \to M$ be the corresponding covering map. Let $\tilde{\mathcal{F}}$ denote a lifting of $\mathcal{F}$ to $\tilde{M}$. Then q induces a map $\tilde{q}(\mathcal{F}) : \tilde{M}/\tilde{\mathcal{F}} \to M/\mathcal{F}$.

The orbit complex $K(\mathcal{F})$ is then defined to be the quotient space $M/\mathcal{F}$ and the structure on $M/\mathcal{F}$ is introduced via the following isomorphism: $K(\mathcal{F}_1)$ and $K(\mathcal{F}_2)$ are *isomorphic* if there exist homeomorphisms

$$h : M/\mathcal{F}_1 \to M/\mathcal{F}_2, \quad \tilde{h} : M/\tilde{\mathcal{F}}_1 \to M/\tilde{\mathcal{F}}_2$$

such that $q(\mathcal{F}_2) \circ \tilde{h} = h \circ q(\mathcal{F}_1)$. Then Theorems 4.4.2, 4.4.3 are valid for the non-orientable surfaces M as well.

4.5 Foliations Without Holonomy

Foliations $\mathcal{F} = (M, \pi^t, \theta)$ without holonomy are represented by the non-wandering flows π^t on M. The latter can be effectively classified by the 'topology' of their separatrix contours; such contours are known to be the boundaries of the 'elementary cells' of non-wandering flows. The equivalence relation adopted to the situation is the so-called *graph–equivalence* relation; roughly speaking, this relation conjugates two non-wandering flows near their separatrix contours and 'forgets' equivalence between their quasiminimal sets. Of course, graph-equivalence of flows is weaker than topological equivalence. We consider both orientable and non orientable surfaces.

Our starting point will be the decomposition theorem, which says that 'elementary cells' of non-wandering flow π^t can border each other either at the separatrix connections of its saddle singular points or at the set of saddle points itself. Here a 1-dimensional CW-complex (similar to the 'Peixoto graph' of the Morse-Smale flow) arises; the vertices of the latter are saddle points and the edges are separatrix connections of the flow π^t. In contrast to the Morse-Smale class these graphs are not 'cellularly' embedded into M (since the quasiminimal sets are non-planar).

4.5.1 Cells

Recall that a point $x \in M$ is said to be a non-wandering point of the flow π^t if, for any neighbourhood U of x and any positive number $T > 0$, $U \cap \pi^t(U) \neq \emptyset$ for a time $|t| > T$. A flow is said to be *non-wandering* if all points of M are non-wandering. For the sake of convenience, we reproduce a decomposition theorem for non-wandering flows taken from an earlier chapter.

Proposition 4.5.1 *Let M be a compact surface, orientable or non-orientable. If π^t is a non-wandering flow on M, then M can be represented as the union $M = M_1 \cup \ldots \cup M_k$ of regions with pairwise disjoint interiors such that the boundary of each M_i is a union of separatrix contours of π^t. The interior of each M_i is either of the following:*

(i) a disc region with centre-type singularity;

(ii) an annulus partitioned by periodic trajectories;

(iii) a Möbius strip partitioned by periodic trajectories;

(iv) a quasiminimal set $\mathcal{Q}_i$.

Moreover, there are four exceptional cases: M_i is a sphere with the equatorial flow (in this case $M = S^2$); a projective plane P^2 with a centre, partitioned by periodic trajectories (in this case $M = P^2$); a torus T^2 partitioned by periodic trajectories (in this case $M = T^2$); or, finally, the Klein bottle K^2 partitioned by periodic trajectories (in this case $M = K^2$).

It is not hard to see that item (iii) happens only when M is a non-orientable surface. Each quasiminimal set of item (iv) can be blown-down (by squeezing its boundary) to a highly transitive flow; since there exist countably many highly transitive flows, the same is true for the 'elementary cells' of non-wandering flows. This fact is in sharp contrast to the Morse-Smale flows, where only a finite number of 'elementary cells' are possible.

Note that non-wandering flows cannot have an exceptional quasiminimal set, i.e., sets homeomorphic to the Cantor set multiplied by $\mathbb{R}$. (The latter can be obtained from a $\mathcal{Q}$-set by the blowing-up of a finite number of trajectories.) If they did, the points lying in the complement of Cantor cells would not satisfy the condition $U \cap \phi^t(U) \neq \emptyset$, so that ϕ^t would have wandering points.

4.5.2 Classification of Elementary Cells

The boundary of a quasiminimal set consists in general of several connected components. By an *elementary cell* of a quasiminimal set one understands a connected component of its boundary. Unlike the Morse–Smale flows, the elementary cells of the quasiminimal set cannot bound a disc region.

To give an idea of the variety of elementary cells, let us consider the following construction. If ϕ^t is a highly transitive flow on M, any of its

$2n$-prong saddles can be blown-up to a circle with the $2n$ saddle points on it. The resulting manifold with the boundary $\partial M \simeq S^1$ can be interpreted as a quasiminimal set $\mathcal{Q}$ of the modified flow $\tilde{\phi}^t$, supported by the region $M \cup \partial M$. The elementary cell of $\mathcal{Q}$ coincides in this case with the boundary ∂M. Moreover, we have to allow the saddle points at ∂M to move freely and merge in different (combinatorial) ways. These operations give us new elementary cells. Roughly speaking, the classification of elementary cells of the $\mathcal{Q}$–sets boils down to a calculation of the blowing-ups and the homotopies of the $2n$-prong saddle points of the highly transitive flows. The result can be expressed by the following lemma.

Lemma 4.5.1 **([226])** *Let $\mathcal{Q}$ be a quasiminimal set obtained from a highly transitive flow on a compact surface M by the blowing-up of the saddle points. Each elementary cell is either a $2n$-gon, $n \geq 1$, or obtained from a $2n$-gon by the contraction of a certain number of its sides.*

Proof. According to item (iv) of Proposition 4.5.1 and the remark above, every $\mathcal{Q}$-set is produced from a highly transitive flow on a surface of genus $g > 0$. Such a flow has $2n$-prong saddles as the singularity set. Taking a certain number of the saddles and then blowing them up, one gets a quasiminimal set whose boundary consists of $2n$-gons (n can vary). A symbolic description of the elementary cells arising in this way is given by the formula:

$$E = E_{\underbrace{11\ldots1}_{2n}}$$

The sides of the $2n$-gons are allowed to contract in different (combinatorial) manners, which gives us a bunch of new elementary cells. To sort the possible cases, we write down these cells in the form

$$E = E_{\underbrace{m_1 m_2 \ldots m_k}_{m_1+m_2+\ldots+m_k=2n}},$$

where $m_1 m_2 \ldots m_k$ is a sequence of positive integers defined up to a cyclic permutation such that every m_j is equal to the 'multiplicity' of vertex of the contracted polygon. The order between m_j corresponds to the order of vertices in the polygon. □

4.5.3 Amalgamation of Elementary Cells

The elementary cells are the building blocks for 'non-elementary' cells, which are obtained as superposition of the elementary cells. Not every pair of elementary cells can be amalgamated together. To answer the question of when two elementary cells can be amalgamated (in at least one way), one needs the notion of a *normal form*.

Definition 4.5.1 ([226]) *The elementary cells $E_{m_1 m_2 \dots m_k}$ and $E_{m'_1 m'_2 \dots m'_k}$ are said to be equivalent if and only if*

$$\begin{aligned} m'_1 &= m_1 \ mod\ 2 \\ m'_2 &= m_2 \ mod\ 2 \\ &\vdots \\ m'_k &= m_k \ mod\ 2. \end{aligned}$$

By a normal form of the elementary cell $E_{m_1 m_2 \dots m_k}$ one understands an equivalent cell whose subscripts are either 0 or 1. Moreover, we drop in the normal form the subscripts 0 and we complete our list by the cell $E = E_{00\dots0}$, which we call a Novikov–Levitt cell [7] *.*

Lemma 4.5.2 ([226]) *Two elementary cells can be amalgamated if and only if they have the same normal form.*

Proof. Let $E_{m_1 m_2 \dots m_k}$ be an elementary cell of Table 1. The vertex $v \in E_{m_1 m_2 \dots m_k}$ is called *semi-stable* if it does not change the direction of flow on the adjacent edges. The vertex is semi-stable if and only if the multiplicity of the vertex is

$$m_j = 0 \ mod\ 2.$$

All other vertices of $E_{m_1 m_2 \dots m_k}$ will be *stable* in the sense that they change the orientation of flow on the adjacent edges. This happens if and only if the multiplicity is

$$m_j = 1 \ mod\ 2.$$

If one eliminates the semi-stable vertices in the elementary cell and assigns the subscript 1 to every stable vertex, then one gets the elementary cell $E_{2n} = E_{11\dots1}$ or the Novikov–Levitt cell.

The amalgamation of two cells $E_{m_1 m_2 \dots m_k}$ and $E_{m'_1 m'_2 \dots m'_k}$ is possible if and only if their normal forms E_{2n} coincide. The latter is a necessary and sufficient condition for the pair of cells to be glued together in a way such that the orientation on the identified sides is retained. □

The cells obtained from the elementary cells by pairwise amalgamations are called *non-elementary cells.* Every non-elementary cell is obtained as a superposition of two elementary cells in which stable vertices of one cell go to the stable vertices of the second cell, while the semi-stable vertices are identified in an arbitrary way – they may be glued to each other, or to the points lying on the edges of the cell.

[7] Such cells were discovered by S. Novikov in the context of the compact leaf theorem for foliations on the three-dimensional sphere. Later G. Levitt studied this type of cell under the name of totally anti-symmetric separatrix cycle. It was shown by the same author that these cycles play an important role in the decomposition of non-wandering flows.

4.5.4 Conley-Lyapunov-Peixoto Graph

By a *ground surface* of the non-wandering flow one understands an (orientable) compact surface $\mathcal{G}$ produced by cutting M along the boundaries $\partial\mathcal{Q}_i$ and gluing the disc regions D into the obtained holes. If $\#$ is a connected sum of manifolds,

$$\mathcal{G} = \left(M \backslash \bigcup_{(i)} \partial\mathcal{Q}_i \right) \#_{(k)} D_k.$$

Remark 4.5.1 *One can 'squeeze' the annuli and Möbius strips of the non-wandering flow by the identification $h : S^1 \to S^1$ of their boundary components S^1 – cf. items (ii), (iii) of Proposition 4.5.1. The 'elementary cells' of the modified flow will be either disc regions or quasiminimal sets.*

By an *orgraph* X of the non-wandering flow one understands an orgraph (disconnected, in general) with vertex set V_X consisting of saddle points lying on the boundaries of $\mathcal{Q}$-sets, and the edge set E_X consisting of the saddle separatrices which connect these saddle points. An (oriented) incidence structure on X is established as follows: a pair of vertices $v_1, v_2 \in V_X$ is joined by an edge $e \in E_X$ if and only if there is a separatrix lying on the boundary of a $\mathcal{Q}$-set and going from the saddle point representing v_1 to the saddle point, representing v_2. The orientation on e coincides with the flow orientation on the respective separatrix. Moreover, one distinguishes the cycles of X obtained as a result of the homotopy of Remark 4.5.1; to each cycle we prescribe the weight $+1$ if the 'squeezed region' was an annulus and the weight -1 if it was a Möbius strip. A *rotation* R on the orgraph X induced by the non-wandering flow ϕ^t is defined to be rotation R corresponding to the embedding of X into the ground surface $\mathcal{G}$.

If $X^\#$ is a dual graph corresponding to the embedding of X into the surface $\mathcal{G}$, then by a *weight function* μ one understands the surjective mapping

$$\mu : V_{X^\#} \to \{\mu_1, \dots, \mu_m\}$$

of the vertex set of the dual graph $X^\#$ onto a finite set of non-negative integer. If $\mathcal{Q}_i$ is a quasiminimal set then function μ takes the same value on each boundary component of $\mathcal{Q}_i$; moreover, this value is equal to the 'genus' $\mu_i > 0$ of $\mathcal{Q}_i$. We set $\mu_i = 0$ if and only if the boundary component bounds a disc. (In other words, μ shows how to glue handles into ground surface $\mathcal{G}$ and what 'genus' each particular handle must have.)

Definition 4.5.2 *A Conley–Lyapunov–Peixoto graph of the non-wandering flow ϕ^t is the orgraph X taken together with the rotation R and the weight function μ. We denote the Conley-Lyapunov-Peixoto graph by X_μ^R.*

By an isomorphism $\cong$ of the Conley-Lyapunov-Peixoto graphs one understands an equivalence relation on the set $\{X_\mu^R\}$ identifying those graphs whose:

(i) oriented incidence structure and collection of the weighted marked cycles are isomorphic;
(ii) rotation systems are equivalent;
(iii) weight functions coincide.

4.5.5 Classification

Equivalence Criterion Let π^t be a non-wandering flow on the surface M and let *Sep* π^t be the separatrix contour bounding the elementary cells. We introduce an equivalence relation identifying those flows which are topologically equivalent on the set *Sep* $\pi^t \subset M$ and arbitrary on the complementary $M\setminus$ *Sep* π^t. In other words, $\pi_1^t \sim \pi_2^t$ if there exists a homeomorphism $h : M \to M$ which sends an π_1^t-orbit through x into π_2^t-orbit through $h(x)$ whenever $x \in$ *Sep* π_1^t. We call this equivalence relation a *graph–equivalence* of non-wandering flows.

Theorem 4.5.1 ([226]) *The non-wandering flows π_1^t and π_2^t are graph–equivalent if and only if their Conley-Lyapunov-Peixoto graphs are isomorphic, $X_\mu^R \cong Y_\mu^R$.*

Proof. Necessary conditions. If the non-wandering flows π_1^t, π_2^t are conjugate by a homeomorphism h, then $X_\mu^R \cong Y_\mu^R$. Indeed, consider the restriction of $h : M \to M$ to the set Sep $\pi_1^t \mapsto$ Sep π_2^t, which establishes an isomorphism of the respective orgraphs, $X \cong Y$. The equivalence of their rotation systems R_X, R_Y follows from the fact that h preserves the 'corner structure' associated with the embedding $i_X : X \to M$ and $i_Y : Y \to M$. Finally, the coincidence of the weight functions μ_X and μ_Y is the consequence of restriction $h : M\setminus G_X \to M\setminus G_Y$, where G_X, G_Y are ground surfaces for π_1^t and π_2^t.

Sufficient conditions. Suppose that $X_\mu^R \cong Y_\mu^R$. Roughly, one must construct a homeomorphism $h : M \to M$ which conjugates π_1^t to π_2^t up to a 'subtle' conjugacy between the corresponding $\mathcal{Q}$-sets. Actually the conjugacy h should be constructed between the 'restriction' of flows π_1^t, π_2^t to the ground surface $G_X \equiv G_Y$ (the last identity follows from $X \cong Y$ and $R_X \sim R_Y$). Now it remains to glue the necessary number of handles of the given genus onto the faces of G under the weight function $\mu_X \equiv \mu_Y$.

(i) First, one constructs an h which establishes the equivalence of the two embeddings $i_X, i_Y : X \to G$, defined by the equivalent rotation systems $R_X \sim R_Y$. But this was already done using the construction of the dual graphs $X^\#$ and $Y^\#$ associated with the embeddings $X \to G$ and $Y \to G$. To finish the proof of this part it remains to 'reverse' the homotopy h_t deforming the homeomorphism $h_0 : X^\# \to Y^\#$ on the dual graphs into a homeomorphism $h_1 : G_0 \to G_0$ of the surface G_0. (This was done detailedly in Ch. 2 to which we refer the reader.) The required homeomorphism h_1 :

$G_0 \to G_0$ is now established. It remains to fix $h = h_1$ and to extend it to the whole G by 'squeezing' the holes $O(V_X)$ to the vertex points of X. The homeomorphism h is the required one, and it conjugates two embeddings $i_X, i_Y : X \to G$ of the graph X.

(ii) Now we can extend the homeomorphism $h : G \to G$ to the surface M using the identity of the weight functions $\mu_X \equiv \mu_Y$. Since the weight function μ (common for both flows) indicates the faces of G to which the handles of a prescribed genus must be glued we construct M on the ground surface $\mathcal{G}$. The genus of M is completely defined by the weight function μ and it is a trivial exercise to prolong h given on the ground surface G to a homeomorphism h' between two surfaces M_X and M_Y having the same genus. □

Realization of a Graph An admissible Conley-Lyapunov-Peixoto graph $\mathcal{Y}_\mu^\mathfrak{r}$ consists of: (i) a finite orgraph $\mathcal{X}$ (disconnected, in general); (ii) a rotation system $\mathfrak{r}$ such that every cycle $C \subset \mathcal{X}$ traced by $\mathfrak{r}$ is isomorphic to one of the elementary or non-elementary cells described before; (iii) a finite number of cycles of weight ± 1; such cycles correspond to the Novikov-Levitt cells; (iv) finally, a weight function μ which is subjected to the following restriction: $k \geq 1 + \frac{1}{2}\sum(n_i - 1)$, where $\mu^{-1}(k) = \{\text{faces} : 2n_i\text{–gons}\}$ is the 'level set' of μ. One should add to admissible graphs those 'exceptional' cases of Proposition 4.5.1: a digraph corresponding to the sphere, a simple loop graph corresponding to the projective plane, a 'positive' one-point graph corresponding to the torus and a 'negative' one-point graph corresponding to the Klein bottle.

Theorem 4.5.2 ([226]) *Let X_μ^R be arbitrary Conley-Lyapunov-Peixoto graph isomorphic to an admissible graph $\mathcal{Y}_\mu^\mathfrak{r}$. Then there exists non–wandering flow on a compact 2-dimensional manifold whose Conley-Lyapunov-Peixoto graph coincides with X_μ^R.*

Proof. Necessary conditions. Let π^t be a non-wandering flow. It is easy to check that its Conley-Lyapunov-Peixoto orgraph X_μ^R satisfies all the required conditions.

Sufficient conditions. We give an outline of the proof, referring the reader to the original work [226]. Let X_μ^R be an orgraph X with the rotation system R and the weight function μ, and isomorphic to an abstract Conley-Lyapunov-Peixoto orgraph $\mathcal{X}_\mu^\mathcal{R}$. Basically, the proof proceeds in the following three stages. First a ground surface G and a 2-cell embedding $X \to G$ is constructed. (Here the Heffter-Edmonds theorem is of fundamental importance.) Next, using the basic properties of the weight function μ, a surface M will be constructed on G, so that we have an embedding $X \to M$ which is not a 2-cell embedding. Starting with the axioms of the definitions above, a non-wandering flow g^t on M is constructed. Finally, one applies (if necessary) the Whitehead homotopy to the 'marked' separatrix cycles of the flow g^t, thus obtaining the flow π^t. □

Application: Hamiltonian Flows If ω is a volume form on the manifold M, then the non-wandering flow is given by a vector field v whose inner product $i_v\omega$ is a closed 1-form. In the case that the inner product is an *exact* form, the flow is Hamiltonian. "Elementary cells" of Hamiltonian flows are especially simple, since no quasiminimal sets may appear in this case. Combining Proposition 4.5.1 with Theorems 4.5.1 and 4.5.2, one extracts the following corollary.

Corollary 4.5.1 *The Hamiltonian flows ϕ_1^t, ϕ_2^t are topologically equivalent if and only if their Conley-Lyapunov-Peixoto graphs are equivalent. Moreover, if X is a connected orgraph whose vertices are of valency $2n, n \geq 1$ with n entering and n exiting edges and there exists a rotation R such that every cycle on X is coherently oriented with some of them having "weight" ± 1, then there exists Hamiltonian flow on a 2-dimensional manifold M (orientable or not) whose Conley-Lyapunov-Peixoto graph coincides with X^R.*

Proof. By the absence of quasiminimal sets for Hamiltonian flows, the weight function μ vanishes on all cycles of X; the first part of corollary follows. Suppose that X and R are as is claimed in corollary. Then by the Heffter theorem, there is a 2-cell embedding of X into an orientable surface M. Each cell is partitioned by periodic trajectories with centre-type singularity; the other singularities are $2n$-separatrix saddles. If certain cycles of X have been 'weighted' one should glue the annulus or the Möbius strip into M – cf Remark 4.5.1. It is an easy exercise to construct a function $H : M \to \mathbb{R}$ such that dH is a 1-form $i_v\omega$ on M. □

4.6 Foliations With Symmetry

4.6.1 Cayley Graph

The standard references are:

W. Burnside, The Theory of Groups of Finite Order. Second Edition, Cambridge, England.

H. S. M. Coxeter and W. O. J. Moser, Generators and Relations for Discrete Groups, Third Edition, Ergebnisse der Mathematik und ihrer Grenzgebiete, Bd. 14, Springer-Verlag, New York-Heidelberg-Berlin (1972).

M. Aschbacher, Finite group theory, Cambridge Univ. Press, (1988).

By G we denote a group with identity. As usual G is called *abelian* if the group operation is commutative and *non-abelian* otherwise. A *group homomorphism* from the group G into the group H is a function $\alpha : G \to H$ of the set G into the set H which preserves the group operations, that is, for all $x, y \in$

G $\alpha(xy) = \alpha(x)\alpha(y)$. The homomorphism α is an *isomorphism* if α is a bijection; an isomorphism of G into itself is called an *automorphism* of the group G. G is said to be *isomorphic* to H if there exists an isomorphism from G to H. This is usually written as $G \cong H$.

Let G be a finite group. We deal with the group G in its 'combinatorial' representation, which defines G as

$$G = \langle g_1, \dots, g_m | r_1, \dots, r_n \rangle$$

where $g_1, \dots, g_m$ are 'generators' while $r_1, \dots, r_n$ is the set of 'relations' on G. (Each 'relation' $r_i = r_i(x_1, \dots x_l)$ is an equation $r_i = 1$ in the group G.) Whether the given representation defines a finite group or not is an open question but we shall always suppose that it does. The 'word problem' for a group given by its combinatorial representation consists in establishing whether two given words are equivalent or not. Below are examples of some often-used finite groups with two generators s and u.

Notation	Name	Relators	Order	Description
K_4	Klein 4-group	$s^2 = u^2 = 1, su = us$	4	abelian
S_3	Symmetric	$s^2 = u^3 = (su)^2 = 1$	6	non-abelian
Q	Quaternion	$s^2 = u^2 = (su)^2$	8	non-abelian

Let $G = \langle g_1, \dots, g_m | r_1, \dots, r_n \rangle$ be a finite group of order $|G|$ with generators $g_1, \dots, g_n$ and relations $r_1, \dots, r_n$. By a *Cayley graph* associated to the group G one understands a regular orgraph Cay G with $|G|$ vertices such that:

- Every vertex of Cay G corresponds to exactly one element $x \in G$ of the group G;
- Two vertices x_1 and x_2 are joined by an edge if and only if there exists a generator g_i such that $x_2 = g_i x_1$;
- The arrow on the above edge goes from x_1 to x_2 [8].

Example 4.6.1 (Dehn) *Let G be the symmetric group S_3 given by two generators s, u and three relations $s^2 = u^3 = (su)^2 = 1$; see the table. To construct the corresponding graph Cay G take an arbitrary point (vertex) to represent the identity of the group G. Then one acts successively on this vertex by the generators s and u so as to obtain a pair of out-going edges with the endpoints representing elements $s, u \in S_3$. If we now take a vertex 's' and act on it by the generator s, then we come back to the identity vertex '1' in accordance with the relation $s^2 = 1$ (thus marking a 2-edge cycle on Cay G). If we take a vertex 'u' and act on it by the generator u we will not obtain the identity 1 but rather a new vertex $u^2 = u^{-1}$ of Cay G. If one now acts on u^2 by the generator u, then in accordance with the relation $u^3 = 1$*

[8] One similarly defines the 'right' Cayley graphs by setting $x_2 = x_1 g$. The right and the left Cayley graphs are isomorphic as orgraphs.

we return to the identity vertex '1'. (Hereby a 3-edge cycle which represents the relator $u^3 = 1$ is traced.) Proceeding similarly with the other vertices of graph Cay G we come to a regular 6-vertex orgraph as pictured in Figure 4.1.

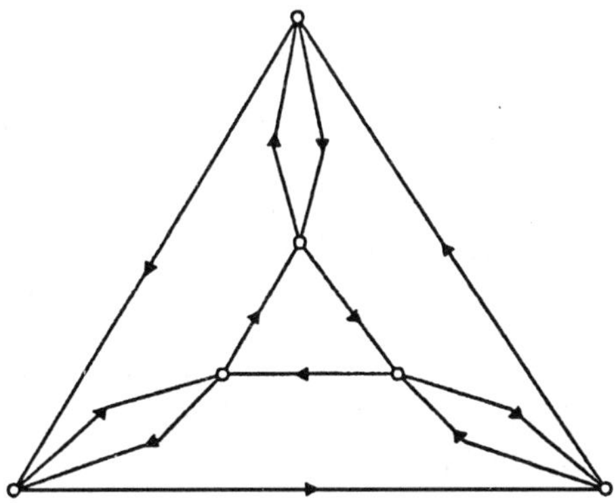

Fig. 4.1. Cayley graph of the group S_3.

4.6.2 Isomorphism

Let π^t be a C^∞ non-wandering flow on the compact surface M. By a G-symmetry of π^t one understands a diffeomorphism $g : M \to M$ which maps the set of trajectories of π^t into itself:

$$g \circ v(x) = v(x), \qquad x \in M, \tag{4.1}$$

provided the g's form the group G by composition. The following two restrictions are assumed:

- Each separatrix cycle of π^t is a Novikov-Levitt cycle, see Section 4.5 for the definition;
- The weight function μ of $\mathcal{F}$ has the form

$$\mu : V_{X\#} \longrightarrow \{1, \ldots, 1\}$$

Definition 4.6.1 *A non-wandering flow π^t is called G-symmetric if there exists a finite group of diffeomorphisms G of the surface M without fixed points in the set Sing π^t such that, for each pair of saddle points $p_1, p_2 \in$ Sing π^t, there exists a unique diffeomorphism $g \in G$ satisfying condition (4.1) and sending the saddle point p_1 into the saddle point p_2.*

Theorem 4.6.1 *([221]) Let π^t be a G-symmetric non-wandering flow on the closed surface M. Then the Conley-Lyapunov-Peixoto orgraph $X = X(\pi^t)$ associated to the flow π^t is isomorphic to the Cayley orgraph Cay G of the symmetry group G. The latter has the two generators s and u. Moreover, the rotation system R of the orgraph X is regular* [9] *with regard to the action of the group G.*

Proof. See [221]. The proof proceeds in the following two stages. First, it is proved that the G-action on M induces the vertex transitive and vertex fixed-point-free automorphisms of the Conley-Lyapunov-Peixoto orgraph X built upon the flow π^t. Secondly, a 'colouring' of the edges of X is established so that in every vertex of X there are exactly two edges coloured by 's' and two edges coloured by 'u'. As far as such a 'colouring' of X by s and u (consistent with the orientation of X) is concerned there exists a finite group G with two generators s and u and a set of relations representing the cycles of the graph X such that the Cayley graph of the group G is isomorphic (as an orgraph) to X; see [221] for details. □

Corollary 4.6.1 *Each inner automorphism of the group G corresponds to a certain diffeomorphism $g : M \to M$ of the surface M which leaves the orbits of $\mathcal{F}$ invariant in accordance with formula (4.1), and vice versa.*

Proof. Every diffeomorphism $g : M \to M$ acting vertex-transitively and vertex-fixed-point-freely on a graph X embedded into M defines an automorphism of X. Since $X \cong \text{Cay } G$ g also defines an inner automorphism of the group G which may be represented canonically as a left L_a or right R_a inner automorphism of the group G.

Conversely, if $L_a : x \mapsto a^{-1}xa$ ($R_a : x \mapsto axa^{-1}$) is a left (right) inner automorphism of G then it generates an automorphism $g : \text{Cay } G \to \text{Cay } G$ of the Cayley orgraph Cay G embedded into M. By formula (1) g can be extended to a diffeomorphism $g : M \to M$ of the surface M. □

4.6.3 Realization

Definition 4.6.2 *Let L be a coherent cycle of an orgraph X. Let $e^* \subset L$ be an arc* [10] *of L with endpoints at the 4-valent vertices of X such that all vertices which lie in the interior of e^* are 2-valent. Then the arc e^* is called a marked edge of the orgraph X.*

Definition 4.6.3 *Let $\mathfrak{G}$ be a finite group with two generators s and u. The group $\mathfrak{G}$ is said to be flow–consistent if:*

[9] The rotation system R is called regular if for each $g \in G$ the local rotation $e_1, \ldots, e_m$ at every vertex $v \in V_X$ maps to the local rotation $g(e_1), \ldots, g(e_m)$ at the corresponding vertex $g(v) \in V_X$.

[10] An arc in the graph X is a 1-1 continuous mapping of the unit interval into X.

(i) There exists a regular rotation system R such that all cycles traced on Cay $\mathfrak{G}$ by R are isomorphic (as orsubgraphs) to a fixed coherent cycle;
(ii) The set E^* of all marked edges $e^* \in E_X$ is $\mathfrak{G}$-invariant. [11]

Theorem 4.6.2 *([221]) Let G be a finite group with two generators s and u. Then the group G can be realized as symmetry group of non-wandering flow on a closed surface M if (and only if) it is isomorphic to a flow–consistent group $\mathfrak{G}$.*

Proof. Necessary conditions. Let $\mathcal{F} : M \times I\!R \to M$ be a G-symmetric $\mathcal{R}$-flow on the closed surface M. Then it is easy to check that its Conley-Lyapunov-Peixoto orgraph X and the associated rotation system R satisfy all the conditions of Definitions 4.6.2 and 4.6.3.

Sufficient conditions. See [221] for details. Let G be a finite group with two generators, s and u, which is isomorphic to a flow-consistent group $\mathfrak{G}$. To prove that there exists a G-symmetric non-wandering flow on a closed surface M one considers the following three steps. First, it must be shown that there exists an appropriate embedding of the Cayley orgraph Cay G of the group G into a (ground) surface $\mathcal{G}$ such that the graph automorphism g : Cay $G \to$ Cay G extends to a diffeomorphism of $\mathcal{G}$. (Thus the action $\mathcal{G} \times G \to \mathcal{G}$ is defined.) Secondly, one should 'attach' to each face of the embedding i : Cay $G \to \mathcal{G}$ a 'standard' cell of genus 1 so as to obtain a surface M. On each cell a Kronecker flow is defined and the resulting surface M is endowed with a flow $\pi^t : M \times R \to M$. Finally, one proves that the flow π^t is G-symmetric and non-wandering. By construction, the orgraph Cay G will be isomorphic to the Conley-Lyapunov-Peixoto orgraph of π^t. □

4.7 Homology and Cohomology Invariants

In this section we consider homology and cohomology invariants of measured foliations. The real one-dimensional homology $H_1(M, I\!R)$ of the closed surface M is a finite dimensional module over $I\!R$. The fundamental group $\pi_1(M)$ is a (non-commutative) group with $2g$-generators satisfying a certain relation. The homology invariants are often calculable and are by far, more constructive than the homotopy invariants.

Transitive foliations have both homotopy and the (co-)homology invariants. The homotopy rotation class is a complete invariant, albeit hardly calculable. In contrast, there are plenty calculable homology invariants which are not complete. This does not diminish their value since they often provide an immediate mean to distinguish between two foliations.

[11] That is, $\mathfrak{g}(E^*) \subseteq E^*$ for each $\mathfrak{g} \in \mathfrak{G}$.

4.7.1 Asymptotic Cycles

Asymptotic cycles are known to be the basic homology (cohomology) invariants of foliations. Let us recall Schwartzman's construction. Let X be a compact metric space. The set of all continuous complex-valued functions on X of absolute value 1 will be denoted by $C(X)$. (In other words, $C(X)$ is the space of continuous functions $f : X \to S^1$ where $S^1 = \{z : |z| = 1\}$ is the unit circle in the complex plane.)

Following S. Eilenberg (*Sur les transformations d'espaces métrique en circonférence, Fund. Math., 24, 1935, 160-176*) let us consider the subset of $R(X) \subset C(X)$ consisting of functions which are homotopic to the constant function $f_0 = 1$. These are called *inessential.* $R(X)$ is a connected subspace of $C(X)$ and is an abelian group under multiplication. Since $\psi \in R(X)$ if and only if $\psi(x) = \exp(2\pi i h(x))$ where $h : X \to R$ is the real-valued function, then $R(X)$ is a normal subgroup in $C(X)$.

Let $l : I \to X$ be a parametrized curve in X and let $f \in C(X)$. Then the value $\Delta_l \; Arg \; f(x)$ measures the change of the argument of $f(x)$ along the curve l. It is easy to see that if l_0 is a closed parametrized curve then $\frac{1}{2\pi}\Delta_{l_0} \; Arg \; f(x)$ is an integer which is equal to the algebraic number of laps made by $f(x)$ as x runs through the curve l_0. Since for $\psi \in R(X)$

$$\frac{1}{2\pi}\Delta_{l_0} \; Arg \; \psi(x) = h(x_0) - h(x_0) = 0,$$

the number

$$\frac{1}{2\pi}\Delta_{l_0} \; Arg \; f(x) = [f](l_0)$$

is determined by the class $[f] \in C(X)/R(X)$.

N. Bruschlinsky (*Stetige Abbildungen und Bettische Gruppen, Math. Ann., 109, 1934, 525-537*) proved that $[f](l_0)$ depends only on the class $[l_0] \in H_1(X,\mathbb{Z})$ of the one-dimensional, integer, homology group. Therefore each element of the factor-group $C(X)/R(X)$ defines a homomorphism

$$[l_0] \longrightarrow f[l_0] \in \mathbb{Z}, \qquad [l_0] \in H_1(X,\mathbb{Z}).$$

Note that by the Poincaré duality, $C(X)/R(X) = H^1(X,\mathbb{Z})$ is a cohomology group. The above construction allowed Schwartzman to define for the flow g^t elements of the group $H_1(X,\mathbb{Z})$ which he called the asymptotic cycles. In fact Schwartzman introduced two cycles: m-cycles and μ-cycles; first we give the definition of an m-cycle.

Denote by $(m, g^t(m))$ an arc of the trajectory between the points m and $g^t(m)$. For a (quasi-regular [262]) point $m \in X$ and each $f \in C(X)$ there exists a limit

$$\lim_{t\to\infty} \frac{1}{2\pi t}\Delta_{(m,g^t(m))} \; Arg \; f = A_m(f).$$

Moreover, $A_m(f)$ depends only on the class $[f] \in C(X)/R(X)$. The correspondence

$$A_m : C(X)/R(X) \to I\!R \qquad [f] \to A_m(f)$$

is a group homomorphism. Therefore $A_m \in H_1(X, I\!R)$ and A_m is called an *m-asymptotic cycle.*

Now let us give the definition of a μ-asymptotic cycle, where μ is an invariant measure of the flow g^t. Take a class $[f] \in C(X)/R(X)$ where $f \in C(X)$ is a function. For almost all $m \in X$ one can define the value $A_m[f]$ as was done earlier. Moreover, the function $A_m[f] : X \to I\!R$ is μ-integrable. The mean value

$$\frac{1}{\mu(X)} \int_X A_m[f] d\mu$$

may be interpreted as an μ-average value of the $[f]$-flow over all m-asymptotic cycles. A functional

$$A_\mu : [f] \to \frac{1}{\mu(X)} \int_X A_m[f] d\mu,$$

defines an element in the homology group $H_1(X, I\!R)$ which is called a *μ-asymptotic cycle* or *asymptotic cycle with respect to the measure* μ. (Roughly speaking, a μ-asymptotic cycle is a μ-average of the m-asymptotic cycles.)

Let $\mathcal{F} = (M, \pi^t, \theta)$ be a foliation with invariant measure μ. Let α be a closed 1-form which defines the cohomology class $[\alpha] \in H^1(M, \mathbb{R})$. Then the integral

$$\int X \lrcorner \alpha d\mu = \int \alpha(X) d\mu$$

depends only on $[\alpha]$, for if $\alpha_2 - \alpha_1 = d\psi$ (ψ is a differential function) then

$$\begin{aligned} \int X \lrcorner \alpha_2 d\mu - \int X \lrcorner \alpha_1 d\mu &= \int (X \lrcorner d\psi) d\mu = \int X(\psi) d\mu \\ &= \int \frac{d}{dt} \psi(\mathcal{F}) d\mu = \frac{d}{dt} \int \psi(\mathcal{F}) d\mu = 0 \end{aligned}$$

due to the invariance of μ. Thus the map

$$A_\mu : [\alpha] \to \int X \lrcorner \alpha d\mu, \qquad \alpha \in [\alpha]$$

defines a linear functional on $H^1(M, \mathbb{R})$ which by duality can be identified with an element $A_\mu \in H_1(M, \mathbb{R})$ of the first homology group.

The integral $\int X \lrcorner \alpha d\mu$ can be interpreted as a "stream" of the invariant measure μ through the leaves of the foliation $F(\alpha)$ defined by the 1-form α. This "physical" interpretation can be realized for the vector fields having nontrivial invariant measure (i.e., a measure which is positive on each open subset of M and such that the measure of any trajectory is zero). In 1973

A.Katok [160] proved that the class of vector fields having nontrivial invariant measure includes the vector fields whose trajectories are non–wandering and whose singularities are all saddles.

Let X be a vector field having the nontrivial invariant measure μ and generating flow π^t. Consider a path $\tau : [0,1] \to M$ of class C^1 in M. Using π^t, let us construct a mapping S of an oriented quadrangle into M given by the formula:

$$S_t : [0,1] \times [0,t] \to M, \quad S(x,y) = \pi^y(\tau(x)), \quad x \in [0,1], \quad y \in [0,t].$$

The measure $\mu(S_t)$ is a 'stream' of the measure which 'crosses' the curve τ in time t. Since μ is non-trivial there exists a finite limit

$$\lambda_\mu(\tau) = \lim_{t\to 0} \frac{1}{t}\mu(S_t).$$

We extend linearly the function λ_μ to all 1-chains of the surface M with coefficients in $\mathbb{R}$.

Lemma 4.7.1 *[160] For every closed curve $C \subset M$ which is contractible, one has $\lambda_\mu(C) = 0$.*

Proof. The curve C splits the surface M. It follows from the definition of invariant measure that the 'inside stream' of measure μ is equal to the 'outside stream' of measure μ. □

It follows from Lemma 4.7.1 that λ_μ defines an element $[\lambda_\mu]$ in the cohomology group $H^1(M,\mathbb{R})$. By Poincaré duality $[\lambda_\mu]$ defines an element (denoted again by A_μ) of the first homology group $H_1(M,\mathbb{R})$ with regard to the nontrivial invariant measure μ which we call a *rotation class* of the flow π^t.

Let us consider the important case of a vector field X preserving an area element ω. By Poincaré duality, A_μ corresponds to an element $B_\mu \in H^1(M,\mathbb{R})$ of the first cohomology group in the following way

$$A_\mu(\alpha) = \int B_\mu \wedge \alpha, \qquad \alpha \in [\alpha] \in H^1(M,\mathbb{R}).$$

Recall that $X\lrcorner\omega = \beta_X$ is a *flux form* of X. A flux form is closed if and only if the vector field X preserves the area element ω. Since

$$(X\lrcorner\alpha)\wedge\omega = X\lrcorner(\alpha\wedge\omega) - \alpha\wedge(X\lrcorner\omega) = (X\lrcorner\omega)\wedge\alpha$$

then

$$A_\mu(\alpha) = \int (X\lrcorner\alpha)d\mu = \int (X\lrcorner\alpha)\omega = \int \beta_X \wedge \alpha.$$

Thus, the flux form $\beta_X = B_\mu$ is a *cohomology* representation of the *asymptotic cycle* which corresponds to A_μ.

4.7.2 Fundamental Class

Let M is a closed, orientable surface of genus $g \geq 1$. Let us consider a class, $S(M)$, of orientable foliations $\mathcal{F} = (M, \pi^t, \theta)$ which obey the following conditions:

- π^t is non-wandering;
- The set Sing $\mathcal{F}$ contains only Morse saddles of total number $2g - 2$.

It follows from Arnold's example that even an analytic foliation of class $S(M)$ may not have a smooth invariant measure. However, a nontrivial invariant measure exists [160]. Let μ be a nontrivial invariant measure of the flow $\pi^t \in S(M)$ and let $A_\mu \in H_1(M, \mathbb{R})$ be a rotation class of the flow π^t with regard to the measure μ.

Denote by $H_1(M,\ Fix\ (\pi^t), \mathbb{R})$ the relative homology group which is introduced in the same way as $H_1(M, \mathbb{R})$ modulo finite set of the 'pinched' points $Fix\ (\pi^t)$. For genus $g \geq 2$ this group is a $(4g - 3)$-dimensional real vector space.

An element $\hat{R}_\mu \in H_1(M,\ Fix\ (\pi^t), \mathbb{R})$ generated by A_μ is called a *fundamental class* relative non-trivial invariant measure μ.

Definition 4.7.1 (A. Katok) *A cone $K(\pi^t)$ in the space of relative homologies $H^1(M,\ Fix\ (\pi^t), \mathbb{R})$ generated by all non-trivial invariant measures μ is called the fundamental class of the flow π^t.*

If π^t is uniquely ergodic (i.e., up to a constant admits a unique invariant measure) then its Katok fundamental class is a ray in the space $H^1(M,\ Fix\ (\pi^t), \mathbb{R})$.

Theorem 4.7.1 (A. Katok) *Let μ be a non–trivial invariant measure of the flow $f^t \in S(M)$ generated by the C^1 vector field v. Then there is a neighbourhood $U(v)$ of v in the space of C^1 vector fields such that the flow $g^t \in S(M)$ generated by vector field $v' \in U(v)$ is topologically equivalent to f^t provided the following conditions hold:*

- *$Fix(f^t) = Fix(g^t)$*
- *there is a nontrivial invariant measure μ' of g^t such that*

$$\hat{R}_\mu(f^t) = \hat{R}_{\mu'}(g^t).$$

In particular, we have

Theorem 4.7.2 (A. Katok) *Let f^t, g^t be two transitive flows on the closed orientable surface M such that*

$$K(f^t) \cap K(g^t) \neq \emptyset, \qquad Fix\ (f^t) = Fix\ (g^t).$$

If the flows f^t, g^t are close enough then f^t, g^t are topologically equivalent.

A. Katok conjectured that Theorem 4.7.1 holds true for any transitive flows (not necessarily "close" in the space of flows). He also conjectured that the condition $K(f^t) = K(g^t)$ instead of $\hat{R}_\mu(f^t) = \hat{R}_{\mu'}(g^t)$ implies the topological equivalence of any flows of the class $S(M)$ (again, not necessarily close to each other). However G. Levitt [185] constructed a counterexample to these conjectures by proving the following theorem.

Theorem 4.7.3 (G. Levitt) *Let M be a closed orientable surface of genus $g \geq 2$, and let $F \subset M$ be a set of $2g-2$ points. Then for almost every cohomology class $\Lambda \in H^1(M, F, I\!R)$ there exist transitive flows $f_i^t, i \in \mathbb{N}$ on M such that*

(i) The set Fix $(f_i^t) = F$ consists of non-degenerate saddles of f_i^t.
(ii) $K(f_i^t) = \{t\Lambda : t > 0\}$.
(iii) The flows f_i^t, f_j^t are topologically distinct for $i \neq j$.

Proof. Let $\Lambda \in H^1(M, F; I\!R)$ be an element of the real homology group with rationally independent periods whose rank is maximal $rk\ \Lambda = 2g$. Each form $\alpha \in \Lambda$ defines the foliation $G(\alpha)$ on M with the transitive components $M_{g_1}, \ldots, M_{g_k}$ so that $g_1 + \ldots + g_k = g$, where g is the genus of M. Therefore, as long as $G(\alpha)$ has no separatrix contours, $G(\alpha)$ is transitive. We shall assume that $G(\alpha)$ is strictly ergodic, i.e., there exists a unique transversal, invariant measure. Note in passing that $G(\alpha)$ has at least one such measure since α is a closed form. From the Masur–Veech theory (see e.g. [287]) it follows that the classes Λ form a set of full measure in the space $H_1(M, F; I\!R)$.

Take an arbitrary disc D which contains all points of F. Each form $\omega \in \Lambda$ in D can be written as $\omega = df$ where f is a smooth function defined on F up to a constant. Since $\Lambda \in H^1(M, F; I\!R)$, f takes pairwise distinct values at the points of F. Let us enumerate the points $s_1, s_2, \ldots, s_{2g-2} \in F$ so that

$$f(s_1) < f(s_2) < \ldots < f(s_{2g-2}).$$

We set

$$\rho = f(s_{2g-2}) - f(s_1) > 0.$$

For the elements $a, b \in H_1(M; \mathbb{Z})$ let us denote by $a \bullet b$ the intersection index of representing curves. (Recall that each element of $H_1(M, \mathbb{Z})$ is representable by a smooth closed curve, since, for two elements of the homology group, the intersection index is independent from the representation chosen.)

By the condition $rk\ \Lambda = 2g$ there exists a basis $\{a_1, b_1, \ldots, a_{2g}, b_{2g}\}$ in $H_1(M; \mathbb{Z})$ such that:

(i) $a_i \bullet a_j = 0,\ b_i \bullet b_j = 0,\ \forall i, j;$
(ii) $a_i \bullet b_i = 1,\ a_i \bullet b_j = 0,\ i \neq j;$
(iii) $\Lambda(a_i) > 0.$

By item (ii), the elements $a_1,\dots,a_g$ can be represented by simple disjoint curves $A_1,\dots,A_g$ such that $M\backslash\cup_{i=1}^g A_i$ is a disc region. We put $a_i^0=a_i, b_i^0=b_i$, $i=1,\dots,g$. Let us build a sequence $\{a_1^n,b_1^n,\dots,a_g^n,b_g^n\}_{n=0}^\infty$ of bases such that:

(i) $a_i^n=a_i, \quad b_i^n=b_i, \quad i=1,\dots,g-2$
(ii) $a_i^n\bullet b_i^n=1, \quad a_i^n\bullet b_j^n=0 (i\neq j), \quad b_i^n\bullet b_j^n=0, \quad a_i^n\bullet a_j^n=0.$
(iii) $a_{g-1}^n=x_n a_{g-1}-y_n b_{g-1}, \quad a_g^n=-x_n a_{g-1}-y_n b_{g-1}+a_g^n+Nb_g$ where x_n, y_n, $n\in\mathbb{N}$, are relatively prime and satisfy the inequalities:

$$0<x_n\Lambda(a_{g-1})+y_n\Lambda(b_{g-1})<K; \tag{4.2}$$

$$\Lambda(a_g)+N\Lambda(b_g)>\Lambda(a_1)+\ldots+\Lambda(a_{g-1})+2[x_n\Lambda(a_{g-1})+y_n\Lambda(b_{g-1})]+\rho \tag{4.3}$$

for some constant $K>0$ and $N\in\mathbb{Z}$.

It follows from the number theory that there exist infinitely many integer pairs (x_n,y_n) satisfying (4.2) for some $K>0$. Next, we choose N satisfying (4.3). Since (x_n,y_n) are relatively prime it follows from the conditions in 3 that there exists a basis for which the conditions in 2 hold true. It also follows from 2 that the elements $a_1^n,\dots,a_g^n$ can be represented as simple, mutually disjoint curves $A_1^n,\dots,A_g^n$ such that $M\backslash\cup_{i=1}^g A_i^n$ is a disc region.

Let us fix n and let us embed the curves $A_1^n,\dots,A_g^n$ into a foliation consisting of periodic leaves and the separatrix connections s_is_i, s_is_{i+1}, $s_{2g-2}s_1$. Apart from the separatrix connections, each leaf is isotopy equivalent to one of the A_i^n. Let us denote by F_n the foliation so obtained.

It follows from condition 3 that the foliations F_n, $n\in\mathbb{N}$, are mutually topologically inequivalent.

Obviously, each foliation F_n is orientable. There are infinitely many closed forms u_n corresponding to F_n. Let us normalize $\int_{b_n}u_n=1$. For every element $\gamma\in H_1(M;\mathbb{Z})$ the form u_n "counts" the intersection index of γ with $a_1^n+\ldots+a_g^n$, i.e.

$$\int_\gamma u_n=\gamma\bullet(a_1^n+\ldots+a_g^n).$$

But

$$a_1^n+\ldots+a_g^n=a_1+\ldots+a_g+Nb_g,$$

i.e., $u_n(\gamma)$ does not depend on n. Therefore the u_n's belong to the same cohomology class, say $U\subset H^1(M;\mathbb{Z})$.

Let us construct for each n a form $\omega_n\in\Lambda(M,\Sigma;\mathbb{R})$ which defines a foliation G_n with $Sing\ G_n=\Sigma$ and which is transversal to F_n except at the points of Σ. We set $\omega_n=df$ in the disc D. Denote by s_is_{i+1}, s_is_i the separatrix connection between s_i, s_{i+1} and the separatrix loop in s_i, respectively. It follows from the inequality $\Lambda(a_i^n)>0, i=1,\dots,g-1$ that df can be extended to the neighbourhoods of all loops s_is_i and handles carrying A_j^n, $j=1,\dots,g-1$, and bounded by the loops. Moreover, the 1-form thus obtained has singular points in Σ and defines a foliation transversal to F_n beyond Σ. Furthermore,

$$\Lambda(s_{2g-2}) = \Lambda(a_g^n) - \sum_{i=1}^{g-1} \Lambda(a_i^n) - \sum_{i=1}^{2g-3} \Lambda(s_i s_{i+1}) = \Lambda(a_g^n) - \sum_{i=1}^{g-1} \Lambda(a_i^n) - \rho > 0,$$

by the inequality (4.3). Therefore, ω_n can be extended to the neighbourhood $s_{2g-2}s_1$ and furthermore to all of M. Since $rk\ \Lambda = 2g$ the foliation G_n is highly transitive. By the choice of the class Λ, G_n is uniquely ergodic. Denote by $\mathcal{F}_n$ a flow whose trajectories coincide with the leaves of G_n. Then

$$K(\mathcal{F}_n) = \{c\Lambda | c > 0\}.$$

Let us show that the flows $\mathcal{F}_n, n \in \mathbb{N}$ are topologically distinct.

Suppose to the contrary that there exists a homeomorphism $\varphi : M \to M$ such that $\varphi(\mathcal{F}_{n_1}) = \mathcal{F}_{n_2}$. Then $\varphi(G_{n_1}) = G_{n_2}$, and therefore $\varphi_*(\omega_{n_2}) = \lambda\omega_{n_1}$ for some $\lambda > 0$. Since $\omega_i \in \Lambda$, we have $\varphi_*(\Lambda) = \lambda\Lambda$. It follows from the equality $rk\ \Lambda = 2g$ that $\varphi_* = \pm id$. But, since $\varphi(G_{n_1}) = G_{n_2}$, we conclude that the foliations $\mathcal{F}_{n_1}, \mathcal{F}_{n_2}$ are equivalent as well. This contradiction proves the theorem. □

4.7.3 Cycles of A. Zorich

Throughout this section M is an orientable, closed surface of genus $g \geq 2$. Let us consider a class of smooth measured orientable foliations such that all singular points are of Morse type. Such foliations are defined by the closed 1-forms on M and the set of non-wandering points is all of M. As a consequence, they have an invariant transversal measure defined via the integral from the corresponding 1-form. We know that in this case M falls into a finite number of periodic (partitioned by periodic trajectories) and transitive components with everywhere dense trajectories. In what follows we confine ourselves to the 'dense' components.

There exists two natural parametrizations of our foliations: a 'Hamiltonian' (which lies beyond our discussion) and parametrization by the natural parameter of a Riemann metric on M. The latter is not unique, but since metrics are non-degenerate, any two of them uniformly bound each other. In the presence of a Riemann metric, the flow given by the closed 1-form can be considered as a Hamiltonian flow with a multi–valued Hamiltonian (which is actually our 1-form). How does one construct a Hamiltonian vector field? One takes a Hamiltonian H and turns dH into a vector field by 'lifting' the indices. If H is no longer given, but we have a closed 1-form, the 'lifting' still defines a vector field but with a 'multi–valued' Hamiltonian. Note that all local properties are valid, e.g., our flow is tangent to the level set of H.

The 'hamiltonian' parametrization is of interest; there are several words on it in the paper

V. Arnold, Topological and ergodic properties of closed 1-forms with incommensurable periods. Funct. Analysis and Its Appl., 1991, 25, 81-90. Transl. from Funk. Anal. i Ego Prilozh., 1991, 25, no 2, 1-12.

For the hamiltonian parametrization only simple zeros are considered; in this typical case, it takes logarithmic time to move near such a point.

Let us return to the main problem. We have in mind the asymptotic behavior of trajectories (leaves of a foliation); in contrast to the homotopy rotation classes considered earlier in this Chapter (where trajectories are approximated by the elements of the fundamental group) we are primarily interested in approximations by the elements of the first homology group. Instead of taking universal covering, an 'abelian' one is taken. In this sense, the expected invariant is cruder.

It is clear now that from this viewpoint, the periodic components are irrelevant. Let us explain how one can pass to the holomorphic 1-differentials (note that, for the moment, no complex structure is present). If one removes all periodic components by squeezing the holes to points and the annuli to cycles, then the resulting flow will be transitive in each component (note that M may be disconnected). On each component, the flow is again given by a closed 1-form. It turns out that one can choose a complex structure on M so that the 1-form is the real part of a holomorphic 1-form. An advantage of such approach is that we start with the infinite-dimensional space of all closed 1-forms, and then reduce the study to a finite-dimensional object – the real parts of holomorphic 1-forms on M. This also yields an exciting link to Teichmuller theory.

A historical note on the representability of closed 1-forms by the real parts of holomorphic 1-forms is in order. There exists (see

E. Calabi, An intrinsic characterization of harmonic one-form. Global Analysis (Papers in Honor of K. Kodaira), Univ. Tokyo Press, Tokyo 1969, pp. 101-117.)

an undeservedly forgotten theorem of Calabi: If a closed 1-form on compact n-dimensional manifold is given, when one can choose a (Riemann) metric on M so that the form is harmonic? Calabi's criterion says that with the assumption that all zeros of the form are non-degenerate, it is necessary and sufficient that through every regular point there exists a closed transversal curve to the form (i.e., the restriction of the form to the curve is non-degenerate). In the case $n = 2$, there always exists a complex structure in which the harmonic 1-form is the real part of a holomorphic 1-form.

Variations of the Calabi theorem have been independently reestablished by Katok and Hubbard–Masur, while Zorich extended the theorem to 1-forms with degenerate zeros (private communication): One can choose a complex structure such that the closed 1-form is harmonic if and only if there are no contractable curves going along the separatrix loops in the positive direction. In particular, if there are no separatrix connections, a complex parametrization is always possible.

There exists an erroneous opinion that the absence of separatrix connections is 'generic' for foliations given by the closed 1-forms on M. This is not true: if there is a periodic component partitioned by the trajectories that contractible, then this component 'survives' all perturbations of the system. Therefore, the respective separatrix connections are 'stable' as well. For foliations given by the real parts of holomorphic 1-forms, there are no separatrix connections at all.

From the above discussion, the 'non-trivial' part of the dynamics is carried by the real parts of holomorphic 1-forms. At this point, we have to consider the family of all such forms, that is, the corresponding moduli space: pairs (complex structure + holomorphic 1-form) factorized by diffeomorphisms.

A general remark on moduli spaces should be made. Actually, the moduli space of pairs (complex structure + holomorphic *quadratic* differential) factorized by diffeomorphisms should be considered. Quadratic means that an object and an atlas are given such that the transformation of coordinates corresponds to the transformation of object according to the law $(dz/dz')^2$. (The usual differential 1-form transforms as $(dz/dz')^1$.) One can (locally) extract a quadratic root of the quadratic differential and obtain (locally) a holomorphic 1-form. But moving around a loop the two branches of the quadratic root can change sign and the 1-form will gain the opposite sign. It follows that the quadratic differentials are 'responsible' for the non-orientable foliations with transversal invariant measure.

The moduli space of holomorphic quadratic differentials is isomorphic to the cotangent bundle over the moduli space of complex structures. The moduli space of holomorphic quadratic differentials is stratified. A quadratic differential may happen to be the 'square' of a globally given holomorphic differential, but it is possible that it is *not* (which is a generic case). Let us consider, however, the first possibility. This chunk of space is isomorphic to the space of moduli of holomorphic 1-differentials. Further stratification depends on the types of zeros. A holomorphic 1-differential admits $2g-2$ zeros, counted with multiplicities. They can either all be different, or two of them can coincide, etc. For instance, when $g = 2$ there are two strata: one of order 2, or two simple zeros. A stratum may have several connected components; recently, M. Kontsevich and A. Zorich classified these connected components (private communication, see also [171]).

Recall that we are working with the subspaces of the cotangent bundle over the moduli space of complex structures. On this bundle there is defined a flow called the *Teichmuller geodesic flow.* This flow preserves stratification, as well as the connected components of each stratum. Moreover, by a theorem proved independently by Masur and Veech, the Teichmuller geodesic flow is ergodic on every connected component of every stratum.

Let us consider the Lyapunov exponents of the Teichmuller flow. By ergodicity, for every connected component of every stratum the Lyapunov exponents are almost everywhere constant. Thus for every connected component of every stratum we obtain a sequence of numbers. These numbers satisfy the following double symmetry:

$$\begin{aligned}-2 &< -(1+\nu_2) \leq -(1+\nu_3) \leq ... \leq -(1+\nu_g) \leq -1 = \dots \\ &= -1(s-1) \leq -(1-\nu_g) \leq ... \leq -(1-\nu_2) < 0 < (1-\nu_2) \leq \dots \\ &\leq (1-\nu_g) \leq 1 = \dots \\ &= 1(s-1) \leq (1+\nu_g) \leq (1+\nu_{g-1}) \leq \dots \leq (1+\nu_2) < 2,\end{aligned}$$

where s is the number of zeros. In practice, everything is defined by g non-negative numbers:

$$1 = \nu_1 > \nu_2 \geq ... \geq \nu_g \geq 0,$$

where g is the genus of the surface. A. Zorich proved that $\nu_2 < 1$ (private communication). The same author used a computer to find out these numbers to nearly 50-80 different strata. In all cases one can clearly observe a simple spectrum: all the Lyapunov exponents are distinct. No general proof of this fact is available (although M. Kontsevich suggested a beautiful formula for μ_i which is not yet entirely proved).

The foliation realized by the real part of a holomorphic differential defines uniquely the stratum as well as its component. From now on, we consider only such foliations. Occasionally, we perturb our foliation, maintaining the types of zeros. Now we are ready to formulate the main result.

Let us consider a typical (in the sense of Lebesgue measure on the component of a stratum) holomorphic 1-differential; consider a foliation given by the real (imaginary) part of it. Such foliations are called horizontal (resp., vertical) foliations. Take an arbitrary leaf of this foliation (or, to avoid simple but tedious operations, take a non-singular leaf). Let us select an arbitrary point x_0 on it. We move along the leaf choosing more and more distant points, say $x(l)$, where l is the natural parameter of the leaf. Each time we consider a cycle $c(l)$ closing the segment $[x(0), x(l)]$, say, by the shortest geodesic.

Theorem 4.7.4 *([316]) The asymptotic cycle*

$$c = \lim \frac{c(l)}{l}$$

is the same for all leaves and all points of M. Moreover, c is proportional to the Poincaré dual of the first cohomology class of the initial closed 1-form defining the foliation. A deviation from the asymptotic cycle has order l^{ν_2}, where

$$1 + \nu_2 < 2$$

is the second Lyapunov exponent. More precisely,

$$\limsup \frac{log||c(l) - c^* l||}{\log l} = \nu_2 < 1.$$

All the limits are uniform.

Proof. The proof uses the interval exchange transformation technique. □

4.8 Smooth Classification

Two C^r-foliations ($r \geq 1$) $\mathcal{F}_1$ and $\mathcal{F}_2$ on the surface M are C^k*-smoothly equivalent* ($r \leq k$) if there exists a C^k-diffeomorphism h of M which sends the leaves of $\mathcal{F}_1$ to the leaves of $\mathcal{F}_2$. This equivalence relation leads to the classification of smooth foliations. The classification smooth foliations is by far more complicate than the one for topological foliations. Even in the "local" setting, smooth equivalence carries a "moduli problem": the invariants are continuous. As in the "topological" case, the smooth classification consists of the following:

(i) Find a constructive invariant which takes the same value on smoothly equivalent foliations.
(ii) To describe all invariants which are admissible, i.e., have a realization in the class of smooth foliations.
(iii) Find a standard representative in each equivalence class, i.e., for a given admissible invariant, construct a foliation with this invariant.

The torus and the Klein bottle are the unique closed surfaces which admit foliations without singularities. For such foliations a solid theory exists already: the Arnold theory of analytic conjugacy of diffeomorphisms of the unit circle. If our surface differs from the torus and the Klein bottle, singularities appear. Their presence adds "local" obstacles to solution of the problem. As a result, not much is known about smooth equivalence in this case, except the theorems of A. Katok and G. Forni concerning foliations given by area-preserving flows.

4.8.1 Torus and Klein Bottle

Lemma 4.8.1 *Let $\mathcal{F}$ be a C^r-foliation without singularities ($r \geq 1$) on the torus T^2 or the Klein bottle K^2. Suppose that $\mathcal{F}$ has no Reeb components. Then $\mathcal{F}$ is C^r-smoothly equivalent to a foliation under certain function over a C^r-diffeomorphism $\phi : S^1 \to S^1$. For the case of the surface T^2 (resp. K^2) the diffeomorphism ϕ is orientation-preserving (resp. reversing).*

Proof. It is sufficient to notice that $\mathcal{F}$ has a global cross-section intersected by each trajectory. □

Corollary 4.8.1 *Suppose the conditions of Lemma 4.8.1 hold. Then $\mathcal{F}$ is C^r-smoothly equivalent to the suspension over a circle C^r-diffeomorphism.*

We recall that an irrational number α is called *Diophantine* if there exist $c > 1$, $d > 1$ such that for any nonzero $p, q \in \mathbb{Z}$ we have

$$|q\alpha - p| > \frac{c}{q^d}.$$

Theorem 4.8.1 *Let $\mathcal{F}$ be a C^∞ (analytic) foliation on T^2 with a Diophantine rotation number. Then $\mathcal{F}$ is C^∞-smoothly (analytically) equivalent to a linear foliation.*

Proof. By lemma 4.8.1, $\mathcal{F}$ is smoothly equivalent to a special flow over a smooth diffeomorphism $\phi : S^1 \to S^1$. The rotation number of ϕ is calculated by an integer unimodular matrix from the rotation number of $\mathcal{F}$. Hence, $rot(\phi)$ is Diophantine as well. According to Yoccoz's result [308], ϕ is C^∞-smoothly or analytically conjugate to a circle rotation. □

Let us consider the torus flows preserving smooth area elements (or having smooth invariant measure). For any analytic flow with analytic invariant measure there exists the following theorem of Kolmogorov [170].

Theorem 4.8.2 *Suppose that the equations $\dot{x} = A(x, y), \dot{y} = B(x, y)$ define an analytic flow $\mathcal{F} = (T^2, f^t, \theta)$ on the torus. Suppose also that $\mathcal{F}$ has an analytic invariant measure with density $U(x, y) > 0$. If $A^2 + B^2 \neq 0$ then there exists an analytical diffeomorphism which transforms the above system to the system of the form*

$$\frac{dx}{dt} = \frac{1}{F(x, y)}, \qquad \frac{dy}{dt} = \frac{\alpha}{F(x, y)},$$

where $F(x, y) > 0$ is a positive function and α is a real constant.

This theorem has the following generalization [162].

Theorem 4.8.3 *Let f^t be a fixed-point-free C^k-flow on torus preserving C^k ($k \geq 1$) area element and let $n = \min(k, r + 1)$. Then f^t is C^n smoothly equivalent to a special flow over a circle rotation.*

Proof. Since f^t preserves a smooth area element then it has no Reeb components. Hence, f^t has a global section. The result follows from the existence of the smooth invariant measure and lemma 4.8.1. □

There exists an interesting result on the degree of regularity of the conjugating mapping.

Theorem 4.8.4 *For any integer $r \geq 0$ there is a C^∞ fixed-point-free flow on the torus that is C^r but not C^{r+1} smoothly equivalent to a linear flow.*

Proof. It is enough to construct a C^∞ diffeomorphism $f : S^1 \to S^1$ which is C^r but not C^{r+1} conjugate to the rotation $R_\alpha : S^1 \to S^1$ for some irrational α. To do this we apply Anosov-Katok's method which was introduced in

D. Anosov and A. Katok, New examples in smooth ergodic theory. Ergodic diffeomorphisms. Transl. of the Moscow Math. Soc., 23, 1970, 1-35.

We shall construct two sequences of C^∞ diffeomorphisms $f_n : S^1 \to S^1$ and their conjugates h_n with rotations R_{α_n} (so, $f_n \circ h_n = h_n \circ R_{\alpha_n}$) for some sequence of rationals α_n such that $f_n \to f$ in the C^∞ topology and $h_n \to h$ in the C^r but not the C^{r+1} topology. To illustrate the idea, we restrict ourselves to the case $r = 1$.

For each n set $h_n = A_1 \circ \ldots \circ A_n$ where $A_i(x) = x + a_i(x)$ is a C^∞ diffeomorphism of the circle S^1. First, we can always choose the C^0 norm of $a_i(x)$ sufficiently small (so that h_{n+1} is C^0 close enough to h_n) and hence $h_n \to h$ in the C^0 topology for some h. Let

$$a_n(x + \frac{1}{q_{n-1}}) = a_n(x)$$

where the sequence of rationals $\alpha_n = \frac{p_n}{q_n}$ will be constructed inductively in the following way. Set $f_n = h_n \circ R_{\alpha_n} \circ h_n^{-1}$ and $\alpha_{n+1} = \alpha_n + \frac{1}{K_n q_n}$. Since A_{n+1} commutes with R_{α_n} then

$$\begin{aligned} f_{n+1,K_n} = h_{n+1} \circ R_{\alpha_{n+1}} \circ h_{n+1}^{-1} &= h_n \circ A_{n+1} \circ R_{\alpha_n} \circ R_{\frac{1}{K_n q_n}} \circ h_{n+1}^{-1} \\ &= h_n \circ R_{\alpha_n} \circ (A_{n+1} \circ R_{\frac{1}{K_n q_n}} \circ A_{n+1}^{-1}) \circ h_n^{-1}. \end{aligned}$$

For sufficiently large K_n the functions $f_{n+1} = f_{n+1,K_n}$ and f_n are C^∞ close. Thus, taking K_n large enough, we can assume that $f_n \to f$ in the C^∞ topology for some C^∞ diffeomorphism f. Note that $f_n \circ h_n = h_n \circ R_{\alpha_n}$ implies $f \circ h = h \circ R_\alpha$ for some irrational $\alpha = \lim_{n\to\infty} \alpha_n$. One can construct a function that is arbitrarily small in the C^1–topology and big in the C^2–topology. Since $h'_{n+1} = h'_n(A_{n+1})(1 + a'_{n+1})$, taking a_{n+1} C^1 small, we have h'_{n+1} as C^0 close to h'_n as we need. Therefore $h_n \to h$ in the C^1 topology and h is a C^1 diffeomorphism. On the other hand,

$$h''_{n+1} = h''_n(A_{n+1})(1 + a'_{n+1}) + h'_n(A_{n+1})a''_{n+1}$$

and we can choose a_{n+1} in such a way as to obtain h''_{n+1} much larger that h''_n. Hence the sequence h''_n is divergent and h is not C^2. Finally, recall that a conjugacy with a minimal circle rotation is defined up to a rotation. □

4.8.2 Surfaces of Genus $g \geq 2$

Presently, substantial results exist only for the special class of foliations given by area preserving flows. Everywhere below in this subsection M is a closed orientable surface of genus $g \geq 2$ endowed with a smooth area 2-form ω. We denote by $S_\omega(M)$ a class of smooth vector fields which satisfy the following conditions:

(i) Every $X \in S_\omega(M)$ preserves ω.
(ii) Each singularity of $X \in S_\omega(M)$ is a generic n-fold saddle (i.e., near a singularity the vector field is Hamiltonian with the Hamiltonian function $\sqcup_{i=1}^{n+1}(\alpha_i x - \beta_i y) + R(x,y)$ in the local coordinates (x,y), and all quotient $\frac{\alpha_i}{\beta_i}$ are different while the $(n+1)$-jet of $R(x,y)$ vanishes at 0).

We fix a set Σ of points $p_1, \dots, p_r \in M$ and natural numbers $n_1, \dots, n_r$ such that

$$\Sigma_{i=1}^r n_i = 2g - 2. \tag{4.4}$$

Denote by $S_\omega(M, \Sigma) \subset S_\omega(M)$ the subset of vector fields such that the point p_i is a generic n_i-fold saddle $(i = 1, \dots, r)$ for every $X \in S_\omega(M, \Sigma)$ and $Fix(X) = \Sigma$. Note that the equation (4.4) agrees with the Poincaré-Hopf formula.

Katok's Local Classification Let $X \in S_\omega(M, \Sigma)$ be a vector field tangent to the smooth flow $\mathcal{F} = (M, f^t, \theta)$ and let μ be a measure defined by ω (i.e. $\omega = d\mu$). Since μ is invariant under f^t, the flux form $X \lrcorner \omega = \beta_X$ of X is closed 1-form. By Poincaré duality, $\beta_X \in H^1(M, I\!R)$ corresponds to an element $A_\mu \in H_1(M, I\!R)$ which is an asymptotic cycle of the flow f^t with regard to the measure μ. Denote by $H_1(M, \Sigma, I\!R)$ the relative homology group. Recall that the element $\hat{R}_\mu \in H_1(M, \Sigma, I\!R)$ corresponding to A_μ is called a fundamental class of the flow f^t relative to the measure μ. The following theorem was proved by A. Katok [160], see also [162].

Theorem 4.8.5 *Let $X_t \in S_\omega(M, \Sigma)$, $0 \leq t \leq 1$, be a smooth family of vector fields such that*

$$\hat{R}_\mu(X_t) = \lambda_t \hat{R}_\mu(X_0)$$

where $\lambda_t > 0$ and $\omega = d\mu$. Then there exists a family $h_t : M \to M$ of Lipschitz homeomorphisms which are C^∞ diffeomorphisms away from Σ, such that $h_t|_\Sigma = id$ and h_t effects a Lipschitz equivalence between the flows generated by X_0 and X_t.

Proof. Consider the 1-form $\beta_t = X \lrcorner \omega$. Since β_t is dual to $R_\mu(X_t)$, $\beta_t = \lambda_t \beta_0$, and therefore, $(\beta_t)' = (\lambda_t)'\beta_0$. Let us fix t and consider the equation

$$d\omega(X_t, V_t) = -(\lambda_t)'\beta_0$$

for the vector field V_t. Since β_0 is closed 1-form, $-(\lambda_t)'\beta_0 = d\psi$ locally for some smooth function ψ. We can assume $\psi|_\Sigma = 0$ because $\beta_0 \in H^1(M, \Sigma; \mathbb{R})$ and ψ is defined up to a constant. It suffices to solve the equation

$$\omega(X_t, V_t) = \psi. \tag{4.5}$$

Since ω is a nondegenerate 2-form the equation (4.5) has a solution defined up to an element from $Ker\ \omega(X_t, \cdot) = X_t$. The solution of (4.5) is uniquely defined if V_t is orthogonal (in a fixed Riemannian metric) to X_t. Due to the local structure of X_t near singularities, V_t is C^∞ away from Σ and is Lipschitz on M. Hence, V_t is uniquely integrable and defines a flow h^t. Denote by h_t time shift along the trajectories of h^t. Obviously, h^t is a Lipschitz homeomorphism which is C^∞ diffeomorphism away from Σ.

Since X_t is an area preserving vector field, 1-form β_t is closed. Hence,

$$\pounds_{V_t}\beta_t = V_t y d\beta_t + d(V_t y \beta_t) = d(V_t y \beta_t).$$

Therefore

$$\frac{d}{dt}(h_t^*\beta_t) = h_t^*(\pounds_{V_t}\beta_t) + h_t^*(\frac{d}{dt}\beta_t) = h_t^*[d(V_t y\beta_t) + \frac{d}{dt}\beta_t] = 0$$

because

$$d(V_t y \beta_t) = d(d(V_t y \mathcal{X}_t y \omega) = d\omega(X_t, V_t) = -(\lambda_t)'\beta_0 = -(\beta_t)'.$$

Hence, $h_t^*\beta_t = \beta_0$ for all $0 \le t \le 1$. The 1-form β_t defines a foliation formed by the trajectories of X_t. It follows that h_t maps trajectories of X_0 onto trajectories of X_t. □

Remark 4.8.1 *Note that*

$$\beta_0 = X_0 y\omega = h_t^*(X_t y \omega) = h_t^* X_t y h_t^* \omega = \lambda_t^{-1} X_0 y h_t^* \omega$$

implies that $h_t^*\omega = \lambda_t \omega$.

If all saddles in Theorem 4.8.5 are Morse then one can get a smooth equivalence on the whole surface.

Theorem 4.8.6 *[160], [162] Let* $X_t \in S_\omega(M, \Sigma)$, $0 \le t \le 1$, *be a smooth family of vector fields such that all saddles of any* X_t *are Morse and*

$$\hat{R}_\mu(X_t) = \lambda_t \hat{R}_\mu(X_0)$$

where $\lambda_t > 0$ *and* $\omega = d\mu$. *Then each* X_t *is smoothly equivalent to* X_0.

Proof. Since any X_t is the area preserving vector field and every singularity is the Morse saddle then, in some neighborhood of each singularity, is locally smoothly equivalent to a standard saddle. We repeat the argument of Theorem 4.8.5. □

One gets smooth equivalence provided the n-fold saddles are *model saddles*, i.e., near a singularity $z = 0$ the foliation is given by

$$Im(z^m dz^2)^{\frac{1}{2}} = Const.$$

Denote by $\mathcal{E}_\omega(M, \Sigma) \subset S_\omega(M, \Sigma)$ a subset of vector fields having only model saddles.

Theorem 4.8.7 *[160], [162] Let $X_t \in \mathcal{E}_\omega(M, \Sigma)$, $0 \le t \le 1$, be a smooth family of vector fields such that*

$$\hat{R}_\mu(X_t) = \lambda_t \hat{R}_\mu(X_0)$$

where $\lambda_t > 0$ and $\omega = d\mu$. Then each X_t is smoothly equivalent to X_0.

Proof. This is much the same as for Theorem 4.8.6. □

Cohomological Equations of G. Forni Cohomological equations play an important role in the smooth classification of dynamical systems. Given diffeomorphism $f : M \to M$ (a dynamical system with a discrete time) a *cohomological equation* is a linear equation:

$$\lambda\phi(f(x)) - \phi(x) = \psi(x)$$

where ψ is a given function on M, λ is a given constant and ϕ is an unknown function. For the flow f^t (a dynamical system with continuous time) generated by the vector field X the cohomological equation is of the form

$$X\phi = \psi.$$

In case of 1–dimensional circle diffeomorphisms, the cohomological equation plays an important role, see for example:

D. V. Anosov, On an additive functional homology equation connected an ergodic rotation on the circle, Math. of USSR, Izvestia, 1973, 7(6), 1257-1271.

V. I. Arnold, Small denominators I. Mapping of the circle onto itself, Translations of AMS (series 2), 46, 1965, 213-284.

V. I. Arnold, Small denominators and problem of stability of motion in classical and celestial mechanics, Russian Math. Surv., 18, 1963, 6, 85-193.

G. R. Belitsky, Equivalence and normal forms of germs of smooth mappings, Russian Math. Survey, 31, 1978, 1, 107-177.

The cohomological equation can be solved for flows on the torus. Concerning the closed orientable surfaces of higher genus, a remarkable paper of G. Forni [109] has recently appeared. Below we give an outline of Forni's results.

Recall that $S_\omega(M)$ is the class of smooth vector fields preserving the area 2-form ω whose singularities are generic n-fold saddles. $\mathcal{E}_\omega(M, \Sigma) \subset S_\omega(M, \Sigma)$ is a subset of vector fields having only model saddles. Let μ be the invariant measure defined by ω. A fundamental class $\hat{R}_\mu(X)$, $X \in S_\omega(M)$ defines the transformation

$$\hat{R}_\mu : \mathcal{E}_\omega(M, \Sigma) \to H^1(M, \Sigma; \mathbb{R}).$$

(In other words, for a given vector field $X \in \mathcal{E}_\omega(M, \Sigma)$ we relate its fundamental class $\hat{R}_\mu(X)$ with regard to the invariant measure μ.) We shall say that a set $\mathcal{F} \subset \mathcal{E}_\omega(M, \Sigma)$ is of *full measure* in $H^1(M, \Sigma; \mathbb{R})$ if $H^1(M, \Sigma; \mathbb{R}) - \hat{R}_\mu(\mathcal{F})$ has zero Lebesgue measure in $H^1(M, \Sigma; \mathbb{R})$.

Finally, let us recall the definition of the Sobolev space. Given area form ω the norm $\| \cdot \|_0$ is defined as

$$\|u\|_0 = \sqrt{\int_M |u|^2 \omega}$$

where $u : M \to \mathbb{R}$ is a L^2 function.

The Sobolev norm $\| \cdot \|_s$, $s \geq 1$, is defined as follows. Let $\{U_i, \phi_i\}$ be a finite covering of M and $\frac{\partial}{\partial x_i}, \frac{\partial}{\partial y_i}$ commuting vector fields supported on U_i. Then

$$\|u\|_s = \max_i \sqrt{\sum_{k+j \leq s} \|\frac{\partial^k}{\partial x_i^k} \frac{\partial^j}{\partial y_i^j} (u \circ \phi_i)\|_0^2}.$$

We set

$$H^s(M) = \{u \in C^\infty(M, \mathbb{C}) : \|u\|_s < +\infty.\}$$

Denote by H_{loc}^{-s} the Sobolev space of functions which are s-times weakly locally differentiable.

Theorem 4.8.8 (Forni) *There exists a full measure set $\mathcal{F}_\omega(M, \Sigma) \subset \mathcal{E}_\omega(M, \Sigma)$ such that for every $X \in \mathcal{F}_\omega(M, \Sigma)$ the following holds: There exists a natural number $r \geq 1$ such that, if $f \in H^{r-1}(M)$ has a compact support in $M - \Sigma$ and satisfies $\int_M f\omega = 0$, then the cohomological equation $Xu = f$ has a distributional solution $u \in H_{loc}^{-s}(M - \Sigma)$.*

Proof. See the details in [109]. □

Using the pair of commuting vector fields X, Y one can develop a Fourier analysis for the metric d_q associated with the holomorphic differential q. Let us introduce a *scalar product:*

$$(u, v)_q = \int_M u\bar{v}\omega_q$$

and the so–called *Dirichlet hermitian form:*

$$\mathcal{Q}(u, v) = (Xu, Xv)_q + (Yu, Yv)_q.$$

There is a constant $C > 0$ such that the following inequality holds:

$$\|u - \int_M u\omega_q\|_0^2 \leq C\mathcal{Q}(u, u)$$

for any $u \in H_q^1(M)$. The *Laplace operator* is defined as:

$$\Delta_q = -(X^2 + Y^2).$$

The following *spectral theorem* is true:

Theorem 4.8.9 *The Dirichlet hermitian form $\mathcal{Q}$ on $L_q^2(M)$ has the following properties:*

(i) $\mathcal{Q}$ is positive semi-define and the set $EV(\mathcal{Q})$ of its eigenvalues is a discrete subset of $[0, +\infty)$.
(ii) Each eigenvalue has finite multiplicity; in particular the eigenvalue 0 is simple and the kernel of $\mathcal{Q}$ consists only of constant functions.
(iii) The space $L_q^2(M)$ decomposes as the orthogonal sum of eigenspaces. Furthermore, all of the eigenfunctions are real analytic on M.

Let q be a holomorphic quadratic differential which is the square of an abelian differential (if this is not the case we may lift q to a double covering ramified over the zeros of q of odd order). Let $\theta \in S^1$. It is possible to associate to q a one-parameter family of holomorphic quadratic differentials $q_\theta = (\exp(-i\theta))^2 q$ and the corresponding horizontal foliations $\mathcal{F}_\theta$. Each $\mathcal{F}_\theta$ can be obtained by the integration of the dual vector field

$$S_\theta = \frac{\exp[-i\theta](X + iY) + \exp[i\theta](X - iY)}{2}.$$

Theorem 4.8.10 (Forni) *There exists a natural number $r \geq 1$ such that the following holds: For almost all $\theta \in S^1$ (with respect to Lebesgue measure) there is a constant $C_\theta > 0$ such that if $f \in H_q^{r-1}(M)$ and $\int_M f\omega_q = 0$ then the cohomological equation $S_\theta u = f$ has a distributional solution $u \in H_q^{-r}(M)$ satisfying the following estimate:*

$$|u|_{-r} \leq C_\theta |f|_{r-1}.$$

Proof. See [109]. □

Theorem 4.8.11 (Forni) *Let $X \in \mathcal{F}_\omega(M, \Sigma) \subset \mathcal{E}_\omega(M, \Sigma)$. Then, for any $s \geq l = 2r-1$, there exists a finite number $n_s > 0$ of X-invariant distributions on $M - \Sigma$, $\mathcal{D}_1^X, \ldots, \mathcal{D}_{n_s}^X \in H_{loc}^{-s}(M - \Sigma)$ such that the following holds. If $f \in H^s(M)$ is supported on a compact set $K \subset M - \Sigma$ and satisfies*

$$\int_M f\omega = 0, \quad \mathcal{D}_i^X(f) = 0, \quad i = 1, \ldots, n_s,$$

there exists a solution $u \in H_{loc}^{s-l}(M - \Sigma)$ of the cohomological equation $Xu = f$. In addition, $u \in H^{s-l}(M)$ if $s > l$ and the following estimate holds:

$$\|u - vol_\omega(M)^{-1} \int_M u\omega\|_{s-l} \leq C_{X,K}^s \|f\|_s.$$

Proof. See [109]. □

Theorem 4.8.12 (Forni) *There exists $l > 0$ such that the following holds. For almost all $\theta \in S^1$ (with respect to Lebesgue measure) there is a constant $C_\theta^s > 0$ and a finite number $n_s > 0$ of S_θ-invariant distributions $\mathcal{D}_1, \ldots, \mathcal{D}_{n_s} \in H_q^s(M)$ such that, $f \in H_q^s(M)$ is compactly supported in $M - \Sigma$ and satisfies*

$$\int_M f\omega_q = 0, \quad \mathcal{D}_i(f) = 0, \quad i = 1, \ldots, n_s,$$

there exists a solution $u \in H_q^{s-l}(M)$ of the cohomological equation $S_\theta u = f$ satisfying the estimate

$$|u - vol_q(M)^{-1} \int_M u\omega_q|_{s-l} \leq C_\theta^s |f|_s.$$

Proof. See [109]. □

Note that even in the case of area-preserving and ergodic flows the solvability of the cohomological equation in the neighbourhoods of fixed points does not imply its "global" solvability. Here cohomological obstacles appear; see [44].

Bibliographic Notes. The idea of studying recurrent flows (and foliations) by 'lifting' them to the universal covering surface $\mathbb{H}$ has a rich history. Apparently André Weil was the first [297], [298] to use this idea to study the rotation numbers for flows without equilibria on the two-dimensional torus. The theory was widely developed by D. Anosov [8], [9],[11],[12],[13]; see also the work of V. Pupko [248]. Consult also our overview of the *Anosov-Weil problem.* This method received development in the works of S. Aranson and Grines [18], [19] and S. Aranson and E. Zhuzhoma [28],[30]. Our exposition is close to the work [18].

5. Curves on Surfaces

Every foliation on the surface is a union of curves. It is therefore logical to pay special attention to curves. On the other hand, the behaviour of curves without self–intersections on surfaces is a problem of independent interest.

5.1 Curves and the Absolute

5.1.1 Notations

Let M be a closed surface of non-positive Euler characteristic $\chi(M)$. Let $l = \{l(t) : t \geq 0\}$ be a semi-infinite simple curve on M, and let $\widetilde{l} = \{\widetilde{l}(t) : t \geq 0\}$ be a connected component of its preimage $\pi^{-1}(l)$ on the universal covering $\widetilde{M}$. (Here $\pi : \widetilde{M} \to M$ is the covering mapping and $\widetilde{M}$ is either the Euclidean plane $\mathbb{R}^2$ or the unit disc Δ [1].) By $\widetilde{d}$ we denote a metric on $\widetilde{M}$. The curve $\widetilde{l}$ *abandons* each compact subset of $\widetilde{M}$, or is *unbounded* if

$$\widetilde{\lim}_{t\to\infty}\widetilde{d}(m_0, \widetilde{l}(t)) = +\infty,$$

where $m_0 \in \widetilde{M}$ is an arbitrary point of the universal covering. A curve lying in a compact subset $\widetilde{M}_0 \subset \widetilde{M}$ is called *bounded.* We say that curve $\widetilde{l}$ *goes to infinity* if

$$\lim_{t\to\infty} \widetilde{d}(m_0, \widetilde{l}(t)) = +\infty.$$

(Evidently the last of the above equations implies the first, the converse being false.) Consider a flow on torus for which the covering flow is given by the system

$$\dot{x} = 1, \quad \dot{y} = \widetilde{f}(x + n, y + k),$$

where $(n, k) \in \mathbb{Z}^2$ are integers. If every trajectory of the covering flow goes to infinity what can be said about its asymptotic behavior? To formalize the latter notion one needs the following definition.

[1] Sometime we replace Δ by the Lobachevsky half-plane $\mathbb{H}$, which should not confuse the reader.

Definition 5.1.1 *Let* $l = \{l(t) : t \geq 0\}$ *be a semi-infinite simple curve, and let* $\widetilde{l} = \{\widetilde{l}(t) : t \geq 0\}$ *be a connected component of its lifting to* $\mathbb{R}^2$ *(if* $\chi(M) = 0$*) or to* Δ *(if* $\chi(M) < 0$*). If* $\widetilde{l}$ *tends to a point* σ *of the absolute* S_∞ *then* $\widetilde{l}$ *is said to have asymptotic direction* σ*.*

(One can also say that l has an asymptotic direction.) Every oriented geodesic line with an 'endpoint' in σ will be called a *representative* of given asymptotic direction. The asymptotic direction is said to be *rational* (resp., *irrational*) if the point $\sigma \in S_\infty$ is a rational (resp., irrational) point of the absolute.

5.1.2 Background

During his stay in India, André Weil became interested in the asymptotic behavior of curves on closed surfaces. Before we turn to the discussion of Weil's heritage, the following empirical remark will be helpful. If one has a covering flow of the fixed-point-free flow on torus, it is possible to introduce the Poincaré rotation number using purely geometric considerations. Denote by $C(p,q)$ a trajectory passing through a point with the integer coordinates $(p,q) \in \mathbb{R}^2$. Let us split the set of rationals $\mathbb{Q}$ into two classes L, R as follows:

$$\frac{p}{q} \in \begin{cases} \mathrm{R} & \text{if } C(p,q) \text{ is ``over'' } C(0,0) \\ \mathrm{L} & \text{if } C(p,q) \text{ is ``under'' } C(0,0) \end{cases}$$

where $p \geq 1$. If $p/q \in R$ and $s/r > p/q$ then $s/r \in R$. Indeed, since $sp > rq$ then the curve $C(rp, sp)$ lies "over" the curve $C(rp, rq)$. Therefore $sp/rp = s/r \in R$. (The same argument is applicable to the set L.) Since each rational number belongs to one of the classes R and L; the closure of the above classes defines a section of the set of real numbers. Every such a section specifies a real number, α, which is equal to the Poincaré rotation number. It can be seen that α is the slope of the straight line defining the asymptotic direction.

Weil proved that the covering trajectories must have an asymptotic direction [298]. Weil's method of the study of asymptotic directions is more geometric than Poincaré's method, which consists in the study of the first return mappings on the global cross-sections to the flows. The Weil method permits the introduction of the rotation numbers and hence to study of the dynamical properties of flows on the torus. What is more important is that Weil apparently inferred that his method works not exclusively for torus flows but also for higher genus flows and is applicable as well to arbitrary families of curves (which are not necessarily integral curves of differential equations). We quote his 1937 talk at the First International Topological Conference in Moscow [298]:

"Dans la présente communication, l'auteur discute deux méthodes pouvant servir à l'etude de la question et d'autres analogues. La première, qui a déjà été développée dans un article du [297], consiste à considérer dans le plan

(x, y) en même temps que la courbe C de la famille, toutes les courbes $C_{p,q}$ qui s'en déduisent par une translation (p, q), p et q étant des entiers: la position relative de ces courbes par rapport à C permet, non seulement de déterminer le nombre de rotation, mais encore la transformation qui ramène la famille étudiée à une forme canonique. La méthode s'applique dans le cas de Poincaré, et plus généralement chaque fois que la famille ne présante pas de 'col à l'infini' (au sens de Niemytzky). [2] *D'ailleurs cette dernière circonstance ne peut vraisemblablement pas se présenter si la famille ne contient pas de courbe fermée.* [3] *À cette méthode se relie encore le théorème suivant, d'ailleurs obtenu par une voie quelque peu différente:*

Soit, sur le tore, une courbe de Jordan, image continue de la demi-droite $0 \leq t < +\infty$; on suppose que cette courbe soit sans point double; alors, si l'image de la courbe dans le plan (x, y), surface de recouvrement universelle du tore, tend vers l'infini avec t, elle y tend avec une direction asymptotique bien déterminée, c'est-à-dire que la rapport $\frac{x(t)}{y(t)}$ tend vers une limite quand t tend vers $+\infty$.

Une généralisation très intéressante du probllème étudié, qui paraît susceptible d'être abordée par la même méthode, est l'étude, sur une surface close de genre p, des solutions d'une équation différentielle du premier ordre n'ayant d'autres points singuliers que de cols, ou en termes topologiques, d'une famille de courbes dont tous les points singuliers sont d'indice négatif. [4] *Un premier résultat est suivant:*

Sur le cercle hyperbolique, surface de recouvrement universelle de la surface étudiée, toute courbe de la famille tend, dans chaque direction, vers un point à l'infinie bien détetminé. (English version is given below.) [5]

[2] This means that the covering flow in the plane does not admit a strip in which the integral curves are homeomorphic to the family of parabolas $y = x^2 + C$. (The remark is ours.)

[3] Magnier obtained the proof of this result. (The remark of A. Weil.)

[4] Magnier obtained very interesting results on this question as well, they will be published shortly. (Remark of A. Weil.)

[5] In the present talk the author discusses two methods which can aid the study of the question, and other analogies. The first method, which was already developed in the paper [297], consists in considering in the plane (x, y) along with the curve C of the family, all the curves $C_{p,q}$ that are obtained by translations (p, q) where p, q are integers: the relative position of these curves with regard to C allows one not only to determine the rotation number, but also the transformation which brings the family to canonical form. The method is applicable to Poincaré's case, and more generally each time when the family does not have 'saddle at infinity' (in the sense of Nemytsky). Anyway, the latter possibility evidently cannot be realized unless the family contains a closed curve. Connected with this method, there is a theorem which can be obtained also in other way:

Let a Jordan curve which is an image of a ray $0 \leq t < +\infty$ on the torus be given; one assumes that this curve has no double points; if the image of the curve on the plane (x, y), which is the universal covering of torus, tends to infinity along with t, it tends in a well-defined asymptotic direction, that is, the quotient $\frac{x(t)}{y(t)}$ tends to a limit when t tends to $+\infty$.

In strict terms, Weil's theorem and Weil's conjecture can be stated as follows.

Theorem 5.1.1 *Let $l = \{m(t) : t \geq 0\}$ be a semi-infinite curve without self-intersections on the torus T^2. Let $\widetilde{l} = \{\widetilde{m}(t) : t \geq 0\}$ be its covering on $D^2 = I\!R^2 \cup \{\infty\}$. Then if $\widetilde{l}$ goes to infinity, it has an asymptotic direction therein.*

Conjecture 5.1.1 *Let $l = \{m(t) : t \geq 0\}$ be a semi-infinite curve without self-intersections on the closed hyperbolic surface M. Let $\widetilde{l} = \{\widetilde{m}(t) : t \geq 0\}$ be its covering on Δ. Then if $\widetilde{l}$ goes to infinity, it has an asymptotic direction therein.*

Neither Weil nor Magnier have ever published a proof of their statements. Weil's method was forgotten until the 60s when D. Anosov and N. Markley, independently, started to involve it in their considerations.

5.1.3 Proof of Weil's Conjectures

Proof of Weil's conjecture. Let l be a semi-infinite, simple curve on a closed surface M of genus $g \geq 2$ and let $\widetilde{l}$ be its lifting to Δ. Suppose to the contrary that $\widetilde{l}$ has no asymptotic direction. Since $\widetilde{l}$ goes to infinity, it must contain in its limit set an interval $I \subseteq S_\infty$. There exists an axis $\widetilde{A}$ of a hyperbolic element $\gamma \in \Gamma$ with endpoints in I. Then $\widetilde{l}$ must intersect $\widetilde{A}$ arbitrarily close (in the Euclidean metric) to the endpoints. But then $\gamma^n(\widetilde{l}) \cap \widetilde{l} \neq \emptyset$ for some $n \in \mathbb{Z}$. This contradicts the assumption that l is simple (i.e., without self-intersections). □

Proof of Weil's theorem. The following lemma plays a key role in the proof:

Lemma 5.1.1 *Let $\widetilde{l}$ be a lifting of curve l to the plane $I\!R^2$. Suppose that $\widetilde{l}$ intersects a straight line $\widetilde{l}_0$ having rational slope at a point which lies infinitely far away (i.e., beyond any compact region). Then $\widetilde{l}$ has a bounded deviation (at least from one side) from $\widetilde{l}_0$.*

Proof of lemma. This is rather technical. The idea is to construct an arc which consists of segments of $\widetilde{l}_0$ and arcs of $\widetilde{l}$. This arc forms a continuous curve $\widetilde{C}$ with respect to the shifts by elements of the lattice $\mathbb{Z}^2$. The curve $\widetilde{C}$ has bounded deviation from $\widetilde{l}_0$ and forms a barrier which cannot be crossed by $\widetilde{l}$. □

A very interesting generalization of the studied problem, which is likely solvable by the same method, is the study on a closed surface of genus p of the differential equations of first order having only singular points of saddle type, or in topological terms, of families of curves with singularities of negative index. The first result is the following:

In the hyperbolic circle which is the universal covering of the studied surface all curves of the family tend in each direction to a well-defined point at infinity.

The rest of the proof of Weil's theorem is transparent. It follows from the above lemma that $\widetilde{l}$ cannot intersect two different straight lines of rational slope at infinitely distant points. (Otherwise there would exist a curve which does not satisfy the lemma.) Now two cases are possible. Either $\widetilde{l}$ does not intersect (infinitely far) no one line with rational slope, and in this case $\widetilde{l}$ is squeezed between two rational slope lines which give in the limit the required asymptotic direction. Or $\widetilde{l}$ satisfies the lemma and then $\widetilde{l}_0$ defines the asymptotic direction. This finishes the proof of Weil's theorem. □

5.1.4 Theorems of D. V. Anosov

From the very beginning D. Anosov singled-out a question which was omitted by A. Weil (dropped by him for the sake of simplicity). Namely, the question was:

Question 1. Under what conditions do the covering trajectories which quit each compact region of the universal covering go to infinity?

(In other words, does the existence of the upper limit imply the existence of the usual limit in the formulas for distances mentioned at the top of this chapter?) A similar question can be asked for the leaves of foliations and laminations:

Question 1'. Under what conditions do the semi-leaves of a covering foliation or lamination which quit every compact region of the universal covering go to infinity?

Example 5.1.1 *Let a flow $\widetilde{f}^t$ be given in a small strip $0 < |y| \leq \varepsilon$ of the axis Ox by equations*

$$\frac{dx}{dt} = \left(\frac{1}{y^2}\sin\frac{1}{y} - \frac{1}{y^3}\cos\frac{1}{y}\right) ye^{-\frac{1}{y^2}}, \qquad \frac{dy}{dt} = -ye^{-\frac{1}{y^2}}.$$

All points of the line $y = 0$ are singular points of the flow. Evidently the flow is extendable to a C^∞ flow on $\mathbb{R}^2$ covering some flow f^t on the torus. The curves $x = \frac{1}{y}\sin\frac{1}{y} + Const$ are integral for the flow $\widetilde{f}^t$ in the strip $0 < |y| \leq \varepsilon$ so that the infinite decreasing of $|y|$ corresponds to $t \to +\infty$.

Let us analyze the above example. The straight line $y = 0$ covers a non-trivial closed curve on the torus. Therefore the flow f^t has a continuum of singular points which form a contractible set on T^2. (It is not hard to deduce that similar flows exist on arbitrary closed surfaces of negative Euler characteristic.) The above example says that in order that the covering trajectories go to infinity, singular points of the underlying flow must not occupy 'too much space' on M. This idea was formalized in the early 60's by D. Anosov.

Theorem 5.1.2 *If the set of singular points of a smooth flow f^t on a closed surface M of non-positive Euler characteristic is finite, then every trajectory of the covering flow $\widetilde{f}^t$ on $\widetilde{M}$ is either bounded or has an asymptotic direction.*

Proof. Suppose that the covering flow $\widetilde{f}^t$ on $\widetilde{M}$ has a semi-trajectory (say positive) $\widetilde{l}^+$ which is unbounded and returns in a neighbourhood of a compact region $U \subset \widetilde{M}$ after arbitrarily big periods of time. Since f^t has a finite number of fixed points, the semi-trajectory $\widetilde{l}^+$ has at least one limit point in U which is different from an equilibrium. Therefore there exists a transversal segment $\widetilde{\Sigma}$ which intersects $\widetilde{l}^+$ in a countable set of points. Since $\widetilde{M}$ is simply connected there exists a monotone sequence of points $\widetilde{m}_i \in \widetilde{l}^+ \cap \widetilde{\Sigma}$ such that the following conditions are satisfied:

(i) The arc $\widetilde{l}_i \subset \widetilde{l}^+$ of $\widetilde{l}^+$ with the endpoints $\widetilde{m}_{2i}, \widetilde{m}_{2i+1}$ and the segment $\widetilde{I}_i \subset \widetilde{\Sigma}$ with the same endpoints form a closed curve which bounds a region $\widetilde{D}_i$ in $\widetilde{M}$.
(ii) The following inclusions hold: $\widetilde{D}_0 \subset \ldots \subset \widetilde{D}_i \subset \widetilde{D}_{i+1} \subset \ldots$

Since the lengths of the arcs grow infinitely, the measure of $\widetilde{D}_i$ does the same as $i \to \infty$. Since the fundamental region of the isometry group Γ has finite measure, there exists $\gamma \in \Gamma$ different from the identity and a region $\widetilde{D}$ such that

$$\gamma(\widetilde{D}) \bigcap \widetilde{D} \neq \emptyset.$$

Let us show that the latter is impossible. Indeed, the transversal segment $\widetilde{I}_i$ can be taken so small that $\gamma(I_i) \cap I_i = \emptyset$. On the other hand, the arc $\widetilde{l}_i \subset \widetilde{l}^+$ cannot have congruent points since otherwise the semi-trajectory must go to infinity. Therefore the set $\gamma(\widetilde{l}_i) \cap \widetilde{I}_i \neq \emptyset$ and consists of a unique point which we denote by $\widetilde{m}$. Since $\widetilde{l}^+$ is unbounded then as time grows infinitely the trajectories of $\widetilde{f}^t$ must leave $\widetilde{D}_i$. Denote by $\widetilde{a}$ an extreme point of I_i whose negative semi-trajectory lies in $\widetilde{D}_i$. Then the negative semi-trajectory through the point $\gamma(\widetilde{m})$ contains the negative semi-trajectory through the point $\widetilde{a}$, and therefore the negative semi-trajectory through $\gamma^2(\widetilde{m})$ contains the negative semi-trajectory through $\gamma(\widetilde{a}) \in \widetilde{D}$. (Evidently the negative semi-trajectory of $\gamma^2(\widetilde{m})$ lies in $\gamma^2(\widetilde{D})$. Therefore the negative semi-trajectory through $\gamma(\widetilde{a})$ lies both in $\widetilde{D}$ and $\gamma^2(\widetilde{D})$. This means that $\widetilde{D} \cap \gamma^2(\widetilde{D}) \neq \emptyset$.

We repeat the above argumentation using the element γ^2 instead of γ, thus obtaining $\widetilde{D} \cap \gamma^4(\widetilde{D}) \neq \emptyset$. Therefore in general $\widetilde{D} \cap \gamma^{2^n}(\widetilde{D}) \neq \emptyset$ for each $n \in \mathbb{N}$. On the other hand, since $\widetilde{D}$ is bounded, there can be only finite number of non-empty intersections $\widetilde{D} \cap \gamma^m(\widetilde{D})$. This contradiction finishes the proof of Anosov's theorem. □

Definition 5.1.2 *A subset $A \subset M$ is called point-contractible if there exists a continuous mapping $\varphi : F \times [0,1] \to M$ such that $\varphi(m,0) = m, \varphi(m,1) = m_0$, where $m \in M$ runs over the set A and $m_0 \in M$ is a fixed point.*

Theorem 5.1.3 *Let f^t be a topological flow on a closed surface M of non-positive Euler characteristic. If the set $Fix\ f^t$ is point-contractible, then every trajectory of the covering flow $\widetilde{f}^t$ on $\widetilde{M}$ is either bounded or has an asymptotic direction.*

Proof. (This theorem was proved by Anosov in 1987; see [10].) The proof is much the same as for Theorem 5.1.2. □

In 1965 at the Topological Symposium in Tiraspol D. Anosov announced Theorem 5.1.2 and proved that the covering unbounded semi-trajectory goes to infinity and has an asymptotic direction. Since the proofs of Weil's conjecture and Weil's theorem had not been published, D. Anosov formulated several questions containing Weil's conjecture and Weil's theorem.

Question 2. Let $l = \{l(t) : t \geq 0\}$ be a semi-infinite simple curve on a closed surface M of non-positive Euler characteristic. Let $\widetilde{l}$ be its lifting to the universal covering M. If $\widetilde{l}$ goes to infinity does it have an asymptotic direction?

Question 3. Which points of the absolute S_∞ can belong to the limit set of $\widetilde{l}$?

Question 4. Suppose that $\widetilde{l}$ tends to the point $\sigma \in S_\infty$. Let an oriented geodesic $\widetilde{g}$ have σ as one of its endpoints. Denote by $\widetilde{d}(t)$ the distance between the point $\widetilde{l}(t)$ and geodesic $\widetilde{g}$. Is the deviation $\widetilde{d}(t)$ is bounded or not? (In other words, does there exist a constant $C > 0$ such that $\widetilde{d}(t) \leq C$ for all $t \in [0, +\infty)$?)

Definition 5.1.3 *Let $\widetilde{l}$ be a curve with asymptotic direction σ. We say that the curve $\widetilde{l}$ has bounded (unbounded) deviation if, for any geodesic $\widetilde{g}$ which has one of its endpoint in σ, the deviation of $\widetilde{l}$ from $\widetilde{g}$ is bounded (unbounded).*

The above definition is correct since if there exists a bounded deviation from one of the geodesics in the direction σ the same is true for any geodesic in the direction σ. Sometime we say that the curve $l \in M$ has bounded (unbounded) deviation, referring to the deviation of $\widetilde{l}$ on $\widetilde{M}$.

5.2 Asymptotic Directions

5.2.1 Of Recurrent Semi-Trajectory

Recall that a non-trivial recurrent semi-trajectory is a semi-trajectory which lies in its own limit set. Such a trajectory can exist only on orientable surfaces of genus $g \geq 1$ or non-orientable surfaces of genus $p \geq 3$. Since the Euler characteristic of these surfaces is non-positive, the question of the existence of asymptotic directions makes sense. It turns out that the asymptotic direction exists for every semi-trajectory with an arbitrary set of fixed points.

Theorem 5.2.1 *Let l be a non-trivial recurrent semi-trajectory of a flow f^t on closed surface of non-positive Euler characteristic. Let $\widetilde{l}$ be a connected component of $\pi^{-1}(l)$ on the universal covering $\widetilde{M}$. Then $\widetilde{l}$ has an asymptotic direction which is irrational.*

Proof. It is enough to consider the orientable surface M. Next we have to consider separately the case of $M = T^2$ a torus and M a surface of genus $g \geq 2$.

For the fixed-point-free torus flows Theorem 5.2.1 was proved by Weil [297] in 1931. For the torus flows with fixed point Theorem 5.2.1 was proved independently by Aranson [16] and Markley [195].

Let $\widetilde{l}$ be a lift of l to the plane $\mathbb{R}^2$. By the Bendixson theorem, $\widetilde{l}$ goes to infinity. Hence, according to the Weil theorem, $\widetilde{l}$ has an asymptotic direction.

A similar speculation yields the proof for the case of a hyperbolic surface where the appeal to the Weil theorem is replaced by one to the Weil conjecture. We give another proof (obtained by Aranson and Grines [18]) in the hyperbolic case which uses properties of the hyperbolic plane.

Since l is a non-trivial recurrent semi-trajectory, it cuts the transversal segment in a countable set of points. The same is true for closed transversal C. Therefore, C cannot split the surface and its complete preimage $\pi^{-1}(C)$ is a countable family of disjoint curves, the endpoints of which lie at the absolute. Each curve from $\pi^{-1}(C)$ is a transversal to the covering flow, and cannot intersect with the trajectory $\widetilde{l}$ more than once. Since $l \cap C$ is a countable set, $\widetilde{l}$ intersects pairwise different curves $\widetilde{C}_i \in \pi^{-1}(C)$. Since the isometry group is discontinuous, the topological limit of the curves $\widetilde{C}_i$ consists of a unique point $\omega(\widetilde{l})$ of the absolute which defines the asymptotic direction of $\widetilde{l}$.

Let us prove the irrationality of $\omega(\widetilde{l})$ following the guidelines of [29]. Suppose to the contrary that $\omega(\widetilde{l}) = \omega$ is rational. Then the stabilizer Γ_ω of ω relative to the isometry group Γ is a non-empty cyclic group generated by an element $g \in \Gamma$. Therefore, the trajectories $g(\widetilde{l}), g^{-1}(\widetilde{l})$ bound a region $\widetilde{D}$ of the Lobachevsky plane so that $\widetilde{l}$ lies in $\widetilde{D}$. Since l is "self-approximating", one can choose an element $g' \in \Gamma$ so that $g'(\widetilde{l})$ lies arbitrarily close to $\widetilde{l}$. By continuity, the ω-limit of $g'(\widetilde{l})$ coincides with ω. Therefore $g' \in \Gamma_\omega$. This contradicts the assumption that g generates Γ_ω. □

(Note that the argument of the above theorem is valid for non-orientable surfaces as well.) In the same vein, one can prove the following statement.

Theorem 5.2.2 *Let l be a positive (negative) semi-trajectory of a flow f^t on closed surface M of non-positive Euler characteristic. Let $\widetilde{l}$ be a connected component of $\pi^{-1}(l)$ on $\widetilde{M}$. If l contains in its $\omega-(\alpha-)$ limit set a non-trivial recurrent trajectory, then $\widetilde{l}$ has an asymptotic direction which is irrational.*

5.2.2 Of Analytic Flow

The Anosov theorem gives the most general sufficient conditions for existence of the asymptotic directions of trajectories of the covering flows. The claim that the set of fixed points is contractible may be dropped because analytic flows imply this propeprty; cf. Anosov [8].

Theorem 5.2.3 (Anosov) *If a flow f^t on a closed surface M of non-positive Euler characteristic is real analytic, then each trajectory of the covering flow $\widetilde{f}^t$ is either bounded, or has an asymptotic direction.*

Proof. The proof is based on the subtle analysis of the flow structure near the set $Fix\ f^t$. In a local chart the set $Fix\ f^t$ is given by the equations $f(x,y) = g(x,y) = 0$, where f, g are analytic functions in a region of $I\!R^2$. It can be shown that $Fix\ f^t$ contains a finite set of "critical" points S, the complement of which (in $Fix\ f^t$) consists of a finite number of pairwise disjoint analytic curves, either closed or ending at S. Despite the fact that $Fix\ f^t$ can be non-contractible, its local structure outside S is well-known (the $Fix\ f^t$ is a segment). It allows to use the same argument as for Anosov's theorem. □

5.2.3 Of Foliation

Theorem 5.2.4 *Let $\mathcal{F}$ be a foliation on a closed surface of negative Euler characteristic. Suppose that the set Sing $\mathcal{F}$ consists of topological saddles of negative Euler-Poincaré index (i.e., m-prong saddles, $m \geq 3$). Then every semi-leaf different from a separatrix has an asymptotic direction.*

Proof. The proof is much akin to those of Theorem 5.2.1. By the assumption, $\mathcal{F}$ has no contractible closed leaves or contours. The limit set of each semi-leaf different from the separatrix is either a closed non-contractible leaf, a contour, or a quasiminimal set. In both cases, there exists a closed transversal C which meets l in a countable number of points. Now the argument of Theorem 5.2.1 must be repeated. The theorem follows. □

5.2.4 Of Curves With Restriction on the Geodesic Curvature

If the geodesic curvature of a curve is vanishing, such a curve looks "like" a geodesic. In this case it is reasonable to expect that deviation of the curve from a geodesic is limited and the curve has an asymptotic direction. For example, a parabola in the plane $I\!R^2$ has the curvature $k/t^{\frac{3}{2}}$, where t is the natural parameter. A parabola deviates from a geodesic arbitrarily far away, but it has an asymptotic direction. The hyperbola has the curvature k/t^3 and has both a limited deviation from a geodesic and an asymptotic direction.

The assumption that the curvature of a curve tends to zero is a strong restriction. Therefore we drop the claim that our curve is a covering curve (for some curve on the underlying surface M). Moreover, we look at curves on a 2-dimensional manifold endowed with metric of non-positive and, generally, non-constant curvature. Such a manifold is called an *Hadamard manifold.* For Hadamard surfaces as well as for the constant curvature surfaces, one can introduce the notion of absolute, cf. [93].

Following the work [176] all curves are supposed to be of class C^2 and parametrized by their length, t. The geodesic curvature in t is denoted by

$k(t)$. It can be shown that if $k(t)$ vanishes as $t \to \infty$, then the curve goes to infinity. We make two assumptions, H_1 and H_2, on "how fast" the curvature tends to zero.

- H_1: There exists $\varepsilon > 0$ such that, $k(t) = O(1/t^{1+\varepsilon})$.
- H_2: There exists $\varepsilon > 0$ such that, $k(t) = O(1/t^{2+\varepsilon})$.

Theorem 5.2.5 *If a curve satisfies hypothesis H_1 then it has an asymptotic direction; moreover, the distance between this curve and the corresponding geodesic is $O(t)$. If a curve satisfies the hypothesis H_2 then it has a bounded deviation from the corresponding geodesic.*

Proof. See [176]. □

A similar result is true for higher dimensional Hadamard manifolds as shown in [176]. Therein one can also find a restriction on curvature such that the curve comes arbitrary close to the geodesic.

A stronger notion of asymptotic direction has been considered in [75] and [120]. Acoording to these works a plane curve l has an *asymptotic direction* if its tangent has a limit direction as the point of tangency tends to infinity. The curve l has an *asymptote* if its tangent has a limit position. If, as before, $k(t) > 0$ is the curvature of l and t is the natural parameter, then a necessary and sufficient condition for existence of asymptotic direction for l is the following:

$$\int_0^\infty k(t)dt < +\infty.$$

In [120] it was shown that necessary and sufficient condition for existence of an asymptote of the curve l is the convergence of the following integral with weight t:

$$\int_0^\infty tk(t)dt < +\infty.$$

In the work [75] these results are generalized to curves in Euclidean space of higher dimensions.

5.3 Approximation of a Curve

One of the beautiful pieces of the theory is the theorem of D. V. Anosov on the approximation of curve by trajectories of a flow. Let us introduce relevant definitions. Let two semi-infinite curves $l_i = \{l_1(t) : t \geq 0\}, l_2 = \{l_2(t) : t \geq 0\}$ on M be given. Fix a metric d on the surface M. We say that the curve l_2 is of distantance less than $r > 0$ from l_1 if there exists a reparametrization $h : I\!R \to I\!R$ of t such that

$$\sup_{0 \leq t < \infty} d(l_1(t), l_2(h(t)) < r.$$

Theorem 5.3.1 *([9]) Let $l = \{l(t) : t \geq 0\}$ be a semi-infinite simple curve on a surface M. For every $r > 0$ there exists a C^∞ flow f^t such that one of its semi-trajectories $l' = \{f^t(m) : t \geq 0\}$ is of distance less than r from l.*

Proof. We shall proceed in several steps. By an ε-homeomorphism one understands a homeomorphism $h : M \to M$ for which $d(m, h(m)) < \varepsilon$ for every $m \in M$. Let us first make a compact arc $l_0 = \{m(t) : 0 \leq t \leq 1\}$ of the curve l 'smooth' by a δ_0-homeomorphism h_0 (the number δ_0 will be specified later on). This implies that the arc $h_0(l_0) = l_1$ of the curve $h_0(l)$ is C^∞ embedded into M. The curves l and $h(l)$ are δ_0-close.

Next, by a δ_1-deformation $h_1 : M \to M$ one 'pushes' the semi-infinite curve $\{m_i(t) : t \geq 2\}$ from a neighbourhood of the arc l_0. (One may suppose $h_1|_{l_0} = Id$.) Notice that the curves $h_0(l) = l_0'$ and $h_1(l_0') = l_1'$ are δ_1-close. Then one 'smoothes' the compact arc $l_2 = \{m_2(t) : 0 \leq 2 \leq 2\} \subset l_1'$ and by a δ_2-deformation one 'pushes' the semi-infinite curve $\{m_3(t) : t \geq 3\}$ from the neighbourhood of the arc $\{m_3 : 0 \leq t \leq 2\}$.

Continuing this process we get a simple C^∞ curve l_∞ which is $(\delta_0 + \delta_1 + \ldots)$-close to l. (Positive numbers $\delta_0, \delta_1, \ldots$ can be chosen such that their sum does not exceed $r > 0$.) Since l_∞ is C^∞ embedded then there exists an open strip $P \subset M$ which contains l_∞. (Formally, there exists a C^∞ embedding $\phi : [0, +\infty) \times (-1, 1) \to M$ such that $l_\infty = \phi([0, +\infty) \times \{0\})$.) We put $P = \phi([0, +\infty) \times (-1, 1))$. Then P can be foliated by a family of curves $\mathcal{F}$ such that $l_\infty \in \mathcal{F}$. Each point $M \backslash P$ is turned into a fixed point of the required flow. By Whitney's theorem (see the end of the chapter) the family $\mathcal{F}$ can be embedded into a C^∞ flow f^t. By the construction f^t has a trajectory $l_\infty = l^+$ which is r-close to the curve l. □

Anosov's theorem says that in order to construct a flow with given asymptotic direction it is enough to construct a simple continuous curve with given asymptotic behaviour. Later Anosov generalized his own theorem by giving it an 'ergodic' interpretation; cf. [14].

Theorem 5.3.2 *Let $l = \{m(t) : t \geq 0\}$ be a semi-infinite simple curve on a surface M. Let μ be a smooth measure on M with everywhere positive density of class C^∞. Then for every real $r > 0$ there exists a C^∞ flow f^t with invariant measure μ such that a trajectory of f^t is r-close to the curve l.*

Proof. When M is orientable, the measure μ is given by a skew-symmetric differential 2-form ω which can be written in the local charts as $\omega = p(x, y) dx \wedge dy$, where $p(x, y)$ is a density function of class C^∞. Let l_∞ and P be as in the proof of Theorem 5.3.1. Take a C^∞ function $H : M \to \mathbb{R}$ such that $H^{-1}(0) = (M \backslash P) \cup l_\infty$, $dH|_{l_\infty} \neq 0$. Then (in a local chart) the system of differential equations

$$\dot{x} = \frac{1}{p(x, y)} \frac{\partial H(x, y)}{\partial y}, \qquad \dot{y} = -\frac{1}{p(x, y)} \frac{\partial H(x, y)}{\partial x},$$

defines a C^∞ flow f^t with an invariant measure μ and trajectory l_∞. In the case of a non-orientable surface an extra argument (using the two-fold covering surfaces) is needed. □

5.4 Limit Sets at the Absolute

Let M be a surface of non-positive Euler characteristic. The universal covering $\widetilde{M}$ of M may be thought of as a unit disc (either $D^2 = \mathbb{R}^2 \cup \{\infty\}$ or Δ). The union $\widetilde{M} \cup S_\infty$ is the closed disc endowed with the natural Hausdorff topology. (We always refer to this topology when we speak about 'nearness' of the points in the disc.)

Let $\widetilde{l} = \{\widetilde{m}(t) : t \geq 0\}$ be a (connected component) of the lifting of a simple closed curve $l \subset M$ to $\widetilde{M}$. The point $m_0 \in S_\infty$ lies in the limit set of $\widetilde{l}$ if in any neighbourhood of m_0 there are the points of $\widetilde{l}$ corresponding to arbitrary large values of the parameter t. We shall focus on Question 3, which concerns possible limit sets of curves (and trajectories) which go to the absolute.

Let $\widetilde{l}$ be as before, and denote its limit set by $\omega(\widetilde{l})$. If $\widetilde{l}$ is unbounded then $\omega(\widetilde{l}) \cap S_\infty \neq \emptyset$. If $\widetilde{l}$ goes to infinity with an asymptotic direction then $\omega(\widetilde{l})$ consists of one point lying at the absolute. Let us consider an example due to Anosov [12] in which all points of the absolute belong to $\omega(\widetilde{l})$.

Theorem 5.4.1 *Let M be a closed surface of non-positive Euler characteristic. There exists a continuous curve $l \subset M$ such that its lifting $\widetilde{l}$ to the universal covering $\widetilde{M}$ contains in its limit set the whole absolute S_∞.*

Proof. Let us take a countable family of neighbourhoods V_n of the set $\widetilde{M} \cup S_\infty$ such that

(i) $\bigcup_n V_n \supset S_\infty$;
(ii) For every point $\sigma \in S_\infty$ and every neighbourhood $U(\sigma)$ of σ there exists V_i such that $V_i \subset U(\sigma)$.

(Evidently, such families exist.) Let us take such a smooth, semi-infinite curve $\widetilde{l}_0 = \{\widetilde{m}_0(t) : t \geq 0\}$ on $\widetilde{M}$ which intersects V_n. Then $\omega(\widetilde{l}_0) \supset S_\infty$ and we are done. The curve $\pi(\widetilde{l}_0) \subset M$ may have self-intersection. We must therefore deform it so as to obtain the required curve. The curve $l_0 = \pi(\widetilde{l}_0) = \{m_0(t) : t \geq 0\}$ can be split into arcs which (after suitable re-parametrization) fit the following conditions:

(i) For $n \geq 0$ each arc $l_{n,n+1} = \{m_0(t) : n \leq t \leq n+1\}$ has no self-intersections;
(ii) $m_0(n+1) \notin \{m_0(t) : 0 \leq t < n+1\}$;
(iii) There exists a sequence $i_n \geq 1$ such that $\widetilde{m}_0(i_n) \in V_n$ for all $n \geq 1$.

Let us modify the arc $l_{1,2}$ so that the new arc $l'_{1,2}$ forms with $l_{0,1}$ an arc without self-intersections, and so that $l_{1,2}, l'_{1,2}$ are homotopic as paths with the same endpoints. We deform similarly the arc $l_{2,3}$ (the role of $l_{0,1}$ is now played by the curve $l_{0,1} \cup l'_{1,2}$). Continuing this process we get a curve l on M which does not have self-intersections. By property (3), the lifting $\widetilde{l}$ to $\widetilde{M}$ will have in its limit set the whole absolute S_∞. □

Corollary 5.4.1 *Let M be a closed surface of non-positive Euler characteristic. Then there exists a C^∞ flow f^t on M such that its covering flow $\widetilde{f}^t$ on $\widetilde{M}$ has a semi-trajectory $\widetilde{l}$ with ω-limit set containing all of the absolute: $\omega(\widetilde{l}) \supset S_\infty$.*

(Roughly speaking, the above flow f^t has a 'wild' trajectory which runs in 'all' asymptotic directions. This is in contrast to the fact that not all points of the absolute are limit points for a fixed flow on M.)

5.5 Geodesic Deviation

In this section we consider the deviation (Question 4) of covering curves from the corresponding geodesics (i.e., geodesics with the same asymptotic directions). V. Pupko [248] stated the restricted deviation property for the curve $l \subset M$ but her prove was incorrect. In 1970s S. Aranson and V. Grines constructed a counter-example (described by D. Anosov in [8]) to Pupko's result. Note that the flow in this example has a continuum of fixed points. Later, D. Anosov [8, 10, 9] constructed counter-examples to the Pupko result on other surfaces including the torus and the Klein bottle.

5.5.1 Deviation Property of Trajectories

It is well known that if f^t is a flow without fixed points on the torus and $\widetilde{l}$ is a semi-trajectory of the covering flow on R^2 then $\widetilde{l}$ has the restricted deviation property ([297], Theorem 6.5 of [197]). For the case $\chi(M) < 0$ the restricted deviation property was proved by N. G. Markley when l is an almost periodic trajectory or l belongs to the closure of some almost periodic trajectory (Theorems 9.14, 9.15, 9.17 [196]). Unfortunately, these results were not published.

In 1995 S. Aranson, V. Grines and E. Zhuzhoma [22, 23] proved the following theorem.

Theorem 5.5.1 *Let f^t be a flow with finitely many fixed points on a closed surface M of non-positive Euler characteristic. Let $\widetilde{l}$ be a semi-trajectory of a covering flow $\widetilde{f}^t$ on $\widetilde{M}$ which has an asymptotic direction. Then $\widetilde{l}$ has the restricted deviation property.*

Proof. Taking the double covering if necessary, we can assume that M is an orientable surface. Consider a semi-trajectory $l = \pi(\widetilde{l})$ on the surface M. If l is a periodic trajectory then l is not homotopic to zero and the result follows (since M is compact). Let l be a non-periodic semi-trajectory. Without loss of generality, we can assume that l is a positive semi-trajectory. Since $\widetilde{l}^+$ has an asymptotic direction then the ω-limit set of l^+ either contains a non-trivial trajectory, consists of a non-trivial homotopic periodic trajectory, or consists of a non-trivial homotopic one-sided contour (i.e., a union of the fixed points and the separatrices). In the last two cases l^+ approaches spirally to a non-trivial homotopic closed curve. This implies the result in these two cases.

So we have to consider the case when $\omega(l)$ is a quasiminimal set. First, we assume that $M = T^2$ is the two-dimensional torus. Then $\widetilde{M} = \mathbb{R}^2$ is the Euclidean plane. The flow f^t is an irreducible one. Hence, there is a blowing-down map $h : T^2 \to T^2$ that sends f^t into a transitive flow f_0^t. Since the Euler characteristic $\chi(T^2) = 0$, any fixed point of f_0^t is a fake saddle, i.e., the degenerate saddle with two saddle sectors.

Without loss of generality we can assume that f_0^t is a linear irrational flow with a certain number of fake saddles. So the integral curves of f_0^t are the geodesics. Therefore the covering flow $\widetilde{f_0^t}$ for f_0^t has the integral curves which are straight lines of the Euclidean plane. Let $\widetilde{h} : R^2 \to R^2$ be a lift of h. Then $\widetilde{h}$ maps a lift $\widetilde{l}^+$ of l^+ into some ray $\widetilde{l}_0$ with the same asymptotic direction as $\widetilde{l}^+$.

Let us show that there is a number $k > 0$ such that $\widetilde{d}(\widetilde{m}; \widetilde{h}(\widetilde{m})) \le k$ for every $\widetilde{m} \in R^2$. Indeed, the unit square $K = \{(x; y) \in R^2 : 0 \le x, y \le 1\}$ is a fundamental domain of the group G of isometries of the covering plane $\pi : \mathbb{R}^2 \to T^2$. Any covering transformation of G is an integer translation and the orbit of K under G covers the plane $\mathbb{R}^2$. Since K is compact and $\widetilde{h}$ is a continuous map then there is $k > 0$ such that $\widetilde{d}(\widetilde{m}; \widetilde{h}(\widetilde{m})) \le k$ for every $\widetilde{m} \in K$. Now take an arbitrary point $\widetilde{m} \in \mathbb{R}^2$. There is a covering transformation $\gamma \in G$ such that $\gamma(\widetilde{m}) \in K$. Since $\widetilde{h}$ is the lift of the map homotopic to the identity then $\widetilde{h} \circ \gamma = \gamma \circ \widetilde{h}$. It follows that $\widetilde{d}(\widetilde{m}; \widetilde{h}(\widetilde{m})) \le k$ because γ preserves distance. Hence $\widetilde{l}^+$ has a restricted deviation from the ray $\widetilde{l}_0$. The result is proved for the case $M = T^2$.

Let us consider the case when M is a closed orientable surface of the genus $g \ge 2$. Then $\widetilde{M} = \mathbb{H}$ is the hyperbolic plane. The main case we have to consider is the following one: the positive semi-trajectory l^+ is non-trivial recurrent and belongs to a trajectory l recurrent in both directions. Let $\widetilde{l}$ be a lift of l to the hyperbolic plane $\mathbb{H}$. Then the limit set of $\widetilde{l}$ consists of two points $\sigma^+, \sigma^- \in S_\infty$. Let $g = g(\widetilde{l})$ be a geodesic with endpoints σ^+, σ^-. Denote by $m(t) \in \widetilde{l}$ the current point such that $m(t) \to \sigma^{+(-)}$ then $t \to +\infty(-\infty)$. Let $[m(t); m_0(t)]$ be the perpendicular from a point $m(t) \in \widetilde{l}$ to the geodesic g where $m_0(t) \in g$. Denote by $d(t)$ the length of $[m(t); m_0(t)]$. We have to prove that the function $d(t)$ is bounded when $t \to +\infty$.

Let us assume the contrary. Then there is a sequence t_n such that $d(t_n) \to +\infty$ provided $t_n \to +\infty$. Passing to a subsequence if necessary, we can assume that the sequence $\pi(m_0(t_n))$ converges to some point $m_0 \in M$ because the surface M is compact. Take a point $\widetilde{m}_0 \in \pi^{-1}(m_0)$. Since $\pi(m_0(t_n))$ converges to m_0 then there exist a sequence $\gamma_n \in G$ such that $\gamma_n(m_0(t_n)) \to \widetilde{m}_0$ provided $t_n \to +\infty$. The closure of the geodesic $\pi(g)$ forms a geodesic lamination Ω which is a geodesic framework of the quasiminimal set $\omega(l) \stackrel{\text{def}}{=} Q$. Therefore the topological limit of geodesics $\gamma_n(g) = g_n$ is a geodesic $g_0 \subset \pi^{-1}(\Omega)$ through the point $\widetilde{m}_0$. Moreover, the endpoints σ_n^+, σ_n^- of the geodesics g_n approaches the endpoints σ_0^+, σ_0^- of g_0 respectively.

Denote by S a geodesic through $\widetilde{m}_0$ which is orthogonal to g_0. Let $\alpha, \omega \in S_\infty$ be the endpoints of S. Since the γ_n is an isometry the sequence of points $m_n = \gamma_n(m(t_n))$ accumulates to at least one of the points α, ω. Without loss of generality we can assume that $m_n \to \omega (n \to +\infty)$. Denote by Γ the topological limit of the sets $\gamma_n(\widetilde{l}) \cup \sigma_n^+ \cup \sigma_n^-$. It is obvious that Γ is a connected set, and $\sigma_0^+, \sigma_0^-, \omega \in \Gamma$. The intersection $\Gamma \cap S_\infty$ does not contain non-trivial open intervals. Therefore there exists an arc $K \subset \Gamma$ with endpoints $k_1, k_2 \in S_\infty$ such that $intK \subset \Delta$ and in at least one of the endpoints k_1, k_2 is different from σ_0^+, σ_0^- (since $\omega \in \Gamma$). We will assume that $k_1 \neq \{\sigma_0^+, \sigma_0^-\}$.

It is obvious that $\pi(K)$ belongs to the quasiminimal set $\omega(l)$. Therefore $\pi(K)$ is neither a closed leaf nor a one-sided contour. Since the flow f^t has finitely many fixed points this implies the existence of a semi-trajectory $\widetilde{l}_1 \subset K$ such that limit set of $\widetilde{l}_1$ contains the point k_1, and $\pi(\widetilde{l}_1) = l_1$ is a non-trivial recurrent semi-trajectory of f^t (Theorem 1 [194]). Then there exists a closed non-contractible transversal C such that $C \cap l_1 \neq \emptyset$.

Since l_1 is a non-trivial recurrent semi-trajectory then it meets C infinitely many times. Hence $\widetilde{l}_1$ intersects the countable family of curves $\widetilde{C}_1, \widetilde{C}_2, \ldots$ of the full preimage $\pi^{-1}(C)$. Since the group G is discontinuous, these curves converge to a point k_1. So we can choose the lift $\widetilde{C}$ of C such that the pair of endpoints C^+, C^- of $\widetilde{C}$ does not separate the pair of points σ_0^+, σ_0^-. Therefore for a sufficiently large n the trajectory $\gamma_n(\widetilde{l})$ has to intersect the curve $\widetilde{C}$ in at least two points. This is impossible because $\widetilde{C}$ is the transversal curve of the flow $\widetilde{f}^t$. □

For the semi-trajectories on torus there exists a stronger version of Theorem 5.5.1 due to Anosov [14]. Namely, the following two theorems have been proved.

Theorem 5.5.2 *Let f^t be a flow on two-dimensional torus T^2 for which at least one of the following conditions holds:*

- *The set of fixed points of the flow f^t is contractible to a point;*
- *The flow f^t is real analytic.*

If $\widetilde{l}$ is a semi-trajectory of the covering flow which has an asymptotic direction, then $\widetilde{l}$ has the bounded deviation property.

Theorem 5.5.3 *Let l be a non-trivial recurrent semi-trajectory of flow on torus. Then l has the bounded deviation property.*

Proof of Theorems 5.5.2 and 5.5.3. If $l = \pi(\widetilde{l})$ is a closed trajectory then it is homotopy non-trivial and theorems follow. It remains to consider the case when l is non-closed.

At the first stage one constructs a simple closed transversal which intersects the semi-trajectory l. (For the non-trivial recurrent semi-trajectories l's this is a well known procedure; to the contrary, when l is not such and the flow is analytic, the construction is more subtle and uses analytical arguments as well as certain assumptions on the set of fixed points.) Next, when the transversal, C_0, is constructed, one can always assume that it coincides with the zero meridian of the torus.

If l intersects C_0 finitely many times, then the covering escapes to infinity, remaining in the strip $k \leq x \leq k+1$. Therefore $\widetilde{l}$ deviates finitely from the axis Oy and Oy defines its asymptotic direction. Otherwise, l intersects C_0 infinitely often and $\widetilde{l}$ has an asymptotic direction $y = ax$ from which it deviates finitely. □

5.5.2 Deviation From the Geodesic Framework

In the previous subsection we considered the deviation property of a trajectory from the corresponding geodesic. It is natural to consider such a property for the family of all one-dimensional trajectories and semi-trajectories of a flow. (For the sake of simplicity we deal only with flows whose fixed points are saddles.)

Let s_0 be a separatrix of the saddle O of $\widetilde{f}^t$. There are two separatrices of O which are adjacent for s_0, i.e., each of them bounds a saddle sector with s_0 near O. Both separatrices are a geometric extension of s_0. In the obvious sense, one of these separatrices, denoted by s_l, is an *extension to the left* and other, denoted by s_r, is an *extension to the right.* Sometimes we say that s_l (s_r) is the Bendixson extension of s_0 to the left (resp. to the right).

A sequence $\ldots, s_n, \ldots$ of separatrices s_n is called a *one-sided sequence* of separatrices, if s_n is the geometric extension of s_{n-1} either to the left or to the right for any n. Note that this sequence may be finite or infinite including both cases $n \in \mathbb{N}$ and $n \in \mathbb{Z}$. The union of separatrices of the one-sided sequence and the corresponding saddles is called a *'generalized' trajectory* of $\widetilde{f}^t$. The image of a 'generalized' trajectory of $\widetilde{f}^t$ under the map π is called a *'generalized' trajectory* of f^t.

Let us take a 'generalized' trajectory or a usual trajectory (not a separatrix) $\widetilde{L}$ of $\widetilde{f}^t$. Then $\widetilde{L}$ has two different endpoints σ_1, σ_2 both belonging to the circle at infinity $\partial\mathbb{H}$. Let $\widetilde{g} = \widetilde{g}(\widetilde{L})$ be a geodesic with endpoints σ_1, σ_2. The geodesic $\widetilde{g}(\widetilde{L})$ is said to be *corresponding* to $\widetilde{L}$.

Denote by $\widetilde{Gf}(\widetilde{f}^t)$ the union of geodesics $\widetilde{g}(\widetilde{L})$ and geodesics $\widetilde{g}(\widetilde{l})$ where $\widetilde{L}$ runs through all of the 'generalized' trajectories $\widetilde{f}^t$ and $\widetilde{l}$ runs through all of the trajectories which are not separatrices of $\widetilde{f}^t$. Then $\widetilde{Gf}(\widetilde{f}^t)$ is a geodesic lamination on $\mathbb{H}$. As a consequence, $Gf(f^t) = \pi(\widetilde{Gf}(\widetilde{f}^t))$ is a geodesic lamination on the surface M which is called *geodesic framework of* f^t. (Note that if F is a transitive flow then this definition of a geodesic framework $Gf(f^t)$ and the earlier definitions coincide.)

Let $\widetilde{L}$ be either a 'generalized' trajectory or usual trajectory of $\widetilde{f}^t$ and let $\widetilde{g}(\widetilde{L})$ be its corresponding geodesic. If there is a constant $k > 0$ such that $\widetilde{d}(\widetilde{m}, \widetilde{g}(\widetilde{L})) \leq k$ for any point $\widetilde{m} \in \widetilde{L}$ then we shall say that $\widetilde{L}$ has a *restricted deviation property*. In this case we denote

$$\widetilde{d}_{\widetilde{L}} = \sup_{\widetilde{m} \in \widetilde{L}} \widetilde{d}(\widetilde{m}, \widetilde{g}(\widetilde{L}))$$

Theorem 5.5.4 *[23] Let f^t be a flow on a closed surface M of negative Euler characteristic. Suppose that all fixed points of f^t are the topological saddles. Let $\widetilde{L}$ be either a 'generalized' trajectory or a usual trajectory of the covering flow $\widetilde{f}^t$. Then $\widetilde{L}$ has the restricted deviation property.*

Proof. This is much the same as that for Theorem 5.5.1. □

Theorem 5.5.5 *[23] Let f^t be a flow on a closed surface M of negative Euler characteristic. Suppose that all fixed points of f^t are topological saddles. Then*

$$\sup \widetilde{d}_{\widetilde{L}} < \infty,$$

where the supremum is taken over all $\widetilde{L}$ which are either 'generalized' trajectories or usual trajectories of the covering flow $\widetilde{f}^t$.

Proof. This is similar to the proof of Theorem 5.5.1. □

In fact, Theorem 5.5.5 assures the uniformity of deviations of all trajectories for any concrete flow of the class of surface flows we are considering. It is natural to call the supremum from Theorem 5.5.5 as the *deviation of the flow from its geodesic framework.*

5.5.3 Ramified Coverings

In this subsection we consider the Anosov-Weil problem for branched coverings. This will allow us to study the questions 1, 2, 4 for surface foliations whose singularities are topological saddles, including thorns. Our method allows us to consider foliations on the sphere and the projective plane as well. Actually, we need the branched covering to 'kill' the thorns on the universal covering $\mathbb{H}$. After such an operation one can consider the asymptotic directions of leaves of foliations even on the sphere and the projective plane.

Denote by $\mathbb{H}^k : z \to z^k$ a mapping of the complex z-plane. A continuous map $\pi : \widetilde{M} \to M$ is called a *branched covering* if any point $m \in M$ has a neighbourhood U such that the complete preimage $\pi^{-1}(U)$ is a union $V_1, V_2, \dots$ of disjoint neighbourhoods, and the restriction $\pi|V_i$ is topologically conjugate to a mapping $\mathbb{H}^k$. The number k is called a *branching index* of the point $z_i = \pi^{-1}(m) \cap V_i$ and is denoted by $k(z_i)$. We shall consider only *regular coverings* such that all points in the complete preimage $\pi^{-1}(m)$ have the same branching index for any $m \in M$. For such coverings the branching order $k(m)$ of $m \in M$ is defined to be the branching index of any point in $\pi^{-1}(m)$. A point $m_0 \in M$ is called a *branch point* if $k(m_0) > 1$. The collection $M_0 \subset M$ of the branch points is called a branched set. It is discrete and finite if M is compact. The number of points in the complete preimage $\pi^{-1}(x)$, $x \in M \backslash M_0$, is called the *multiplicity* of the covering. For an arc-wise connected M this number is independent from the point $x \in M \backslash M_0$.

In this section we consider foliations on surfaces whose singularities are topological saddles (including thorns). It is well known that the sum of indices of all singularities is equal to the Euler characteristic of the surface, which is an integer. Hence the number of singularities with a semi-integer index is even. This useful remark allows us to apply the special branched coverings to 'kill' the thorns.

First we consider an orientable closed surface M of genus $g \geq 0$. Let n be an even natural number such that $n \geq 4$ if $g = 0$ and n is arbitrary if $g \geq 1$. Fix a set Σ_n of n points on the surface M. Σ_n is said to be a *marked set.* Since n is even, there exists a two-fold branched covering $p : \hat{M} \to M$ with the branch set $M_0 = \Sigma_n$ [30]. Note that the order of any branch point is equal to 2. By the condition $n \geq 4$ if $g = 0$ the surface $\hat{M}$ is closed, orientable, and of genus $g \geq 1$. So $\hat{\pi} = p \circ \pi : \widetilde{M} \to M$ is the universal covering map where $\widetilde{M}$ is either the Euclidean plane $\mathbb{R}^2$ or the hyperbolic plane $\mathbb{H}$ where $\pi : \widetilde{M} \to \hat{M}$ is a non-branched universal covering map.

If M is a non-orientable surface of genus $g \geq 1$ we require that $n \geq 2$ if $g = 1$ and n is any natural even number if $g \geq 2$. Again fix a set Σ_n of n points on the surface M. There is a non-branched double covering $q : M' \to M$ where M' is an oriented closed surface. The complete preimage $q^{-1}(\Sigma_n)$ consists of $2n$ points and $2n \geq 4$ if the genus of M' equals to 0. So there is a two-fold branched covering $p : \hat{M} \to M'$ with the branch set $q^{-1}(\Sigma_n)$. The map $\hat{\pi} = q \circ p \circ \pi : \widetilde{M} \to M$ is the universal covering where $\widetilde{M}$ again is either the Euclidean plane $\mathbb{R}^2$ or the hyperbolic plane $\mathbb{H}$.

Let us denote by $\mathcal{F}_n$ the set of foliations on the surface M with the marked set Σ_n such that the set of singularities of a semi-integer index of any $F \in \mathcal{F}_n$ coincides with Σ_n. The last condition implies that any foliation $F \in \mathcal{F}_n$ has a lift under the covering $\hat{\pi}$, i.e., the foliation $\widetilde{F}$ is lifted to the universal covering space $\widetilde{M}$. So we can formulate questions 1,2,3 of the Anosov-Weil problem (see the Introduction) for the leaves and the semi-leaves of the foliations of the class $\mathcal{F}_n$. We shall follow the notations and the definitions of the Introduction.

Theorem 5.5.6 *[23] Let $F \in \mathcal{F}_n$ be a foliation on a closed surface M with the marked set Σ_n. Let $\widetilde{l}$ be a semi-leaf of a covering foliation $\widetilde{F}$ on $\widetilde{M}$ which is not a separatrix. Then $\widetilde{l}$ has an asymptotic direction and $\widetilde{l}$ has the restricted deviation property.*

Proof. Since the branching order of any branch point of two-fold branched covering $p : \hat{M} \to M(M')$ is equal to 2, the covering foliation $\widetilde{F}$ has no thorns. Hence any singularity (if one exists) of $\widetilde{F}$ is a saddle of non-positive index. Then the semi-leaf $\widetilde{l}$ has an asymptotic direction.

Let $\hat{F}$ be a lift of F under p if M is the orientable or under $p \circ q$ if M is non-orientable. So $\hat{F}$ is the foliation on the closed orientable surface $\hat{M}$. Again since the order of any branch point of p equals 2 then any singularity (if one exists) of $\hat{F}$ is a saddle of non-positive index (note that if M is the sphere with $n = 4$ or M is the projective plane with $n = 2$ then $\hat{F}$ is a foliation on the torus without singularities). Let $\hat{l}$ be a lift of l to the surface $\hat{M}$. It is obvious that the restriction of a deviation of $\widetilde{l}$ can be considered for $\widetilde{l}$ as a lift of $\hat{l}$ for a non-branched universal covering $\pi : \widetilde{M} \to \hat{M}$. The result follows from Theorem 5.5.5. □

5.5.4 Swing of Trajectories

In this subsection we study the unbounded deviation property and a swing of the trajectory near the corresponding hyperbolic line.

Let $\widetilde{m}(t) \in \widetilde{l}$ be a point on the curve $\widetilde{l} \subset \widetilde{M}$ such that $\widetilde{m}(t) \to \sigma \in S_\infty$ as $t \to +\infty$. Let γ be a hyperbolic line (either a geodesic that represents the direction σ or an equidistant curve with the endpoint σ). This line divides the hyperbolic plane Δ into two domains $\widetilde{D}_1$, $\widetilde{D}_2$. We will say that $\widetilde{l}$ *swings near the hyperbolic line γ in the positive direction* if there are two sequences of points $\widetilde{m}(t_{2n+1}) \in \widetilde{D}_1$, $\widetilde{m}(t_{2n}) \in \widetilde{D}_2$ such that $t_0 < t_1 < \ldots < t_{2n} < t_{2n+1} < \ldots \to +\infty$. As a consequence, $\widetilde{l} \neq \gamma$ and there is a sequence of points $\widetilde{m}(t_k) = \widetilde{l} \cap \gamma$ approaching to σ as $t_k \to +\infty$. If $t_k \to -\infty$ then we get a *swing in the negative direction*. If, moreover, $\widetilde{l}$ has the *unbounded deviation property*, i.e.,

$$\widetilde{\lim}_{t \to +\infty} d(t) = +\infty$$

where $d(t) = \widetilde{d}(m(t), \gamma)$ is the distance between $\widetilde{m}(t)$ and the line γ, then we shall say that $\widetilde{l}$ has a *resonance swing* near the hyperbolic line γ.

Let l be a non-trivial trajectory of the flow f^t. For such a trajectory there is *corresponding geodesic* $g(l)$ with the same asymptotic directions. If we take any lift $\widetilde{l}$ of l then there is a lift $\widetilde{g}(\widetilde{l})$ of $g(l)$ such that $\widetilde{l}$ and $\widetilde{g}(\widetilde{l})$ have the same endpoints; one of them $\sigma \in S_\infty$ represents the asymptotic positive direction of $\widetilde{l}$. We shall say that $\widetilde{g}(\widetilde{l})$ is the *corresponding geodesic* for $\tilde{l}$. It is obvious that if a covering for a non-trivial recurrent trajectory $\tilde{l}$ has unbounded deviation from $\tilde{g}(\tilde{l})$ on both sides then $\tilde{l}$ swings near $\tilde{g}(\tilde{l})$. But if $\tilde{l}$ has unbounded deviation from $\tilde{g}(\tilde{l})$ only on one side then we can guarantee that $\tilde{l}$ swings near some equidistant curve of the geodesic $\tilde{g}(\tilde{l})$.

Theorem 5.5.7 *[23] Let f^t be a flow on a closed orientable surface M of genus $g \geq 2$ and let l be a non-trivial trajectory of f^t. Let $\tilde{l}$ be a lift of l and $\tilde{g}(\tilde{l})$ be the geodesic corresponding to $\tilde{l}$. If a positive (negative) semi-trajectory of l has the unbounded deviation property then it has a resonance swing near the equidistant curve of the geodesic $\tilde{g}(\tilde{l})$ in the positive (negative) direction.*

Proof. We shall only prove the theorem for a positive semi-trajectory $\tilde{l}^+$ of $\tilde{l}$. Denote by $\sigma_+(\sigma_-) \in \partial\mathbb{H}$ the asymptotic direction of $\tilde{l}$ and $\tilde{g}(\tilde{l})$ in the positive (negative) direction. Let $\tilde{m}(t) \in \tilde{l}$ be a current point on $\tilde{l}$ such that $\tilde{m}(t) \to \sigma_+$ as $t \to +\infty$. Let $d(t) = \tilde{d}(m(t), \tilde{g}(\tilde{l}))$ be the distance between $\tilde{m}(t)$ and the geodesic $\tilde{g}(\tilde{l})$. Denote by $\tilde{p}(t)$ the perpendicular dropped from the point $\tilde{m}(t)$ to the geodesic $\tilde{g}(\tilde{l})$ (if $\tilde{m}(t) \in \tilde{g}(\tilde{l})$ then $\tilde{p}(t) = \tilde{m}(t)$). This perpendicular is a geodesic segment, whose endpoints are $\tilde{m}(t)$ and $\tilde{m}_0(t) in \tilde{g}(\tilde{l})$. It is evident that the length of $\tilde{p}(t)$ equals $d(t)$. By the assumption, $\limsup_{t\to+\infty} d(t) = +\infty$. To prove the theorem it is enough to show that $\liminf_{t\to+\infty} d(t) < +\infty$. We give a proof by contradiction. Suppose that

$$\liminf_{t\to+\infty} d(t) = +\infty.$$

The geodesic $\tilde{g}(\tilde{l})$ divides the hyperbolic plane Δ into two domains $\tilde{D}_1$, $\tilde{D}_2$. By the assumption there is a point $\tilde{m}(t_0) \in \tilde{l}^+$ such that $\tilde{m}(t_0) \notin \tilde{g}(\tilde{l})$. The above equality implies that we can assume $\tilde{l}^+ \cap \tilde{g}(\tilde{l}) = \emptyset$. We denote by σ^- the first point of the intersection of $\tilde{l}^-(\tilde{m}(t_0))$ with $\tilde{g}(\tilde{l})$ in the direction of decreasing time (in the case of $\tilde{l}^-(\tilde{m}(t_0)) \cap \tilde{g}(\tilde{l}) \neq \emptyset$) and $\sigma^- = \sigma_-$ otherwise. We put $\tilde{l}^+(\sigma^-) = \tilde{l}$ if $\sigma^- = \sigma_-$.

Denote by $\tilde{D}_0$ an open domain bounded by $\tilde{l}^+(\sigma^-)$ and $[\sigma^-;\sigma_+)$ where $[\sigma^-;\sigma_+)$ means the positive geodesic ray of the geodesic $\tilde{g}(\tilde{l})$ starting at the point σ^-. Without loss of generality we can assume that $\tilde{D}_0 \subset \tilde{D}_1$. Let $\tilde{E} \subset \tilde{D}_1$ be a curve whose points are equidistant from the geodesic $\tilde{g}(\tilde{l})$ such that the hyperbolic distance between $\tilde{g}(\tilde{l})$ and $\tilde{E}$ equals $2d(t_0)$. Denote by $\tilde{D}$ the open domain bounded by $\tilde{g}(\tilde{l})$ and $\tilde{E}$. Because of the assumption there is a parameter $\tilde{t}$ such that

$$\tilde{l}^+(\tilde{m}(\tilde{t})) \cap \tilde{D} = \emptyset$$

(i.e., the semi-trajectory $\tilde{l}^+(\tilde{m}(\tilde{t}))$ leaves the domain $\tilde{D}$). Since l is a non-trivial trajectory then there a sequence $\gamma_k \in G$ such that $\tilde{d}(\gamma_k(\tilde{m}_0(t_0))$; $[\tilde{m}_0(\tilde{t}),\sigma_+)) \to 0, \gamma_k(\tilde{m}_0(t_0)) \in \tilde{D}$ as $k \to \infty$. Hence $\gamma_k(\tilde{p}(t_0)) \in \tilde{D}$ if k is large enough, because γ_k is an isometry and the length of $\tilde{p}(t_0)$ equals $d(t_0)$

(we recall that the hyperbolic distance between $\widetilde{g}(\widetilde{l})$ and $\widetilde{E}$ equals to $2d(t_0)$). Then the equality written above implies that $\gamma_k(\widetilde{p}(t_0)) \in \widetilde{D}_0$; in particular, $\gamma_k(\widetilde{m}(t_0)) \in \widetilde{D}_0$. As a consequence, the trajectory $\gamma_k(\widetilde{l})$ passes through the point $\gamma_k(\widetilde{m}(t_0)) \in \widetilde{D}_0$. According to [18], $\gamma_k(\sigma_+) \neq \sigma_+$, $\gamma_k(\sigma_-) \neq \sigma_-$. We note that $\gamma_k(\widetilde{g}(\widetilde{l})) \subset D_1$ because $\gamma_k(\widetilde{m}_0(t_0)) \in \widetilde{D} \subset D_1$. As a consequence, the geodesic $\gamma_k(\widetilde{g}(\widetilde{l}))$ intersects $\widetilde{D}$ and has to leave the domain $\widetilde{D}$. Hence the trajectory $\gamma_k(\widetilde{l})$ has to intersect the trajectory $\widetilde{l}$. This is impossible. □

Now let us give the definition of a *weak swing* of a non-trivial trajectory near the corresponding geodesic $g(l)$. Since $g(l)$ is not closed and does not divide the surface M into open domains, we need to apply the covering hyperbolic plane. The geodesic $g(l)$ divides $\mathbb{H}$ into two domains $\widetilde{D}_1$, $\widetilde{D}_2$. We shall say that l *swings weakly under the geodesic* $\widetilde{g}(l)$ *in the positive direction* if there are a sequence of points $\widetilde{m}(t_{2n+1}) \in \widetilde{l}$, $\widetilde{m}(t_{2n}) \in \widetilde{l}$ and a sequence of hyperbolic maps $\gamma_n \in G$ such that $t_0 < t_1 < \ldots < t_{2n} < t_{2n+1} < \ldots \to +\infty$ and $\gamma_n(\widetilde{m}(t_{2n+1})) \in \widetilde{D}_1$, $\gamma_n(\widetilde{m}(t_{2n})) \in \widetilde{D}_2$. (There is a similar definition of swing in the negative direction.)

A non-trivial trajectory is called *semi-proper* if it does not approach itself from one side. Such a side is called a *proper side.* A non-trivial trajectory is *interior* if it is not semi-proper. In other words, a non-trivial trajectory is interior if it is a limit of itself from both sides.

Theorem 5.5.8 *[23] Let f^t be a flow on a closed orientable surface M of genus $g \geq 2$ and l be an interior non-trivial trajectory of f^t. Let $g(l)$ be a geodesic corresponding to l. If some positive (negative) semi-trajectory of l does not coincide with any positive (negative) ray of $g(l)$ then l swings weakly under $g(l)$ in the positive (negative) direction.*

Proof. We shall only prove the theorem for a positive semi-trajectory l^+, the proof for a negative one being similar. Let $\widetilde{l}$ be a lift of l and $\widetilde{g}(\widetilde{l})$ a geodesic corresponding to $\widetilde{l}$. The geodesic $\widetilde{g}(\widetilde{l})$ divides the hyperbolic plane $\mathbb{H}$ into two domains $\widetilde{D}_1$, $\widetilde{D}_2$. Denote by $\sigma_+(\sigma_-) \in \partial\mathbb{H}$ an asymptotic direction of $\widetilde{l}$ and $\widetilde{g}(\widetilde{l})$ in the positive (negative) direction. By the hypothesis, for any number $N > 0$ there is a time $t_0 > N$ such that the current point $\widetilde{m}(t_0) \in \widetilde{l}^+$ and $\widetilde{m}(t_0) \notin \widetilde{g}(\widetilde{l})$. To prove the theorem it is enough to find a map $\gamma \in G$ and two parameters t_1, t_2 both more than t_0 such that $\gamma(\widetilde{m}(t_1)) \in \widetilde{D}_1$, $\gamma(\widetilde{m}(t_2)) \in \widetilde{D}_2$.

If $\widetilde{l}^+(\widetilde{m}(t_0)) - \widetilde{m}(t_0)$ intersects both domains $\widetilde{D}_1$, $\widetilde{D}_2$ then the result follows ($\gamma = id$). Suppose that either $\widetilde{l}^+(\widetilde{m}(t_0)) \subset \mathbb{H} - D_1$ or $\widetilde{l}^+(\widetilde{m}(t_0)) \subset \mathbb{H}\backslash D_2$. If $\widetilde{l}^+(\widetilde{m}(t_0)) \cap \widetilde{g}(\widetilde{l}) \neq \emptyset$ then we denote by σ^+ the first point of the intersection of $\widetilde{l}^+(\widetilde{m}(t_0))$ with $\widetilde{g}(\widetilde{l})$. If $\widetilde{l}^+(\widetilde{m}(t_0)) \cap \widetilde{g}(\widetilde{l}) = \emptyset$ then we put $\sigma^+ = \sigma_+$. In the same way we denote by σ^- the first point of the intersection of $\widetilde{l}^-(\widetilde{m}(t_0))$ with $\widetilde{g}(\widetilde{l})$ in the direction of decreasing time (in the case of $\widetilde{l}^-(\widetilde{m}(t_0)) \cap \widetilde{g}(\widetilde{l}) \neq \emptyset$) and $\sigma^- = \sigma_-$ otherwise. Let us denote by $\widetilde{D} \subset \mathbb{H}$ the domain bounded by $\widetilde{g}(\widetilde{l})$ and the arc of $\widetilde{l}$ between the points σ^+, σ^-.

Since the non-trivial trajectory l is interior then the geodesic $\pi(\widetilde{g}(\widetilde{l}))$ is the limit of itself from both sides. Hence there is a hyperbolic isometry $\gamma^{-1} \in G$, $\gamma^{-1} \neq id$, such that $\gamma^{-1}(\widetilde{g}(\widetilde{l})) \cap \widetilde{D} \neq \emptyset$. Moreover, we can assume that the point $\widetilde{m}(t_0)$ and the geodesic $\widetilde{g}(\widetilde{l})$ belong to different components into which the geodesic $\gamma^{-1}(\widetilde{g}(\widetilde{l}))$ divides the plane Δ. According to [18], $\gamma^{-1}(\sigma_+) \neq \sigma_+$, $\gamma^{-1}(\sigma_-) \neq \sigma_-$. As a consequence, the geodesic $\gamma^{-1}(\widetilde{g}(\widetilde{l}))$ has to leave the domain $\widetilde{D}$ at a point $\widetilde{m}(t') \in \widetilde{l}^+(\widetilde{m}(t_0))$ corresponding to the parameter $t' > t_0$. Hence there are two parameters $t_1, t_2 > t_0$ such that $\gamma(\widetilde{m}(t_1)) \in \widetilde{D}_1$, $\gamma(\widetilde{m}(t_2)) \in \widetilde{D}_2$. This proves the theorem. □

If a nontrivial recurrent trajectory l is not an interior one then, generally speaking, l does not swing near $g(l)$. However, if l has the unbounded deviation property then it swings weakly under $g(l)$.

Theorem 5.5.9 *[23] Let f^t be a flow on a closed orientable surface M of genus $g \geq 2$ and l be a non-trivial trajectory of f^t (possibly semi-proper). Let $g(l)$ be the geodesic corresponding to l. If some positive (negative) semi-trajectory of l has the unbounded deviation property then l swings weakly near $g(l)$ in the positive (negative) direction.*

Proof. We shall use the notions of the proof of Theorem 5.5.8 and prove the theorem only for a positive semi-trajectory l^+, the proof for a negative one being similar. If l is an interior non-trivial trajectory then the result follows from Theorem 5.5.8. So we suppose that the trajectory l is semi-proper. In this case, for the point σ_+ there are two geodesics having σ_+ as endpoint and projecting onto non-trivial recurrent geodesics each without self-intersections on the surface M [27]. We shall denote these geodesics by $\widetilde{g}_1(\sigma_+)$, $\widetilde{g}_2(\sigma_+)$. One of these geodesics, for example $\widetilde{g}_1(\sigma_+)$, corresponds to the geodesic $\widetilde{g}(\widetilde{l})$. The geodesics $\widetilde{g}_1(\sigma_+)$, $\widetilde{g}_2(\sigma_+)$ divide $\mathbb{H}$ into three domains $\widetilde{D}_1$, $\widetilde{D}$, $\widetilde{D}_2$, see [23].

If $\widetilde{l} \cap D_1 \neq \emptyset$ then the proof is similar to the proof of Theorem 5.5.8 because the domain $\widetilde{D}_1$ contains geodesics of the form $\gamma(\widetilde{g}_1(\sigma_+))$, $\gamma \in G$, arbitrarily close to $\widetilde{g}_1(\sigma_+)$. Therefore we assume that $\widetilde{l} \cap D_1 = \emptyset$. Since the positive semi-trajectory of l has the unbounded deviation property then $\widetilde{l} \cap D_2 \neq \emptyset$. Again the rest of proof is similar to the proof of Theorem 5.5.8 because the domain $\widetilde{D}_2$ contains geodesics of the form $\gamma(\widetilde{g}_1(\sigma_+))$, $\gamma \in G$, arbitrarily close to $\widetilde{g}_2(\sigma_+)$. □

If l is not a non-trivial trajectory then Theorem 5.5.9 is not correct in general. There is an example of non-closed, one-dimensional trajectory l that has both positive and negative asymptotic directions but does not oscillate under $g(l)$ (where $g(l)$ is the corresponding geodesic) even if l has the unbounded deviation property.

5.6 Unbounded Deviation

This section is a collection of examples of different types of behavior of curves on surfaces. These examples illustrate theorems proved in previous sections.

5.6.1 Surfaces of Genus $g \geq 2$

First we give a construction showing how to obtain a foliation by the special gluing of two foliations, one transitive and the other of Denjoy type.

Let M_1, M_2 be two orientable surfaces endowed with the foliations F_1, F_2 respectively. Suppose that F_1 is a transitive foliation (for example, an irrational foliation of the torus or sphere [256]). Let us assume that F_2 has an exceptional minimal set $\Omega(F_2)$, i.e., a nowhere dense minimal set locally homeomorphic to the product of the Cantor set and a segment. Suppose also that the complement $M_2 \backslash \Omega(F_2)$ of $\Omega(F_2)$ contains a simply connected domain W with accessible boundary consisting of two leaves L_1, L_2 (for example, F_2 is the Denjoy foliation of the torus). Then the complement $M_2 \backslash \Omega(F_2)$ consists of simply connected domains. Such foliations exist on the sphere as well; see [256].

Let D_1 be a disc on M_1 the boundary ∂D_1 of which is transverse to F_1 except at two points $a_1, c_1 \in \partial D_1$. Suppose there are no singularities of F_1 in D_1. We can think of ∂D_1 as a circle with the natural orientation induced by M_1. Let $(x; y)$, $x, y \in \partial D_1$, denote an interval passing from x to y in the positive direction. Two points $a_1, c_1 \in \partial D_1$ divide ∂D_1 into intervals $(a_1; c_1)$ and $(c_1; a_1)$. We denote by $\phi_1 : (a_1; c_1) \to (c_1; a_1)$ a leaf-wise map induced by the restriction of the foliation F_1 on D_1. Take some point $d_1 \in (a_1; c_1)$ and let $b_1 = \phi_1(d_1)$.

Let $D_2 \subset W$ be an open disc on M_2 whose boundary ∂D_2 is transverse to F_2 except at two points $b_2, d_2 \in \partial D_2$. Moreover, let $\partial D_2 \cap \Omega(F_2) = b_2 \cup d_2$ and $b_2 \in L_1, d_2 \in L_2$. Then the disc D_2 divides the domain W into two parts W^+, W^-. Without loss of generality we can suppose that there are no singularities of F_2 in D_2. Again we can think of the boundary ∂D_2 of D_2 as a circle with the natural orientation induced by M_2. Let us take two points $a_2 \in (b_2; d_2), c_2 \in (d_2; b_2)$. Let $\Theta : \partial D_1 \to \partial D_2$ be an orientation-reversing homeomorphism such that $\Theta(a_1) = c_2, \Theta(b_1) = b_2, \Theta(c_1) = a_2, \Theta(d_1) = d_2$. Let us consider a homeomorphism

$$\phi_2 = \Theta \circ \phi_1 \circ \Theta^{-1}|_{[a_2;c_2]} : [a_2; c_2] \to [c_2; a_2]$$

(It easy to see that $\phi_2(a_2) = a_2, \phi_2(c_2) = c_2, \phi_2(d_2) = b_2$.) Let $x_0 = c_2, x_1, \ldots, x_n, \ldots$ be a sequence of points $x_i \in (c_2; b_2) \subset \partial D_2$ such that $x_n \to b_2$ monotonically as $n \to \infty$. Then the sequence $\{x_{-n} = \phi_2^{-1}(x_n)\}_0^\infty$ of points $x_{-i} \in (d_2; c_2)$ is monotone and $x_{-n} \to d_2$ monotonically as $n \to \infty$. Now we replace the foliation F_2 restricted to W^+ by a special foliation $F(W^+)$. The points x_n, x_{-n} belong to the separatrices of tripod O_n which has a separatrix connection with a thorn T_n. Moreover, we can construct the foliation

$F(W^+)$ in such a way that the leaf-wise map $(d_2; c_2) \to (c_2; b_2)$ induced by $F(W^+)$ coincides with ϕ_2.

Before we specify the foliation $F(W^+)$ let us take a transversal segment T of the foliation F_2 through points b_2, d_2. An interval $(b_2; d_2) \subset T$ divides T into two subsegments T^-, T^+. Suppose that the subsegment T^+ contains the point d_2. Since leaves L_1, L_2 are non-trivially recurrent, W^+ intersects both subsegments T^-, T^+ infinitely many times. Let W_1^+ be a first interval of the intersection W^+ with T^+. We define a sequence W_n^+ of intervals of $W^+ \cap T^+$ as follows: W_{n+1}^+ is the first interval of the intersection $W^+ \cap T^+$ which lies between the point d_2 and the interval W_n^+.

Now we construct the foliation $F(W^+)$ more accurately. One can deform if necessary the foliation $F(W^+)$ in such a way that the thorn T_n and the tripod O_{n+1} will be located near the interval W_n^+. Moreover we require that the singularities T_n, O_{n+1} approach the point d_2 like a "wedge". An analogous foliation $F(W^-)$ is constructed in the domain W^-. The thorns and tripods of $F(W^-)$ accumulate (in a "wedge") to the point b_2.

We identify the manifolds $M_1 \backslash D_1$ $M_2 \backslash D_2$ via the homeomorphism Θ. As a consequence, we obtain an orientable closed surface M. The leaves of foliations $F_1|M_1 - D_1$, $F(W^+)$, $F(W^-)$ and $F_2|M_2 - (D_2 \cup W)$ form a family of curves denoted by F'. We declare the points $a_2 = \Theta(c_1)$, $c_2 = \Theta(a_1)$, $b_2 = \Theta(b_1)$, $d_2 = \Theta(d_1)$ to be the singularities. Then the family F' gives a foliation (denoted by F) because the points $b_2 = \Theta(b_1)$, $d_2 = \Theta(d_1)$ of accumulation of thorns and tripods are singularities and the segments between T_n and O_{n+1} at which tangents of the curve of F are transversal to the leaves of the foliation F_2 approach the singularity d_2.

The surface M is a connected sum of the surfaces M_1, M_2 and we shall wright $M = M_1 \# M_2$. The foliation F on M will be called a *special connected sum of the foliations* F_1, F_2. Note that any leaf $l(m)$ through a point $m \in M_1 - D_1$ is non-trivially recurrent, provided $l(m)$ is not a separatrix. In this case the limit set of $l(m)$ is equal to $(M_1 - D_1) \cup \Omega(F_2)$ (such leaves exist because the set of singularities of F is countable). Now we are ready to present some concrete examples.

Let $M_1 = M_2 = T^2$ be the two-dimensional torus, F_1 a minimal foliation, and F_2 a Denjoy foliation. Then $M = M_1 \# M_2$ is an orientable closed surface of genus $g = 2$. The special connected sum F of the foliations F_1, F_2 has a non-trivial recurrent leaf l which has the limit set $(M_1 - D_1) \cup \Omega(F_2)$. Let $\widetilde{l}$ be a lifting of l to the hyperbolic plane $\mathbb{H}$. Since the leaf-wise map $(d_2; c_2) \to (c_2; b_2)$ induced by $F(W^+)$ coincides with $\phi_2 = \Theta \circ \phi_1 \circ \Theta^{-1}$ where ϕ_1 is the leaf-wise map induced by the restriction of the foliation F_1 on D_1, then both of the semi-leaves $\widetilde{l}^+$, $\widetilde{l}^-$ of $\widetilde{l}$ have asymptotic directions. Each of the semi-leaves l^+, l^- of l intersects the interval (a_1, c_1) infinitely many times and contains both leaves L_1, L_2 of F_2 in its limit set. Since L_1, L_2 are non-trivial recurrent then they have an asymptotic direction. Therefore both semileaves $\widetilde{l}^+$, $\widetilde{l}^-$ of $\widetilde{l}$ have unbounded deviation. Moreover, since l intersects both intervals (a_1, d_1), (d_1, c_1) infinitely many times, both semileaves of each of the leaves L_1, L_2 belong to the limit set of l. Hence $\widetilde{l}$ swings and has unbounded deviation on both sides. Modifying in the obvious way the foliation $F(W^+)$ or $F(W^-)$ one can ensure that $\widetilde{l}$ has unbounded deviation to one side only.

Now we can construct a flow with a semi-trajectory having unbounded deviation. Denote by $[O_n, T_n]$ the union of the singularities O_n, T_n and the separatrix of $F(W^+)$ which connects these singularities. Let F' be the foliation F restricted to the domain $M \setminus (\Omega(F_2) \cup_n [O_n, T_n])$. Then the leaves of the foliation F' form the orientable family of curves which satisfies the conditions of the Whitney theorem on the embedding of a family of curves into a flow [6]. Hence, there is a flow f^t on M such that $Fix\,(f^t) = \Omega(F_2) \cup_n [O_n, T_n]$. It is easy to see that f^t has a semi-trajectory with unbounded deviation.

Let $M_1 = S^2$ be the two-dimensional sphere, F_1 a transitive foliation [256], M_2 an orientable closed surface of genus $g \geq 1$, and F_2 a Denjoy type foliation. Then $M = M_1 \# M_2$ is an orientable closed surface of genus $g \geq 1$. The special connected sum F of the foliations F_1, F_2 yield an example of a foliation for which question 1 has a negative solution.

5.6.2 Irrational Direction on Torus

This construction is again due to Anosov [12] and is based on the approximation properties of irrational numbers. Let $l : y = ax, x \geq 0$ be a ray with an irrational slope α in the Euclidean plane $\mathbb{R}^2$. Let p_i/q_i be rational approximations to α which form an alternating sequence converging to α:

$$\frac{p_0}{q_0} < \frac{p_2}{q_2} < \ldots < \frac{p_{2n}}{q_{2n}} \ldots \to \alpha \leftarrow \ldots \frac{p_{2n-1}}{q_{2n-1}} < \ldots < \frac{p_1}{q_1}.$$

In this case the line

$$P_i : y = \frac{p_i}{q_i} x + \frac{(-1)^i}{q_i}$$

meets l at a point t_i. Denote by l_i the segment of the ray l between the origin and t_i. The idea consists in the "stretching" of l_i near the point t_i in the direction P_i. Here is a formal expression of the above deformation:

$$\widetilde{F}_i(x, y) = (x, y) + c_i \varphi_{\varepsilon_i}(q_i y - p_i x)(q_i, p_i)$$

where φ_{ε_i} is a specially chosen approximation function; see [12] for the details. It is easy to check that $\widetilde{F}_i$ is the covering homeomorphism for a homeomorphism on torus. (Real numbers ε_i are chosen such that $\widetilde{F}_i$ would not shift the points of l_{i-1} near the point t_{i-1}.) Then the curve $L_n = \widetilde{F}_1 \circ \ldots \circ \widetilde{F}_n(l_n)$ is a compact curve without self-intersections. Moreover, the union $L = \cup L_n$ has the same property.

[6] Such a theorem will be proved in the next Section.

The numbers c_i define the size of "needles" lying at l after l is transformed by the homeomorphisms $\widetilde{F}_i$'s. The c_i's can be chosen such that "needles" deviate arbitrary far out from l, on the one hand, and on the other hand would lie in a region bounded by a parabola with symmetry axis l. It can be seen that L has both an asymptotic direction and infinite deviation from the ray l.

5.6.3 Rational Direction on Torus

Finally, let us mention an example of unbounded deviation of a curve from a geodesic with rational slope. We refer the interested reader to the original works [10], [11]. We remind the reader that a covering of a curve without self-intersections which has rational slope can deviate only from one side; cf. previous sections. Such examples on the torus and the Klein bottle were constructed in 1988 by Anosov [10], [11].

5.7 Family of Curves

It is known that an arbitrary flow $\phi^t : M \times I\!R \to M$ on a surface M defines a family of (disjoint) curves in M corresponding to the set of orbits of ϕ^t. Let us consider an inverse problem: when can an arbitrary family of curves on a surface M be represented as the set of orbits of some flow on M? This question was studied by H. Whitney [301] who defined local (necessary and sufficient) conditions which the arbitrary family should obey.

Denote by γ_1 and γ_2 a pair of arcs lying in a surface M endowed with a metric Dist (x, y) where $x, y \in M$. Take a homeomorphism $h : \gamma_1 \to \gamma_2$ which sends the endpoints of γ_1 to the endpoints of γ_2. By the *span* of two (oriented) arcs γ_1 and γ_2 one understands the positive function

$$\sigma(\gamma_1, \gamma_2) = \inf_{h \in H} \sup_x Dist(x, h(x)),$$

where $x \in \gamma_1, h(x) \in \gamma_2$, and H is the set of all homeomorphisms between the arcs γ_1, γ_2.

Definition 5.7.1 (Whitney) *A family F of curves in M is called regular if the following conditions are satisfied:*

(i) no two curves $\Gamma_1, \Gamma_2 \in F$ intersect each other;
(ii) for each $\Gamma \in F$ and each arc $\gamma \subset \Gamma$ and any $\varepsilon > 0$ there exists a $\delta > 0$ such that if $x' \in M$ lies on some curve $\Gamma' \in F$ and Dist $(x, x') < \delta$ for $x \in \gamma$, then $\sigma(\gamma, \gamma') < \varepsilon$ for some $\gamma' \subset \Gamma'$.

(The first condition can be weakened to the statement that two curves of family F have at most two common points. The second condition means that the family F locally 'fills-up' a space 'uniformly'.) Below two examples of regular families of curves are given.

Example 5.7.1 *Take a family of parabolas $y = a(1 + x^2)$ in the Euclidean plane $\mathbb{R}^2$. Each value $a \in \mathbb{R}$ defines an element (curve) of the family.*

Example 5.7.2 *The family F is formed by curves which are equidistant from a given segment in the Euclidean plane. Every value $\alpha > 0$ of the function Dist yields an element (closed curve) of the family F. (This example can be modified by replacing the segment with an open ray coming out a fixed point $x \in \mathbb{R}^2$. In this case the equidistant family defines a singularity of thorn type placed at x.)*

Let now assign to each curve $\Gamma \in F$ of a regular family F a positive direction. An arc $\gamma \subset \Gamma$ with the endpoint in $x \in M$ is *positively oriented* if it lies in the positive direction from x. Otherwise it is *negatively oriented.* Let $U_\gamma \subset M$ be a neighbourhood of an arc $\gamma \subset \Gamma$ of curve Γ in a regular family F. We say that the curves of F are *similarly oriented in the neighbourhood U_γ* if, for a sequence $\gamma_1, \gamma_2, \dots$ of oriented arcs in F approaching γ,

$$\lim_{k \to \infty} \sigma(\gamma_k, \gamma) = 0$$

the arcs γ_k and γ have the same orientation.

Definition 5.7.2 (Whitney) *A regular family of curves F is called orientable if it is possible to assign a direction to every curve of the family so that they are all similarly oriented in the neighbourhood of each arc of F.*

(Note that in Example 5.7.1 the regular family F is orientable, while in Example 5.7.2 it is non-orientable.)

It is natural to consider along with the regular families of curves (both orientable and non orientable) the so-called *families of paths* in M. Such families are extensions of regular families of curves by certain sets of singular points. [7]

Definition 5.7.3 (Whitney) *Let $R \subset M$ be a closed subset of locally compact separable space M. Suppose that F is a regular family of curves in the complement $M \backslash R$ of R in M. Then the curves of F taken together with the point of R are called a (regular, orientable) family of paths in M. The points of R are called singular points of the family.*

Following [301] denote by $f(x,t)$ a function given on a family of paths. The statement below establishes a criterion of (local) embeddability of a family of paths on M into a flow $\phi^t : M \times \mathbb{R} \to M$.

Theorem 5.7.1 (Whitney) *Let F be a regular orientable family of paths on a surface M. Then there exists a function $f(x,t)$ on F which satisfies the following properties:*

[7] Whitney [301] calls it *invariant points* of a family F.

(i) *for every point* $x \in F \subset M$ *and every real number* $t \in \mathbb{R}$ *there exists a unique point* $y = f(x,t)$ *which lies on the curve* Γ *through* x, *a singular point in which case it coincides with* x; *moreover, for each* $y \in \Gamma$ *there is a* $t \in \mathbb{R}$ *such that* $y = f(x,t)$;

(ii) *the function* $f(x,t)$ *is continuous in both variables;*

(iii) *every point* $x \in M \backslash R$ *moves in the positive (negative) direction along the curve* Γ *through* x *as* t *increases (decreases);*

(iv) *for every* $x \in F$ *and any real numbers* t_1 *and* t_2, $f(f(x,t_1),t_2) = f(x,t_1+t_2)$ *and* $f(x,0) = x$.

In other words, F *defines a flow* $\phi^t : M \times \mathbb{R} \to M$ *on the surface* M.

Proof. To prove the above theorem some additional notions and auxiliary statements will be helpful.

Measure function μ. Let M be a separable metric space, and let $x_1, x_2, \dots$ be a sequence of points dense in M. To each of these points we associate the following functions

$$f_i(y) = \frac{1}{1 + Dist(x_i, y)}, \quad i = 1, 2, \dots$$

(Note that $0 < f_i(y) \leq 1$ and $|f_i(y_1) - f_i(y_2)| \leq \text{Dist}\,(y_1, y_2)$.) Let A be a non-empty subset of M. For fixed i consider the various values of $f_i(y)$ where $y \in A$ runs through the set A. We introduce the following two functions on A:

$$\mu_i(A) = \sup_{y_1, y_2 \in A} |f_i(y_1) - f_i(y_2)|, \quad \mu(A) = \sum_{i=1}^{\infty} \frac{\mu_i(A)}{2^i}.$$

Functions g and g'. Let x be a point on a curve $\Gamma \subset F$ of a regular family F, and consider an arbitrary point $y \in \Gamma$. For each arc $xy \subset \Gamma$ there is a real number $\mu(xy)$ which increases continuously from zero as y moves away from x. For a given real number α, we can consider an 'inverse' function of μ, that is a function

$$y = g(x, \alpha), \qquad \mu(x, g(x, \alpha)) = \alpha.$$

Note that g is in general a two-valued function. To avoid this ambiguity, one introduces a function g'. Again take a point $x \in \Gamma$, and fix a positive real number $\alpha > 0$. If y is a point of Γ such that $\mu(xy) = \alpha$, we define a function $y = g'(x, \alpha)$ or a function $y = g'(x, -\alpha)$ according to whether y lies in the positive or the negative direction from x along the arc pq. It is easy to see that g' is a single-valued function of α.

Tubes $\mathfrak{T}$. A set of arcs is called *self-compact* if any sequence chosen from these arcs contains a subsequence $\gamma_1, \gamma_2, \dots$ approaching an arc γ from the above set: $\lim_{n \to \infty} \sigma(\gamma, \gamma_n) = 0$. A set of arcs of a regular family of oriented paths form *a tube* if the following conditions are satisfied:

(1) The arcs are self-compact;
(2) No two distinct arcs have common points;
(3) The set of points in $\mathfrak{T}$ contains inner points.

Denote by S_1 and S_2 the set of points which are the negative and positive ends of a tube $\mathfrak{T}$, respectively. These will be called the *negative* and *positive ends* of $\mathfrak{T}$. Evidently, the sets $\mathfrak{T}, S_1$ and S_2 are compact and closed. Moreover, S_1 and S_2 are connected provided $\mathfrak{T}$ is so.

Lemma 5.7.1 *If $\mathfrak{T}$ is a tube with ends S_1 and S_2 and $\gamma \subset F$ is an arc of F containing at least two points of $S_1 \cup S_2$, then Diam $(\gamma) \geq$ Dist $(S_1, S_2) > 0$.*

Proof. This is certainly true if γ contains points of both S_1 and S_2. Suppose that γ contains two points x', x'' of S_1 (or, S_2). Suppose x'' lies in positive direction from x'. If γ' is an arc of $\mathfrak{T}$ containing x' then $\gamma' \subset \gamma$ and the arc γ also contains a point of S_2. Therefore, Diam $(\gamma) \geq$ Dist (S_1, S_2). □

Lemma 5.7.2 *The ends S_1 and S_2 of a tube $\mathfrak{T}$ are cross-sections of the family F.*

Proof. Let $S = S_1$ be a cross-section of F through a point $x \in M$. To prove the lemma one needs to establish that: (i) each point $x \in S$ lies within an arc $\gamma \subset F$ such that for some $\varepsilon > 0$ each arc in the ε-vicinity $U_\varepsilon(\gamma)$ of γ contains at most one point of S; (ii) $x \in S$ lies within an arbitrary small arc $\gamma' \subset F$ such that for some $\varepsilon > 0$ each arc of $U_\varepsilon(\gamma')$ contains exactly one point of S; and (iii) S is closed. We do it step-by-step as follows.

If $x \in S$ then one takes an arc γ of Diam $(\gamma) < \varepsilon, \varepsilon = \frac{1}{3}$ Dist (S_1, S_2) which contains the point x. Then by Lemma 5.7.1 any arc in $U_\varepsilon(\gamma)$ contains at most one point of S, and the condition (i) is satisfied. Suppose that $x \in S$ lies on an arc γ of $\mathfrak{T}$ which contains the inner points of $\mathfrak{T}$. Then all points of γ except the endpoints, p and q, are inner points of $\mathfrak{T}$. Denote by q_0q_1 an arc of diameter $< \frac{1}{2}$ Dist (S_1, S_2) and containing p in its interior. Suppose that q_1 lies in the interior of pq. Now we choose ε small enough that $U_\varepsilon(q_0) \cap \mathfrak{T} = \emptyset$ and $U_\varepsilon(q_1) \subset \mathfrak{T}$. Then any arc of $U_\varepsilon(q_0q_1)$ contains exactly one point of S, and condition (ii) is satisfied. We have already noted that (iii) S is closed. Therefore, S is a cross-section of the family F. □

Lemma 5.7.3 *If S is a compact cross-section of a regular family F passing through a point x, then for some real number $\alpha > 0$ the set of arcs y_1, y_2 for which $y_1 = g'(x, -\alpha), y_2 = g'(x, \alpha)$ form a tube $\mathfrak{T}$ containing x as an inner point.*

Proof. A modification of the proof of Lemmas 5.7.1 and 5.7.2. □

First sequence of tubes. By Lemma 5.7.3 for a compact cross-section S through a point $x \in M$ there exist tubes $\mathfrak{T}_1$ and $\mathfrak{T}_2$ containing x bounded by the cross-sections $S_1 = g'(S, \alpha), S_2 = g'(S, -\alpha)$ and $S_3 = g'(S, 2\alpha), S_4 = g'(S, -2\alpha)$, respectively. Likewise we can define an infinite sequence of tubes $\mathfrak{T}_1, \mathfrak{T}_2, \ldots$

such that Diam $(\mathfrak{T}_k) < 1/i$ for $k > i$. If the above sequence satisfies the following conditions:

(1) $\mathfrak{T}_1, \mathfrak{T}_2, \ldots$ covers $M \backslash R$;
(2) Any compact and closed set in $M \backslash R$ has common points with a finite number of tubes;
(3) The diameter of the tubes vanishes: $\lim_{n \to \infty}$ Diam $(\mathfrak{T}_n) = 0$,

then it is called *a (first) sequence of tubes.*

First sequence of pseudo-tubes. Any set of arcs which satisfies properties (1) and (2) of the definition of a tube is called *a pseudo-tube.* We can formally replace the sequence of tubes $\mathfrak{T}_1, \mathfrak{T}_2, \ldots$ by a sequence of pseudo-tubes $\mathfrak{T}'_1, \mathfrak{T}'_2, \ldots$ provided that, in addition to (1)-(3), they satisfy:

(4) No two ends of pseudo-tubes have common points;
(5) Any pseudo-tube $\mathfrak{T}'_i$ contains points of at most one of the ends of any preceding pseudo-tube $\mathfrak{T}'_k, k < i$.

(The reader is encouraged to verify that it is always possible to construct the first sequence of pseudo-tubes starting from a sequence of tubes; see [301].)

Second sequence of pseudo-tubes. It is possible to replace the first sequence of pseudo-tubes by a sequence of pseudo-tubes $\mathfrak{T}''_1, \mathfrak{T}''_2, \ldots$ which satisfy properties (1)-(3) of the definition of a sequence of tubes, and also the following property:

(6) If any inner point z of an arc xy of a pseudo-tube $\mathfrak{T}''_k$ lies in a preceding pseudo-tube $\mathfrak{T}''_l, l < k$, then *all* points of xy lie in the preceding pseudo-tubes.

(For the construction of function $f(x,t)$ the pseudo-tubes of the second sequence will be used. We leave the reader to check that such sequences always exist; see [301] for details.)

Function τ'. With each motion (xy) [8] of family F we associate a real number $\tau'(xy)$ with the following properties:

(i) $\tau'(xy)$ is greater than, less than, or equal to zero according to whether (xy) is oriented positively, negatively, or is a null motion;
(ii) τ' is continuous, that is, if $(x_iy_i) \to (xy)$, then $\tau'(x_iy_i) \to \tau'(xy)$;
(iii) $\tau'[(xy) + (yz)] = \tau'(xy) + \tau'(yz)$, and
(iv) $\tau'(yx) = -\tau'(xy)$.

[8] That is a parametrization of a curve Γ by the unit segment I.

The function τ' is defined successively through the second sequence of pseudo-tubes $\mathfrak{T}_1'', \mathfrak{T}_2'', \ldots$ of the family F. Consider first the tube $\mathfrak{T}_1''$. If z is any point of an arc $xy \subset \mathfrak{T}_1''$ which determines the motion (xz) we set

$$\tau'(xz) = \mu(xz).$$

If z' is another point of xy we put $\tau'(zz') = \tau'(xz') - \tau'(xz)$.

Using induction, suppose now that $\tau'(xy)$ is defined for all motions lying entirely in the set $\mathfrak{T}_1'' \cup \ldots \cup \mathfrak{T}_{i-1}''$. One must extend its definition to the tube $\mathfrak{T}_i''$. Indeed, if the motion (xy) is interior to $\mathfrak{T}_i''$ then $\tau'(xy)$ is defined as before. Consider now any motion (xy) which lies in $\mathfrak{T}_1'' \cup \ldots \cup \mathfrak{T}_i''$. Let $xx_1, x_1x_2, \ldots, x_my$ be a set of arcs defining the motion (xy) such that each x_kx_{k+1} lies in some tube $\mathfrak{T}_l''$. For every arc x_kx_{k+1} there is a motion (x_kx_{k+1}) for which the function $\tau'(x_kx_{k+1})$ is already defined. [9] Clearly, $(xy) = (xx_1) + (x_1x_2) + \ldots + (x_my)$ and we set

$$\tau'(xy) = \tau'(xx_1) + \tau'(x_1x_2) + \ldots + \tau'(x_my).$$

(It is not hard to see that the above definition determines $\tau'(xy)$ uniquely, and the resulting function τ' is continuous.)

Functions τ and f. Now we replace the function τ' by a function τ which has all the properties (i)-(iv) of τ', and an additional one:

(v) If (xy) is an arc of a pseudo-tube of a second sequence $\mathfrak{T}_i''$ and (xy) is the corresponding motion, then $\tau(xy) \geq 1$.

To introduce τ, first let us define a continuous function $\phi(x) > 0$ for each $x \in M \backslash R$. For this any positive function can be chosen, and we take such ϕ for which $\phi(x) \leq \Delta_i$, where Δ_i is a minimum for $\tau'(xy)$ chosen among all arcs in $\mathfrak{T}_i''$. (It is not hard to see that such a ϕ always exists.)

Take any motion (xy) and represent it by arcs $xx_1, \ldots, x_my$. If we let x' move along these arcs from x to y, we will obtain sub-motions (xx') of the motion (xy). The corresponding numbers increase from 0 to $\tau'(xy)$. Therefore, for any value of τ' between 0 and $\tau'(xy)$, there is a corresponding point $x_{\tau'}$ for which $\tau'(xx_{\tau'}) = \tau'$. We set

$$\tau(xy) = \int_0^{\tau'(xy)} \frac{d\tau'}{\phi(x_{\tau'})}.$$

Moreover, if xy is any arc in a tube $\mathfrak{T}_i''$ then the following estimate is valid:

$$\tau(xy) \geq \int_0^{\Delta_i} \frac{d\tau'}{\Delta_i} = 1.$$

[9] To accurately define $\tau'(x_k, x_{k+1})$ more space is needed and we omit here some details for which we refer the reader to the original paper of Whitney [301].

Finally, let us define the function f. For this let us take a point $x \in M \backslash R$ and a real number $t \in \mathbb{R}$. Let us consider the motion (xy) in the family F for which $\tau(xy) = t$. This yields us a function

$$y = f(x, t).$$

In this case $x \in R$ belongs to a set of singular points, it is assumed $f(x,t) = x$. We claim that the function $f(x,t)$ is the required one.

Indeed, for every x and $t \in \mathbb{R}$, there is a unique point $y = f(x,t)$ such that property (i) holds. By the construction of the sequence of pseudo-tubes $\mathfrak{T}_i''$ property (iii) of Theorem 5.7.1 is satisfied. Also it is evident that $f(x,t)$ is continuous; see property (ii). Finally, property (iv) follows from the 'additivity' of motions: $(xy) = (xz) + (zy)$ for any z in the arc xy. Since the function $f(x,t)$ satisfies all of the conditions (i)-(iv), Theorem 5.7.1 is proven. □

Bibliographic Notes. The history of the Anosov-Weil problem is rather curious; see a remarkable overview of the question in [13]. A. Weil [297],[298] first posed this problem for flows without equilibrium points on the torus. He conjectured that the method which uses the universal covering is applicable to the study of flows on higher genus surfaces. Then the idea was abandoned for almost 40 years until it was revived by D. V. Anosov. Stimulated by the latter V. I. Pupko [248] considered the problem for arbitrary non-closed curves on compact surfaces. A few years later the question was profoundly studied by S. Kh. Aranson and V. Z. Grines [18] leading to the generalization of the Poincaré rotation numbers for flows on higher genus surfaces. Finally D. V. Anosov, in the series of works [8],[9],[11],[12],[13], established the basic facts of the theory by considering its different aspects. Theorem 5.7.1 was formulated and proved by H. Whitney [301].

6. Non-compact Surfaces

There exists no advanced theory of foliations on non-compact (open) surfaces comparable to that of the compact case. This is due to the greater "freedom" of behaviour permitted to the leaves. Foliations on open surfaces are close to the line (of dimension 1) foliations in three–dimensional spaces; this is especially true for open surfaces of infinite genus. To a large extent, the theory of such foliations is a collection of examples of different effects which cannot take place on compact surfaces.

6.1 Foliations in the Plane

6.1.1 Non Singular Case

The simplest non-compact surface is the Euclidean plane $\mathbb{R}^2$, and the simplest foliation on it is a fixed-point-free flow. Historically, this was the first case studied. Object more general than fixed-free-point flows, namely, foliations in the plane without singular points, have been studied by W. Kaplan [154]. The leaves of the foliation can be thought of as a curve-family which Kaplan calls *regular*. H. Whitney showed in 1933 that every regular, oriented family of curves in a paracompact space is flow-embeddable. Later on, Kaplan specified that in the case of the plane the orientability is no longer needed. This result of Kaplan is a consequence of the following theorem.

Theorem 6.1.1 *Let $\mathcal{F}$ be a foliation of $\mathbb{R}^2$ without singular points. Then there exists a continuous function $f : \mathbb{R}^2 \to \mathbb{R}$ such that:*

(i) for each $c \in \mathbb{R}$ the equation $f(x, y) = 0$ defines at most countable set of leaves of $\mathcal{F}$;
(ii) in a neighbourhood of any point (x_0, y_0) there are points (x, y) for which $f(x, y) > 0$, and there are points (x, y) for which $f(x, y) < 0$.

In particular, $f(x, y)$ cannot have local extrema.

Proof. This theorem is a consequence of Urysohn's Lemma; see Kaplan [154] for the details. □

Later Kaplan generalized this theorem (we return below to his generalization). To prove Theorem 6.1.1 Kaplan introduced the so–called *chordal system of curves* which allows to obtain the topological classification of singularity-free foliations (and point-fixed-free flows) on the plane. Without touching the technical details, we consider the simplest chordal system formed by three curves.

Let $C_1, C_2, C_3 \subset I\!R^2$ be three simple, pairwise disjoint curves, each tending at both ends to infinity. It follows from the Jordan theorem that each curve C_i divides the plane into two distinct regions of which C_i is the common boundary. So there are actually five different subdivisions of $I\!R^2$ by these curves, which are codified as:

$$C_1|C_2|C_3; \quad C_2|C_1|C_3; \quad C_2|C_3|C_1; \quad |C_1, C_2, C_3|^+; \quad |C_1, C_2, C_3|^-.$$

Definition 6.1.1 *By an (abstract) chordal system (CS) in the Euclidean plane $I\!R^2$ one understands a non-void family of disjoint curves $a, b, c...$ on which the two types of relations: $a|b|c$ and $|a, b, c|^\pm$, are defined in such a way that the following axioms are satisfied:*

(i) $a|b|c$ *is equivalent to* $c|b|a$;
(ii) $|a, b, c|^+$ *is equivalent to* $|b, c, a|^+$ *(and, hence, to* $|c, a, b|^+$*);*
(iii) $|a, b, c|^+$ *and* $|a, c, d|^+$ *imply* $|a, b, d|^+$ *and* $|b, c, d|^+$;
(iv) $|a, b, c|^\pm$ *and* $a|b|d$ *imply* $c|b|d$ *and* $|a, d, c|^\pm$, *whereby* $[a, b, c] \sim [a, d, c]$; [1]
(v) $a|b|c$ *and* $b|c|d$ *imply* $a|b|d$ *and* $a|c|d$;
(vi) of the three relations $b|a|c; b|a|d; c|a|d$ *at most two can hold.*

Let $\mathcal{F}$ be a non-singular foliation of $I\!R^2$. Kaplan constructs the "canonical" chordal system $CS(\mathcal{F})$ for $\mathcal{F}$ in the following way. All curves of $\mathcal{F}$ form the chordal system at the first step. Next, there exists a regular procedure (see Theorems 29, 30 of [154]), called a *normal subdivision* of $\mathcal{F}$ which decomposes $I\!R^2$ into a countable set of non-overlapping regions partitioned by the leaves of foliation $\mathcal{F}$.

We present the main idea of the procedure. The simplest foliation is that by parallel lines. In every foliation one can extract subfoliations homeomorphic to parallel lines: e.g., a subfamily of curves meeting at a single cross-section. It is therefore natural to try to decompose the whole foliation into such pieces. The decomposition formed by curves crossing at an infinite cross-section which starts at a finite point and tends to the infinity. Such a family of curves is homeomorphic to parallel lines partitioning the half-plane. The normal subdivision consists precisely in the decomposition of the foliation into such non-overlapping pieces. The basic statement of the procedure is the following

[1] The relation $[a, b, c] \sim [a', b', c']$ means that the relation holding for a, b, c is also true for a', b', c'.

Theorem 6.1.2 *(Theorems 29, 30 of [154]) Suppose the curves l_1 and l_2 of F are joined by a cross-section p_1p_2 with $p_i \in l_i$. Let S be the set of curves crossing p_1p_2, excluding p_1 and p_2. Then S forms an open set and the condition $l_1|l_3|l_2$ is equivalent to the condition that l_3 is in S. Moreover, $S \cup l_1 \cup l_2$ can be mapped homeomorphically on a strip $0 \leq y \leq 1$, $-\infty < x < +\infty$ in such a way that the curves of $S \cup l_1 \cup l_2$ become the straight lines $y = Const$.*

Taking (any) one representative in each component, we obtain a (countable) system of non-intersecting curves which form the normal chordal system $CS(\mathcal{F})$. Kaplan showed that $CS(\mathcal{F})$ does not depend on the normal subdivision of $\mathcal{F}$.

In the paper [155] Kaplan obtained a topological classification of the singularity–free foliations in the plane using normal chordal systems. Every normal chordal system $CS(\mathcal{F})$ can be represented by the isomorphic set of non-intersecting chords in the circle, which is denoted by $K(\mathcal{F})$.

Theorem 6.1.3 *Let $\mathcal{F}_1, \mathcal{F}_2$ be two foliations in the plane $I\!R^2$ without singularities. Then $\mathcal{F}_1$ is topologically equivalent to $\mathcal{F}_2$ under an orientation-preserving homeomorphism if and only if $K(\mathcal{F}_1)$ is isomorphic either to $K(\mathcal{F}_2)$ or to its mirror reflection [2] $K'(\mathcal{F}_2)$. Moreover, to any $\mathcal{F}$ there corresponds a normal chordal system $CS(\mathcal{F})$ and to every (abstract) normal chordal system $\mathfrak{CS}$ there corresponds a foliation $\mathcal{F}$ for which $CS(\mathcal{F})$ is isomorphic to $\mathfrak{CS}$.*

Proof. See [155]. □

Note that Theorem 6.1.3 gives also the classification of foliations on the sphere which admit one singular point. For completeness, we also quote the 1942 work of Kaplan [156] (*The structure of a curve-family on a surface in the neighborhood of an isolated singularity, Amer. J. Math 64, 1-35*) where a classification of such foliations with two singularities was obtained.

6.1.2 Singular Case

In this section the results of L. Markus [198] are reviewed. A collection S of curves filling the plane region R is *parallel* if S is mapped by an orientation preserving homeomorphism (also called an o-preserving homeomorphism) to either the plane partitioned by parallel lines, or to the punctured plane partitioned by concentric circles, or by rays issued from the deleted point. For brevity, we say that R is parallel and refer to the three cases as *strip, annular* and *radial (spiral)*, respectively.

A trajectory l is a *separatrix in the sense of Markus* if l is not embedded into a parallel neighbourhood N such that:

[2] Relatively to the diameter of the circle containing $K(\mathcal{F})$.

(i) all the trajectories in N have the same α- and ω-limit sets, $\alpha(l)$ and $\omega(l)$;
(ii) N is bounded by $\alpha(l) \cup \omega(l)$ and exactly two trajectories l_1, l_2 for which $\omega(l_1) = \omega(l_2) = \omega(l)$ and $\alpha(l_1) = \alpha(l_2) = \alpha(l)$.

(Note that the sets l_1, l_2 and $\alpha(l) \cup \omega(l)$ are always disjoint; furthermore, N can be taken to be either a strip or an annulus.)

The following special cases are immediate consequences of the above definition. Every fixed point is a separatrix in the sense of Markus. A closed trajectory is a separatrix in the sense of Markus if and only if every neighbourhood of l intersects a semi-trajectory which is a spiral. The separatrices in the sense of Markus include all usual separatrices and the bounding cycles of Kaplan's classification; the latter form barriers to extension of the transversals.

Let $\sigma(f^t)$ be the union of all separatrices in the sense of Markus of flow f^t on $\mathbb{R}^2$. Each component of the complement to $\sigma(f^t)$ is called a *canonical region* of f^t.

It is important for Markus consideration that, although there may be infinitely many separatrices, they cannot cluster about some trajectory, which would then be a limit separatrix.

If a separatrix $l_0 \subset \; Clos\ (\sigma(f^t)\backslash l_0)$, then l_0 is called a *limit separatrix.*

Following [180], Markus chooses an arbitrary trajectory in each canonical region to form a "topological picture" of the flow. More precisely, for each flow f^t on $\mathbb{R}^2$ one defines a *separatrix configuration*, $S\sigma(f^t)$, as a collection of separatrices (in the sense of Markus) taken together with one representating trajectory chosen in each canonical region.

Theorem 6.1.4 *Let f_1^t, f_2^t be flows in $\mathbb{R}^2$ having no limit separatrices other than fixed points. A necessary and sufficient condition for the topological o-equivalence of f_1^t and f_2^t, is the existence of an o-preserving homeomorphism $\varphi : \mathbb{R}^2 \to \mathbb{R}^2$ such that $\varphi[\sigma(f_1^t)] = \sigma(f_2^t)$ and $\varphi[S\sigma(f_1^t)] = S\sigma(f_2^t)$.*

Proof. The central part of the proof is consisting in showing that every canonical region is a parallel region. □

Corollary 6.1.1 *Let f_1^t, f_2^t be the two flows in $\mathbb{R}^2$ having no limit separatrices other than the fixed points. Then, f_1^t is topologically equivalent to f_2^t if and only if $S\sigma(f_1^t)$ and $S\sigma(f_2^t)$ are equivalent under a homeomorphism φ which maps $\sigma(f_1^t)$ onto $\sigma(f_2^t)$.*

Corollary 6.1.1 is similar to the Leontovich-Maier Theorem 4.4.1. However, the "singular" trajectories of Leontovich-Maier and Markus theorems are essentially different. The singular trajectories introduced by Leontovich and Maier are tailored for compact surfaces. On noncompact surfaces there are "too many" such trajectories. (One easily constructs the relevant examples.)

For further reading, we recommend the thorough book of A. Beck [43] *(Continuous Flows in the Plane, Springer-Verlag, 1974)* where the Kaplan-Markus theory is given in the terms of Kaplan diagrams.

6.1.3 Level Set of Harmonic Functions

A remarkable generalization of Theorem 6.1.1 has been made by Kaplan [157] *(Topology of level curves of harmonic functions, Trans. Amer. Math. Soc. 63 (1948), 514-522)*. Before we plunge into the details, let us fix our vocabulary.

The complex plane will be denoted by $\mathbb{C}$. The function $f : D \to \mathbb{C}$, $D \subset \mathbb{C}$, is *holomorphic* in D if the first derivative $\frac{df}{dz}$ (and hence all subsequent derivatives) exist for every $z \in D$. Equivalently,

$$\frac{\partial f}{\partial \bar{z}} \equiv 0, \quad \text{where} \quad \frac{\partial}{\partial \bar{z}} = \frac{1}{2}(\frac{\partial}{\partial x} + i\frac{\partial}{\partial y}).$$

A function f is said to be *harmonic in the domain* D if

$$\Delta f \equiv 0, \quad \text{where} \quad \Delta = \frac{\partial^2}{\partial x^2} + \frac{\partial^2}{\partial y^2} = 4\frac{\partial^2}{\partial z \partial \bar{z}}$$

is the Laplace operator.

Recall that a *Riemann surface* is a one-dimensional, complex, analytic manifold S with maximal atlas $\{U_\alpha, z_\alpha\}$ such that the transition functions $t_{\alpha\beta} = z_\alpha z_\beta^{-1}$ are holomorphic whenever they are defined. This atlas is also called a *conformal structure* on S. The Riemann surfaces S_1, S_2 are *conformally equivalent* if there exists a homeomorphism $\varphi : S^1 \to S^2$ which preserves the conformal structure.

Let S be a Riemann surface with fixed conformal structure $\{U_\alpha, z_\alpha\}$. A real-valued function V on S is *harmonic* on S if and only if, for each α, the composition $V \circ z_\alpha^{-1}$ is harmonic on $z_\alpha(U_\alpha)$.

Now we are ready to formulate Kaplan's theorem.

Theorem 6.1.5 *Let $\mathcal{F}$ be a foliation in the plane without singular points. Then there exists a conformal structure which turns $\mathbb{R}^2$ into a Riemann surface S such that $\mathcal{F}$ is the level curve family of a harmonic function on S.*

Proof. By Theorem 6.1.1, there exists a continuous function $f(x, y)$ without local extrema such that $\mathcal{F}$ is the level set of $f(x, y)$. Since $\mathcal{F}$ is oriented, there exists an orthogonal foliation $\mathcal{F}'$ which is the level set of some function $g(x, y)$.

Neither f nor g have local extrema; therefore the mapping

$$T_1 : (x, y) \mapsto (u = f(x, y), v = g(x, y))$$

is a local homeomorphism of $z(= x + iy)$-plane to $w(= u + iv)$-plane. T_1 is either the orientation-preserving or the orientation-reversing homeomorphism. By taking $-g$ instead of g, if necessary, it is possible to ensure that T_1 preserves the orientation.

The surface S is defined as follows. Its points are all pairs (z, w), where $w = T_1(z)$. The mapping $T_2 : z \mapsto (z, w)$ is one-to-one and defines a topology on S. The mapping $T_3 : (z, w) \mapsto w$ is locally a homeomorphism. This allows us to define local conformal coordinates on S. It is easy to see that if two local charts overlap, the mappings T_i's coincide on the common part and, hence, the transition function $\equiv Id$.

Let H be a preimage of $\mathcal{F}$ under the mapping T_2. Then, under T_3, H goes to the family of lines $u = f(x, y) = Const$; thus u is a harmonic function. □

It is well known (the Uniformization Theorem) that S is conformally equivalent either to the open unit disc $D_1 : x^2 + y^2 < 1$, or to the entire xy-plane D_∞. If in Theorem 6.1.5 S can be chosen as D_1, $\mathcal{F}$ will be called *hyperbolic*. If S can be chosen as D_∞, $\mathcal{F}$ will be called *parabolic*. The cases are not mutually exclusive, as the following theorem shows.

Theorem 6.1.6 *Every non-singular foliation $\mathcal{F}$ in the plane $\mathbb{R}^2$ is hyperbolic. There exists infinitely many topological types of $\mathcal{F}$ which are not parabolic.*

Proof. To prove that S can always be chosen as D_1, it is enough to ensure that $|w|$ is bounded on S. The level set functions $f(z), g(z)$ are replaced by the functions

$$\bar{f}(z) = tanh\ f(z), \quad \bar{g}(z) = tanh\ g(z)$$

to ensure that this is the case. For then $\bar{f}, \bar{g}$ are level-curve functions, since *tanh* is a monotone continuous function and $|\bar{f}| < 1, |\bar{g}| < 1$. Thus $|w| = |\bar{f} + i\bar{g}|$ will be bounded on S; T_3 is a conformal mapping on the domain D which is bounded. By the Liouville Theorem, D goes to D_1, but not to D_∞. Hence every $\mathcal{F}$ is hyperbolic.

Consider the level-curve families $f_n(x, y)$:

$$\begin{aligned} f_n &= e^x \sin y, & 0 \le y \le 2\pi n, \\ f_n &= y, & -\infty < y \le 0, \\ f_n &= y - 2\pi n, & 2\pi n \le y < \infty. \end{aligned}$$

They represent the infinity of distinct foliations, none of which is parabolic. □

Kaplan's student William M. Boothby generalized Theorems 6.1.1, 6.1.5 to foliations with saddle type oriented singularities [57]. [3] An analogy to a chordal system is introduced, which along with one-dimensional leaves containes also *trees*, i.e., the union of saddles and separatrices. (The allowence for such a term is justified, since this union will give a graph containing no cycles.)

Theorem 6.1.7 *Let $\mathcal{F}$ be a foliation in the plane with oriented saddle singularities. Then on $\mathbb{R}^2$ there exists a continuous function f such that:*

---

[3] Boothby calls such saddles *multiple*, i.e., given by the level curves of the function $Re\ z^n, n > 1$. Another term for this is *branch point* of multiplicity n.

(i) for each $c \in \mathbb{R}$ the equation $f(x, y) = c$ defines, unless it is empty, a finite or at most countably infinite collection of trees (and regular curves) of $\mathcal{F}$;

(ii) in every neighbourhood of any point $p \in \mathbb{R}^2$ there exist points q's for which $f(q) > f(p)$ and points $\tilde{q}$'s for which $f(\tilde{q}) < f(p)$.

Proof. See [57]. □

Corollary 6.1.2 *Every foliation $\mathcal{F}$ mentioned in Theorem 6.1.7 is flow-embeddable.*

Theorem 6.1.8 *Let $\mathcal{F}$ be a foliation in the plane with oriented saddle singularities. Then there exists a conformal structure which turns $\mathbb{R}^2$ into a Riemann surface S and such that $\mathcal{F}$ is the level curve family of a harmonic function on S.*

Proof. The proof is much the same as for Theorem 6.1.5 but with an extra study of the singularity points; cf. Boothby [58]. □

The necessity of conditions in Theorem 6.1.8 follows from the well-known theorem of harmonic function theory.

Theorem 6.1.9 *Let f be a harmonic function on the Riemann surface S. Then the level curve family of f yields a foliation on S whose singularities are oriented saddles.*

Digression: the level set of pseudo-harmonic functions. The importance of the study of level set of harmonic functions has been forecast by M. Morse and J. Jenkins in the work [205] *(The existence of pseudoconjugates on Riemann surfaces, Fund. Math. 39 (1952), 269-287)* which we quote in below.

"Among the characteristic of a function U which is harmonic on a Riemann surface G are the topological interrelations of the level lines of U < ... >. It seems likely that outstanding problems in Riemann surface theory, such as the type problem, the nature of essential singularities, the existence of functions on the Riemann surface with restricted properties cannot be thoroughly understood in the absence of a complete analysis of the topological characteristics of these level lines..."

If a homeomorphism acts on the family of level curves, then the topological type of the family does not change. This remark leads to the pseudo-harmonic functions; cf. *M. Morse and M. Heins, Topological methods in the Theory of Functions of a Complex Variable, II, Annals of Math. 46 (1945), 625-666.*

Namely, let $u(x, y)$ be a non-constant, harmonic function defined in the neighbourhood N of a point (x_0, y_0). If N is subjected to an arbitrary homeomorphism T which fixes point (x_0, y_0) and moves the point (x, y) to the point (x', y'), then the function $U(x', y') = u(x, y)$ is said to be *pseudo-harmonic.* In other words, the pseudo-harmonic functions are obtained locally from harmonic ones by composition with a topological transformation. The question naturally arises if the same is true for pseudo-harmonic functions "globally". The answer is yes for the plane $\mathbb{R}^2$; cf. Morse and Jenkins [205].

Theorem 6.1.10 *Let U be a pseudo-harmonic function in a simply connected domain D_z lying in the z-plane. Then there exists a homeomorphism φ from D_z onto a domain D_w in the w-plane and a function U^* harmonic on D_w such that $U = U^*\varphi$. Moreover, if D_z is of hyperbolic type, the mapping φ can be chosen so as to map D_z onto itself.*

To conclude, let us again cite the 1952 work of Morse & Jenkins [205]: *"The reader familiar with the geometric theory of dynamical systems as initiated by Poincaré with its attention to singularities, periodicity, recurrence, etc., will recognize that the underlying topological theory of the level lines of a pseudo-harmonic function is a form of 2-dimensional topological dynamics..."*

6.2 Structural Stability

Morse-Smale flows are not, in general, structurally stable on open manifolds; cf. Peixoto and Pugh [238]. To "control" the structural stability of such flows, more severe restrictions must be imposed. The major obstacles here are the so-called saddle points at "infinity". Below we treat the subject in more detail following the 1982 memoir of J. Kotus, M. Krych and Z. Nitecki [172] *(Global structural stability of flows on open surfaces, Mem. Amer. Math. Soc. 261, V.37).*

Let M be an open surface. By a *strong C^r-topology* of vector fields f on M one understands a topology given by neighbourhoods of the form $||f_1(x) - f_2(x)|| < \varepsilon(x), x \in M$ and their derivatives up to the order r. A flow f^t is said to be *globally C^r structurally stable* if for every flow g^t in a strong C^r neighbourhood of f^t there exists a homeomorphism $h : M \to M$ close to the identity which sends f^t-orbits to g^t-orbits and respects the time order.

If $O_+(x)(O_-(x))$ is a positive (negative) semi-trajectory through x, then one distinguishes three possible types of behaviour:

- $O_\pm$ is *bounded* if it is contained in a compact set $C \subset M$;
- $O_\pm$ *escapes to infinity* if for each compact set C there exists a point $y \in O_\pm$ such that $O_\pm \cap C = \emptyset$;
- $O_\pm$ *oscillates* if it is neither bounded nor escapes to infinity.

Suppose that $O_\pm(x)$ is an orbit of flow f^t which escapes to infinity. The (first) *positive (negative) prolongational limit set* of $x \in M$ is defined to be the set $J^\pm = \{y \in M |\ f^t(x_n, t_n) \to y\}$, for a sequence $x_n \to x$ and $t_n \to \pm\infty$.

Definition 6.2.1 *Two unbounded semi-trajectories $O_+(x)$ and $O_-(y)$ are said to form a saddle at infinity, if each escapes to infinity and $y \in J^+(x)$ or $x \in O^-(y)$.*

For a flow f^t on M we denote by $W^+(f^t)$ $(W^-(f^t))$ the union of all stable (resp. unstable) separatrices of saddle points, including those at infinity. By $Per\ f^t$, as usual, one understands the set of periodic points (closed orbits and fixed points) of f^t. Here is a sufficient condition for global structural stability:

Theorem 6.2.1 *Suppose that M is an open surface and f^t is a $C^r, r \geq 1$ flow on M, satisfying the following conditions:*

(i) there are no oscillating orbits and there are no non-trivial minimal sets;
(ii) every orbit in $Per\ f^t$ is hyperbolic;
(iii) $Clos\ W^-(f^t) \cap Clos\ W^+(f^t) \subset Per\ f^t$

then (a) $\Omega(f^t) = Per\ f^t$ and (b) f^t is globally structurally stable.

Proof. The proof is based on the filtration construction, which is rather standard in the field. It also uses decomposition theorems similar to those which we discussed in earlier chapters; see [172] for details. □

The question of necessary conditions for global structural stability is rather delicate. It is settled only in the case of the Euclidean plane $\mathbb{R}^2$:

Theorem 6.2.2 *Suppose that $M = \mathbb{R}^2$ is the Euclidean plane and let f^t be a globally structurally stable flow on M. Then for f^t the conditions (i)-(iii) of Theorem 6.2.1 are satisfied.*

Proof. See [172]. □

The problem of genericity for Morse-Smale flows on closed surfaces has been discussed earlier. Genericity fails, in general, for Morse-Smale flows on open surfaces, as does structural stability. One of the known criteria applies only to flows in the plane.

Theorem 6.2.3 *Let f^t be a flow in the Euclidean plane $\mathbb{R}^2$, and suppose F^t satisfies the following conditions:*

(i) every orbit $Per\ f^t$ is hyperbolic;
(ii) there are no saddle connections between finite saddle points;
(iii) there are no oscillating orbits;
(iv) $W^-(f^t) \cap W^+(f^t) = \emptyset$.

Then f^t's are generic (open and dense) in the space of flows for all $r \geq 1$.

Proof. See [172]. □

6.3 Examples

6.3.1 Depth of the Centre

The depth of the centre of flows on a compact surface cannot exceed 3. As was noticed by D. Neumann [211], this is no longer true for open surfaces (namely, surfaces of infinite genus), where the depth can be arbitrary big. In this respect, flows on surfaces of infinite genus are the same as flows in three-dimensional space.

Before we pass to Neumann's example, let us recall different concepts of central sequence, due to Birkhoff [50], Birkhoff & Smith and Maier:

G. D. Birkhoff and P. A. Smith, Structure analysis of surface transformations, J. Math. Pures et Appl. 7 (1928), 345-379.
A. Maier, Sur un problème de Birkhoff, Dokl. Acad. Sci. USSR 55 (1947), 473-475; Sur les trajectoires dans l'espace à trois dimensions, ibid., 579-475.

Let f^t be a flow on a metric space M.

Birkhoff's central sequence. Let NW denote a non–wandering set of points of flow f^t. Define by the recursion:

1. $\Omega_0 = M$;
2. $\Omega_{\alpha+1} = NW(f^t|\Omega_\alpha)$, if α is an ordinal and Ω_α is defined;
3. $\Omega_\alpha = \cap_{\beta<\alpha}\Omega_\beta$, if α is a limit ordinal and Ω_β is defined for all $\beta < \alpha$.

The transfinite sequence $\alpha \to \Omega_\alpha$ is called a *1-central sequence* of f^t. The depth of the 1-central sequence, $d_1(f^t)$, is the least ordinal α such that $\Omega_{\alpha+1} = \Omega_\alpha$.

We recall that the limit set $\lim(f^t)$ is the union of $\omega-$ and α-limit sets of all trajectories of f^t.

Birkhoff-Smith's central sequence. Define recursively:

1. $\omega_0 = M$;
2. $\omega_{\alpha+1} = \ Clos\ [\lim(f^t|\omega_\alpha)]$, if α is an ordinal and ω_α is defined;
3. $\omega_\alpha = \cap_{\beta<\alpha}\omega_\beta$, if α is a limit ordinal and ω_β is defined for all $\beta < \alpha$.

We call the transitive sequence $\alpha \to \omega_\alpha$ a *2-central sequence* of f^t. The depth of 2-central sequence, $d_2(f^t)$, is the least ordinal α such that $\omega_{\alpha+1} = \omega_\alpha$.

Maier's central sequence. Define recursively:

1. $M_0 = M$;
2. $M_{\alpha+1} = \lim(f^t|M_\alpha)$, if α is an ordinal and M_α is defined;
3. $M_\alpha = \cap_{\beta<\alpha}M_\beta$, if α is a limit ordinal and M_β is defined for all $\beta < \alpha$.

Transfinite sequence $\alpha \to M_\alpha$ is called a *3-central sequence* of f^t. The depth of 3-central sequence, $d_3(f^t)$, is the least ordinal α such that $M_{\alpha+1} = M_\alpha$. Since $M_\alpha \subset \omega_\alpha \subset \Omega_\alpha$ for any ordinal α, we have $d_1(f^t) \geq d_2(f^t)$.

The question of the existence of flows (in $\mathbb{R}^n$) with 1-central sequence of an arbitrary (countable) depth was asked by Birkhoff (*Some unsolved problems of theoretical dynamics, Science, 94, 1941, 598-600*). A satisfactory answer has been obtained by Maier *(On the ordinal number of central trajectories, Dokl. Acad. Nauk SSSR 59 (1948), 1393-1396)*, who proved that it is possible to construct a flow on any n-manifold ($n \geq 3$) such that $d_1 \geq d_3$ are arbitrarily chosen countable ordinals. Neumann [211] proved that it is also true for flows on open manifolds.

Theorem 6.3.1 *Let α, β and γ by any countable ordinals satisfying $\alpha \geq \beta \geq 2, \gamma \geq 2$. Then there exists an orientable open surface M and a C^∞-flow f^t such that $d_1(f^t) = \alpha, d_2(f^t) = \beta$ and $d_3(f^t) = \gamma$.*

Proof. We give a sketch of the construction for a fixed countable ordinal which is not a limit ordinal. Let U be an open unit square $|x| < 1, |y| < 1$ with a fixed orientation. Let A and B denote the open subsets of $I_0 = \{x = 1, 0 < |y| < 1\}$ and $J_0 = \{x = -1, 0 < |y| < 1\}$, respectively. A and B are oriented by the given orientation of U. We set $S = U \cup A \cup B$. If $f : A \to B$ is any orientation reversing homeomorphism, then the quotient space $M = S/f$ is an oriented, open surface. If we consider S to be fibered by its intersection with horizontal lines, so that each fiber is oriented from the left to the right, then we obtain an oriented 1-dimensional foliation of M embedded into the flow $f^t|M$. Let us put $E_1 = U$. Notice that if we place a fixed point on the segment $U \cap \{y = 0\}$, then for the flow f_1^t we would have $\alpha(f_1^t) = 1$. First, we construct E_2. For each $i \in \mathbb{N}$ fix $0 < \varepsilon_i < 1/3i(i+1)$, and let

$$K_i = \{x = 1,\ \frac{1}{i} + \varepsilon_i < y < -\frac{1}{i+1} - \varepsilon_i\}, J_i = \{x = -1,\ \frac{1}{i+1} + \varepsilon_i < y < \frac{1}{i} - \varepsilon_i\}.$$

Let $A_2 = \cup K_i, B_2 = \cup J_i$, both oriented by fixed orientation of U, and let $f : A_2 \to B_2$ be an orientation reversing diffeomorphism that takes K_i onto J_i for each $i \in \mathbb{N}$. Set $E_2 = (U \cup A_2 \cup B_2)/f$ and let g_2^t denote a C^∞-flow f_2^t on E^2. Note again that if we place a fixed point at $U \cap \{y = 0\}$, then $\alpha(f_2^t) = 2$.

In the same (inductive) manner one constructs the set E_3 with the flow f_3^t, such that $\alpha(f_3^t) = 3$. In general, we obtain the flow f_n^t with $\alpha(f_n^t) = n$ for an arbitrary natural number n. The same schema leads to the construction of a flow f_α^t with the prescribed depths d_1, d_2 and d_3. □

6.3.2 Minimal Sets

Recall that a *minimal set* of a flow is a closed, invariant, non-empty set which has no proper subsets with the same property.

P. Schweitzer [263] (item 34.1 of *Some problems in foliation theory and related areas, Lect. Notes Math. 652, 240-252*) posed the following problem: "Does there exist any manifold M^n with a flow that has no minimal set?" (Note, that M^n is necessarily a non-compact manifold.)

After almost 20 years T. Inaba [151] solved this problem by proving the following theorem.

Theorem 6.3.2 (Inaba) *There exists a flow on an open surface which admits no minimal set.*

Proof. Take a countable family of surfaces, each topologically equivalent to Neumann's surface with a "simplest" foliation. The first copy is glued together with the second one along the countable family of strips, accumulating to a leaf L_1, lying in the second copy. The first copy is then deformed into a long strip, and like a "tinsel garland" is thrown over to the second copy. Thus, a leaf L_2 lying at the closure of all leaves in the first copy is obtained; however, $L_1 \not\subset \ Clos\ (L_2)$. We repeat the procedure a second time, so that $L_2 \not\subset \ Clos\ (L_3)$. We continue by induction infinitely many times. The foliation has no minimal set because in the closure of any leaf there exists a leaf with a "smaller" closure. Take sequences $\{a_k\}, \{b_k\}, \{c_k\}, \{d_k\}, \{e_k\}, (k = 1, 2, 3, \cdots)$ such that:

$$a_1 = 9/10, \quad a_k > b_k > c_k > d_k > e_k > a_{k+1}, \quad (k = 1, 2, 3, \cdots)$$

and $\lim_{k\to\infty} a_k = 0$. Put

$$P = \{(x, y) \in I\!R^2 \mid -1 < x < 0,\ x - 2 < y < \min\{x + 1,\ 9/10\}\},$$

$$Q_k = \{(x, y) \in I\!R^2 \mid 9/10 < x < 21/10,\ c_k < y < a_k\}$$

$$R_k = \{(x, y) \in I\!R^2 \mid 29/10 < x < 41/10,\ b_{k+1} < y < d_k\}$$

$k = 1, 2, 3, \cdots$, and consider on each of them a foliation by the horizontal lines $y =$ constant.

Let N be the quotient space obtained from the disjoint union $P \cup \bigcup_{k=1}^{\infty} Q_k \cup \bigcup_{k=1}^{\infty} R_k$ by the following identifications: $(x, y) \in Q_k\ (9/10 < x < 1)$ is identified with $(x - 1, y - 2) \in P$, $(x, y) \in Q_k\ (2 < x < 21/10)$ with $(x + y - 3, y) \in P$, $(x, y) \in R_k\ (29/10 < x < 3)$ with $(x - 3, y) \in P$ and $(x, y) \in R_k\ (4 < x < 41/10)$ with $(4 - x, y - 1) \in P$. Denote by $\mathcal{F}_N$ the foliation on N induced from the horizontal foliations on the pieces.

Let $(N_n, \mathcal{F}_n)\ (n = 1, 2, 3, \cdots)$ be countably many copies of $(N, \mathcal{F}_N)$. Put

$$R^{(0)} = \{(x, y) \in I\!R^2 \mid 49/10 < x < 61/10,\ 0 < y < 9/10\}$$

and let $(N_0, \mathcal{F}_0)$ be the quotient space obtained from the union $P \cup R^{(0)}$ by identifying $(x, y) \in R^{(0)}\ (49/10 < x < 5)$ with $(x - 5, y) \in P$ and $(x, y) \in R^{(0)}\ (6 < x < 61/10)$ with $(6 - x, y - 1) \in P$.

For each $k = 1, 2, 3, ...$, put

$$S_k^{(n)} = \{(x, y) \in P \subset N_n \mid a_{k+1} - 1 < x < e_k - 1\} \quad (n = 0, 1, 2, \cdots)$$

and

$$T_k^{(n)} = \{(x,y) \in P \subset N_n \mid \frac{d_k - e_k}{1 - e_k}x + d_k < y < \frac{b_k - c_k}{1 - c_k}x + b_k\} \quad (n = 1, 2, 3, \cdots).$$

For each $n \geq 0$ and $k \geq 1$, choose a foliation preserving diffeomorphism $\varphi_k^{(n)}$ from $S_k^{(n)}$ onto $T_k^{(n+1)}$ so that

$$\varphi_k^{(n)}((1-t)a_{k+1} + te_k - 1, a_{k+1}) = ((1-t)(c_k - 1), c_k)$$

and that

$$\varphi_k^{(n)}((1-t)a_{k+1} + te_k - 1, e_k - 3) = ((1-t)(d_k - 1), d_k)$$

for $0 < t < 1$.

Now, let M be the quotient space obtained from the disjoint union $\bigcup_{n=0}^{\infty} N_n$ by identifying $S_k^{(n)}$ with $T_k^{(n+1)}$ via $\varphi_k^{(n)}$ for all $n \geq 0$ and $k \geq 1$, and let $\mathcal{F}_M$ be the foliation on M induced from $(N_n, \mathcal{F}_n)$. We will show that $(M, \mathcal{F}_M)$ has the desired property.

It suffices to show that the closure $\overline{F}$ is not a minimal set for any leaf F of $\mathcal{F}_M$. Let L_n $(n = 0, 1, 2, \cdots)$ be the leaf of $\mathcal{F}_M$ defined by

$$L_n = (-1, 0) \times \{0\} \subset P \subset N_n \subset M.$$

Then, by the construction, one can observe that $\overline{L_n} = \bigcup_{i \geq n} L_i$ for all $n \geq 0$. Thus, $\overline{L_n} \supsetneq \overline{L_{n+1}}$, and hence $\overline{L_n}$ is not a minimal set. Next let us consider a leaf F of $\mathcal{F}_M$ such that $F \neq L_n$ for any n. Then, by construction, either (maybe both) of the two ends of F, say e, is of the form $(-1, s] \times \{y\} \subset P \subset N_m$ for some $m \geq 0$ and some s, $y < 0$. In N_{m+1} the end e behaves similarly to L_m and is asymptotic to L_{m+1}, that is, $\overline{F} \supset L_{m+1}$. This implies that $\overline{F}$ is not a minimal set because $\overline{F} \supsetneq \overline{L_{m+1}}$. This proves the non-existence.

The foliation $\mathcal{F}_M$ is transversely orientable, but not tangentially orientable. If one wants to get an example of a tangentially orientable foliation (or a flow), one only has to take a suitable double covering of M. This completes the proof of the theorem. □

A generalization of Inaba's theorem was obtained by Jean-Christophe Beniere and Gael Meigniez; cf. [47]

Theorem 6.3.3 (Beniere-Meigniez) *Every noncompact manifold carries a flow without minimal set, except the real line and surfaces of finite genus.*

Proof. The canvas of the proof is the construction of a flow without minimal set on a 2-dimensional open surface similar to Inaba's example. This flow is then "adapted" to higher dimensional manifolds. □

6.3.3 Minimal Flows

A complement to Theorem 6.3.3 was proved by J.-Ch. Benière and G. Hector [46].

Theorem 6.3.4 (Beniere-Hector) *Every non-compact surface of non-zero genus admits a minimal flow.*

Proof. Every non-compact surface S is derived from a compact one, say M, after removing a subset Σ of a totally discontinuous set which is equivalent to Cantor set, i.e., $S = M \setminus \Sigma$. In case Σ is finite or countable the theorem follows easily after one considers a highly transitive flow on M and one removes fixed points and (the necessary number of) fake saddles. In case Σ is uncountable, the proof requires more care. For example, if $M = T^2$ one faces the following task: Find an irrational α and a Cantor set $K \subset S^1$ such that $R_\alpha^n \cap K = \emptyset, \forall n \in \mathbb{Z}$, where R_α is rotation through the angle α of the unit circle S^1. One requires K. Falconer's results on Hausdorff measures to prove that such a Cantor set exists. For M's of genus $g \geq 2$ one considers a similar problem for the interval exchange transformations. □

For non-compact orientable surfaces of finite genus Theorem 6.3.4 can be formulated more precisely.

Theorem 6.3.5 *Let S be a non-compact orientable surface of finite genus $g > 0$. Then for each $2 \leq r \leq 2g$ there exists a minimal foliation $\mathcal{F}$ given on S by a closed 1-form ω. Moreover, the rank of the group Per ω is equal to r.*

Proof. As in Theorem 6.3.4, the required $\mathcal{F}$ is derived from a highly transitive foliation $\mathcal{F}'$ on a compact M. $\mathcal{F}'$ is the suspension of either a homeomorphism or an interval exchange transformation over S^1. Such an $\mathcal{F}'$ is always given by a closed 1-form. □

6.3.4 Transitive Flows

In this section we construct a transitive flow f^t on a surface M of infinite genus such that f^t has a nontrivial recurrent trajectory which is not dense in M. Thus, Maier's theorem (see Chapter 3) is not valid for a surface of infinite genus. This example is due to S. Aranson and E. Zhuzhoma; see [229]

Let f_0^t be an irrational flow on the torus T^2 and $m_0 \in T^2$ some point. Since any semitrajectory of f_0^t is recurrent in the sense of Birkhoff, there is a sequence of pairwise disjoint arcs $a_n b_n$ of some trajectory l of f_0^t such that $\cup_{n=1}^{\infty}[a_n b_n]$ is dense in T^2 and $a_n, b_n \to m_0$ as $n \to \infty$. Moreover, we can suppose that $m_0 \notin l$.

Putting a fake saddle in m_0 and then deleting m_0, we get a transitive flow f_1^t on the punctured torus $T^2 - m_0$. Starting with f_1^t, one can get a flow f_2^t on the surface M_2 of infinite area with the following properties:

(1) M_2 is homeomorphic to the torus $T^2 - m_0$,
(2) the ideal boundary of M_2 consists of a unique end E_2,
(3) the flow f_2^t has a unique quasiminimal set Q_2 whose complement is a unique simply connected domain W,
(4) there is a sequence of flow boxes B_n bounded by transversal segments Σ_n^+, Σ_n^- and arcs l_1^n, l_2^n of trajectories of f_2^t such that

(a) $Int\ (B_n) \subset W$,
(b) $l_1^n, l_2^n \subset Q_2$,
(c) the segments Σ_n^+, Σ_n^- approach E_2 as $n \to \infty$,
(d) the union $\cup_{n=1}^{\infty}(l_1^n \cup l_2^n)$ is dense in Q_2.

Roughly speaking, f_2^t is a Denjoy-type flow obtained by blowing up the trajectory l and unfolding the point m_0. After Denjoy's surgery, the arc $a_n b_n$ becomes the flow box B_n. We denote by $\theta_2^n : \Sigma_n^+ \to \Sigma_n^-$ the Poincare map induced by f_2^t.

Again starting with f_1^t, one can obtain a transitive flow f_3^t on the surface M_3 of infinite area with the following properties:

(1) M_3 is homeomorphic to the torus $T^2 - m_0$,
(2) the ideal boundary of M_3 consists of a unique end E_3,
(3) there is a sequence of flow boxes P_n bounded by transversal segments C_n^+, C_n^- and arcs s_1^n, s_2^n of trajectories of f_3^t such that P_n approach E_3 as $n \to \infty$.

We can assume that the restriction of f_3^t to P_n induces the Poincare map $C_n^+ \to C_n^-$ denoted by θ_3^n. Now we cut M_2 along the segments Σ_n^+, Σ_n^-. This cutting along Σ_n^+ (resp., Σ_n^-) gives a boundary component consisting of two intervals Σ_{n1}^+, Σ_{n2}^+ (resp., Σ_{n2}^-, Σ_{n2}^-), one of which Σ_{n1}^+ (resp., Σ_{n1}^-), belongs to the boundary of B_n and $\Sigma_{n1}^+ \cap \Sigma_{n2}^+ =$ the endpoints of Σ_n^+ (resp. $\Sigma_{n1}^- \cap \Sigma_{n2}^- =$ the endpoints of Σ_n^-). We put fixed points on $\Sigma_{n2}^+ \cup \Sigma_{n2}^-$, $n = 1, 2, \ldots$. The surface that we have obtained we denote by M_4 and the flow on M_4 by f_4^t.

Now let us remove all of the P_n from M_3. We put fixed points on $s_1^n \cup s_2^n$, $n = 1, 2, \ldots$. We denote the surface and the flow by M_5 and f_5^t, respectively.

Let $\eta_n : C_n^+ \to \Sigma_{n1}^+$ be an arbitrary homeomorphism, $n = 1, 2, \ldots$. Let us glue M_4 and M_5 by homeomorphisms η_n and

$$\eta_n' = \theta_3^n \circ \eta_n^{-1} \circ (\theta_2^n)^{-1} : \Sigma_{n1}^- \to C_n^-, \quad n = 1, 2, \ldots.$$

We obtain a surface M_6 of infinite genus. The flows f_4^t, f_5^t induce a flow f_6^t on M_6. By construction, the flow f_6^t is transitive, and therefore a quasiminimal set of the flow f_6^t is M_6. On the other hand, there is a nontrivial recurrent trajectory l_0 of the quasiminimal set Q_2 which does not intersect the segments Σ_n^+, Σ_n^-, $n = 1, 2, \ldots$. Hence, l_0 is nowhere dense on M_6. Thus, flow f_6^t is the required one.

Bibliographic Notes. The development of the qualitative theory of flows and foliations on open surfaces is connected with the names of H. Whitney [301], W. Kaplan [154], L. Markus [198], J. Kotus, M. Krych & Z. Nitecki [172], D. Neumann [211], T. Inaba [151], J. -C. Benière & G. Meigniez [47], G. Hector [46], S. Aranson & E. Zhuzhoma [229].

Part II

Applications

7. Ergodic Theory

7.1 Notations

We shall consider the collection of *Borel sets*, B_i, of a topological space X, which contains the open subsets and is closed under taking complements and countable unions. Such a collection is called a σ-algebra. The Borel sets are always assumed to be *measurable.* A function μ which associates to each measurable set a non-negative real number so that for each countable collection of disjoint Borel sets B_i one has

$$\mu\left(\bigcup B_i\right) = \sum_i \mu(B_i)$$

is called a *measure.* The measure μ is a *probability measure* if $\mu(X) = 1$ and a σ*-finite measure* if X can be written as a countable union of sets with finite measure. (Finite measure means that for all B_i, $\mu(B_i) < \infty$.)

Let f^t be a topological flow on X. The measure μ is *invariant* under f^t if

$$\mu(A) = \mu(f^t(A)) = \mu(f^{-t}(A)),$$

for each measurable set A and all $t \in \mathbb{R}$; it is easy to see that if A is measurable then the sets $f^t(A), f^{-t}(A)$ are measurable as well. If X is compact then according to the Bogolyubov-Krylov theorem every flow f^t must have at least one probability measure invariant under f^t; see

N. N. Bogolyubov and A. N. Krylov, La théorie génerale de la mesure dans son application à l'étude des systemes dynamiques de la mécanique non-linéaire. Ann. of Math. II, Ser. 38, 1937, 65-113.

This measure in general may be badly distributed between isolated points and has no relation to the smooth structure. If ω is a differential nondegenarate 2-form on a two-dimensional manifold M then it defines a *smooth measure* which is given by

$$\mu_\omega(A) = \int_A \omega.$$

In local coordinates

$$\mu_\omega(A) = \int_A p(x_1, \ldots, x_n) dx_1 \ldots dx_n$$

where $p(x_1, \dots, x_n)$ is a differential of non-negative function (the so–called *density function*).

7.2 Existence of Invariant Measures

7.2.1 Liouville's Theorem

Let the form ω be given by a density function $p(m)$ on a n-dimensional manifold M so that

$$d\mu_\omega = p(x_1, \dots, x_n) dx_1 \dots dx_n.$$

Theorem 7.2.1 (Liouville) *Suppose that a flow f^t is defined in a local chart $(x_1, \dots, x_n)$ by the system of differential equations*

$$\frac{dx_1}{dt} = X_1(x_1, \dots, x_n), \quad \dots, \quad \frac{dx_n}{dt} = X_n(x_1, \dots, x_n).$$

Then the measure μ (finite or σ-finite) given by $d\mu = p(x_1, \dots, x_n) dx_1 \dots dx_n$ is invariant under f^t if and only if

$$\sum_{k=1}^{n} \frac{\partial}{\partial x_k} (pX_k) = 0. \tag{7.1}$$

Proof. See for example *I. P. Cornfeld, S. V. Fomin, Ya. G. Sinai, Ergodic Theory, Springer-Verlag, Ch.2, §2, p.48.* □

Consider the system

$$\frac{dx_1}{dt} = w(x_1, \dots, x_n) X_1(x_1, \dots, x_n), \quad \dots,$$
$$\frac{dx_n}{dt} = w(x_i, \dots, x_n) X_n(x_1, \dots, x_n),$$

where $w(m)$ is a positive C^∞ function. Such a system defines a flow f_w^t on M. The flow f_w^t is obtained from f^t by time-rescaling by the function $w(m)$.

Proposition 7.2.1 *Suppose that a flow f^t is defined in a local chart $(x_1, \dots, x_n)$ by the system of differential equations*

$$\frac{dx_1}{dt} = X_1(x_1, \dots, x_n), \quad \dots, \quad \frac{dx_n}{dt} = X_n(x_1, \dots, x_n).$$

If the flow f^t admits a smooth invariant measure given by the C^∞ density function $p(m) > 0$ then the function

$$p_w(m) = \frac{p(m)}{w(m)}$$

is the density function of an invariant smooth measure for the flow f_w^t defined by the system

$$\frac{dx_1}{dt} = w(x_1, \dots, x_n) X_1(x_1, \dots, x_n), \qquad \dots,$$
$$\frac{dx_n}{dt} = w(x_i, \dots, x_n) X_n(x_1, \dots, x_n),$$

Proof. This follows from Liouville's theorem. □

Remark 7.2.1 *Poincaré understands by invariant measure an integral invariant, that is, an expression of the form*

$$\int\limits_M p(x_1, \dots, x_n) dx_1 \dots dx_n$$

with the property

$$\int\limits_D p(x_1, \dots, x_n) dx_1 \dots dx_n = \int\limits_{f^t(D)} p(x_1, \dots, x_n) dx_1 \dots dx_n \tag{7.2}$$

for any $t \in \mathbb{R}$ and $D \subset M$. It is easy to see that condition (7.2) is equivalent to the Liouville condition (7.1).

Consider a 2-dimensional dynamical system

$$\frac{dx}{dt} = A(x, y), \quad \frac{dy}{dt} = B(x, y). \tag{7.3}$$

Let us obtain necessary and sufficient conditions under which the standard Euclidean area

$$\mu_s(D) = \int\int\limits_D dx dy, \quad D \subset \mathbb{R}^2$$

($p \equiv 1$) is an integral invariant. From (7.1) it follows that

$$\frac{\partial A}{\partial x} + \frac{\partial B}{\partial y} = 0.$$

The latter means that expression

$$B\ dx \ - \ A\ dy \ = dH(x, y)$$

is an exact differential. System (7.3) can be written as

$$\frac{dx}{dt} = -H_y', \quad \frac{dy}{dt} = H_x'. \tag{7.4}$$

7.2.2 Torus

Let us consider a linear flow on the torus which is given by the system of differential equations

$$\frac{dx}{dt} = \lambda_1, \qquad \frac{dy}{dt} = \lambda_2.$$

Evidently the Lebesgue measure is an invariant measure for such a flow (with the density $p(m) = Const$). If one applies an analytic substitution of time and an analytic transformation of coordinates, then by Proposition 7.2.1 we shall obtain a flow with an analytic invariant measure. It turns out that the converse is also true. This was first proved in 1953 by A. N. Kolmogorov [170]. He showed that each analytic fixed-point-free torus flow with an analytic invariant measure can be obtained from a linear one by time-rescaling after an analytic transformation of coordinates.

Theorem 7.2.2 (Kolmogorov) *Suppose that the system $\dot{x} = A(x,y), \dot{y} = B(x,y)$ defines an analytic flow f^t on the torus. Suppose also that f^t has an analytic invariant measure with density $U(x,y) > 0$. If $A^2 + B^2 \neq 0$ then there exists an analytical diffeomorphism which transforms the above system to a system of the form*

$$\frac{dx}{dt} = \frac{1}{w(x,y)}, \qquad \frac{dy}{dt} = \frac{\alpha}{w(x,y)},$$

where $w(x,y) > 0$ is a positive function and α is a real constant.

Proof. Since the density function is analytic the flow f^t cannot have invariant annulus with Reeb components (that is, the integral curves cannot give rise to a Reeb foliation in the annulus). Therefore, f^t must have a global cross-section C which is not contractable. One can suppose C to be analytically embedded. Since C is non-contractable there exists an analytic transformation $\phi : T^2 \to T^2$ which sends C into the zero meridian of the torus. The flow f^t will be sent to an analytic flow g^t with cross-section coinciding with the zero meridian C_0. Evidently the first-return time for each trajectory hitting C_0 is an analytic function. It follows therefore that g^t is given by the formula displayed in the statement of the theorem. □

It is not hard to see that the flow f^t_α defined by the system from (Kolmogorov) Theorem 7.2.2 has the rotation number α. As a consequence, one can deduce that the rotation number of the original flow can be calculated from the right-hand sides of the dynamical system and the density function of its invariant measure.

Corollary 7.2.1 *Let the conditions of Theorem 7.2.2 hold. Then the rotation number of the flow f^t is*

$$\lambda = \frac{\int\int_{T^2} UBdxdy}{\int\int_{T^2} UAdxdy}.$$

In 1957 S. Sternberg [271] generalized Theorem 7.2.2.

Theorem 7.2.3 (Sternberg) *Suppose that the system* $\dot{x} = A(x,y), \dot{y} = B(x,y)$ *defines a* C^k *flow* f^t *on the torus. Suppose also that* f^t *possesses an integral invariant* $U(x,y) > 0$ *of class* C^k. *If* $A^2 + B^2 \neq 0$ *then there exists a* C^k *diffeomorphism which transforms the above system to the system*

$$\frac{dx}{dt} = \frac{1}{w(x,y)}, \qquad \frac{dy}{dt} = \frac{\alpha}{w(x,y)},$$

where $w(x,y) > 0$ *is a positive function and*

$$\alpha = \frac{\int\int_{T^2} UBdxdy}{\int\int_{T^2} UAdxdy}$$

is a real constant equal to the rotation number of the flow f^t. *Moreover, in new coordinates the covering trajectories are represented by the straight lines.*

Proof. The proof is similar to the proof of Theorem 7.2.2. □

In the new coordinates, the time is "non-linear". The linearization of time depends on the arithmetic properties of the rotation number and becomes an issue of small denominators. In Kolmogorov's work [170] sufficient "arithmetic" conditions for the linearization of the flow f^t are given. We formulate this result without proof, referring the reader to [170]. See also the papers

A. Winter. The linear difference equation of first order for angular variables, Duke Math. J. 12 (1945), 445-449.

and [271] where similar questions are considered.

Theorem 7.2.4 *Suppose the conditions of Theorem 7.2.2 hold. Suppose that there exist* $C > 0$ *and* $h > 0$ *such that*

$$|m - n\alpha| \geq Ch^n,$$

for all integers $m, n > 0$. *Then there exists an analytic transformation of coordinates which sends the flow* f^t *to the linear flow with* $\lambda_2 = \alpha\lambda_1$.

Kolmogorov's results give rise to a natural question: Does any analytical flow without fixed points on the torus and with non-wandering trajectories, always have an invariant analytical measure? (Note that such a flow is minimal provided that it has no periodic trajectories).

In 1961 V. I. Arnold [15] showed that that this is not true even in the case of minimal flows. He constructed an analytical minimal fixed-point-free flow with a unique invariant measure which is not absolutely continuous with respect to Lebesgue measure. We omit Arnold's construction here (although we wish to discuss some details), since there is a comprehensive description of it in the literature. See for example:

I. P. Cornfeld, S. V. Fomin and Ya. G. Sinai, Ergodic Theory, in: Grundlehren der mathematischen Wissenschaften, Bd.245, Springer-Verlag, Berlin-Heidelberg-New York (1982).

To obtain Arnold's example it is enough to consider a corresponding analytic diffeomorphism of the circle S^1 and then take a suspension. Arnold's example is the limit of a sequence of analytic diffeomorphisms with a double-asymptotic periodic orbit. The required diffeomorphism has an irrational rotation number which is approximated 'unusually fast' by rational numbers.

M. Herman [144] rediscovered 'Arnold's example' in the form

$$f_b(x) = x + a \sin 2\pi x \ + \ b \ mod \ 1.$$

He showed that for fixed $|a| < 1/2$ there exists a residual set with regard to the real parameter b for which the analytic diffeomorphism $f_b(x)$ has a unique invariant measure which is not absolutely continuous with respect to Lebesgue measure.

7.2.3 Surfaces of Genus $g \geq 2$

Flows on the torus with the smooth invariant measure which we considered above have no equilibria and consist of non-wandering trajectories. Moreover, the measure of each trajectory is necessarily equal to zero. If we pass to the surfaces of higher genus the difficulties grow, since every such flow must have fixed points. Euler's formula implies that there are fixed points of negative index. The simplest among such points is (in a sense) a non-degenerate (Morse) saddle. It is natural therefore to consider the class of flows with non-wandering trajectories whose fixed points are Morse saddles. In 1973 such flows have been considered by A. Katok, who proved the existence of an invariant measure such that each trajectory has measure 0 and is positive on each open subset of M [160].

An invariant measure μ of the flow f^t is called *nontrivial* if it positive on each open subset of M and the measure of each trajectory is 0. In the case of μ is σ-finite we will assume that $\mu(M - U) < \infty$ for any neighbourhood U of the set $Fix(f^t)$.

Let us consider a class $S(M)$ of flows on orientable closed surfaces M of genus $g \geq 1$ which obey the following conditions:

1. $NW(f^t) = M$;
2. The set $Fix\ f^t$ consists of Morse saddles (the total number of which is $2g - 2$).

Theorem 7.2.5 *Every flow $f^t \in S(M)$ has a nontrivial invariant measure.*

Proof. Since $f^t \in S(M)$ the surface M can be presented as a union of of sets $M_1 \cup \ldots \cup M_k$ with pairwise disjoint interiors. The boundaries of such sets consist of the separatrix connections and saddle points. Moreover, one can assume that $Int\ M_i$ consists of periodic trajectories or of non-trivial recurrent trajectories which are everywhere dense in M_i. It easily follows now that μ exists on $Int\ M_i$. The measure of separatrix contours is 0. □

Remark 7.2.2 *The proof of Theorem 7.2.5 shows that the theorem is true for flows with non–wandering trajectories and fixed points which are arbitrary saddles, not necessarily of Morse type.*

Arnold's example shows that it is impossible to prove the existence of smooth nontrivial invariant measures for $f^t \in S(M)$. Let us consider an important subclass $SA(M) \subset S(M)$ of smooth flows having at least one smooth, non-trivial, ergodic, invariant measure. Such flows are necessarily transitive. If $M = T^2$ then every $f^t \in SA(M)$ is uniquely ergodic. For a closed orientable surface M of genus ≥ 2 A. Katok [160] proved the following result.

Theorem 7.2.6 *Let M be a closed orientable surface of genus $g \geq 2$. Then every $f^t \in SA(M)$ has at most g non-trivial, ergodic, invariant measures.*

Proof. A short and elegant proof of this theorem is given in [162], Theorem 14.7.6. E. Sataev [259] proved that the upper boundary in Theorem 7.2.6 is exact.

7.3 Ergodicity

Let us consider dynamical systems with an invariant measure μ. A dynamical system is said to be *ergodic* with respect to μ if any invariant set A has either zero measure $\mu(A) = 0$ or full measure $\mu(A) = \mu(M)$. Clearly ergodicity depends on the invariant measure μ. We shall be interested in the ergodicity of dynamical systems relative to smooth measures. Since an ergodic flow with smooth measure on a closed surface must be transitive there are no ergodic flows on the sphere, projective plane or the Klein bottle. Note that if the invariant measure is not smooth, then f^t is not necessarily transitive.

7.3.1 Torus

Non Singular Case First, let us consider the ergodicity of fixed-point-free flows on the torus. Such a flow is defined by the system of differential equations (7.3):

$$\dot{x} = A(x,y), \quad \dot{y} = B(x,y)$$

where $A^2 + B^2 > 0$ and A, B are periodic of period one in both variables. The ergodicity of such flows is equivalent to transitivity. Therefore, according to Denjoy's theorem [88], if $A(x,y) > 0$ or $B(x,y) > 0$ and the flow f^t defined by (7.3) is C^2 then ergodicity of f^t is equivalent to the irrationality of the rotation number of f^t. Ergodic fixed-point-free flows on the torus were studied in [89]; the deviation of the covering trajectories from the asymptotic direction has also been estimated there.

A linear minimal flow on the torus gives an example of an ergodic flow with respect to the standard Euclidean area form $dx \wedge dy$ (which is evidently invariant for the above flow). It turns out that up to topological equivalence this is the only possibility for fixed-point-free C^1-flows on tori. In 1951 T. Saito [258] proved the following theorem.

Theorem 7.3.1 *Let f^t be a C^1 fixed-point-free flow on torus defined by the equations (7.3):*

$$\frac{dx}{dt} = A(x,y) \quad \frac{dy}{dt} = B(x,y)$$

where

$$A = \sum a_{mn} \exp 2\pi i(mx+ny), \quad B = \sum b_{mn} \exp 2\pi i(mx+ny)$$

are real-valued periodic functions of period 1. Suppose that f^t has an integral invariant

$$\int\int_D dxdy.$$

Then f^t is ergodic with respect to the measure defined by the above integral if and only if

$$a_{00} \neq 0, \quad b_{00} \neq 0 \quad \text{and } \frac{a_{00}}{b_{00}} \text{ is irrational}.$$

Theorem 7.3.1 can be reformulated as follows.

Theorem 7.3.2 *Let f^t be a C^1 fixed-point-free flow on the torus with an invariant standard Lebesgue measure μ_s. Then f^t is ergodic with respect to μ_s if and only if the rotation number $rot(f^t)$ of f^t is irrational.*

Proof. Suppose the rotation number $rot(f^t)$ is irrational. Since the divergence of f^t is zero, f^t is minimal, and hence, ergodic. Suppose that f^t is ergodic with respect to μ_s. Assume that $rot(f^t)$ of f^t is rational. Then f^t has at least one periodic trajectory. Since f^t preserves standard Lebesgue measure then all trajectories are periodic. Obviously such a flow is not ergodic. □

For the curious reader, we wish to give a proof which does not appeal to rotation numbers. The Liouville condition (7.1) implies that $ma_{mn} + nb_{mn} = 0, \quad m,n = 0, \pm1, \pm2, \ldots$ Hence we can find a sequence c_{mn} such that

$$a_{mn} = nc_{mn}, \quad b_{mn} = -mc_{mn}, \quad (m,n) \neq (0,0).$$

Therefore we can write

$$A = a_{00} + \sum nc_{mn} \exp 2\pi i(mx+ny), \quad B = b_{00} - \sum mc_{mn} \exp 2\pi i(mx+ny).$$

The function

$$H(x,y) = a_{00}y - b_{00}x + \frac{1}{2\pi i}\sum c_{mn}\exp 2\pi i(mx+ny)$$

is an integral of our flow. In general, H is not a single-valued function because

$$H(m,n) = a_{00}n - b_{00}m + H(0,0).$$

If $a_{00} = b_{00} = 0$ then $H(x,y)$ is a single-valued continuous integral and therefore the flow is not ergodic. If $a_{00} \neq 0$ (the case $b_{00} \neq 0$ is similar) and $\frac{b_{00}}{a_{00}} = \frac{p}{q}$ is a rational then the function

$$\exp 2\pi i \frac{p}{a_{00}} H$$

is a single-valued continuous integral. Hence f^t is not ergodic. □

Kolmogorov's Theorem 7.2.2 implies that the analytical flow f^t on the torus with $Fix\ f^t = \emptyset$ is ergodic with respect to the analytical invariant measure if and only if its rotation number is irrational. Indeed, up to a time change f^t is analytically equivalent to a linear flow. If the linear flow has an irrational rotation number then it is minimal and therefore ergodic. If the rotation number is rational then all of the trajectories are periodic and the linear flow is not ergodic.

Singular Case If a smooth invariant measure is given, are there ergodic flows with fixed points on the torus? By transitivity, the fixed points must be fake saddles. An example of such a flow is in:

V. Stepanov, Sur une extension du théorème ergodique, Compositio Math. 3 (1936), 239-253.

Stepanov considered flows $f^t_{\alpha,0}$ defined by the system of differential equations

$$\frac{dx}{dt} = A(x,y), \quad \frac{dy}{dt} = \alpha A(x,y) \tag{7.5}$$

where α is an irrational number and A is a 1-periodic, non-negative, continuous function which satisfies the Lipschitz condition and vanishes at the unique point of the torus. It follows from Proposition 7.2.1 that

$$\mu(D) = \iint_D \frac{dxdy}{A(x,y)}, \quad D \subset T^2$$

is an integral invariant of system (7.5). Stepanov showed that the $f^t_{\alpha,0}$ are ergodic with respect to the Borel measure $\mu(D)$. In general, this measure may be either finite or infinite. But if the function $A(x,y)$ is of class C^1 then $\mu(T^2) = \infty$.

Following J. C. Oxtoby [233], f^t is said to be a *Stepanov flow* if it is topologically equivalent to the flow defined by a system of the form (7.5) with irrational α and a unique fixed point, say $m_0 \in T^2$. Oxtoby investigated the ergodicity and strict ergodicity of Stepanov flows.

Theorem 7.3.3 *Let f^t be a Stepanov flow with fixed point m_0. Suppose that f^t admits a normalized invariant measure μ for which $\mu(m_0) = 0$. Then μ is ergodic with respect to f^t. Moreover, μ is unique in the class of such measures.*

Another of Oxtoby's theorem describes the set of normalized invariant measures of Stepanov flows.

Theorem 7.3.4 *Let f^t be a Stepanov flow with fixed point m_0. Then one of the following holds:*

1. *The only normalized invariant measure is the trivial one μ_0; for the latter $\mu_0(m_0) = 1$.*
2. *f^t has a unique ergodic invariant measure $\mu \neq \mu_0$; moreover, μ is the unique invariant measure for which $\mu(T^2) = 1$, $\mu(m_0) = 0$. Every finite invariant measure of f^t is a linear combination of μ_0 and μ.*

Are there analytic flows with a fixed point on the torus and which are ergodic with respect to a smooth normalized invariant measure? (Note that the flow defined by (7.5) does not give an answer since the invariant measure is infinite.) J. C. Oxtoby constructed an example of such flows:

$$\frac{dx}{dt} = \alpha[1 - \cos 2\pi(x - y)] + (1 - \alpha)(1 - \cos 2\pi y),$$
$$\frac{dy}{dt} = \alpha[1 - \cos 2\pi(x - y)].$$

The above flow, denoted by $f^t_{Ox,\alpha}$, coincides with Stepanov's for irrational α. $f^t_{Ox,\alpha}$ is analytic with zero divergence, hence it preserves the standard Lebesgue measure. Oxtoby proved that $f^t_{Ox,\alpha}$ is topologically equivalent to a flow defined by (7.5) but the function $A(x, y)$ violates Lipschitz conditions.

For C^∞ flows with fixed points on the torus, A. Blohin [51] proved the following theorem.

Theorem 7.3.5 (Blohin) *Let f^t be a C^∞-flow on the torus with finitely many fixed points. Suppose that f^t preserves a smooth (positive) measure μ. Then f^t is ergodic relative to μ if and only if each point $p \in$ Fix f^t has Euler index 0 and the rotation number of the flow is irrational.*

Proof. The sum of the indices of the fixed points on the torus is zero. Therefore, if there are singularities of non-zero index then there exists a singularity of positive index (center-type singularity). Such a flow is evidently non-ergodic. If the index of each fixed point is 0 one can 'remove' all of them while retaining the smooth invariant measure μ as well as the first-return mapping on the global cross-section to f^t (better to say an 'extended' return mapping). The rest of the proof follows from theorems on fixed-point-free flows. □

7.3.2 Surfaces of Genus $g \geq 2$

Theorem 7.3.6 *Let M be a closed surface different from the sphere, the projective plane and the Klein bottle. Then there exists a C^∞-flow f^t on M with a smooth invariant measure on it with respect to which f^t is ergodic.*

Proof. Take a minimal linear flow f_0^t on the torus (the irrational flow), and add to it two fake saddles S_1, S_2 lying on distinct orbits of f^t. Next let us 'blow-up' each fake saddle, so that instead of S_i there appears a disc D_i bounded by separatrices $l_i^{(1)}, l_i^{(2)}$ and saddles $S_i^{(1)}, S_i^{(2)}$. Let us remove the interiors of D_1, D_2. The flow so obtained we denote by f_1^t. Since the flow f_0^t is analytic with invariant Lebesgue measure, it is possible to carry out the above construction so that f_1^t is C^∞ with smooth invariant measure with respect to which it is ergodic.

We identify the boundaries of the discs D_1, D_2 so that $S_1^{(1} \sim S_2^{(2)}, S_1^{(2)} \sim S_2^{(2)}$. For the separatrix there are two possibilities: either $l_1^{(1)} \sim l_2^{(1)}, l_1^{(2)} \sim l_2^{(2)}$ or $l_1^{(1)} \sim l_2^{(2)}, l_1^{(2)} \sim l_2^{(1)}$. The first identification gives an orientable surface (pretzel), the second one gives a non-orientable surface of Euler's characteristic -2. Taking care with time-scaling one can get a C^∞-flow f_2^t with a smooth invariant measure. Because the fake saddles s_1, s_2 belong to different trajectories, the flow f_t^2 is ergodic with regard to the above invariant measure μ.

Similarly on can construct the required flows on all surfaces of the even negative Euler characteristic. It remains to do the same for the surfaces of odd negative Euler characteristic (such surfaces are non-orientable). It follows from the above construction that it is enough to do so on the non-orientable closed surface of the Euler characteristic -1 (a torus with a cross-cap).

This is done by setting a fake saddle s_0 onto a trajectory of a linear irrational flow on torus, then blowing-up s_0 to the disc bounded by a loop separatrix. Next we remove the interior part of the above disc and glue a cross-cap into the hole. Reparametrizing the flow on the loop, we identify opposite points on it. □

7.4 Mixing

A flow f^t on the space X with invariant measure μ is said to be *mixing* if, for a given subset $A \subset X$ of positive measure and any two subsets $B, C \subset X$ of a finite positive measure,

$$\frac{\mu\left(f^t(A)\cap B\right)}{\mu\left(f^t(A)\cap C\right)} \longrightarrow \frac{\mu(B)}{\mu(C)}, \qquad \text{as} \quad t \to 0.$$

Ergodicity follows from the mixing property. Indeed, if we assume that there exists an invariant set A such that $\mu(A) > 0, \mu(X\backslash A) > 0$, then setting $C = A, B = X\backslash A$ we get

$$\frac{\mu(f^t(A)\cap B)}{\mu(f^t(A)\cap C)} = \frac{\mu(A\cap(X\backslash A))}{\mu(A)} = 0.$$

Conversely, by the mixing property,

$$\frac{\mu(f^t(A)\cap B)}{\mu(f^t(A)\cap C)} = \frac{\mu(B)}{\mu(C)} = \frac{\mu(X\backslash A)}{\mu(A)} > 0,$$

a contradiction which proves the claim.

There exists a 'physical' interpretation of mixing: If one takes a set A_1 of a positive measure, then any other set A_2 of positive measure which moves along the flow starts from some definite moment of time (and up to infinity) to intersect A_1. Moreover, the measure of $A_1 \cap f^t(A_2)$ at the moment t is asymptotically (as $t \to \infty$) proportional to $\mu(A_1)$. Roughly speaking, while moving, any set of positive measure 'paints' uniformly all of the phase space of the dynamical system.

7.4.1 Torus

Non Singular Case From Kolmogorov's Theorem 7.2.2 it follows that an analytic fixed-point-free torus flow with smooth invariant measure is not mixing. In 1972 A. V. Kochergin [166] proved that topological flows may be mixing; if, however, the flow is "Lipschitz" mixing is excluded:

Theorem 7.4.1 (Kochergin) *Suppose the flow f^t on the torus has no fixed points, and let f^t be given by the system*

$$\dot{x} = A(x,y), \qquad \dot{y} = B(x,y),$$

where A, B are periodic functions in each argument. If A and B obey the Lipschitz condition then f^t cannot be mixing.

Proof. The flow f^t has no periodic trajectories. Since there are no fixed points and no periodic trajectories, one can construct a cross-section C. By the assumption of the theorem it follows that the first-return mapping $\phi : C \to C$ is transitive and has irrational rotation number. Therefore f^t can be represented as a special flow built upon the irrational rotation of the circle and a function $f(x)$ which satisfies Lipschitz's condition. It remains to prove the following lemma, which has independent interest.

Lemma 7.4.1 *Let f^t be a special flow built upon rotation of S^1 by irrational number α and function $f(x)$ of bounded variation on S^1. Then f^t is not mixing.*

Proof of lemma 7.4.1. Let us denote $V = Var_{S^1} f$ and let us normalize $\int_{S^1} f(x)dx = 1$. Let $\delta > 0$ and let us take an interval $\Delta \subset S^1$ of length $|\Delta| < 1/(6V+8)$. We set

$$A = \bigcup_{-3V \le t \le 3V+8} f^t(\Delta).$$

Since the length of Δ is small we have $\mu(\alpha) > 0, \mu(X \backslash A) > 0$. Take an arbitrary interval lying strictly in the interior of Δ. Let

$$B = \bigcup_{a \le t \le \delta} f^t(\Delta_1).$$

From the irrationality of α it follows that there exists an infinite sequence $q_n \to +\infty$ such that $R_\alpha^{q_n} \subset \Delta$. Then $f^{q_n}(B) \subset A$, which proves the absence of mixing. This proves the lemma and the theorem. □

On the other hand, applying continuous time-rescaling of an irrational flow, Kochergin constructed a topological fixed-point-free flow with the mixing property. In the case of fixed-point-free flows the situation is similar to the Denjoy theory for non-trivial minimal sets: there are C^1 but not $C^r, r \ge 2$ flows with such sets. According to Kochergin, there are topological fixed-point-free flows that are mixing, but $C^r, r \ge 1$ flows cannot be mixing. Thus 'smooth' and 'mixing' properties are related.

Recall that a flow f^t with an invariant measure μ is said to be *aperiodic* (with respect to μ) if the set of periodic trajectories is of measure 0. In 1973 Kochergin [167] proved that any ergodic aperiodic flow on a compact Lebesgue space can be transformed by continuous time-rescaling to a mixing flow. As a consequence we have

Corollary 7.4.1 *Every closed surface different from the sphere, the projective plane and the Klein bottle admits a mixing topological flow.*

Proof. The idea of the proof is to modify time in the vicinity of an arbitrary neighbourhood of positive measure so that each element of a sufficiently small partition is 'stretched'. □

Singular Case Let us discuss mixing for smooth flows with fixed points on the torus. We have to look at ergodic flows (since mixing implies ergodicity), i.e., transitive flows with fake saddles. All such flows are special flows. Recall the definition. Let $\theta : S^1 \to \mathbb{R}^1 \cup \{+\infty\}$ be a strictly positive function and $\varphi : S^1 \to S^1$ be an orientation-preserving homeomorphism of the circle S^1. We normalize $\int_{S^1} \theta dx = 1$. "Vertical motion" on the set

$$T_{\varphi,\theta} = \{(x,t): \quad x \in S^1, \quad 0 \le t \le \theta(x)\}/(x,\theta(x)) \simeq (\varphi(x),0)$$

determines a flow $Sus_\theta^t\,(\varphi)$ called a *special flow* under θ built over φ. Note that $T_{\varphi,\theta}$ is a torus and the fixed points of $Sus_\theta^t\,(\varphi)$ are fake saddles corresponding to the *singularities* of θ. Denote by μ_θ a restriction of the Lebesgue measure $dxdt$ on $T_{\varphi,\theta}$. Then μ_θ is a normalized invariant measure with respect to $Sus_\theta^t\,(\varphi)$.

If $\varphi = R_\alpha$ is a rotation on the irrational angle α then the mixing or non-mixing of $Sus_\theta^t\,(\varphi)$ depends on the asymptotic behavior of function θ near singularities. Fix an irrational α and define the following class $Sus_{\theta,\alpha}^t(D)$ of special flows:

1. The function θ has a finite number of singularities $x_1, \ldots, x_r \in S^1$.
2. θ is a C^2 function outside the singular points and $\theta'' > 0$ in a punctured neighbourhood of each singularity.
3. There exist constants $A_i > 0$, $B_i > 0$, $i = 1, \ldots, r$, such that

$$\lim_{x \to x_i^+} \frac{\theta(x)}{\ln(x - x_i)} = -A_i, \qquad \lim_{x \to x_i^-} \frac{\theta(x)}{\ln(x - x_i)} = -B_i.$$

4. $D = |\sum_{i=1}^r A_i - \sum_{i=1}^r B_i|$.

In 1976 Kochergin [169] proved that if $D = 0$ and α admits approximation by fractions with the rate $\frac{1}{n^2 \ln n}$, i.e.,

$$|\alpha - \frac{p}{q}| < \frac{C}{q^2 \ln q}, \qquad C > 0,$$

then the special flow $Sus_{\theta,\alpha}^t(D)$ is not mixing. In 1992 Ya. Sinai and K. Khanin [266] proved the following theorem.

Theorem 7.4.2 *Let $\alpha = [a_1, \ldots, a_n, \ldots]$ be expansion of α into a continued fraction. If $D \neq 0$ and $a_n \le n^{1+\gamma}$ for some $\gamma > 0$ then the special flow $Sus_{\theta,\alpha}^t(D)$ is mixing.*

A smooth area preserving flow f^t on the torus has at most one transitive component, say K. The restriction $f^t|K$ is a special flow. At the end of the paper

V. Arnold, Topological and ergodic properties of closed 1-forms with incommensurable periods. Funct. Analysis and Its Appl. 25 (1991), 81-90. (Transl. from Funk. Anal. i Ego Pril., 1991, 25, no 2, 1-12.

the author remarks that in the typical case the flow $f^t|K$ satisfies the conditions of the Sinai-Khanin Theorem 7.4.2. Thus we have:

Theorem 7.4.3 *A transitive component of a generic area-preserving torus flow has the mixing property.*

7.4.2 Surfaces of Genus $g \geq 2$

We say that a C^∞ function $\theta : (a,b) \to \mathbb{R}$ is of *class* $\mathcal{K}(a,b)$ if

1. $\theta'' > 0$.
2. There is a non–increasing function $M(x)$, $x \in (0,1]$, such that $M(x) \geq 1$ and
$$\frac{\theta''(x)}{M(t)} \leq \theta(a + t(x-a)) \leq M(t)\theta''(x)$$
for any $t \in (0,1]$ and $x \in (a,b)$.
3. There is a non–decreasing function $\omega(x)$, $x > 0$, such that $\lim_{x\to 0+} \omega(x) = 0$ and
$$\theta''(x) \geq \frac{1}{(x-a)^2\omega(x-a)}$$
for any $x \in (a,b)$.

Recall that the definition of a special flow $Sus_\theta^t\ (\psi)$ under a function
$$\theta : S^1 \to \mathbb{R}^1 \cup \{+\infty\}$$
built over an interval exchange transformation $\psi : S^1 \to S^1$ is similar to that of a special flow over a homeomorphism. In this case the singularities of ψ are the points of discontinuity.

Theorem 7.4.4 (Kochergin) *Let $\psi : S^1 \to S^1$ be an ergodic interval exchange transformation and $x_1, \ldots, x_r \in S^1$ points of discontinuity of ψ. Suppose that $\theta(x) \geq c > 0$ and*
$$\theta(x) = \theta_0(x) + \sum_{i=1}^{r} [\theta_i^+(x - x_i + \theta_i^-(x_i - x)]$$
where θ_i^+, $\theta_i^- \in \mathcal{K}(0,1)$ and $\theta_0(x)$ has a bounded second derivative. Then the special flow $Sus_\theta^t\ (\psi)$ is mixing.

Thus if the singularities of the function θ are of "maximal power" then the special flow has the mixing property. As a consequence of this theorem, we get

Theorem 7.4.5 (Kochergin) *On each compact surface different from the sphere, projective plane and the Klein bottle there exist smooth mixing flows with an invariant measure having a smooth positive density.*

For the verification of mixing property and the explicit construction of mixing flows Kochergin introduced the notion of a *mixing saddle* [168]. Roughly speaking, a mixing saddle s_0 is a saddle whose time of passage along the trajectories near separatrices is of class $\mathcal{K}(0,\delta)$. Kochergin [168] proved

Theorem 7.4.6 (Kochergin) *If all fixed points of flow f^t on a compact surface are mixing saddles then f^t is mixing.*

Example 7.4.1 *Let f^t be a flow in the plane with the Hamiltonian $H(x, y) = y(x^2 + y^2)$. The origin $(0, 0)$ is fake saddle with two separatrices l_α, $_\omega$ lying on the axis $y = 0$. The circle $S_0^1 = \{x^2 + y^2 = 1\}$ is transversal to $l_{\alpha,\ \omega}$. Let $A = S_0^1 \cap l_\alpha$ and $B = S_0^1 \cap l_\omega$. Denote by $\Sigma_A \subset S_0^1$ (cor. $\Sigma_B \subset S_0^1$) a transversal segment through A (cor. B). As a coordinate on Σ_A, Σ_B we take the values of H. Since f^t is defined by the equations*

$$\frac{dx}{dt} = x^2 + 3y^2, \quad \frac{dy}{dt} = -2xy$$

then the passage–time between Σ_A and Σ_B is

$$t(H) = -\int_{\sqrt{H}}^{\sqrt[3]{H}} \frac{dy}{2xy} = \int_{\sqrt{H}}^{\sqrt[3]{H}} \frac{dy}{\sqrt{Hy - y^4}}.$$

One can prove that $t(H) \in \mathcal{K}(0, 1)$, i.e. $(0, 0)$ is a mixing saddle (see details in [168]). Inserting this saddle into the phase space of an irrational linear flow on torus we get a smooth mixing flow satisfying Theorem 7.4.4.

Example 7.4.2 *The fixed point $(0, 0)$ is a mixing saddle of flow with the Hamiltonian*

$$H(x, y) = (x^2 + y^2)^k Re(x + iy)^n, \quad n \geq 1, k \geq 1, n + k \geq 3.$$

This saddle is of use in constructing other examples of smooth mixing flows on surfaces.

7.5 Entropy

Entropy is a measure of the complexity of dynamical system. It was introduced by A. N. Kolmogorov:

New metric invariant of transitive dynamical systems and automorphisms of Lebesgue's spaces, Dokl. Akad. Nauk SSSR, 119, 1958, 5, 861-864.
On the unit-time entropy as metric invariant of automorphisms, Dokl. Akad. Nauk SSSR, 124, 1959, 4, 754-755.

Kolmogorov's entropy is usually called the metric entropy. Suppose that ξ is a finite partition $M_1 \cup \ldots \cup M_n$ of the space M endowed with a measure μ. The *entropy of the partition* ξ is the number

$$H_\mu(\xi) = -\sum_{i=1}^{n} \mu(M_i) \log \mu(M_i).$$

(If $\mu(M_i) = 0$, we set $\mu(M_i) \log \mu(M_i) = 0$.) The *entropy of an automorphism* $T : M \to M$ is the number

$$h_\mu(T) = \sup_{\xi \in M} H(\xi),$$

where ξ varies over all partitions of M. Let f^t be a flow with invariant measure μ. The *metric entropy* of f^t is defined as the entropy of the time–1 shift f^1 along the trajectories of the flow, that is, $h_\mu(f^t) = h_\mu(f^1)$. The definition of the metric entropy as the entropy of the time–1 mappings was motivated by the formula

$$h_\mu(F^t) = |t| h_\mu(f^1),$$

where F^t is the time–t shift along the trajectories of f^t; the later formula is taken from

L. M. Abramov, On the entropy of a flow, Dokl. Akad. Nauk SSSR 128, 1959, 5, 873-876; transl. in AMS Transl. 49, 1966, ser.2, 167-170.

L. M. Abramov also proved an important formula which connects the entropy of a special flow built upon the ergodic automorphism $T : X \to X$ with the measurable functions $f : M \to \mathbb{R}$ separated from 0:

$$h_\mu(f^1) = \frac{h_\nu(T)}{\int f d\nu},$$

where ν is a T-invariant measure.

Along with the metric entropy, in the work

R. Adler, A. Konheim, M. McAndrew, Topology entropy, Trans. Amer. Math. Soc. 114, 1965, 309-319

a topological entropy is introduced. Let A be a covering of the topological space X. Let $N(A)$ denote the number of sets in a subcovering of minimal cardinality. A subcovering of a covering is *minimal* if no other subcovering contains fewer elements. The *entropy of the covering* A is a number

$$H(A) = \log N(A).$$

Let $\phi : X \to X$ be a mapping of the topological space X, and let A be a covering of X. The *entropy of ϕ with respect* to A is defined as

$$h(\phi, A) = \lim_{n \to \infty} \frac{H(A \cup \phi^{-1} A \cup \ldots \cup \phi^{-n+1} A)}{n}.$$

The *topological entropy of the mapping ϕ* is given by

$$h(\phi) = \sup_{A \in X} h(\phi, A)$$

where A varies over all possible coverings of X. Finally, the *topological entropy of the flow f^t* is defined as

$$h(f^t) = h(f^1),$$

which is actually the topological entropy of the time–1 shifts along the flow f^t.

Both topological and metric entropies have been studied by L. S. Young [309]. She showed that the entropy of surface flows is trivial.

Theorem 7.5.1 (Young) *Let f^t be an ergodic flow on a compact orientable surface M with invariant measure μ. Then $h_\mu(f^t) = 0$.*

Proof. From ergodicity it follows that the measure μ is concentrated on the closure of a recurrent trajectory of f^t. If μ is concentrated on a fixed point or periodic trajectory the claim is evident. Therefore let us assume that the support of μ belongs to a quasiminimal set $\Omega(f^t)$. Then there exists a closed transversal C where f^t induces a piecewise monotone transformation ϕ. By Gutierrez's theorem, ϕ is semi-conjugate to an interval exchange transformation. Therefore $h_\nu(\phi) = 0$ where ν is measure induced on C by μ. By Abramov's formula,

$$h_\mu(f^t) = \frac{h_\nu(\phi)}{\int d\nu} = 0.$$

□

The fact that

$$h(f^t) = \sup_\mu h_\mu(f^t)$$

where the supremum is taken over ergodic measures μ, was established by

E. Dinaburg, On the relations among various entropy characteristics of dynamical systems. Math. USSR, Izvestia 5, 1971, 337-378.

and

L. Goodwyn, Topological entropy bounds measure-theoretic entropy, Proc. Amer. Math. Soc. 23 (1969), 679-688.

Theorem 7.5.2 (Young) *Let f^t be a flow on compact surface M. Then its topological entropy is zero, $h(f^t) = 0$.*

Proof. It suffices to prove the theorem for an orientable surface because if M is nonorientable, one can consider an orientable covering $\widetilde{M}$ and a covering flow $\widetilde{f}^t$ on $\widetilde{M}$. The corresponding diagram commutes, and

$$h(f^t) \leq h(\widetilde{f}^t).$$

The result follows from Theorem 7.5.1 and the Dinaburg-Goodwyn formula.

□

Bibliographic Notes. Ergodic theory studies the statistical properties of group actions and measure-preserving transformations of spaces with a measure. A good introduction to the subject is: *Ya. Sinai. Introduction to Ergodic Theory, Princeton Univ. Press, 1976.* Although flows on surfaces exhibit weak statistical properties, there are several interesting results in this direction due to A. N. Kolmogorov, V. I. Arnold, A. Blohin, S. Schwartzman, A. V. Kochergin, E. A. Sataev, L. S. Young and others.

8. Homeomorphisms of the Unit Circle

Important classes of foliations can be obtained by suspension over homeomorphisms and endomorphisms of the unit circle. Thus the methods of one-dimensional dynamics play a distinguished role in the theory of foliations. A modern encyclopaedia of the subject is the monograph:

W. de Melo & S. van Strien, One-dimensional dynamics, Ergebnisse der Mathematik und ihrer Grenzgebiete, 25, Springer-Verlag, 1993.

8.1 Denjoy Flow

Let T^2 be a two-dimensional torus and let $\pi^t : T^2 \times I\!R \to T^2$ be a flow on T^2 of class $C^r, r \geq 0$. A flow π^t is said to be *Denjoy* if it admits no singular points and closed orbits.

Theorem 8.1.1 (Denjoy) *Let $\pi^t : T^2 \times I\!R \to T^2$ be a Denjoy flow on the torus, generated by a vector field $v(x)$ of the class C^r. If $r \geq 1$ and the first derivative Dv of the vector field v has bounded variation on T^2 then each orbit of π^t is dense in T^2.*

Remark 8.1.1 *In particular, the Denjoy theorem holds for the C^2 vector fields, since existence and continuity of the second derivative implies $Var_{x \in T^2} Dv < \infty$.*

Let S^1 be a global cross-section of the Denjoy flow $\pi^t : T^2 \times I\!R \to T^2$ (such a cross-section always exists on the torus). Denote by $\phi : S^1 \to S^1$ a homeomorphism of S^1, induced by the first return mapping on S^1, which acts on $x \in S^1$ by the formula $x \mapsto \pi^{\tau(1)}(x)$, where $\tau(1)$ is minimal positive time such that $\pi^{\tau(1)} \in S^1$. The following chain of equations

$$D_x\phi(x) = D_x\pi^{\tau(1)}(x) = \int_0^{\tau(1)} D_t[D_x\pi^t(x)]dt = \int_0^{\tau(1)} D_x[D_t\pi^t(x)]dt =$$
$$= \int_0^{\tau(1)} D_x v(x)dt = \tau(1)D_x v_x$$

allows us to estimate a differentiability class of ϕ. This one turns out to coincide with differentiability class of the vector field $v(x)$. Moreover, if $Var_{x\in T^2}v(x)$ is bounded such will be $Var_{x\in S^1}\phi(x)$.

Let $p \in S^1$ be a point on the circle S^1. Denote by $p_n = \phi^n(p), n \in \mathbb{Z}$ the image of p under the action of homeomorphism $\phi : S^1 \to S^1$ and set $p_0 = p$. The following auxiliary statement is valid.

Lemma 8.1.1 *Let $p, q \in S^1$ be a pair of distinct points on S^1. For every positive integer N, there exists an integer $n > N$ such that either of two conditions are satisfied:*

(i) n arcs $p_{-k}q_{n-k}(k = 1, 2, ..., n)$ are pairwise disjoint, or

(ii) n arcs $p_{n-k}q_{-k}(k = 1, 2, ..., n)$ are pairwise disjoint.

Proof. Consider $2N$ arcs $pp_i, i = \pm 1, \pm 2, ..., \pm N$. Choose a minimal arc pp_m such that $p \prec p_m \prec p_i$ lies between p and p_i for all $i \neq m$. Since $\phi : S^1 \to S^1$ has no periodic points, there exists an integer $h, |h| > N$, such that $p \prec p_h \prec p_m$. In addition, one chooses $|h|$ to be the smallest possible. Then $p \prec p_m \prec p_j$ for all $0 < |j| < |h|, j \neq m$ and, henceforth,

$$p \prec p_h \prec p_j, \quad 0 < |j| < |h|.$$

The number $n = |h|$ is the required one.

Indeed, suppose that neither (i) nor (ii) are satisfied. It means that among numbers $1, 2, ..., n$ there exists such $k, l (k \neq l)$ that

$$p_{-k} \prec p_{-l} \prec q_{h-k}.$$

Since $\phi : S^1 \to S^1$ does not reverse the ordering of points, one applies ϕ^k to the above ordered sequence by setting $k - l = j, 0 < |j| < |h|$. This yields us $p \prec p_j \prec p_h$, what is a contradiction with the initial assumption. □

Proof of the theorem. Let $M = \overline{O(x)}$ be a closure of the orbit $O(x) = \{\phi^n(x) | n \in \mathbb{Z}\}$ of a point $x \in S^1$. Suppose to the contrary, that $M \neq S^1$. Choose an arc $pq \subset S^1 \backslash M$ in the complement of M. Arcs p_kq_k are disjoint and the series $\sum_{k=-\infty}^{k=+\infty} l_k$ is convergent, where $l_k = ||q_k - p_k||$ is the length of arc. In particular, l_k vanishes as k tends to $\pm\infty$.

On the other hand, by the *Mean Value Theorem*, $l_1 = ||q_1 - p_1|| = ||\phi(q) - \phi(p)|| = D\phi(\xi)||q - p|| = l_0 D\phi(\xi)$, where ξ is a point on the arc pq. Proceeding by the induction, one easily obtains

$$\frac{l_n}{l_0} = \prod_{k=1}^{n} D\phi(\xi_{n-k}), \qquad \frac{l_{-n}}{l_0} = \prod_{k=1}^{n} \frac{1}{D\phi(\zeta_{-k})}.$$

Bringing these expressions together, and taking their logarithm, one obtains

$$\log \frac{l_0^2}{l_n l_{-n}} = \log \prod_{k=1}^{n} \frac{D\phi(\zeta_{-k})}{D\phi(\xi_{n-k})} \leq \sum_{k=1}^{n} |\log D\phi(\zeta_{-k}) - \log D\phi(\xi_{n-k})|.$$

On account of Lemma 8.1.1, we have a variation of the function $\log D\phi(x)$ to the right of the above formula. By assumption of the theorem, $Var_{x\in S^1} D\phi(x)$ is bounded and, henceforth, such will be $Var_{x\in S^1} \log D\phi(x)$. Finally, this yields us that the quotient $\frac{l_0^2}{l_n l_{-n}}$ is bounded regardless what n is chosen. This is a contradiction with $l_{\pm n} \to 0$. □

Conditions of the Denjoy theorem cannot be relaxed.

Example 8.1.1 (Denjoy) *There are Denjoy homeomorphisms $\phi : S^1 \to S^1$ of the circle S^1 which belong to the class C^1 and such that the orbit $O(x) = \{\phi^n(x) | n \in \mathbb{Z}\}$ of $x \in S^1$ is nowhere dense in S^1.*

Proof. The idea of the construction is as follows. One starts with an irrational homeomorphism ϕ, which is a rotation through the angle $\alpha \in \mathbb{R}\backslash\mathbb{Q}$. Then taking a rational approximation $\{p_n/q_n\}$ of the point α, one amalgamates at the points $\{p_n/q_n\}$ the segments of *finite* total length. Mapping ϕ is extended to these segments in such a eay that they are invariant by ϕ. Since ϕ 'squeezes' the pasted segments to 0, there are orbits of ϕ, which are nowhere dense in S^1.

Indeed, denote by $I_m = [0, l_m], m \in \mathbb{Z}, l_m > 0$ a set of segments such that

(i) $\sum_{m\in\mathbb{Z}} l_m = l < \infty$

(ii) $\lim_{m\to\pm\infty}(l_{m+1}/l_m) = 1$

(as such one can take $l_m = (1 + m^2)^{-1}$). Let α be an irrational number, and let $\alpha_m, m \in \mathbb{Z}$ are chosen to satisfy the conditions

$$0 \le \alpha_m \le 1, \qquad m\alpha - \alpha_m \in \mathbb{Z}.$$

Let $[0, 1]$ be a unit segment, containing α. For every $m \in \mathbb{Z}$, instead of α_m one attaches the segment I_m and obtains a segment of the total length $1 + l$. By identifying its endpoints, one gets a circle $S^1 = [0, 1] \cup (\cup_{m\in\mathbb{Z}} I_m)$. The required homeomorphism $\phi : S^1 \to S^1$ is established as follows.

(1) If $x \in [0, 1]$, then ϕ acts by the formula $x \mapsto x + \alpha$,

(2) Otherwise $x \in \cup_{m\in\mathbb{Z}} I_m$ and ϕ brings I_m to I_{m+1}. The function ϕ is piecewise linear and smooth at the points ε and $l_m - \varepsilon$. Clearly, $\phi \in C^1$ and $\phi^m(I_0)$ vanishes as $m \to \pm\infty$. In other words, there exist nowhere dense orbits of ϕ. □

8.2 Cherry Class

8.2.1 Cherry Example

Theorem 8.2.1 (Cherry) *There exists a flow $\pi^t : T^2 \times \mathbb{R} \to T^2$ given by a C^∞ vector field v on the torus, which has the following structure of orbits:*

(i) *Sing v consists of two singularities σ and α of the saddle and node (source) types, both of which are hyperbolic;*
(ii) *the unstable manifold $W^u(\alpha) = \{x \in T^2 | \lim_{t\to-\infty} \pi^t(x) = \alpha\}$ of the source α is dense in T^2, and a compact set $\Lambda = \overline{T^2 \backslash W^u(\alpha)}$ is transitive (that is there exists $x \in \Lambda\backslash\sigma$ such that $\overline{O(x)} = \Lambda$);*
(iii) *if $x \in T^2\backslash(\Lambda \cup \alpha)$, then $\alpha(x) = \alpha$ and $\omega(x) \subset \Lambda$, where $\alpha(x)$ and $\omega(x)$ are α and ω-limit sets of the point x;*
(iv) *there exists a closed transversal $S^1 \subset T^2$, such that $S^1 \cap \Lambda$ is the Cantor set.*

Proof. Our proof will proceed in the three steps. First, a geometrical idea of the Cherry flow on torus will be presented. Next an analytical construction, due to Dos Anjos [6], will be exposed. Finally, properties (i)-(iv) will be established.

(1) One starts with the irrational flow on the torus, given by a closed 1-form $\omega = d\phi + \alpha d\psi$ of the maximal rank with $\alpha \in I\!R\backslash\mathbb{Q}$. We have already mentioned that each orbit of the flow is dense in T^2. Take any orbit and 'open' this one as it is shown in Fig. 8.1 trying to squeeze the 'tongue' of the opening in a narrow stripe, not hurting the nearby trajectories. Next one fills up the stripe with the orbits of unstable node (source), placed inside the stripe. The obtained irrational flow with the saddle and node is called a *Cherry flow* on the torus.

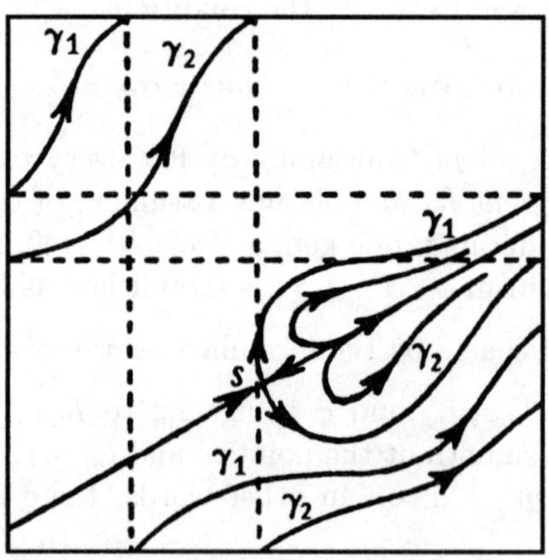

Fig. 8.1. The Cherry cell on torus.

(2) Let us consider $I\!R^2$ as the universal covering of the torus $T^2 \simeq I\!R^2/\mathbb{Z}^2$ and a family of periodic (covering) flows $\pi^t_\alpha : I\!R^2 \times I\!R \to I\!R$, given by a system of differential equations

$$\frac{dx}{dt} = 1 + \cos 2\pi x + \sin 2\pi y, \quad \frac{dy}{dt} = \alpha(1 + \sin 2\pi x) + \cos 2\pi y,$$

where $\alpha \in [0, 1]$ is a real parameter of the family π_α^t. In view of the periodicity of π_α^t, it is enough to consider a unit square $Q = I \times I$.

(i) Suppose that $\alpha = 0$. Then π_0^t has two equilibria in Q with coordinates $\sigma(1/2, 3/2)$ and $\alpha(3/2, 3/2)$. The linearization of π_0^t in σ and α gives us

$$A(\sigma) = \begin{pmatrix} -1 & 0 \\ 0 & 1 \end{pmatrix}, \quad A(\alpha) = \begin{pmatrix} 1 & 0 \\ 0 & 1 \end{pmatrix}.$$

Clearly, σ is a hyperbolic saddle and α is a hyperbolic node of the flow π_0^t.

(ii) Suppose that $\alpha = 1$. Then again, π_1^t has the equilibria $\sigma(1/2, 1/2)$ and $\alpha(3/2, 3/2)$ in Q. The linearization in the above points yields us

$$A(\sigma) = \begin{pmatrix} 0 & -1 \\ -1 & 0 \end{pmatrix}, \quad A(\alpha) = \begin{pmatrix} 1 & 0 \\ 0 & 1 \end{pmatrix}.$$

Thus, σ is a (hyperbolic) saddle, and α is a (hyperbolic) node of the flow π_1^t. Note that the rotation number α for the flows π_0^t and π_1^t is, correspondingly, 0 and 1. Varying parameter α in the family π_α^t, from $\alpha = 0$ to $\alpha = 1$, one obtains infinitely many Cherry flows π_α^t, which correspond to the irrational values of α.

(3) It is a straightforward exercise to check that singular points σ and α of the flow $\pi_\alpha^t, \alpha \in I\!R\backslash\mathbb{Q}$, are hyperbolic. By *irrationality* of α the Cherry flow has no periodic orbits.

Let us show that $W^u(\alpha)$ is dense in T^2. For this it is enough to show that a compact set $K = S^1\backslash W^u(\alpha)$ has an empty interior. Let to the contrary, K has a *non* empty interior $J \subset K$, where J is a maximal connected interval. Let $\phi : S^1 \to S^1$ be a homeomorphism induced on a transversal circle S^1 and consider the intervals $J_n = \phi^n(J)$. Since there are no periodic points, intervals J_n are pairwise disjoint, and their length grows with n, $|J_{n+1}| > |J_n|$. This is a contradiction with J chosen maximal.

By the same argument, $\Lambda \cap S^1$ is a Cantor set (a set with the empty interior) and each trajectory of $x \in \Lambda\backslash\sigma$ is dense in Λ. It is clear also, that the orbits (which are not in Λ or stationary) have its α-limit set in the source α, and its ω-limit set in the Cantor set Λ. This concludes the proof of the Cherry theorem. □

8.2.2 Flows With One Cell

Cherry example constructed in the previous section is by no means isolated. There exists a wider class of flows with the similar behaviour.

Definition 8.2.1 *Let π^t be a Cherry flow on the torus. The flow π^t is said to belong to a $\mathcal{B}$-class if it has precisely one hyperbolic sink and one hyperbolic saddle point p. Moreover the divergence $div_p \pi^t$ in the saddle point p is strictly positive.*

Theorem 8.2.2 (Cherry) *Let π^t be a C^∞ Cherry flow of the class $\mathcal{B}$. Then there exists an infinite set $L \subset T^2$ such that every point $x \in L$ is recurrent: $\omega(x) = \alpha(x) = L$. Moreover, the following is true:*

- *For every point $x \in T^2$ which is not contained in the unstable manifold of the saddle point one has $\alpha(x) = L$;*
- *For every point $x \in T^2$ which is not contained in the stable manifold of the sink or the saddle one has $x \in L$.*

Proof. As it was already mentioned the proof of the theorem is based on the study of the order-preserving mappings of the unit circle S^1 which is a cross-section for the Cherry flow π^t. Below we introduce necessary terminology and reformulate our theorem in the terms of circle mappings.

Let $S^1 = \mathbb{R} \bmod 1$ be a unit circle. Let $\phi : S^1 \to S^1$ be a continuous, order-preserving, degree one map which is possibly constant on a finite number of intervals. Since ϕ is order preserving it has a correctly defined rotation number $\alpha(\phi)$. This number is irrational if and only if ϕ has no periodic points. If $\alpha(\phi)$ is irrational, then ϕ is semi-conjugate to the rotation through the angle α. The semi-conjugacy is continuous, monotone, has degree one and maps the orbits of ϕ to the orbits of rotation α.

Assume that $\phi : S^1 \to S^1$ is continuous, order-preserving, degree one and everywhere C^1 except possibly in a finite number of points. Denote by K_ϕ the closure of the set of points $x \in S^1$ such that either of the following conditions holds:

- ϕ is not C^1 in x, or
- f is constant on some neighbourhood of x, or else
- the first derivative $D\phi(x)$ vanishes in the point x.

In order to reformulate our main theorem in terms of the circle mapping ϕ one introduces the following restrictions (further it will be clear that all of them are satisfied for the flows of class $\mathcal{B}$). Namely we say that a circle mapping $\phi : S^1 \to S^1$ is an $\mathcal{A}$-mapping if:

(i) Mapping ϕ is everywhere strictly increasing except a unique interval K_ϕ;
(ii) For each point $x \in S^1 \backslash K_\phi$ there exists a neighbourhood U of x such that ϕ is a local C^1-diffeomorphism on U and $D\phi$ has a bounded variation on U;

(iii) At the boundary points a and b of the interval $K_\phi = [a, b]$ the one-sided derivative $D\phi(a) = D\phi(b) = \infty$; moreover, there exists $\varepsilon > 0$ such that

$$D\phi|_{[a-\varepsilon,a)}, \qquad D\phi|_{(b,b+\varepsilon]}$$

are monotone.

Definition 8.2.2 *An interval $I \subset S^1$ is called a wandering interval of the mapping $\phi : S^1 \to S^1$ if*

- *$\phi^m(I) \cap \phi^n(I) = \emptyset$ for all $m, n \in \mathbb{N}$ unless $m = n$;*
- *For all $n \in \mathbb{N}$ it holds $\phi^n \not\subset \mathcal{B}_\phi$.*

Clearly, Theorem 8.2.2 is an implication of the following basic lemma.

Lemma 8.2.1 *Let $\phi : S^1 \to S^1$ be a circle mapping of class $\mathcal{A}$ with the irrational rotation number. Then ϕ has no wandering intervals.*

Proof of Lemma. Assume that $\phi \in \mathcal{A}$. This means that ϕ has the degree 1, is everywhere increasing except the interval K_ϕ where it is constant and $\phi|_{S^1\setminus K_\phi}$ is of the class C^1. Moreover, the conditions (ii) and (iii) on $\log D\phi$ are satisfied. We subdivide the proof of Lemma 8.2.1 into several auxiliary statements. First let us set

$$\Lambda = S^1 \setminus \bigcup_{i\geq 0} \phi^{-i}(K_\phi).$$

Proposition 8.2.1 *Let $\phi \in \mathcal{A}$ and the rotation number $\alpha(\phi)$ is irrational. Then for each $x \in \Lambda$ there exists $N \in \mathbb{N}$ such that*

$$D\phi^N(x) > 1.$$

Proof of Proposition 8.2.1. Fix $x \in \Lambda$. It can be shown [199] that there exist $j \in \mathbb{N}$ such that $x \in \phi^j(U)$ and the intervals

$$Int\ U, \quad Int\ \phi(U), \quad \ldots, \quad Int\ \phi^{-j}(U)$$

are pairwise disjoint, where U is the smallest interval containing K_ϕ. Let $\varepsilon > 0$ be a small positive number which is mentioned in item (iii) of the definition of the $\mathcal{A}$-class of mappings. Let us denote

$$V = Var_{x\in U\setminus K_\phi} \log D\phi, A = Max_{x\in U\setminus K_\phi} \log D\phi, B = Min_{x\in U\setminus K_\phi} \log D\phi.$$

(We may always assume $D\phi|_{U\setminus K_\phi} > 1$.) Furthermore, since $K_\phi, \phi^{-1}(K_\phi), \ldots,$ $\ldots, \phi^{-j}(K_\phi)$ are disjoint we may assume that

$$\log \frac{|K_\phi|}{|f^{-j}(K_\phi)|} > V + 2A + 3|B|.$$

Now we choose $y \in f^{-j}(K_\phi)$ such that

$$D\phi^{-j}(y) = \frac{|K_\phi|}{|f^{-j}(K_\phi)|}.$$

Suppose that T is the smallest interval containing both x and y. Notice that since $U \supset K_\phi$ we have $y \in \phi^{-j}(K_\phi)$ and therefore $T \subset \phi^{-j}(U)$ is a proper subset of $\phi^{-j}(U)$. It follows immediately from the disjointness of preimages of U that the intervals

$$T, \quad \phi(T), \ldots, \quad \phi^{-j}(T)$$

are pairwise disjoint. Since ϕ is an orientation-preserving mapping this implies that the points

$$\{x, \phi(x), \ldots, \phi^j(x)\} \qquad \{y, \phi(y), \ldots, \phi^j(y)\}$$

are alternatively distributed at S^1. (That is between each two x's there is an y's, and *vice versa*.) Now we have a rather standard chain of estimates:

$$\log D\phi^N(x) = \sum_{i=0}^{j} \log D\phi(x_i) \geq \sum_{i=0}^{j} \log D\phi(x_i) - \sum_{i=0}^{j} \log D\phi(y_i) + V + 2A + 3|B|,$$

where $N = j + 1$, see [199] for details. The most 'narrow' place in the above chain is the behaviour of sum near the points close to the set K_ϕ. But it was assumed earlier that $\log D\phi$ is decreasing on the semi-interval $(b, b + \varepsilon]$ and increasing on the semi-interval $[a - \varepsilon, a)$. Near the 'endpoints' x_s, y_{s+1} and x_{r-1}, y_r, x_r of the above intervals the following two estimates are true:

$$\log D\phi(x_s) - \log D\phi(y_{s+1}) \geq Min(0 - A, B - A) \geq -|B| - A,$$
$$\log D\phi(x_{r-1}) - \log D\phi(y_r) + \log D\phi(x_r) \geq Min(B - A + B, B - A) \geq -2|B| - A.$$

Bringing now all parts together we get

$$\log D\phi^N(x) \geq (-|B| - A) - V + (-2|B| - A) + V + 2A + 3|B| = 0.$$

Clearly this implies that $D\phi^N(x) > 1$. Proposition 8.2.1 is proved. □

Proposition 8.2.2 *Suppose that all the assumptions of Proposition 8.2.1 are satisfied. Then there exists $N \in \mathbb{N}$ such that for all $x \in \Lambda$ it holds*

$$D\phi^N(x) > 1.$$

Proof of Proposition 8.2.2. Let us set

$$\Lambda' = S^1 \setminus \bigcup_{i \geq 0} \phi^{-i} Int K_\phi.$$

We claim that for all $x \in \Lambda'$ there exists an interval U of x and a number $N \in \mathbb{N}$ such that $D\phi^N(y) > 1$ for all $y \in U \cap \Lambda$. Indeed, if $x \in \Lambda$ then it is true by Proposition 8.2.1.

If $x \in \Lambda'\backslash\Lambda$ then the proof is more complicated. Namely there exists $n \geq 0$ such that $\phi^n(x) \in \partial K_\phi$. Let $N = n+1$. Since $F\phi^N(y) \to \infty$ as $y \to x$ the claim also holds for $x \in \Lambda'\backslash\Lambda$. Since Λ' is compact one covers it by the intervals U_i as above and such that for every i there exists N_i so that the following holds:

$$D\phi^{N_i}|_{U_i \cap \Lambda} > 1.$$

Let us introduce the following notation:

$$N' = \max_i\{N_i\},$$
$$\rho_1 = \min_i\{\min\{\log D\phi^{N_i}|_{U_i\cap\Lambda}\}\}, \quad \rho_2 = \min_{0\leq l\leq N'}\{\min\{\log D\phi^l|_\Lambda\}\}$$

Since $\rho_1 > 0$ N can be chosen such that

$$\rho_1 \frac{N - N'}{N'} + \rho_2 > 0.$$

We are to show that $D\phi^N(x) > 1$ for all $x \in \Lambda$. But it is rather technical. One takes a subdivision of intervals $U_{i_1}, U_{i_2}, \ldots, U_{i_n}$ such that

$$x \in U_{i_1},\ \phi^{N_{i_1}}(x) \in U_{i_2},\ \phi^{N_{i_1}+N_{i_2}}(x) \in U_{i_3}, \ldots, \phi^{N_{i_1}+N_{i_2}+\ldots+N_{i_l}}(x) \in U_{i_{l+1}}.$$

By the definition of N' it holds

$$N - (N_{i_1} + N_{i_2} + \ldots + N_{i_l}) \leq N'.$$

The standard estimates of the $\log D\phi^N(x)$ (see [199] for details) yields us

$$\log D\phi^N(x) \geq \rho_1 \frac{N - N'}{N'} + \rho_2 > 0.$$

Proposition 8.2.2 follows. □

Proof of Lemma 8.2.1. Suppose to the contrary that there is a wandering interval $I \subset S^1$. Since $\phi^n(I)$ are disjoint it leads to a contradiction with Proposition 6.1.2 which states that $\phi^N|_\Lambda$ is an expanding mapping. Lemma is proved. □

Proof of Theorem 8.2.2. On order to apply Lemma 8.2.1 let us establish first that the mapping ϕ induced by the vector field mentioned in the theorem is of the class $\mathcal{A}$. Indeed, properties (i) and (ii) are easily verified. Property (iii) follows from the fact that the transition (Dulac) function near the hyperbolic saddle is defined by the formula

$$x \longmapsto |x|^\alpha, \qquad \alpha = -\lambda_1/\lambda_2,$$

where λ_1, λ_2 are the eigenvalues of the saddle point p. Clearly, the condition $div_p\pi^t = \lambda_1 + \lambda_2 > 0$ implies that $0 < \alpha < 1$. Conditions (iii) of the definition of $\mathcal{A}$-class follow.

Since there are no wandering intervals of the first return mapping ϕ defined on the global cross-section S^1, the complementary

$$S^1 \backslash \bigcup_{n \geq 0} \phi^{-n}(B_\phi)$$

has a non-zero Lebesgue measure. Theorem 8.2.2 follows. □

8.2.3 Flows With Several Cells

Statement Let $\pi^t : M \times \mathbb{R} \to M$ be a C^∞ Cherry-type flow. That is, the Cherry cell is no longer unique, but there is a finite number $N \geq 2$ of "codirected" cells.

Definition 8.2.3 *The flow π^t is said to be of a $\mathcal{B}$-class if the following conditions are satisfied:*

(i) Flow π^t has a finite number of singularities all of which are hyperbolic;
(ii) There are no periodic orbits and saddle connections;
(iii) There are no sources (i.e. all 'cells' are codirected), and
(iv) The divergence of π^t in each singularity p is non-positive:

$$div_p\ \pi^t \leq 0.$$

Theorem 8.2.3 (Martens-van Strien-de Melo-Mendes) *If $\pi^t : T^2 \times \mathbb{R} \to T^2$ is a C^∞ $\mathcal{B}$-flow on torus, then there exists an infinite set $L \subset T^2$ such that every point of it is recurrent being dense in L: $\alpha(x) = \omega(x) = L, x \in L$. Moreover, the following is true:*

- *For every point $x \in L$ which is not contained in the unstable manifold of the saddle point one has $\alpha(x) = L$;*
- *For every point $x \in L$ which is not contained in the stable manifold of the sink or the saddle one has $x \in L$.*

Keeping the notation K_ϕ, B_ϕ introduced earlier for the continuous, order-preserving, degree one mappings $\phi : S^1 \to S^1$ of the unit circle S^1, let us reformulate Theorem 8.2.3 in terms of the circle mappings ϕ.

Definition 8.2.4 *A circle mapping $\phi : S^1 \to S^1$ is said to be of a class $\mathcal{A}$ if it obeys the following conditions:*

(i) K_ϕ has a finite number of components;
(ii) For each point $x \in S^1 \backslash K_\phi$ there exists a neighbourhood U of x such that ϕ is a local C^1-diffeomorphism on U and $D\phi$ has a bounded variation on U;

(iii) For every boundary point $x_0 \in \partial K_\phi$ there exist neighbourhoods U, V of $x_0, \phi(x_0)$ respectively, a constant $\alpha \geq 1$ and a pair of C^1 'normalizing' diffeomorphisms $f_r : I \to U, f_l : I \to V$ such that Df_r, Df_l have a bounded variation and

$$f_l \circ \phi \circ f_r(x) = \pm |x|^\alpha,$$

or is equal to 0.

(Further it will be shown that $\mathcal{B}$-flows induce precisely $\mathcal{A}$-mappings, and vice versa.) Theorem 8.2.3 is a consequence of the following basic lemma.

Lemma 8.2.2 *Let $\phi : S^1 \to S^1$ be a circle mapping of class $\mathcal{A}$. Then ϕ has no wandering intervals.*

Circle Transformations Here we develop some analytic tools which give the estimates of the 'shape' of non-linearity of the n-th iteration $\phi^n(T)$ of an interval $T \subset S^1$ provided intervals $T, \phi(T), \dots, \phi^n(T)$ are disjoint.

Definition 8.2.5 *Let $\phi : S^1 \to S^1$ be a continuous order-preserving mapping and $I, T \subset S^1$ be intervals in S^1 with Clos $I \subset$ Int T. We introduce the following notation:*

$$D(T, I) = \frac{|T||I|}{|L||R|}, \qquad B(\phi, T, I) = \frac{D(\phi(T), \phi(I))}{D(T, I)},$$

where L and R are 'left' and 'right' components of $T \backslash I$.

It may be checked directly that the operator B is multiplicative:

$$B(\phi \circ \psi, T, I) = B(\phi, \psi(T), \psi(I)) B(\psi, T, I).$$

Definition 8.2.6 *Let $\phi : S^1 \to S^1$ be a mapping of the unit circle S^1 which is of class C^3 in the point $x \in S^1$. Then the operator*

$$S\phi(x) = \frac{\phi'''(x)}{\phi'(x)} - \frac{3}{2} \left(\frac{\phi''(x)}{\phi'(x)} \right)^2$$

is called the Schwarzian derivative in the point x.

Proposition 8.2.3 *Let $\phi : T \to T$ be a C^3 diffeomorphism of the open interval T such that for every $x \in T$ it holds $S\phi(x) < 0$. Then*

$$B(\phi, T, I) > 1$$

for every interval I such that Clos $I \subset$ Int T.

Proof. See [86]. □

Let $\mathcal{I}, \mathcal{U}$ be the interval neighbourhoods of K_ϕ such that Clos $\mathcal{I} \subset$ Int $\mathcal{U}$. Moreover, let the number of components of $\mathcal{I}$ and $\mathcal{U}$ be the same as for the set K_ϕ.

Proposition 8.2.4 *Let ϕ be an $\mathcal{A}$-mapping. Let $\Xi = \{T_1, T_2, \dots, T_n\}$ be a collection of n intervals in S^1 such that:*

(i) every point of S^1 is contained in at most three intervals of Ξ;
(ii) $T_i \subset T_j$ implies $i = j$ and
(iii) none of the intervals in Ξ contains a component of $\mathcal{U}\backslash\mathcal{I}$ or points of K_ϕ.

Then there exists $V < \infty$ such that the following estimate is true:

$$\sum_{i=1}^{n} \log B(\phi, T_i, I_i) \geq -5V.$$

Proof. Denote by $V_\phi = \text{Var}\ (\log D\phi|_{S^1\backslash\mathcal{I}})$. Since on $S^1\backslash\mathcal{I}$ the mapping ϕ is C^1 and $D\phi$ has bounded variation one gets $V_\phi < \infty$. Moreover, on each $\mathcal{U}_i$ ϕ has the form $f_l^i \circ (\pm|x|^\alpha) \circ f_r^i$ where f_l^i, f_r^i are C^1 diffeomorphisms such that $\log D\phi_l^i, \log D\phi_r^i$ have bounded variation on $\mathcal{U}_i$. Therefore

$$V_l = \sum_i Var \log D\phi_l^i, \qquad V_r = \sum_i Var \log D\phi_r^i,$$

are both finite. Let us set $V = V_\phi + V_l + V_r$.

Let $\Omega_1 = \{i | T_i \cap \mathcal{I} = \emptyset\}$ and $\Omega_2 = \{i | T_i \subset \mathcal{U}\}$ be the index sets. Since the intervals T_i do not contain components of $\mathcal{U}\backslash\mathcal{I}$ we have $\Omega_1 \cup \Omega_2 = \{1, 2, \dots, n\}$. Consider both possibilities step by step.

(i) First let us assume that $i \in \Omega_1$. Let R_i and L_i be 'right' and 'left' components of $T_i \backslash I_i$. Because ϕ is C^1 on $S^1\backslash\mathcal{I}$ we may apply the Mean Value Theorem on T_i. Taking this fact into account and using the definition of $B(\phi, T_i, I_i)$ we find $l_i \in L_i, r_i \in R_i, m_i \in I_i$ and $t_i \in T_i$ such that

$$\log B(\phi, T_i, I_i) = \log \left(\frac{D\phi(m_i) D\phi(t_i)}{D\phi(l_i) D\phi(r_i)} \right),$$

where $m_i \in (l_i, r_i)$. Two estimates follow immediately:

$$|\log B(\phi, T_i, I_i)| \leq |\log D\phi(m_i) - \log D\phi(l_i)| + |\log D\phi(t_i) - \log D\phi(r_i)|,$$
$$|\log B(\phi, T_i, I_i)| \leq |\log D\phi(m_i) - \log D\phi(r_i)| + |\log D\phi(t_i) - \log D\phi(l_i)|.$$

Let l_i, m_i, r_i, t_i be the increasing ordering on the set of thus far defined points. Applying either of the above inequalities one gets

$$\begin{aligned}|\log B(\phi, T_i, I_i)| &\leq |\log D\phi(m_i) - \log D\phi(l_i)| + |\log D\phi(t_i) - \log D\phi(r_i)| \\ &\leq Var \log D\phi|_{T_i},\end{aligned}$$

and henceforth

$$\log B(\phi, T_i, I_i) \geq -Var \log D\phi|_{T_i}.$$

(ii) Now let us assume that $i \in \Omega_2$. Then T_i is contained in some component $\mathcal{U}_i$ of $\mathcal{U}$ so as ϕ has the form $f_l^i \circ (\pm x^\alpha) \circ f_r^i$. By the multiplicativity

$$B(\phi, T_i, I_i) = B(f_l^i, T_i'', I_i'')B(\pm x^\alpha, T_i', I_i')B(f_r^i, T_i, I_i).$$

(Here $T_i' = f_r^i(T_i), I_i' = f_r^i(I_i), T_i'' = \pm|f_r^i(T_i)|^\alpha$ and $I_i'' = \pm|f_r^i(I_i)|^\alpha$.) Because $\alpha \geq 1$ the Schwarzian derivative of $\pm|x|^\alpha$ is less or equal to 0 so as one gets $B(\pm|x|^\alpha, T_i, I_i') \geq 1$. Simple calculations yield

$$\log B(\phi, T_i, I_i) \geq -\left(Var \log Df_l^i|_{T_i''} + Var \log Df_r^i|_{T_i}\right).$$

Since every point of S^1 is contained in at most three intervals one can write $\{T_i | i = 1, 2, \ldots, n\} = A_1 \cup \ldots \cup A_5$ where each A_j consists of collection of mutually disjoint intervals. The same is true for the intervals T_i''. Hence from the above inequalities one comes to

$$\sum_{i=1}^{n} \log B(\phi, T_i, I_i) \geq -5(V_\phi + V_l + V_r).$$

The roposition is proved. $\square$

Proposition 8.2.5 *Let $\phi : S^1 \to S^1$ be an $\mathcal{A}$-mapping of the unit circle. Then there exists $A_0 > 0$ and intervals $I, W \subset S^1$ such that*

(i) $I \subset W$ and $W \backslash I$ consists of a unique component H;
(ii) $|H| \leq |I|$;
(iii) $I \cap K_\phi = \emptyset$,

then the following inequality holds:

$$\frac{|\phi(I)|}{|\phi(H)|} \geq A_0 \frac{|I|}{|H|}.$$

Proof. We shall argue towards a contradiction. So suppose that there exists a sequence of intervals $I_n, W_n, H_n = W_n \backslash I_n$ satisfying (i)-(iii) and such that

$$\lim_{n \to \infty} \frac{|\phi(I_n)|/|I_n|}{|\phi(H_n)/|H_n|} \longrightarrow 0.$$

Since $|H_n| \leq |I_n|$ this implies that $|W_n| \to 0$ and $W_n \to K_\phi$. Taking a subsequence we may assume that $W_n \to p$ where $p \in K_\phi$ is a point in the set K_ϕ.

Without loss of generality we may assume that p is a critical point of ϕ so that near p $\phi = f_l \circ \pm|x|^\alpha \circ f_r$ where we take $|x|^\alpha$ or 0 if $x \geq 0$ and $-|x|^\alpha$ or 0 if $x \leq 0$. Therefore

$$\frac{|\phi(I_n)|/|I_n|}{|\phi(H_n)|/|H_n|} \geq \left(\frac{\min Df_l}{\sup Df_l}\right)\left(\frac{\min Df_r}{\sup Df_r}\right)\left(\frac{|I_n'|^\alpha/|I_n'|}{|H_n'|^\alpha/|H_n'|}\right),$$

where $H_n' = f_r(H_n)$ and $I_n = f_r(I_n)$. Since f_r is a C^1-diffeomorphism denoting $K_n = |H_n'|/|I_n'|$ one gets $\limsup K_n < \infty$ and therefore there exists $0 < K < \infty$ such that $K_n < K$ for all $n \geq 0$. Here three possibilities arise.

(i) I_n lies between p and H_n. In this case rather technical calculations [199] yield us

$$\frac{|I_n'|^\alpha/|I_n'|}{|H_n'|^\alpha/|H_n'|} \geq \frac{K}{(1+K)^\alpha - 1} > 0.$$

But this is a contradiction with the initial assumption.

(ii) H_n lies between p and I_n so that $H_n \cap K_\phi = \emptyset$. Since the derivative of x^α is increasing we obtain in this case

$$\frac{|I_n'|^\alpha/|I_n'|}{|H_n'|^\alpha/|H_n|} \geq 1.$$

This is again a contradiction with the assumption at the top of the proof.

(iii) Let, finally, H_n lies between p and I_n but $H_n \cap K_\phi$ is non-empty. If H_n intersects an interval component of K_ϕ then we can choose the maximal interval $\tilde{W}_n$ with $I_n \subset \tilde{W}_n \subset W_n$ such that $\tilde{W}_n \cap K_\phi = \emptyset$. Let $\tilde{H}_n = \tilde{W}_n \backslash I_n$ and $\tilde{H}_n' = f_r(H_n')$. By the construction $|\tilde{H}_n'| \leq |H_n'|$ and $|\tilde{H}_n'|^\alpha = |H_n'|^\alpha$. Therefore

$$\frac{|I_n'|^\alpha/|I_n'|}{|H_n'|^\alpha/|H_n'|} \geq \frac{|I_n'|^\alpha/|I_n'|}{|\tilde{H}_n'|^\alpha/|\tilde{H}_n'|}.$$

As in the case (ii) the last term is bounded away from 0 and we come to a contradiction as before. Therefore we may assume that $H_n \cap K_\phi$ is an isolated point. But here again after a calculation we come to

$$\frac{|I_n'|^\alpha/|I_n'|}{|H_n'|^\alpha/|H_n'|} \geq \frac{1}{(1+K)^\alpha},$$

what is a contradiction with the initial assumption. Since all possible cases are considered Proposition is proved. □

Proposition 8.2.6 *Let $\phi \in \mathcal{A}$ and $I, T \subset S^1$ be the intervals of S^1 such that Clos $I \subset$ Int T. Suppose as before that L and R are 'left' and 'right' components of $T \backslash I$. Let $0 < \gamma < 1$ be a real number. Suppose finally that*

$$\frac{|\phi(I)|}{|\phi(R)|} \geq \gamma \frac{|I|}{|R|}, \qquad \frac{|\phi(I)|}{|\phi(L)|} \geq \gamma \frac{|I|}{|L|}.$$

Then

$$B(\phi, T, I) \geq \gamma^2.$$

Proof. This consists in a stepwise substitution of the above inequalities into the formula for $B(\phi, T, I)$. □

Proposition 8.2.7 *Let $\phi \in \mathcal{A}$ and let $I, T \in S^1$ be the intervals in S^1 such that Clos $I \subset$ Int T. If L and R are components of $T \backslash I$ such that*

(i) Either $|L| \leq |I|$ or $|R| \leq |I|$;
(ii) $I \cap K_\phi = \emptyset$,

then there exists a constant $A_1 > 0$ such that

$$B(\phi, T, I) \geq A_1.$$

Proof. Let A_0 be as in Proposition 8.2.5. (We may assume that $0 < A_0 < 1$. We are going to prove Proposition 8.2.7 taking $A_1 = A_0^2/3$. Renaming possibly L and R we may consider the case $|R| \leq |I|$. By Proposition 8.2.5 we get

$$\frac{|\phi(I)|/|I|}{|\phi(R)|/|R|} \geq A_0,$$

and therefore

$$B(\phi, T, I) \geq A_0|L|/|T|.$$

(i) Suppose that $|L| \geq |I|$. Then on account of $|R| \leq |I|$ it gives us

$$B(\phi, T, I) \geq \frac{A_0|L|}{|L| + |I| + |R|} \geq \frac{A_0|I|}{2|I| + |R|} \geq A_0/3 \geq A_1,$$

and we are done.

(ii) Assume that $|L| < |I|$. Then again applying Proposition 8.2.5 we get

$$\frac{|\phi(I)|/|I|}{|\phi(L)|/|L|} \geq A_0.$$

It follows immediately from the above estimates that $B(\phi, T, I) \geq A_0^2 = A_1$. Since all cases are considered Proposition 8.2.7 is proved. □

Further we consider the family $\mathcal{I} \subset \mathcal{U}$ of the neighbourhoods of K_ϕ as it was introduced in Proposition 8.2.4.

Lemma 8.2.3 *Let $\phi \in \mathcal{A}$ and let $I, T \subset S^1$ be the intervals in S^1 such that $I \subset$ Clos $I \subset$ Int T. Let L and R be 'left' and 'right' components of $T \backslash I$. Then there exists $\varepsilon > 0$ and B_0 such that for any $n \geq 0$ satisfying the following conditions:*

(i) $|L| < \varepsilon|I|$ and $|R| < \varepsilon|I|$;
(ii) Every point $x \in S^1$ is contained in at most three elements of the sequence $T, \phi(T), \ldots, \phi^{n-1}(T)$; and
(iii) $\phi^i(I) \cap K_\phi = \emptyset$ for $0 \leq i \leq n-1$,

it holds

$$B(\phi^n, T, I) \geq B_0.$$

Proof. Let $A_1 < 1$ be chosen as in Proposition 8.2.7. Let V be as in Proposition 8.2.4, for which we assume $e^{-5V} < 1$. Let, finally, N be the total number of components of the set K_ϕ. We set $B_0 = (A_1 e^{-5V})^{12N+1}$ and $\varepsilon = \sqrt{B_0/3}$.

Let $0 \le k_1 \le k_2 \le \ldots \le k_s \le n-1$ be the integers such that either $\phi^i(T)$ contains a component of $\mathcal{U}\backslash\mathcal{I}$ or $\phi^i(T) \cap K_\phi$ is non-empty. From the disjointness of the images of T we get $s \le 12N$. From the definition of t_1 one has that $\phi^i(T)$ does not contain a component of $\mathcal{U}\backslash\mathcal{I}$ for $i < k_1$ and since ϕ is aperiodic we have that $\phi^i(T) \subset \phi^j(T)$ implies $i = j$. It follows now from the Proposition 8.2.4 that

$$B(\phi^{k_1}, T, I) \ge e^{-5V}.$$

Similarly, for the number k_2 we get (see [199] for details)

$$B(\phi^{k_2}, T, I) \ge \left(e^{-5V}\right)^2 A_1 \ge B_0$$

To finish the proof we repeat the above procedure $12N$ times and we obtain at the last step:

$$B(\phi^{k_s+1}, T, I) \ge \left(e^{-5V} A_1\right)^{12N}.$$

Bringing together all the above estimates gives us the inequality of Lemma 8.2.2. □

Proof of Theorem *Proof of Lemma 8.2.2.* Let us assume to the contrary that the circle mapping ϕ has a wandering interval I. We may suppose that I is a maximal wandering interval, i.e. not included into a bigger wandering interval. Since the forward orbit of I consists of disjoint intervals and K_ϕ has at most finite number of components we can always assume that $\phi^i(I) \cap K_\phi = \emptyset$.

To obtain a contradiction we have to show that there exists a sequence of integers q_n and neighbourhoods T_n of I such that ϕ^{q_n} is contracting on I but strongly expanding on $T_n \backslash I$. We shall do this by showing that one can control the non-linearity of $\phi^{q_n}|_{T_n}$. From the previous section it follows that this is possible if the intervals $T_n, \phi(T_n), \ldots, \phi^{q_n}(T_n)$ are more or less disjoint. (This will follow from the fact that ϕ preserves the ordering on S^1.)

Let us define sequence $\{q_n\}$ inductively. Let $R = R_\alpha$ be a rigid rotation through the angle α which is conjugate to ϕ. Take an $x \in S^1$ and set $q_0 = 1$. Define q_n as the time of the n-th closest approach:

$$q_{n+1} = \min_{dist(x, R^i(x)) < dist(x, R^{q_n}(x))} i > q_n.$$

(Note that the above numbers q_n depend only on the rotation R.) It can be easily seen that

Proposition 8.2.8 *Dist* $(I, \phi^{q_n}(I)) \longrightarrow 0$ *as* $n \longrightarrow \infty$.

In the next three propositions the ordering of the sequence $\phi^{q_n}(I)$ on S^1 is described.

Proposition 8.2.9 *The intervals $\{\phi^{q_{2n}}(I)|n \geq 0\}$ approach I from the right and the intervals $\{\phi^{q_{2n+1}}(I)|n \geq 0\}$ approach I from the left (or vice versa).*

Proposition 8.2.10 *Let $Q(2n)$ be the set of integers $i \in \mathbb{N}$ such that $q_{2n} \leq i \leq q_{2n+2}$ and such that $\phi^i(I)$ lies between $\phi^{q_{2n+2}}(I)$ and $\phi^{q_{2n}}$. Then $i \in Q(2n)$ if and only if there exists $k \in \mathbb{N}$ such that i has the form $i = q_{2n} + kq_{2n+1}$ and $q_{2n} \leq i \leq q_{2n+2}$. In particular, there exists $a_{2n+2} \in \mathbb{N}$ such that $q_{2n+2} = q_{2n} + a_{2n+2}q_{2n+1}$.*

Proposition 8.2.11 *Let T_n be the smallest interval containing $\phi^{-q_n}(I), I$ and $\phi^{q_n}(I)$. Then every point in S^1 belongs to at most three intervals of the sequence*

$$T_n, \phi(T_n), \ldots, \phi^{q_{n+1}-1}(T_n).$$

Proof of Propositions 8.2.9-8.2.11. See [144]. □

Proposition 8.2.12 *Let $\phi \in \mathcal{A}$ be a circle mapping and assume that I is its wandering interval. Then there exists $n_0 < \infty$ such that if $n > n_0$ then*

$$|\phi^{q_{2n}+a_{2n+2}q_{2n+1}}(I)| = |\phi^{q_{2n+2}}(I)| > |\phi^{q_{2n+2}-q_{2n+1}}(I)|.$$

Proof. Let T_n be the smallest interval containing $\phi^{-q_{2n+2}}(I), I$ and $\phi^{-q_{2n+1}}(I)$. Since $T_n \subset [\phi^{q_{2n+1}}(I), \phi^{-q_{2n+1}}(I)]$ (the last notation means an interval containing both intervals in bracket) one deduces from Proposition 8.2.11 that each point of S^1 is contained in at most three intervals $T_n, \phi(T_n), \ldots,$ $\phi^{q_{2n+2}-1}(T_n)$. Moreover, for the components L_n and R_n of $T_n \backslash I$ one has $|L_n|, |R_n| \to 0$ as $n \to \infty$.

Therefore we can apply Lemma 8.2.3 and get that there exists $B_0 > 0$ and $n_0 < \infty$ such that for all $n \geq n_0$ it holds

$$B(\phi^{q_{2n+2}}, T_n, I) \geq B_0.$$

Since $\phi^{q_{2n+2}}(R_n) \supset \phi^{q_{2n+2}-q_{2n+1}}(I)$ one gets

$$\frac{|\phi^{q_{2n+2}-q_{2n+1}}(I)|}{|\phi^{q_{2n+2}}(I)|} \leq \frac{|\phi^{q_{2n+2}}(R_n)|}{|\phi^{q_{2n+2}}(I)|} \leq \frac{|L_n||R_n|}{B_0|I|^2} \frac{|\phi^{q_{2n+2}}(T_n)|}{|\phi^{q_{2n+2}}(L_n)|} \leq \frac{|L_n||R_n|}{B_0|I|^3}.$$

For n large the last term tends to 0 and since $q_{2n+2} = q_{2n} + a_{2n+2}q_{2n+1}$ Proposition 8.2.12 follows. □

Proposition 8.2.13 *Let $\phi \in \mathcal{A}$ and let I be a wandering interval of ϕ. Then there exists sequences $\{n_k\}$ and $\{i_{n_k}\}$ obeying the inequality $0 < i_{n_k} < a_{2n_k+2}$ and such that for $i = i_{n_k}$ it holds*

$$|\phi^{q_{2n_k}+iq_{2n_k+1}}(I)| < |\phi^{q_{2n_k}+(i-1)q_{2n_k+1}}(I)|,$$
$$|\phi^{q_{2n_k}+iq_{2n_k+1}}(I)| < |\phi^{q_{2n_k}+(i+1)q_{2n_k+1}}(I)|.$$

(In particular, if ϕ has a wandering interval then there are infinitely many integers $n \geq 0$ such that $a_{2n} > 1$.)

Proof. On account of Proposition 8.2.12 for $n \geq n_0$ the following argument is true. Suppose that for some n we cannot find an integer $0 < i < a_{2n+2}$ such that the above inequalities are satisfied. Then in view of inequality of Proposition 8.2.12 for all $0 < i \leq a_{2n+2}$ one has

$$|\phi^{q_{2n}+iq_{2n+1}}(I)| > |\phi^{q_{2n}+(i-1)q_{2n+1}}(I)|.$$

Because $q_{2n+2} = q_{2n} + a_{2n+2}q_{2n+2}$ it follows that

$$|\phi^{q_{2n+2}}(I)| > |\phi^{q_{2n}}(I)|.$$

In view of disjointness of $\phi^i(I)$ the sum $\sum_i |\phi^i(I)| \leq 1$ so that there must be a sequence $n_k \to \infty$ such that the above inequality is violated. Therefore the sequence $\{n_k\}$ is the required one for which the inequalities of Proposition 8.2.13 holds. □

Now we are in position to finish the proof of Lemma 8.2.2. Let n_k and i_{n_k} be as in Proposition 8.2.13, $q_n = q_{2n_k} + i_{n_k}q_{2n_k+1}$ and K_n be the smallest interval containing $\phi^{q_{2n_k+1}}(I), I$ and $\phi^{-q_{2n_k+1}}(I)$. Let L_n and R_n be as before. Proposition 8.2.12 guarantees the disjointness of intervals needed to apply Lemma 8.2.3 for the mapping $\phi^{q_n}|_{K_n}$. From the inequalities of Proposition 8.2.13, we get

$$3\frac{|L_n||R_n|}{|I|^2} \geq B(\phi^{q_n}, K_n, I) \geq B_0.$$

But this is a contradiction since $|L_n|, |R_n| \to 0$ as $k \to \infty$ by the assumption that I is a wandering interval. Lemma 8.2.2 is proven. □

Proof of Theorem 8.2.3. First let us show that the flows $\pi^t : T^2 \times \mathbb{R} \to T^2$ of class $\mathcal{B}$ generate the circle mappings $\phi : S^1 \to S^1$ of class $\mathcal{A}$. Indeed, properties (i) and (ii) are evident. Property (iii) follows from the assumption that $div_p \pi^t \leq 0$ in the saddle points of π^t. Indeed by the results of §6.2.4 the Dulac's mapping near the separatrices of the saddle point p is C^∞ conjugate to the mapping

$$x \longmapsto |x|^\alpha, \qquad \alpha = -\lambda_1/\lambda_2,$$

where λ_1 and λ_2 are the eigenvalues in the hyperbolic saddle p. By the assumption $div_p \pi^t = \lambda_1 + \lambda_2 \leq 0$ and therefore $\alpha \geq 1$. So the condition (iii) is satisfied.

Now we apply Lemma 8.2.2. Since the $\mathcal{A}$-mapping has no wandering intervals then for every

$$x \in S^1 \setminus \bigcup_{n \geq 0} \phi^{-n}(B_\phi),$$

where B_ϕ is are 'black' intervals of ϕ, both the forward and the backward orbits are everywhere dense in the above defined subset. (In other words, the 'grey' intervals exist.) In particular, there exists a non-finite recurrent orbit of ϕ which is suspended to a recurrent trajectory of the flow π^t. Theorem 8.2.3 is proved. □

Appendix: A Dulac Function We recall some known results concerning the normal forms of vector fields near hyperbolic singularities, which allow us to estimate the transition (Dulac) functions near the critical points of the circle mappings ϕ. All the exposed results are rather classical and may be found in [273].

Let X be a C^∞ vector field with a singularity at 0 which is a saddle (or admits a saddle sector). Let p_1 and p_2 be the points in $W^s(0)$ and $W^u(0)$, respectively. Furthermore, let S_1 and S_2 be a C^2 curves through p_1 and p_2 transversal to X.

Definition 8.2.7 *By the Dulac (or transition) function near the saddle in* 0 *one understands a mapping along the orbits of* X

$$T : S_1 \backslash p_1 \longrightarrow S_2 \backslash p_2,$$

defined between S_1 *and* S_2 *with* p_1 *and* p_2 *deleted.*

In this section we want to show that T is equal to a mapping

$$\phi(\alpha) = \pm |x|^\alpha, \alpha = |\lambda_1|/\lambda_2,$$

where $\lambda_1 < 0 < \lambda_2$ are the eigenvalues in the saddle point 0.

Lemma 8.2.4 *Let* X *be a* C^∞ *vector field on* $\mathbb{R}^2$ *which has a hyperbolic singularity at* 0 *of saddle type with eigenvalues* $\lambda_1 < 0 < \lambda_2$. *Let* $\alpha = |\lambda_1|/\lambda_2$. *There exists mappings* ϕ, ψ *of* $\mathbb{R}$ *of class* C^1 *such that* $D\phi, D\psi$ *have bounded variation and such that the mapping* T *has the form*

$$T(x) = \phi \circ \phi_\alpha \circ \psi(x).$$

Proof.

(I) First let us establish the above formula when X is a linear vector field

$$X = \lambda_1 x \frac{\partial}{\partial x} + \lambda_2 y \frac{\partial}{\partial y}$$

(this case is trivial but gives a right understanding of how the whole thing works). As S_1 and S_2 in this case one chooses the lines $x = 1$ and $y = 1$. By an explicit integration of the above vector field:

$$|x|^{\lambda_1} |y|^{\lambda_2} = C,$$

one comes after the appropriate choice of the constant C to the required formula for ϕ_α.

(II) X is non-resonant $|\lambda_1| \neq |\lambda_2|, 2|\lambda_2|, 1/2|\lambda_2|$. In this case Sternberg's Linearization Theorem [273] claims that X is C^2-linearizable in 0 and we get the case (I).

(III) Suppose that X is resonant, that is either $|\lambda_1| = 2|\lambda_2|$ or $|\lambda_1| = |\lambda_2|$ (The case $|\lambda_1| = 1/2|\lambda_2|$ is similar to the first possibility.) We may always assume that in these two cases vector field X takes the following 'normal' forms:

$$X(1:2) = x\frac{\partial}{\partial x} - 2y\frac{\partial}{\partial y} + O(x^2+y^2), \qquad X(1:1) = x\frac{\partial}{\partial x} - y\frac{\partial}{\partial y} + O(x^2+y^2).$$

Proposition 8.2.14 *Let X be a C^∞ vector field with a hyperbolic singularity in 0. Then the following is true.*

(i) If X has the normal form $X(1:2)$ then there exists a C^∞ change of coordinates ϕ such that it conjugates $X(1:2)$ to the vector field

$$\phi_*(X) = x(1+ax^2y)\frac{\partial}{\partial x} - 2y(1+bx^2y)\frac{\partial}{\partial y} + O(x^6+y^6);$$

(ii) If X has the normal form $X(1:1)$ then there exists a C^∞ change of coordinates ϕ such that it conjugates $X(1:1)$ to the vector field

$$\phi_*(X) = x(1+axy)\frac{\partial}{\partial x} - 2y(1+bxy)\frac{\partial}{\partial y} + O(x^4+y^4).$$

Proof. According to the general theory we search for a vector field Y such that the Lie product $[X,Y]$ vanishes. Taking the particular normal forms we come to the following (called *homological*) equations:

$$\left[x\frac{\partial}{\partial x} - \beta y\frac{\partial}{\partial y}, x^n y^m\frac{\partial}{\partial x}\right] = (n - \beta m - 1)x^n y^m\frac{\partial}{\partial x},$$
$$\left[x\frac{\partial}{\partial x} - \beta y\frac{\partial}{\partial y}, x^n y^m\frac{\partial}{\partial y}\right] = (n - \beta m + \beta)x^n y^m\frac{\partial}{\partial y},$$

where $\beta \in \{1,2\}$. We can delete all monomials by an appropriate choice of Y, except the resonant ones. It follows from the above formulas that such are precisely two monomials x^2y if $\beta = 2$ and two monomials xy if $\beta = 1$.

(IV) Now we will show that even resonant monomials in the normal forms $\phi_*X(1:2)$ and $\phi_*X(1:1)$ can be deleted by C^1+ 'bounded variation' diffeomorphism (although by the price of appearing certain monomials of higher degree).

(a) Indeed, take the normal form $\phi_*X(1:2)$ and apply to it the coordinate change:

$$\tilde{x} = x - ax^3y\log|x|, \qquad \tilde{y} = y - bx^2y^2\log|y|$$

(Note that the smoothness class of the above diffeomorphism is indeed C^1+ 'bounded variation'.) After the direct computations we obtain the normal form

$$\tilde{X}(1:2) = \tilde{x}\frac{\partial}{\partial \tilde{x}} - 2\tilde{y}\frac{\partial}{\partial \tilde{y}} + O(x^6 + y^6).$$

(b) Take now the normal form $\phi_* X(1:1)$ and let us make a C^1+ 'bounded variation' substitution

$$\tilde{x} = x - ax^2 y \log|x|, \qquad \tilde{y} = y - bxy^2 \log|y|.$$

It may be verified directly that it yields us the following normal form:

$$\tilde{X}(1:1) = \tilde{x}\frac{\partial}{\partial \tilde{x}} - \tilde{y}\frac{\partial}{\partial \tilde{y}} + O(x^4 + y^4).$$

Lemma 8.2.4 follows. □

8.3 Foliations on the Sphere

An unexpected application of the unit circle transformations, is that to the non-orientable foliations on the sphere. If foliation is orientable, then by the theorem of Bendixson (see the *Addendum* to this chapter) there are no recurrent leaves which are non trivial. However, if one drops the claim of orientability, there are plenty of such leaves (cf. labyrinths). A delicate question is when the non-orientable foliations in a simply connected domain still cannot have non trivial recurrent leaves (the so-called *problem of Rosenberg* [256]). In this section a class of non-orientable foliations on the sphere is considered, for which the problem of Rosenberg can be solved.

8.3.1 Notations

Let $\mathcal{F} = (M, \pi^t, \theta)$ be a C^∞ foliation given on a compact surface M. Let $N = M/\theta$ be the 'underlying' surface which carries the 'non-orientable' foliation $F = \pi^t/\theta$. Specifically, we are interested in the case $N = S^2$. (The 2-dimensional sphere as an underlying surface.) Moreover, the set Sing F of a C^∞ foliations on N is fixed by an even number of non-orientable singularities of the thorn, sun-set or tripod type.

Definition 8.3.1 *If a Morse-Smale foliation $F \in \mathcal{F}_0^r$ has no cycles (closed 1-leaves), it is called a regular* [1] *foliation. (In other words, all leaves of F are compact and tend from a singularity to a singularity.)*

Definition 8.3.2 *Foliation F is called arational if it has no compact 1-leaves.*

[1] Or, the gradient like, see [202]

We introduce a class, an A-class, of foliations on the 2-sphere which are produced from a 'standard foliation with 4 thorns on S^2' by a finite number of homotopy operations of an 'opening of a leaf' and 'deformation of a thorn into a sun-set'.

• *Standard foliation of S^2.* Consider a 2-dimensional torus T^2 as a covering surface M of a sphere with the ramification set in 4 (distinct) points (apply the Riemann-Hurwitz formula (1) for $k = 2$). Take an irrational flow π^t : $T^2 \times I\!R \to T^2$ on T^2 such that chosen points lie on the different recurrent orbits. These points are declared the equilibria (two-separatrix saddles) of the flow π^t. Clearly, the flow π^t on T^2 covers a (non-orientable) foliation, F_4, on S^2 such that Sing F_4 consists of 4 thorns. All 1-leaves of F_4 are recurrent and we call it a 'standard foliation with 4 thorns'. (Further we deal with a covering representation of F_4 which is given by the triple $\mathcal{F}_4 = (T^2, \pi^t, \theta)$.)

• *Opening of a leaf.* Take a recurrent leaf, γ, which is not a separatrix leaf of a foliation produced from F_4. Making an infinite cut along γ with the banks γ_+ and γ_-, we 'open' γ and then we glue-up between γ_+ and γ_- an infinite narrow stripe F_γ. Then one attaches sun-set and tripod singularities to γ_+ and γ_-, foliating F_γ as it is shown in Fig. 8.2.

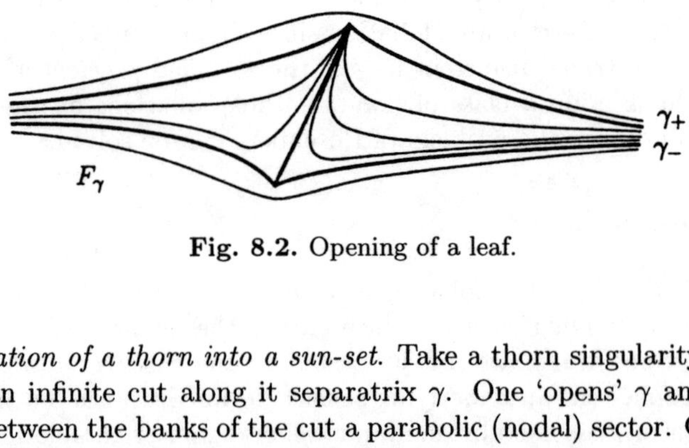

Fig. 8.2. Opening of a leaf.

• *Deformation of a thorn into a sun-set.* Take a thorn singularity of F_4 and produce an infinite cut along it separatrix γ. One 'opens' γ and then one glues-in between the banks of the cut a parabolic (nodal) sector. Clearly, one obtains a foliation on the sphere with an extra sun-set singularity instead of a thorn.

Fix a natural metric on S^2. The 'tongues' which appear as a result of the above homotopy operations can follow two different patterns at the infinity. They either 'expand', covering more and more area on S^2, so that each separatrix of an A-foliation is 'caught' by a parabolic (nodal) sector of a sun-set. Or else, the tongues can be squeezed sufficiently fast, so that no one separatrix of an A-foliation goes to a sun-set. (Note that by the irrationality of F_4 A-foliations admit no cycles.)

Corollary 8.3.1 *A-foliations are either arational or regular.*

(Class A is an important class, which includes the Rosenberg labyrinths in the disc and annulus [256], generalized measured foliations [280] and the Plykin attractors on the sphere [245], etc.)

Let F be a C^∞ A-foliation on the 2-sphere. Let $\mathcal{F} = (M, \pi, \theta)$ be a covering representation of F on the surface M of genus $g = 1/2|\text{ Sing } F| - 1$. For every $p \in$ Sing $\mathcal{F}$ one defines a divergence of $\mathcal{F}$ in p as follows.

Definition 8.3.3 *Denote by E_1 and E_2 the flow boxes of a thorn and a sun-set which has been obtained by a deformation of a thorn. Denote by E_3 a (cylindric) flow box which contains a tripod and a sun-set, obtained by a homotopy of the opening of a leaf. All of them are shown in Fig. 8.3. Shaded regions correspond to the orbits which enter or emerge from the nodal sectors of the respective singularities. The divergence in the above singularities is given by the table.*

Singularity	*Flow box*	*Normal form*	*Divergence*
thorn	E_1	(3)	$div_p\mathcal{F} = 0$
sun-set	E_2	(2)	$div_p\mathcal{F} = \|\alpha - \beta\|$
tripod	E_3	(2)	$div_p\mathcal{F} = \beta - \alpha$
sun-set	E_3	(2)	$div_p\mathcal{F} = \beta - \alpha$

The 'eigenvalues' α, β of the singularities in the flow box E_3 are chosen such that the in-going (into E_3) separatrix always corresponds to an α-eigenvalue, while out-going separatrix corresponds to a β-eigenvalue.

8.3.2 Main Result

Theorem 8.3.1 *Let $\mathcal{F}$ be a C^∞ A-foliation on the 2-sphere with at least one sun-set singularity. If in every point $p \in Sing\mathcal{F}$ the inequality $div_p\mathcal{F} \leq 0$ holds, then all leaves of $\mathcal{F}$ are compact.*

Proof. Proof will consist of the two parts. First we show that for the A-foliations there exists a global cross-section such that the corresponding return map does not admit interval exchange transformations (IET), see [162] for the definition. Next, we shall prove that under the restrictions, imposed on the divergence of the foliation in singular points, there exists an ergodic measure, invariant under the return map (or, what is the same, there are no 'wandering intervals'). Thus far, the theorem will follow.

Let $\mathcal{F}$ be an A-foliation on the 2-sphere.

Lemma 8.3.1 *There exists a simple closed curve S^1 which is transversal to $\mathcal{F}$ and each orbit of $\mathcal{F}$ (except for the singularities and the compact separatrices) hits S^1. Moreover, the return mapping induced on S^1 by the orbits of $\mathcal{F}$ preserves the order of the intervals of continuity.*

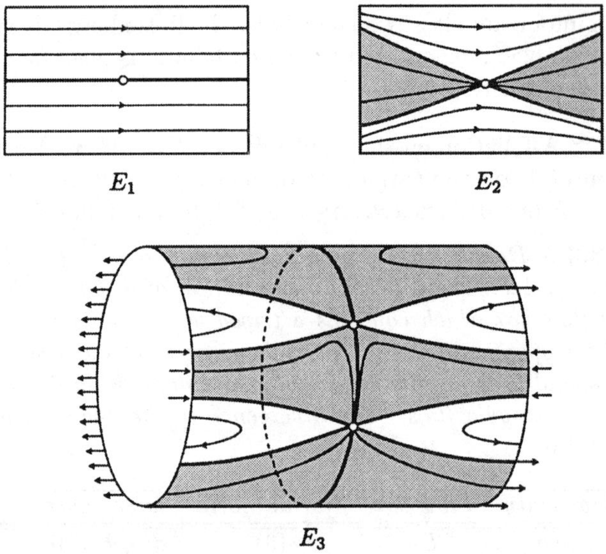

Fig. 8.3. Flow boxes $E_1 - E_3$.

Proof. By the definition, A-foliations can be obtained from $\mathcal{F}_4 = (T^2, \pi^t, \theta)$ by a finite number of homotopies.

(i) Since $\pi^t : T^2 \times I\!R \to T^2$ is an irrational flow on torus, one can easily construct a global cross-section, S^1, of π^t (as such one of the meridians of T^2 can be chosen). The return mapping $\phi : S^1 \to S^1$, induced by the orbits of π^t, is defined everywhere except for 4 distinct points $P = \{p_i | i = 1, ..., 4\}$ in which the separatrices of the 2-saddles hit S^1. The set P splits S^1 into 4 disjoint intervals. However, the permutation $i \mapsto \pi(i)$ defined by ϕ is trivial (i.e. it keeps the orientation and the cyclic order of the intervals on S^1).

(ii) Let now a homotopy of the deformation of a thorn into a sun-set be applied to one of the singularities of $\mathcal{F}_4$. It means that in the above mentioned set P a point $p \in P$ must be replaced by a closed segment $A_p \in S^1$, on which the mapping ϕ is not defined. (It corresponds to the substitution $E_1 \to E_2$ of the flow boxes, shown in Figure 8.3). Clearly, the cross-section S^1 remains the same. Moreover, by the arguments of (i), ϕ does not permute the continuity intervals on S^1.

(iii) Let, finally, the homotopy of the opening of a leaf be applied to an $\mathcal{F}_4$ foliation. (The general case of an arbitrary A-foliation is treated similarly and can be included, with slight modifications, in the below scheme.) Let γ be a recurrent orbit of a C^∞ irrational flow π^t different from the separatrices of the 2-saddles. Let us 'open' γ to a narrow stripe $\gamma_\varepsilon = \gamma \times \varepsilon$. Notice that the cross-section S^1 for a such flow will remain the same, as well as the continuity intervals at S^1 will preserve their order with regard to the ϕ-actions.

Now split γ_ε into two parts, $\gamma_\varepsilon^+ \cup \gamma_\varepsilon^-$, and move aside the banks of the cut. This procedure yields us a torus with a hole, D, which breaks the stripe γ_ε.

Let $\theta : T^2 \to T^2$ be an involution, defined by the triple $\mathcal{F}_4 = (T^2, \pi^t, \theta)$. Denote by $\widetilde{\gamma}$ a recurrent orbit, which stays in the involution with γ. Likewise, let $\widetilde{\gamma}_\varepsilon, \widetilde{\gamma}_\varepsilon^+$ and $\widetilde{\gamma}_\varepsilon^-$ be the respective stripes on the torus. Denote by $\widetilde{D}$ a hole which breaks the stripe $\widetilde{\gamma}_\varepsilon$. What remains to be done is to glue-up a 'handle' into T^2 which connects D and $\tilde{D}$. The handle must contain a pair of singularities covering the tripod and the sun-set and equivalent to the flow box E_3 shown in Fig. 8.3. We claim that the above surface, $T^2 \# T^2$, is a double covering representation surface for a 'standard' foliation on S^2 with one leaf opening.

Indeed, all the orbits in the stripe γ_ε^+ enter the stable parabolic sector of the singularity in the flow box E_3. All orbits in the stripe γ_ε^- emerge from the unstable sector of the same singularity. Clearly, the above operations does not change the position of the global cross-section S^1. And, as it was noticed at the top of this item, the continuity intervals on S^1 will preserve their cyclic order under the ϕ-actions of the flow on $T^2 \# T^2$. Lemma is proved. □

Let $\phi : S^1 \to S^1$ be a return mapping defined by an A-foliation on a cross-section S^1. As it was mentioned earlier ϕ is defined on $S^1 \setminus \cup_{k \in K} A_k$, where A_k are closed sets in S^1. The set A_k is either an isolated point or a segment which corresponds to the points where orbits that go to (go out) a sink (source) section of a saddle-node type singularity representing the sun-set, hit S^1. Our nearest purpose is to extend ϕ to the set $\cup_{k \in K} A_k$.

(i) Suppose a separatrix of a 2-saddle hits S^1, as given by the flow-box E_1. Transition map ϕ_0 in the neighbourhood of 2-saddle is $y \approx \exp(\pi\beta/\alpha)x + \ldots$ Since $\phi = \phi_R \circ \phi_0$, where ϕ_R is a diffeomorphism along 'regular' part of the flow, the diffeomorphism ϕ near $A_k = p$ has the form $x \mapsto a_0 e^{\pi\beta/\alpha} x + \ldots$ We extend ϕ in an evident way:

$$\phi(p) = 0.$$

(ii) Suppose that A_k is a segment in S^1, where 'blind' orbits of the flow in the flow-box E_2 hit S^1. The transition mapping ϕ_0 near the saddle-node type singularity has the form $y \approx |x|^{\alpha/\beta} + \ldots$ By the condition $div_p \mathcal{F} \leq 0$ one gets immediately $\alpha = \beta$. Taking advantage of the regular component, the diffeomorphism ϕ near the ends of A_k is given by $|x| \mapsto a_0|x| + \ldots$ Note that by the symmetry of the flow a_0 has the same value for the 'right' and 'left' ends of A_k. Therefore we extend ϕ to A_k linearly:

$$\phi = a_0 x, \quad x \in A_k.$$

(iii) Finally, let us consider the flow-box E_3. Suppose that the transition map near the 6-saddle p_1 is $\phi_0^{(r)}$ and near the saddle-node p_2 is $\phi_0^{(l)}$. They have the form $\phi_0^{(r)}(x) = |x|^{\alpha_1/\beta_1} + \ldots$ and $\phi_0^{(l)} = |x|^{\alpha_2/\beta_2} + \ldots$, where $\alpha_1, \beta_1, \alpha_2, \beta_2$ are the local values in the corresponding points. On account of the regular components a_0, b_0, we prolong ϕ to $A_k = [0,1]$ as follows:

$$\phi = \begin{cases} a_0|x-1|^{\alpha_1/\beta_1}, & x \geq 1 \\ -a_0|x-1|^{\alpha_1/\beta_1}, & 1-\varepsilon_1 \leq x < 1 \\ \phi \in C^\infty \text{and monotone}, & \varepsilon_2 < x < 1-\varepsilon_1 \\ b_0|x|^{\alpha_2/\beta_2}, & 0 < x \leq \varepsilon_2 \\ -b_0|x|^{\alpha_2/\beta_2}, & x \leq 0. \end{cases}$$

By now, we have ϕ extended to the entire S^1. Note that by Lemma 8.3.1 ϕ is a homeomorphism and by the condition *div* $\mathcal{F} \leq 0$ it is also a diffeomorphism of the circle.

Definition 8.3.4 *A diffeomorphism $\phi : S^1 \to S^1$ is said to have a critical point $x \in S^1$ if $D\phi(x) = 0$.*

Clearly, there are no critical points of ϕ in A_k, and there are exactly two critical points $x(p_1) = 0$ and $x(p_2) = 1$ of ϕ at the complementary to A_k.

Lemma 8.3.2 *Let $\mathcal{F}$ be a C^∞-smooth* A*-foliation with $div_p\mathcal{F} \leq 0$ in each $p \in Sing\mathcal{F}$. Let $\phi : S^1 \to S^1$ be the corresponding return map on the global cross-section S^1, prolonged in the way described above. Then ϕ has no wandering intervals.*

Proof. For the proof below we use the Denjoy arguments [88] refined by the technique of Yoccoz [307]. Let $I \subset S^1$ be an interval in S^1 with the ends a and b and let $|I|$ be the length of I. Suppose $\phi : S^1 \to S^1$ be an orientation preserving C^1-homeomorphism with a finite set of critical points K. Along with [307] we consider the function

$$M(\phi, I) = \begin{cases} \frac{|\phi(I)|}{|I|}[D\phi(a)D\phi(b)]^{-\frac{1}{2}}, & a,b \notin K \\ \infty, & a,b \in K. \end{cases}$$

The function M is multiplicative, $M(\phi \circ \psi, I) = M(\phi, \psi(I))M(\phi, I)$. The strategy one should follow now is to estimate $M(\phi, I)$ from below. It will be done step by step as follows.

(i) Suppose that I does not lie in the vicinity of K. By the Mean Value Theorem, one has

$$M(\phi, I) = D\phi(\xi)[D\phi(a)D\phi(b)]^{-\frac{1}{2}}, \quad \xi \in I.$$

By assumption, ϕ is $C^{1+\varepsilon}$ and therefore the variation of $D\phi$ on I is finite

$$M(\phi, I) \geq \exp(-\frac{1}{2}Var_{\{I\}}D\phi).$$

(ii) Let $x_i \in K$ be a critical point of ϕ. Suppose $I \subseteq (x_i - \varepsilon, x_i)$ or $I \subseteq (x_i, x_i + \varepsilon)$.

(a) Let, in addition, $x_i \notin \overline{I}$. In the vicinity of x_i the diffeomorphism ϕ is given by (6): $\phi = \pm A|x|^\delta$, where $A \in \{a_0^{(i)}, b_0^{(i)}\}_{i=1}^K, \delta \in \{\alpha_i/\beta_i\}_{i=1}^K$. Yoccoz [307] demands $(D\phi)^{-1/2}$ to be convex in I (or, what is the same, $S\phi \leq 0$ on I, where $S\phi = \phi'''/\phi' - \frac{3}{2}(\phi''/\phi')$). We leave it to the reader to check that the Schwarzian derivative S of ϕ is non-positive. Then the inequality $M(\phi, I) \geq 1$ follows [307].

We give an independent estimate. Since $D\phi = \pm A\delta|x|^{\delta-1}$, we calculate directly $M^2(\phi, I) = (\xi^2/ab)^{\delta-1}$, where $\xi \in I$ is the Mean Value Point. For the convex functions $\xi \approx \frac{a+b}{2}$ and therefore $M^2(\phi, I) \geq 1$ (recall that $\delta \geq 1$ by the assumption of the theorem). Hence, $M(\phi, I) \geq 1$.

(b) Suppose, finally, that $x_i \in \overline{I}$. In this case [307] one has $0 \leq x_i - a \leq b - x_i$. We denote $I_1 = [x_i, b]$ and, in view of (6), $D\phi(a)D\phi(b) \leq A^2\delta^2|b - x_i|^{2(\delta-1)}$ and $|\phi(I_1)| = \int_{x_i}^b D\phi(t)dt = |b - x_i|^\delta$. Now it follows that

$$M(\phi, I) \geq \frac{|\phi(I_1)|}{2|I_1|}[D\phi(a)D\phi(b)]^{-\frac{1}{2}} \geq \frac{1}{2A\delta}.$$

The last step to prove Lemma 8.3.2 is to establish the following proposition.

Proposition 8.3.1 *Let ϕ be a homeomorphism of S^1 specified in lemma 6.3.2. Then there exists a positive constant C such that for all $N \geq 1$, all $0 \leq k \leq N$ and all $I \subset S^1$, not degenerate to a point, the inequality $M(\phi^k, I) \geq C$ holds.*

Proof follows from the estimates above, and we refer the reader to [307]. □

To finish the proof of Lemma 8.3.2, let us suppose the contrary: there exists a wandering interval $I \subset S^1$ with the endpoints a and b. In particular, it yields that $\lim_{k\to\infty} |\phi^k(I)| = 0$. On account of the order of points a_k, b_k on S^1, it means that $M(\phi^k, I) \to 0$ as $k \to \infty$. This is a contradiction with Proposition 8.3.1 proving Lemma 8.3.2. □

Proof of Theorem 8.3.1. Denote by $A = \cup_{k\in K} A_k$ a set of all points $x \in S^1$ where the orbits going-out from the source section of a saddle node k hit S^1 first time, and by $\Omega = \cup_{k\in K} \Omega_k$ those points of $x \in S^1$, where the orbits going-in a sink section of a saddle-node k hit S^1 first time. Take a forward orbit $O(A) = \{\phi^{n_1}(A)|n_1 \in \mathbb{N}\}$ of A and a backward orbit $O(\Omega) = \{\phi^{-n_2}(\Omega)|n_2 \in \mathbb{N}\}$ of Ω. It follows from Lemma 8.3.2 that there exist a finite n_1 and n_2 such that $O(A)$ and $O(\Omega)$ cover $S^1\backslash(A \cup \Omega)$. It means that each orbit of $\mathcal{F}$, different from the equilibria, goes from a source section of a saddle-node to a sink section of a saddle-node. In particular, all leaves of $\mathcal{F}$ are compact.

Theorem is proven. □

8.3.3 Application to the Labyrinths

In the paper [256], p.29, H. Rosenberg has formulated several problems concerning the labyrinths in discs and annuli (these are special cases of $\mathcal{A}$-foliations on S^2, the simplest of which is an $\mathcal{A}$-foliation with 3 thorns and 1 sun-set singularity). One of the proposed questions is to find a criterion to discern between two possible types of behaviour of leaves in the labyrinth: they can all be compact or they can loose itself inside the labyrinth. This problem admits a simple solution in terms of the divergence of the foliations in singular points. Namely, one readily derives from Theorem 8.3.1 the following

Corollary 8.3.2 *Let $\mathcal{L}$ be a Rosenberg labyrinth in the disc. Then all leaves of $\mathcal{L}$ are compact if $div\mathcal{L} \leq 0$ in all its singular points.*

8.3.4 Appendix: The Dulac Functions

In this section we calculate the relevant normal forms of the 'hyperbolic' non-orientable singular points. We estimate the relevant 'transition mappings' near such singularities what delivers us a justification to the formulas used in the previous section.

C^0 Normal Forms As it was shown earlier, C^0 change of coordinates and time brings the vector field in singularity to the respective 2-jet:

$$w(u,v) = [\alpha u^2 + (\beta-1)uv]\frac{\partial}{\partial u} + [(\alpha-1)uv + \beta v^2]\frac{\partial}{\partial v}$$

$$w(u,v) = [\alpha u^2 + \beta uv - v^2]\frac{\partial}{\partial u} + [(\alpha+1)uv + \beta v^2]\frac{\partial}{\partial v}$$

Both of the above equations are integrable and admit the first integrals (see also [218]) given, respectively, by

$$|u|^\beta |v|^\alpha |u-v|^{1-\alpha-\beta} = C$$

$$\frac{\alpha+1}{2}\log|\frac{u^2}{v^2}+1| + \beta \arctan\frac{u}{v} + \log|v| = C.$$

It is well-known that the germ $T(0)$ of the transition mapping $T: S_1 \to S_2$ does not depend on the particular choice of S_1 and S_2. Therefore we consider $S_1 = (-1,x), S_2 = (y,1)$ and $S_1 = (-1,x), S_2 = (1,y)$ in the above two cases. The transition mapping $y = T(x)$ is implicitly given by

$$|x|^\alpha |1+x|^{1-\alpha-\beta} = |y|^\beta |1+y|^{1-\alpha-\beta},$$

$$\frac{\alpha+1}{2}\log|1+x^2| - \beta\arctan\frac{1}{x} - \alpha\log|x|$$
$$= \frac{\alpha+1}{2}\log|1+y^2| + \beta\arctan\frac{1}{y} - \alpha\log|y|.$$

It readily follows now that the main part of $T(0)$ of the first normal form is

$$y \approx |x|^{\alpha/\beta} + ...,$$

and of the second normal form is

$$y \approx e^{\pi\beta/\alpha}|x| + ...$$

Of course, the homeomorphism h which has been used for the normalization does not keep C^1+'bounded variation'$\simeq C^{1+\varepsilon}$-structure of the foliation. Still it is clear that the restriction of transformations to the class $C^{1+\varepsilon}$ will 'add' some higher degree monomials in the above normal forms, and these monomials will not influence much on the transition mappings obtained. This point of view will be formalized in the next section.

$C^{1+\varepsilon}$ Normal Forms

Lemma 8.3.3 *The transition mappings T_1 and T_2 near the singular points defined by the general (non-reduced) normal forms are $C^{1+\varepsilon}$– equivalent to the transition mappings calculated for the reduced normal forms.*

Proof. Let $\mathcal{F} \in \mathcal{F}^r(M)$ and let 0 be an equilibrium of $\mathcal{F}$ given in some local charts by the first or the second normal form. If we look at $\phi(u,v), \psi(u,v)$ as at the formal series of monomials of even degrees, then, after an appropriate change of coordinates and time, in typical case ϕ and ψ take the form $\phi(u,v) = 0, \psi(u,v) = \sum_{m=2}^{\infty} C_m v^{2m}$ for the first normal form, and $\phi(u,v) = \sum_{m=2}^{\infty} C_m v^{2m}, \psi(u,v) = 0$ for the second normal form. In fact, as it was proven in §1.3, the above are $C^{1+\varepsilon}$ finitely determined, so that there exists N such that they are $C^{1+\varepsilon}$-equivalent to

$$w(u,v) = [\alpha u^2 + (\beta - 1)uv]\frac{\partial}{\partial u} + [(\alpha - 1)uv + \beta v^2 + \sum_{m=2}^{N} C_m v^{2m}]\frac{\partial}{\partial v},$$

$$w(u,v) = [\alpha u^2 + \beta uv - v^2 + \sum_{m=2}^{N} C_m v^{2m}]\frac{\partial}{\partial u} + [(\alpha + 1)uv + \beta v^2]\frac{\partial}{\partial v}.$$

It is impossible to estimate directly the mapping T for the above germs, because they are not integrable. However, T can be evaluated after the blowing-up procedure, the result of which is given in the charts $(u,\eta), (\theta, v)$, where $\eta = u/v, \theta = v/u$, by the equations

$$\frac{d\theta}{dt} = -\theta + \theta^2 - \theta \sum_{m=2}^{N} C_m v^{2m-2}, \quad \frac{dv}{dt} = \beta v + (\alpha - 1)\theta v + \sum_{m=2}^{N} C_m v^{2m-1}$$

in the first case, and by

$$\frac{du}{dt} = \alpha u + \beta u\eta - u\eta^2 + \sum_{m=2}^{N} C_m u^{2m-1}\eta^{2m},$$

$$\frac{d\eta}{dt} = \eta + \eta^3 - \sum_{m=2}^{N} C_m u^{2m-2}\eta^{2m+1},$$

$$\frac{d\theta}{dt} = -1 - \theta^2 + \sum_{m=2}^{N} C_m v^{2m-2},$$

$$\frac{dv}{dt} = \beta v + (\alpha+1)\theta v,$$

in the second case. The basic idea now is to normalize the obtained vector fields locally, then integrate it and glue together the T-mappings, defined piecewise in the vicinity of the blowing-up circle. C^∞ orbital normal forms in the neighbourhood of the hyperbolic saddle are well understood [37]. They are given in (x, y)-coordinates by

$$\frac{dx}{dt} = x + \varepsilon(x^{pk+1}y^{qk} + ax^{2pk+1}y^{2qk}), \quad \frac{dy}{dt} = -\omega y,$$

where $\varepsilon = 0$, if ω is not resonant and $\varepsilon = \pm 1$, if $\omega = p/q$ is resonant, $p, q, k \in \mathbb{N}$.

Comparing the above normal forms, one finds that by an orbital C^∞ transformation they can be reduced to

$$\frac{du}{dt} = \alpha u, \quad \frac{d\eta}{dt} = -\eta + \varepsilon u^{2m-2}\eta^{2m}, \quad \alpha \in \left\{\frac{2m-1}{2m-2}\right\}_{m=2}^{N}$$
$$\frac{d\theta}{dt} = -\theta, \quad \frac{dv}{dt} = \beta v,$$

in the first case and to

$$\frac{du}{dt} = \alpha u, \quad \frac{d\eta}{dt} = \eta + \varepsilon u^{2m-2}\eta^{2m+1}, \quad \alpha \in \left\{\frac{m}{1-m}\right\}_{m=2}^{N}$$
$$\frac{d\theta}{dt} = -1, \quad \frac{dv}{dt} = 0.$$

in the second case.

First normal form. Step 1. α is not resonant. We have $\alpha \neq (2m-1)(2m-2)^{-1}$ and $\varepsilon = 0$. The transition mapping $T = T_\beta \circ T_t \circ T_\alpha$, where T_α and T_β are the transition mappings near the saddle sections, while T_t is the transition mapping along the saddle connection which joins the saddle sections. These normal forms are easily integrable, so that for T_α and T_β one obtains $T_\alpha : |x| \mapsto |x|^\alpha, T_\beta : |x| \mapsto |x|^{1/\beta}$. Mapping T_t is the transition function in the 'standard flow-box' so that $T_t : x \mapsto a_0 x + ...$, where $a_0 \neq 0$. Finally, we come to $T : |x| \mapsto a_0|x|^{\alpha/\beta} + ...$, which differs from the required normal form only by a positive multiplier.

First normal form. Step 2. α is resonant. In this case $\alpha = (2m-1)(2m-2)^{-1}$ and $\varepsilon = \pm 1$. This normal form is still integrable with the first integral given by

$$-\eta^{1-2m}u^{2-2m} + 2\varepsilon(m-1)\log|u| = C.$$

It follows now that the transition mapping T_α near the resonant saddle is $T_\alpha : |x| \mapsto |x|^{(2m-1)(2m-2)^{-1}}(1+...)$. The mappings T_β and T_t remain as before, so that in the resonant case one obtains

$$T : |x| \mapsto a_0|x|^{\frac{1}{\beta}(\frac{2m-1}{2m-2})} + ...$$

It is evident now that T does not differ from the required form if we substitute $\alpha = (2m-1)(2m-2)^{-1}$.

Second normal form. Step 1. α is not resonant. In this case one gets $\alpha \neq m(1-m)^{-1}$ and $\varepsilon = 0$. The transition mapping $T = T_{\alpha'} \circ T_t \circ T_\alpha$, where T_α and $T_{\alpha'}$ are transitions near the saddle, while T_t is a 'standard flow-box' transition mapping. It is easy to see that $T_\alpha : |x| \mapsto |x|^{1/\alpha}, T_{\alpha'} : |x| \mapsto |x|^\alpha$ and $T_t : x \mapsto a_0 x + ...$ One comes to $T : |x| \mapsto \widetilde{a}_0|x| + ...$, which coincides with the required form up to the choice of a positive constant $\widetilde{a}_0$.

Second normal form. Step 2. $\alpha = m(1-m)^{-1}$ is resonant. This implies that $\varepsilon = \pm 1$. In this case the form is integrable with the first integral

$$-\eta^{-2m}u^{2-2m} + 2\varepsilon(m-1)\log|u| = C.$$

As an easy consequence, one obtains the following estimates:

$$T_{\alpha'} : |x| \mapsto |x|^{\frac{m}{m-1}}(1+...), \qquad T_\alpha : |x| \mapsto |x|^{\frac{m-1}{m}}(1+...)$$

The mapping T_t is as before and we obtain $T : |x| \mapsto \widetilde{a}_0|x| + ...$ Note that in the resonant case T coincides, up to a multiplier, with the expression postulated by the lemma.

Now all cases are considered and Lemma 8.3.3 is proven. □

8.4 Addendum: Bendixson's Theorem

Theorem 8.4.1 (Bendixson) *Let $\mathcal{F}$ be orientable foliation in a disc D, given by a C^r vector field v, which is transversal to ∂D being directed inside D. Assume also $|Sing\ \mathcal{F}| < \infty$. Let $x \in D$ be a point in the D and let $\omega(x)$ be ω-limit set of the orbit, passing through x. Then either of the following possibilities is realized:*

(i) $\omega(x)$ is a singular point, or
(ii) $\omega(x)$ is a closed orbit, or

(iii) $\omega(x)$ *consists of singular points* $p_1, ..., p_n$ *and regular orbits* $\gamma \subset \omega(x)$ *for which* $\alpha(\gamma) = p_i$ *and* $\omega(\gamma) = p_j$.

Proof. (i) If $\omega(x)$ does not contain regular points, then it is reduced to the unique singular point. This follows from the fact that $\omega(x)$ is connected set and the set Sing v is finite.

(ii) Suppose that $\omega(x)$ does not contain singular points. Let $f \in \omega(x)$ be a regular point. Let us show that the orbit $O(y)$ of the point y, is closed. Indeed, let τ_y be a transversal segment, passing through the point y. Consider points $\{y_n\} = \tau_y \cap O(x)$, where $O(x)$ is the orbit of x and let (y_n, y_{n+1}) be a loop made by $O(x)$ around y (region bounded by this loop and the transversal is often called a *'Bendixson sack'*, since the inside orbits cannot leave it). Clearly, the sequence $\{y_n\}$ is monotone and converges to y.

Suppose to the contrary, that $O(y)$ is *not* closed. Then y^* is a point where $O(x)$ hits the first time τ_y (it is necessarily so, because $O(y)$ cannot leave the Bendixson sack). Thus far, one obtains a monotone sequence $\{y_n\}$, converging to $y^* \in \omega(x)$. It is a contradiction, because the monotone bounded sequence converges to the unique limit.

(iii) Suppose that $\omega(x)$ contains both regular and singular points. First let us show that if $p_1, p_2 \in \omega(x)$, then there exists a unique orbit $\gamma \subset \omega(x)$, such that $\alpha(\gamma) = p_1$ and $\omega(\gamma) = p_2$.

Indeed, suppose to the contrary, that there are two distinct orbits γ_1 and γ_2 such that $\alpha(\gamma_i) = p_1, \omega(\gamma_i) = p_2$. Take a small transversals τ_1 and τ_2 passing through regular points of γ_1 and γ_2, and let l be a part of the orbit $O(x)$ between points on τ_1 and τ_2, where $O(x)$ hits τ_1 and τ_2. Region Ω bounded by $l, \tau_1, \tau_2, \gamma_1$ and γ_2 is a 'sack' for $O(x)$, so that $\omega(x) \subset \Omega$.

It is a contradiction, since $p_1 \cap \omega(x) \neq \emptyset$. Let now $\gamma \subset \omega(x)$ be an orbit in $\omega(x)$. Then $\omega(\gamma)$ must be a singular point in $\omega(x)$. Indeed, if $\omega(\gamma)$ contains a regular point $y \in \omega(\gamma)$, then drawing a small transversal τ_y as in the item (ii), one obtains γ to be a closed orbit and $\omega(x) = \gamma$. This is a contradiction, since $\omega(x)$ contains also singular points.

Case $\alpha(\gamma)$ is similar. This concludes the proof of the Bendixson theorem. □

Corollary 8.4.1 *If* $\mathcal{F}$ *is an orientable foliation in the disc* D *with* ∂D *transversal to* $\mathcal{F}$, *and* $|Sing\mathcal{F}| < \infty$, *then* $\mathcal{F}$ *admits no recurrent leaves which are non trivial.*

(As it was noticed by Aranson and Grines [21], the claim $|\text{Sing}\mathcal{F}| < \infty$ cannot be omitted. Otherwise $\mathcal{F}$ can readily have non trivial recurrent leaves.)

Bibliographic Notes. The Denjoy theorem [88] appeared as a counterexample to the Poincaré conjecture on the existence of analytic flows on the torus with limit set of the Cantor type. Our exposition is close to the original one; cf. [88]. An example of C^1 flow on the torus with the limit set of the Cantor type has been constructed by Denjoy [88] and is exposed following Tamura [277]. The Cherry example of C^∞ smooth singular flow on the torus is due to Cherry [78] and is presented in the version of Dos Anjos [6]. Cherry's example remained unrevised until the work of M. Martens, P. Mendes, W. de Melo and S. van Strien [199] of 1990 where a class of the Cherry examples was considerably enlarged. J. C. Yoccoz considered the above problem from the point of view of one-dimensional dynamics; cf. [307]. He proved that if a diffeomorphism is analytic in the critical point then there are no discontinuous examples of such diffeomorphisms. S. Kh. Aranson, M. I. Malkin and E. V. Zhuzhoma establish sufficient conditions for the flows with 'oppositely directed' Cherry cells to be 'regular'; cf. [24].

9. Diffeomorphisms of Surfaces

9.1 A–diffeomorphisms

Basic sets[1] of the pseudo-Anosov (A–) diffeomorphisms and homeomorphisms generate an interesting class of non-orientable foliations. R. V. Plykin, and later on V. Z. Grines and A. Yu. Zhirov used such foliations to obtain a classification of A–diffeomorphisms on closed surfaces. As an introductory reading we recommend the classical survey of

S. Smale, Differentiable dynamical systems, Bull. Amer. Math. Soc. 73 (1967), 747-817.

9.1.1 Attractors of R. V. Plykin

The basic sets of A-diffeomorphisms of the two-dimensional sphere may be either of dimension 0 or 1. By an *attractor* (*repeller*) one understands a basic set B of the dimension 1. In the example below the construction of ϕ is carried out for the two-dimensional plane $\mathbb{R}^2$ which is next compactified to the sphere.

Example 9.1.1 (Plykin) *Let $\phi : \mathbb{R}^2 \to \mathbb{R}^2$ be a diffeomorphism of the plane which maps the 4-connected region D shown in Fig. 9.1 to itself. (Region D is composed from the three semi-discs with three smaller semi-discs deleted.) Let the image $\phi(D)$ of the region D is as it is shown in Fig. 9.1. If one fixes the generators $\gamma_1, \gamma_2, \gamma_3$ of the fundamental group $\pi_1(D)$ then ϕ acts on π_1 as follows:*

$$\begin{aligned}
\phi_*\gamma_1 &= \gamma_1^{-1}\gamma_3^{-1}\gamma_1\gamma_3\gamma_1, \\
\phi_*\gamma_2 &= \gamma_2\gamma_1^{-1}\gamma_3\gamma_1\gamma_2\gamma_1^{-1}\gamma_3^{-1}\gamma_1\gamma_2^{-1}, \\
\phi_*\gamma_3 &= \gamma_2\gamma_1^{-1}\gamma_3\gamma_1\gamma_2^{-1}.
\end{aligned}$$

[1] The term 'basic set' is an invention of S. Smale; see *Differentiable dynamical systems, Bull. Amer. Math. Soc. 73 (1967), 747-817.* Every diffeomorphism satisfying an axiom A can be spectrally decomposed into a finite union of basic sets. The classification of diffeomorphisms depends on the classification of basic sets, see the above cited paper of S. Smale.

The diffeomorphism ϕ can be prolonged to a structural stable diffeomorphism of the sphere, the non-wandering set of which consists of the hyperbolic one-dimensional attractor Λ and four point repellers. (Three of which correspond to the 'holes' of D and one corresponds to the 'exterior' of D.)

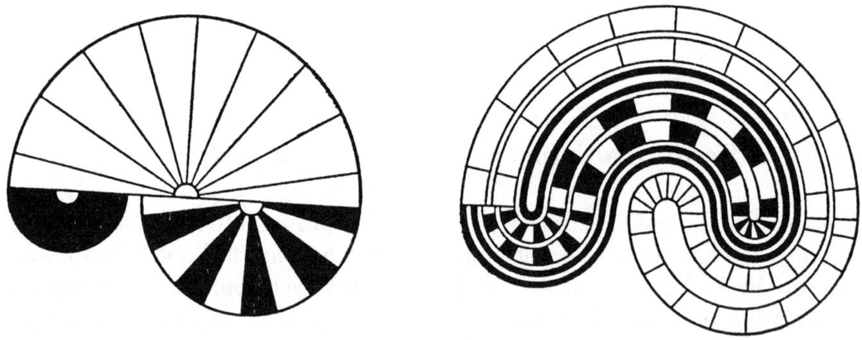

Fig. 9.1. The Plykin attractor

(It is easily recognizable that Fig. 9.1 'corresponds' to the simplest Rosenberg's labyrinth in the disc which possesses three thorns and one tripod type singularities, see Fig. 0.2.) Further we shall see, that the above representation of the diffeomorphisms of the sphere with one-dimensional basic sets is typical.

9.1.2 One–Dimensional Basic Sets on the Sphere

Theorem 9.1.1 (Plykin) *Let $\mathcal{K} = \bigvee_{i=1}^{N} S_i$ be an amalgamated product (bouquet) of circles S_i. Then every one-dimensional hyperbolic attractor Λ of the diffeomorphism $\phi : S^2 \to S^2$ with the cohomology rank N is a generalized Williams solenoid (see [246],[302] for the definition) induced by the self-mappings of the bouquet $\mathcal{K}$.*

Proof. We give an idea of the proof only, referring to [246] for the details. Let us choose a minimal natural number k such that there exists k-periodic point $y \in \Lambda$ and the mapping ϕ^k keeps the orientation of $W^s(y)$. Suppose that y is not a boundary point. Then there exists a segment $[x_0, x_1]^s$ containing y and such that x_0, x_1 belong to the unstable manifold of a boundary periodic point of the attractor Λ. Moreover, the segments $[x_0, x_1]^s, [x_0, x_1]^u$ intrsect only at the end points and make a loop, C_{x_0,x_i} which is called a *type I loop*. Extending the segment $[x_0, x_1]^u$ up to the points x_{-1}, x_2 where the extension meets the segment $[x_0, x_1]^s$, one comes to a pair of loops $C_{x_0,x_{-1}}, C_{x_1,x_2}$ of the type I. In the similar way one constructs a finite set of the type I loops $C_{y_{2i},y_{2i+1}} = [y_{2i}, y_{2i+1}]^u \cup [y_{2i}, y_{2i+1}]^s, i = 0, \dots, 2m + 1$, see [246]. The

irreducible region of the attractor Λ is a maximal connected region which is bounded by the above system of loops.
Let G_0 be an irreducible region which lies beyond D and has with D the common boundary. Splitting the boundary ∂G_0 into a finite number of components one may obtain a sequence of irreducible regions $G_j, j = 1, \ldots, l$, see [246]. There exists an integer m_j such that

$$\phi^{m_j} G_j \bigcap G_j \neq \emptyset.$$

It can be easily seen now that each region $D_j = \cup_{t \geq 0} \phi^{t m_j} G_j$ is a connectivity component of the set $S^2 \backslash \Lambda$ and therefore $t = n$. Let us show that the set

$$D \backslash \bigcup_{j=1}^{n} G_j,$$

can be represented as a union of the segment $[x_0, x_1]^s$ and n quadrangles each of which has no periodic boundary points in its interior.
Indeed, let us take for each of the segments $[x_0, x_1]^u, [x_0, x_{-1}]^u, [x_1, x_2]^u$ the nearest *main arc* [2] lying in the region G. Since at the interior of the quadrangles bounded by the above pairs and the pairs of segments of the stable manifold of $[x_0, x_1]^s$ there are no periodic boundary points, the interior of each constructed quadrangle cannot have common points with the irreducible regions G_j. Continuing the above process in the same way [246] we shall obtain the series of the quadrangles $\Pi_1, \Pi_2, \ldots, \Pi_n$ the number of which coincides with the number n of the main arcs of the region G. Now we set

$$\Phi = \bigcup_{i=1}^{n} \Pi_i.$$

Let us define a mapping $\psi : \Phi \to \mathcal{K}$, where $\mathcal{K} = \vee_{i=1}^{n} S_i$ is a bouquet of the circles S_i. To do so we identify all the points of the segments of stable manifold in Π_i and the segment $[x_0, x_1]^s$ is identified with the 'common' point of the bouquet $\mathcal{K}$. It can be easily verified that the above mapping ψ is defined correctly and is the required one. Theorem 9.1.1 is proved. □

9.1.3 Surfaces of Genus $g \geq 1$

Attractors of R. V. Plykin correspond to the quasiminimal sets of non-orientable foliation on a surface. Such sets are not spacious and therefore any surface may carry an arbitrary number of Plykin's attractors.
A classification scheme for the basic sets of A–diffeomorphisms with such attractors has been suggested in the series of papers of A. Yu. Zhirov, see for

---

[2] That is an arc $l_{0,j}^u = [t_0^j, t_{2k_j+1}]^u$ where $t_0^j, \ldots, t_N^j$ is the boundary partition of G_j.

example [311], [312]. On an arbitrary compact surface, the attractors can be codified so that one arrives to a constructive invariant of A–diffeomorphisms. Both the equivalence and realization problems for such invariants can be positively solved. We refer the interested reader to the original papers cited above.

9.2 Singularity Data

Let $\phi : M \to M$ be a pseudo-Anosov homeomorphism of an orientable surface M of genus g with n punctures. We always assume that ϕ has one-dimensional (hyperbolic) basic sets. It was shown earlier that ϕ determines the stable and unstable transverse measured foliations $\mathcal{F}$. These foliations have the common p-pronged saddle singularities at the finite number of points $x_1, \ldots, x_n$. H. Masur and J. Smillie in [201] established a list of the 'local' data restrictions on an arbitrary measured foliation $\mathcal{F}$ to be foliation given by one-dimensional basic sets of a pseudo-Anosov homeomorphism.
Let ϕ be a pseudo-Anosov homeomorphism defined as before and let p_i be the number of prongs of the singularity at the point $x_i \in M$. Let ε be either $+1$ or -1 depending on whether the foliation $\mathcal{F}$ is orientable or is non-orientable.

Definition 9.2.1 *The $n+1$ dimensional vector $(p_1, p_2, \ldots, p_n; \varepsilon)$ where p_i is the number of prongs emerging from the singular point $x_i \in$ Sing $\mathcal{F}$ and ε is the orientability of $\mathcal{F}$, is called a singularity data of the foliation $\mathcal{F}$.*

The following theorem due to Masur and Smillie [201] [3] establishes the restrictions on the data vector which describe the 'impossible' vectors.

Theorem 9.2.1 (Masur-Smillie) *Let $\phi : M \to M$ be a pseudo-Anosov homeomorphism of an orientable surface M of genus g. Let $\mathcal{F}$ be a measured foliation defined by the homeomorphism ϕ. Then the singularity data $(p_1, p_2, \ldots, p_n, \varepsilon)$ of $\mathcal{F}$ obeys the following restrictions:*

(i) $\sum_{i=1}^{n}(p_i - 2) = 4(g-1)$,
(ii) $\varepsilon = -1$ *if and only if* p_i *is odd,*
(iii) The singularity data $(6; -1), (3,5; -1), (1,3; -1)$ *and* $(\emptyset; -1)$ *are prohibited.*

Proof. A pair of transverse measured foliations with the singularity data $(p_1, \ldots, p_n; \varepsilon)$ determines a meromorphic quadratic differential $\phi(z)dz^2$ with simple poles having the singularity data $(k_1, \ldots, k_n; \varepsilon)$ where $k_i \geq -1$ is the order of the singular points of ϕ and ε is the orientability index of the foliation $\mathcal{F}$ defined by the horizontal trajectories of ϕ. Theorem 9.2.1 is an implication of the following basic statement.

[3] In fact Masur and Smillie have proved more by establishing the necessary and sufficient conditions for realization of the data vector.

Lemma 9.2.1 (Masur-Smillie) *Let $\phi(z)dz^2$ of a meromorphic quadratic differential defined on a compact surface M of genus g. Let $(k_1, k_2, \ldots, k_n; \varepsilon)$ be the singularity data of $\phi(z)dz^2$. Let $\sigma = \sum_{i=1}^{n} k_n$ be the sum of the orders of its singular points. Then the singularity data vector must obey the following list of restrictions:*

(i) $\sigma \equiv 0$ *Mod* 4 *and* $\sigma \geq -4$;
(ii) $\varepsilon = -1$ *if any* k_i *is odd;*
(iii) The singularity data $(4; -1), (1, 3; -1), (-1, 1; -1)$ *and* $(\emptyset; -1)$ *are prohibited.*

Proof of Lemma 9.2.1. Conditions (i) and (ii) are simple implications of the Index Theorem applied to the singular points of the quadratic differential $\phi(z)dz^2$. The main difficulty is the case (iii) to which we restrict our attention. We shall consider the restrictions (iii) consequently.

(a) Let us suppose to the contrary that the singularity data $(4; -1)$ and $(3, 1; -1)$ can be realized for a quadratic differential $\phi(z)dz^2$. Then M must be a surface of genus 2. Let us endow M with a Riemannian structure. Every Riemann surface of genus 2 is hyperelliptic. It means that M can be represented as a two-fold ramified covering of the Riemann sphere. Denote by z the corresponding covering map and let $p_1, \ldots, p_6$ be the branch points of M. Assume also that $z(p_i) \neq \infty$. Then there is a map w of M to the Riemann sphere which satisfies the equation:

$$w^2 = \Pi_{i=1}^{6} (z - z(p_i)) .$$

Moreover, there exists an involution $\theta : M \to M$ which satisfies the conditions $z \circ \theta = z$ and $w \circ \theta = -w$ and which fixes the points p_i. It may be seen [101] that the abelian differentials

$$\omega_1 = \frac{dz}{w}, \qquad \text{and} \qquad \omega_2 = \frac{zdz}{w},$$

form a basis for the vector space of the abelian differentials on M. Clearly, $\theta_*(\omega_i) = -\omega_i$. On the other hand the quadratic differentials

$$\omega_1^2, \qquad \omega_2^2 \qquad \text{and} \qquad \omega_1\omega_2,$$

form a basis for the vector space of quadratic differentials on M. Therefore every quadratic differential $\phi(z)dz^2$ satisfies the equality $\theta_*(\phi) = \phi$.
Suppose that $\phi(z)dz^2$ is a quadratic differential with a single zero of the order 4 at the point $p \in M$. We have to show that ϕ must realize the data $(4; +1)$ rather then $(4; -1)$.
Indeed, since $\theta_*(\phi) = \phi$ we have $\theta(p) = p$. Therefore p is one the ramification points p_i. Each p_i is the Weierstrass point of M, see [101]. It means that for each p_i there exists an abelian differential on M with a single zero of the order 2 at p_i. Let ω_p be such a differential at p. The quotient ϕ/ω_p^2

is therefore a meromorphic function on M without zeros or poles. Therefore $\phi = C$ is a constant function. Thus $\phi = C\omega_p^2$ and ϕ determines the orientable foliation $(4; +1)$.

Next we have to show that there is no quadratic differential with a zero of order 3 at a point $p_1 \in M$ and a zero of order 1 at a point $p_2 \in M$. If such a $\phi(z)dz^2$ exists we again have $\theta_*(\phi) = \phi$ and therefore $\theta(p_i) = p_i$ and so the p_i's are the Weierstrass points. Then the function $\phi/\omega_{p_1}^2$ is meromorphic with a simple pole at p_1 and a simple zero at p_2. Thus $\phi/\omega_{p_1}^2$ defines a degree 1 covering mapping of M to the Riemann sphere. This yields a contradiction.

(b) Suppose that $\phi(z)dz^2$ is a quadratic differential on a torus T^2 which realizes the singularity data $(-1, 1; -1)$. There exists a quadratic differential $\psi(z)dz^2$ on T^2 which defines a flat structure [4] without singularities. (In other words, the corresponding singularity data is $(\emptyset; +1)$.) Then ϕ/ψ defines a degree 1 covering of T^2 to the Riemann sphere, which is a contradiction.

(c) Suppose finally that $\phi(z)dz^2$ defines the data $(\emptyset; -1)$ for a foliation on torus T^2. Here again, the quotient ϕ/ψ must be a constant, which is impossible. Lemma 9.2.1 is proved. □

Proof of Theorem 9.2.1. The singularity data $(p_1, p_2, \ldots, p_n; \varepsilon)$ and $(k_1, k_2, \ldots, k_n; \varepsilon)$ are connected by the formula $k_i = p_i - 2$. Moreover, every quadratic differential may be recovered from two transversal foliations defined by the pseudo-Anosov diffeomorphisms. Thus the items (i),(ii) and (iii) of the theorem follow. □

9.3 Isotopy Classes of Diffeomorphisms

Let M be a surface, which for brevity is supposed to be orientable, possibly with a boundary ∂M. Denote by $\mathcal{J}(M)$ the set of isotopy classes of simple closed curves on M, not bounding a disc and not isotopic to ∂M. Denote by $\mathcal{PJ}(M)$ a projective completion of $\mathcal{J}(M)$, taken in a natural 'nearness' topology [5] in the space of curves. $\mathcal{PJ}(M)$ is homeomorphic to a sphere $S^{6g+2b-7}$, where g and b are genus and number of the boundary components of M. One can think of $\mathcal{J}(M)$ consisting of a dense set of 'rational' points on this sphere.
$\mathcal{PJ}(M)$ has two other interpretations. First, it is the space of all equivalence classes of measured foliations on M. Second, it forms a boundary (*Thurston boundary*) for the Teichmuller space of M, to which the action of the group of diffeomorphisms $f : M \to M$ extends.

[4] A Riemannian structure with zero Gaussian curvature.
[5] This topology reflects how close are two curves in the sense of their 'general' direction on M.

(i) *Measured foliations.* Let $\mathcal{F}$ be a measured foliation and let $a \in \alpha$ be a closed curve on M, representing the isotopy class $\alpha \in \mathcal{J}$. One defines $\int_a \mathcal{F}$ to be a total variation of the y-coordinates of a, as measured locally with respect to any chart. If $\beta \in \mathcal{J}$, one sets

$$i(\mathcal{F}, \beta) = \inf_{b \in \beta} \int_b \mathcal{F},$$

where $b \in \beta$ is a simple closed curve, representing β. Two measured foliations $\mathcal{F}_1$ and $\mathcal{F}_2$ are *measure equivalent*, if for all $\beta \in \mathcal{J}, i(\mathcal{F}_1, \beta) = i(\mathcal{F}_2, \beta)$. Two measured foliations $\mathcal{F}_1$ and $\mathcal{F}_2$ are *projectively equivalent*, if there is a $\lambda \in \mathbb{R}$, such that for all $\alpha \in \mathcal{J}$, it holds $i(\mathcal{F}_1, \alpha) = \lambda i(\mathcal{F}_2, \alpha)$.

Proposition 9.3.1 *For every measured foliation $\mathcal{F}$ there is a $\beta \in \mathcal{J}$ such that $i(\mathcal{F}, \beta) \neq 0$.*

Denote by $\mathcal{M}$ the set of measured foliations on M. The map $i : \mathcal{M} \times \mathcal{J} \to \mathbb{R}$ defines a map $i_* : \mathcal{M} \to \mathbb{R}^{\mathcal{J}}$ from the set of measured foliations to $\mathbb{R}$. One defines projective equivalence classes of measured foliations,

$$\mathcal{PF} = \{p \circ i_*(\mathcal{F}) | \mathcal{F} \in \mathcal{M}\}$$

where $p : \mathbb{R} \to \mathbb{R}$ is projective equivalence on $\mathbb{R}$.

Lemma 9.3.1 *$\mathcal{PF} = \mathcal{PJ}$. In other words, the topological sphere $\mathcal{PJ}(M)$ given by the projective completion of the isotopy classes of simple closed curves on M, can be identified with the set of all projective equivalence classes of measured foliations on M.*

(ii) *Teichmuller space.* Let M be a surface of genus g, where g for the sake of simplicity is supposed ≥ 2. *Teichmuller space* is the space of conformal structures on the topological surface M, where two are equivalent if there is a conformal map between them which is homotopic to the identity. This space is denoted by $\mathcal{T}(M)$ and is identified with the set of isotopy classes of Riemannian metrics on M of curvature -1 (such metric is called *hyperbolic*). If $\gamma \in \mathcal{T}(M)$ is a Riemannian metric on M and $\alpha \in \mathcal{J}(M)$ is an isotopy class of simple closed curves on M, one defines $i(\gamma, \alpha)$ to be the infimum of the length of a simple closed curve $a \in \alpha$, representing α. (It is well-known result, that there is a unique closed geodesic representing α).
From this viewpoint two hyperbolic metrics g_1 and g_2 are isotopic, if for every $\alpha \in \mathcal{J}, i(g_1, \alpha) = i(g_2, \alpha)$. $\mathcal{T}(M)$ is endowed with the minimal topology such that $i(\gamma, \alpha)$ is continuous. As before, the map $i : \mathcal{T} \times \mathcal{J} \to \mathbb{R}$ defines a map $i_* : \mathcal{T} \to \mathbb{R}^{\mathcal{J}}$, and let now $p \circ i_* : \mathcal{T}(M) \to \mathrm{P}(\mathbb{R}^{\mathcal{J}})$ be a continuous projective completion of the Teichmuller space.

Lemma 9.3.2 *The closure $\overline{p \circ i_*(\mathcal{T}(M))} = p \circ i_*(\mathcal{T}(M)) \cup \mathcal{PJ}$. Moreover, $\mathcal{T}(M) \cup \mathcal{PJ}(M)$ can be endowed with a natural topology, in which it is homeomorphic to a closed disc $D^{6g+2b-6}$, where again, g and b are genus and number of the boundary components of M.*

The closure $\mathcal{T}(M) \cup \mathcal{PJ}(M)$ of the Teichmuller space $\mathcal{T}(M)$ is denoted by $\overline{\mathcal{T}(M)}$, and its boundary is called *Thurston boundary* of the Teichmuller space. Attaching Thurston boundary to $\mathcal{T}(M)$ is called *Thurston compactification* of $\mathcal{T}(M)$.
It is known that any diffeomorphism $f : M \to M$ induces an isometry of M. The group of isometries induced by the group of diffeomorphisms f, is called *modular group* and is denoted by Mod f. The main idea of Thurston's theory is that the action of Mod f can be continuously extended to $\overline{\mathcal{T}(M)}$. This allows to apply the Brower fixed point theorem to the compact topological space $\overline{\mathcal{T}(M)}$, and obtain the following

Corollary 9.3.1 *Let $f : M \to M$ be a diffeomorphism of the surface M. Then, either*

(i) f fixes an element of the Teichmuller space, or
(ii) f fixes a projective class of measured foliations.

In case (i) there exists a Riemannian metric g, such that diffeomorphism f is isotopic to f', where f' has a finite order. [6]

In case (ii) further classification is possible. Diffeomorphism $f : M \to M$ is called *reducible* by Γ, if Γ is a system of disjoint simple closed curves $\Gamma_1, ..., \Gamma_n$, such that Γ_i belongs to the distinct elements of the space $\mathcal{J}(M)$ and is invariant by the action of f (however, Γ_i can be permuted). In case f is reducible, one cut M along Γ_i obtaining a simpler surface (disconnected, in general). The following statement (whose proof is omitted) is a central result on the isotopy classification of surface diffeomorphisms, see [100], [280].

Theorem 9.3.1 (Thurston) *Any diffeomorphism $f : M \to M$ of a surface M is isotopic to a diffeomorphism f', such that either*

(i) f' has finite order, or
(ii) f' is pseudo-Anosov diffeomorphism, or
(iii) f' is reducible by a system of curves Γ surrounded by small tubular neighborhoods $N(\Gamma)$, such that on $M \backslash N(\Gamma)$ f' satisfies either (i) or (ii).

Proof. Follows from Lemma 9.3.2 and Corollary 9.3.1. □

Bibliographic Notes. Anosov diffeomorphisms have been introduced in [7] under a different name. They were called the *U-diffeomorphisms* derived from the Russian *Usloviya* meaning the *Conditions*. Pseudo-Anosov diffeomorphisms were studied by Laudenbach [100] as one of isotopy classes of the surface diffeomorphisms. The one-dimensional basic sets of pseudo-Anosov homeomorphisms have been investigated by R. V. Plykin [245],[246],

[6] That is there exists $n \geq 1$ such that $(f')^n = \mathrm{Id}$.

A. Yu. Zhirov [311],[312], H. Masur and J. Smillie [201]. Thurston's results on the isotopy classification of diffeomorphisms are exposed accordingly with the classical work [100]. For further development of Thurston's theory, see the bibliography to [280].

10. C*-Algebras

10.1 Irrational Rotation Algebra

C^*-algebras are the (non-commutative) Banach algebras with an involution. Such algebras appear naturally in a dynamical context; see Pedersen [235]. The ones we shall be considering will either be irrational rotation algebras or algebras generated by interval exchange transformations of the unit circle. (The latter we call the *Artin rotation algebras*; they are canonically isomorphic to the C^*-algebras of minimal flows on a surface of genus $g > 1$.)

We start with the irrational rotation algebras A_α; see Pedersen [235]. Let α be an irrational number and let $\phi : S^1 \to S^1$ be a rotation through the angle $2\pi\alpha$ on the unit circle S^1. Then the group of integers $\mathbb{Z}$ acts on S^1 as a transformation group by mean of powers of ϕ. We can form the corresponding *crossed product* C^*-algebra:

$$A_\alpha = C(S^1) \rtimes_\phi \mathbb{Z}, \tag{10.1}$$

where $C(S^1)$ denotes the algebra of continuous complex-valued functions on S^1.

Fix an irrational number α. The canonical model for A_α is a norm-closed C^*-algebra generated by two unitary operators in the Hilbert space $L^2(S^1)$:

$$Uf(t) = z(t)f(t), \qquad Vf(t) = f(t - \alpha),$$

that is, multiplication by the unimodular function $z(t)$ and the rotation operator, cf. Davidson [83]. It can easily be derived that $UV = e^{2\pi i\alpha}VU$. A_α is a simple , unital C^*-algebra with a unique normalized trace; see Pedersen [235].

10.1.1 Dimension Groups

General references to the subject are Effros [95] and Davidson [83]. An *AF (approximately finite-dimensional)* algebra is defined to be the norm closure of an ascending sequence of finite dimensional algebras M_n; M_n is an algebra of $n \times n$ matrices with entries in $\mathbb{C}$. The index $n = (n_1, \ldots, n_k)$ will represent a *multimatrix* algebra $M_n = M_{n_1} \oplus \ldots \oplus M_{n_k}$. Let $M_1 \xrightarrow{\varphi_1} M_2 \xrightarrow{\varphi_2} \ldots$ be a chain of algebras and homomorphisms. The set-theoretic limit $A = \lim M_n$ has a natural algebraic structure given by the formula $a_m + b_k \to a + b$, where $a_m \to a, b_k \to b$ for the sequences $a_m \in M_m, b_k \in M_k$. The homeomorphisms of the above (multimatrix) algebras admit a canonical description; see Effros [95]. Suppose that $p, q \in \mathbb{N}$ and $k \in \mathbb{Z}^+$ are such numbers that $kq \leq p$. Let us define a homomorphism $\varphi : M_q \to M_p$ by the formula

$$a \longmapsto \underbrace{a \oplus \ldots \oplus a}_{k} \oplus 0_h,$$

where $p = kq + h$. More generally, if $q = (q_1, \ldots, q_s), p = (p_1, \ldots, p_r)$ are vectors in $\mathbb{N}^s, \mathbb{N}^r$, respectively, and $\Phi = \phi_{ij}$ is a $r \times s$ matrix with the entries in $\mathbb{Z}^+$ such that $\Phi(q) \leq p$, then the homomorphism φ is defined by the formula:

$$\begin{aligned} a_1 \oplus \ldots \oplus a_s \longrightarrow & \underbrace{(a_1 \oplus a_1 \oplus \ldots)}_{\phi_{11}} \oplus \underbrace{(a_2 \oplus a_2 \oplus \ldots)}_{\phi_{12}} \oplus \ldots \oplus 0_{h_1} \\ \oplus & \underbrace{(a_1 \oplus a_1 \oplus \ldots)}_{\phi_{21}} \oplus \underbrace{(a_2 \oplus a_2 \oplus \ldots)}_{\phi_{22}} \oplus \ldots \oplus 0_{h_2} \oplus \ldots \end{aligned}$$

where $\Phi(q) + h = p$. We say that φ is a *canonical* homomorphism between M_p and M_q. By an appropriate choice of the matrix units any homomorphism $\varphi : M_q \to M_p$ can be made canonical; see Effros [95]. It was noticed by Elliott [97] that a canonical system of homomorphisms $M_{p_1} \xrightarrow{\varphi_1} M_{p_2} \xrightarrow{\varphi_2} \ldots$ admits further simplification. He proved that φ_i's determine (and are determined by) positive [1] homomorphisms $\varphi : \mathbb{Z}^s \to \mathbb{Z}^r$ on the ordered lattice of integer.

Definition 10.1.1 *By a dimension group P of a canonical system of homomorphisms associated to an AF C^*-algebra, one understands the inductive limit of the sequence $\mathbb{Z}^{r_1} \xrightarrow{\varphi_1} \mathbb{Z}^{r_2} \xrightarrow{\varphi_2} \ldots$ of (partially) ordered abelian groups.*

We wish to apply dimension groups to the study of a concrete object: an irrational rotation algebra A_α. A formal definition of this algebra can be found in [83]. Let us introduce the class of ordered groups on the lattice $\mathbb{Z}^2$ endowed with positive cone $K^+ = \{(p, q) \in \mathbb{Z}^2 : p\alpha + q \geq 0\}, \quad \alpha \in \mathbb{R}^+ \backslash \mathbb{Q}$. It can be verified directly that K^+ induces a total ordering on $\mathbb{Z}^2$; indeed, since α is irrational for each $(p, q) \in \mathbb{Z}^2$ either $p\alpha + q \geq 0$ or $p\alpha + q < 0$. The ordered group $P_\alpha = (\mathbb{Z}^2, K^+)$ plays a key role in classification of the irrational rotation algebras A_α's.

Lemma 10.1.1 *Let A_α be an irrational rotation C^*-algebra. Let $\lim_{n\to\infty}(\mathbb{Z}^{r_n}, \varphi_n)$ be the dimension group associated to A_α. Then the latter is order isomorphic to P_α.*

Proof. Effros and Shen[96]. □

[1] A homomorphism φ is *positive* if $\varphi(a) \geq 0$ whenever $a \geq 0$ is a non-negative vector.

P_α 'measures' the 'sizes' of projections (or 'dimensions') in the algebra A_α. The *Unimodular Conjecture* suggests that P_α must be a limit of the direct sequence of positive unimodular operators φ_n acting on the lattice $\mathbb{Z}^2$: $\mathbb{Z}^2 \xrightarrow{\varphi_1} \mathbb{Z}^2 \xrightarrow{\varphi_2} \ldots$, where $\varphi_n = \begin{pmatrix} a & b \\ c & d \end{pmatrix}, a, b, c, d \in \mathbb{Z}^+, ad - bc = \pm 1$. For P_α [2] the Unimodular Conjecture is true; see Effros [95].

10.1.2 Continued Fractions

If $a_0, a_1, a_2, \ldots$ is a (finite or infinite) sequence of integers which are positive with the possible exception of a_0, then the aggregate

$$\alpha = a_0 + \cfrac{1}{a_1 + \cfrac{1}{a_2 + \ldots}} \stackrel{\text{def}}{=} [a_0, a_1, a_2, \ldots]. \tag{10.2}$$

is called a (regular) *continued fraction.* For *finite* continued fractions the following two classical statements are valid.

Lemma 10.1.2 *Every rational number p/q is represented by a finite continued fraction $[a_0, \ldots, a_n]$ in one and only one way, provided $a_n \neq 1$. If $a_n = 1$ then p/q has two (and only two) equivalent representations $[a_0, a_1, \ldots, a_n] = [a_0, a_1, \ldots, a_{n-1}, a_n - 1, 1]$.*

Proof. See Hermann [144]. □

Lemma 10.1.3 (Minkowski) *Suppose that $\begin{pmatrix} a & b \\ c & d \end{pmatrix}$ is a unimodular matrix with positive integer entries. Then there exist non-negative integers $a_0, a_1, \ldots, a_n$, such that $\begin{pmatrix} a & b \\ c & d \end{pmatrix} = \begin{pmatrix} a_n & 1 \\ 1 & 0 \end{pmatrix} \cdots \begin{pmatrix} a_0 & 1 \\ 1 & 0 \end{pmatrix}$. Moreover, if $b > d$, then $a/b = [a_0, \ldots, a_n]$.*

Proof. See Minkowski [204], p.159. □

In the case of an *infinite* continued fraction, the rationals $A_n/B_n = [a_0, \ldots, a_n]$ can be regarded as 'convergents' to a real number $\xi \in \mathbb{R}$. The following 'existence and uniqueness' lemma is true. The proof can be found in

O. Perron, Die Lehre von den Kettenbruchen, Teubner, Leipzig, 1929. Reprinted by Chelsea, N.Y., 1950.

Lemma 10.1.4 *Every infinite continued fraction converges to an irrational number. Conversely, every irrational number is decomposable in one and only one way into a continued fraction which is necessarily infinite.*

[2] As well as for the wider class of all finitely generated dimension groups.

Two irrational numbers ξ_1 and ξ_2 are said to be equivalent, $\xi_1 \sim \xi_2$, if they are modular equivalent, i.e., there exist integers a, b, c, d obeying the condition $ad - bc = \pm 1$ such that $\xi_2 = (a\xi_1 + b)/(c\xi_1 + d)$. The statement below reflects the principal property of continued fractions. (In fact, it makes the apparatus of continued fractions applicable to the key questions of hyperbolic geometry; Artin [38].)

Lemma 10.1.5 *Let $\xi = [a_0, a_1, \ldots]$ and $\eta = [b_0, b_1, \ldots]$ be two irrational numbers given by their continued fraction expansion. Then $\xi \sim \eta$ if and only if their continued fraction expansion has the form $\xi = [a_0, a_1, \ldots, a_{n-1}, g_0, g_1, \ldots], \eta = [b_0, b_1, \ldots, a_{m-1}, g_0, g_1, \ldots]$. (In other words, their continued fraction expansions coincide everywhere except possibly at a finite number of terms.)*

Proof. See the treatise of O. Perron indicated above. □

10.1.3 Effros-Shen's Theorem

Theorem 10.1.1 *Let A_α be an irrational rotation C^*-algebra. Suppose that the real number α has a continued fraction expansion $\alpha = [a_0, a_1, a_2, \ldots]$. Let φ_n be a composition of isometries of the lattice $\mathbb{Z}^2 \subset \mathbb{R}^2$: $\varphi_n = \begin{pmatrix} a_0 & 1 \\ 1 & 0 \end{pmatrix} \cdots \begin{pmatrix} a_n & 1 \\ 1 & 0 \end{pmatrix}$. Then the dimension group P_α of the irrational rotation algebra A_α is isomorphic to the direct limit $P_\alpha = \lim_{n\to\infty}(\mathbb{Z}^2, \varphi_n)$. Moreover, if $\alpha = [a_0, a_1, \ldots]$ and $\beta = [b_0, b_1, \ldots]$ are two irrational numbers, then A_α and A_β are isomorphic if and only if $a_{m+k} = b_m$ for an integer $k \in \mathbb{Z}$. (In other words, the irrational numbers α and β are modular equivalent: $\beta = (a\alpha + b)/(c\alpha + d)$, $\quad ad - bc = \pm 1$, where $a, b, c, d \in \mathbb{Z}$ are integers.)*

Proof. See Effros and Shen [96]; we wish to give an outline of the proof. Let us consider the commutative diagram:

$$
\begin{array}{ccccccc}
\mathbb{Z}^2 & \xrightarrow{\varphi_{02}} & \mathbb{Z}^2 & \xrightarrow{\varphi_{24}} & \mathbb{Z}^2 & \longrightarrow & \\
\downarrow \theta_0 & & \downarrow \theta_1 & & \downarrow \theta_2 & & \\
(\mathbb{Z}^2, P_\alpha) & \xrightarrow{\mathrm{Id}} & (\mathbb{Z}^2, P_\alpha) & \xrightarrow{\mathrm{Id}} & (\mathbb{Z}^2, P_\alpha) & \longrightarrow &
\end{array}
$$

where $\theta_0 = Id$ and $\theta_m = \varphi_{0,2m}^{-1}$. Suppose that $\varphi_{m,m+1}$ is as stated in the theorem. Elementary calculations with matrices and continued fractions yield (see Minkowski's Lemma)

$$\varphi_{0,2m} = \begin{pmatrix} a_{2m-1} & 1 \\ 1 & 0 \end{pmatrix} \cdots \begin{pmatrix} a_0 & 1 \\ 1 & 0 \end{pmatrix} = \begin{pmatrix} A_{2m-1} & B_{2m-1} \\ A_{2m-2} & B_{2m-2} \end{pmatrix},$$

where $m \geq 1$ and $\frac{A_{2m-2}}{B_{2m-2}}, \frac{A_{2m-1}}{B_{2m-1}}$ are the convergents of α. First, let us notice that the θ_m are positive (i.e. their entries are positive integers); see Effros-Shen. Let us show that the θ_m are also order-preserving isomorphisms of $\mathbb{Z}^2$. For this, it is enough to show that if $(\tau_1, \tau_2) \in P_\alpha$ then $(\tau_1, \tau_2) \in \theta_m(\mathbb{Z}^2_+)$. Indeed,

$$\theta_m^{-1} = \varphi_{0,2m}(\tau_1, \tau_2) = (A_{2m-1}\tau_1 + B_{2m-1}\tau_2, A_{2m-2}\tau_1 + B_{2m-2}\tau_2) \in \mathbb{Z}^2_+.$$

This finishes the proof of the first part of the theorem. The second part is a consequence of the modular property of continued fractions. □

10.1.4 Projections of A_α

Let $C(S^1)$ be a space of the complex-valued continuous functions on the unit circle S^1. For convenience we shall view the elements of $C(S^1)$ as continuous functions on the real line $\mathbb{R}$ periodic of period 1. Therefore ϕ becomes the shift

$$\phi(f)(t) = f(t - \alpha),$$

for $f \in C(S^1)$ and $t \in \mathbb{R}$. Notice that for the operators ϕ and M_f, $\phi \circ M_f = M_{\phi(f)} \circ \phi$. An operator $p \in A_\alpha$ is called a *projection* if $p^2 = p = p*$. The following theorem concerns the connection between the traces of the algebra A_α and its projections p.

Theorem 10.1.2 (Rieffel) *For each $\beta \equiv \alpha$ Mod $\mathbb{Q}$ lying in the unit interval $[0, 1]$ there exists a projection $p \in A_\alpha$ such that $\tau(p) = \beta$.*

Proof. An element of A_α is *supported* on $\{-1, 0, 1\}$ if it has the form

$$M_h \circ \phi^{-1} + M_f + M_g \circ \phi,$$

for $f, g, h \in C(S^1)$. The theorem follows from the statement below.

Lemma 10.1.6 *For each $\beta \in (\mathbb{Z} + \alpha\mathbb{Z}) \cap [0, 1]$ there is a projection $p \in A_\alpha$, supported on $\{-1, 0, 1\}$ such that $\tau(p) = \beta$.*

Proof of lemma. Let p be a projection supported on $\{-1, 0, 1\}$ and expressed in terms of f, g, h as before. Since p is self-adjoint the function f must be real-valued, and $h = \phi^*(\bar{g})$. Because p is idempotent one finds that

(1) $g(t)g(t - \alpha) = 0$;
(2) $g(t)[1 - f(t) - f(t - \alpha)] = 0$;
(3) $f(t)[1 - f(t)] = |g(t)|^2 + |g(t + \alpha)|^2$,

for each $t \in \mathbb{R}$. Conversely, if $f, g \in C(S^1)$ satisfy (1)-(3) and $h = \phi^*(\bar{g})$, then the corresponding element of A_α will be a projection. Since translation by α on $C(S^1)$ is equivalent to translation by $\alpha - [\alpha]$ we may assume $\alpha \in [0, 1]$. Furthermore, since ϕ^* is translation by $1 - \alpha$ (so that $A_\alpha \cong A_{1-\alpha}$) we can assume that $\alpha \in [0, 1/2]$.

Let us show how to construct a projection p such that $\tau(p) = \alpha$. For this, let f be 'almost' the characteristic function $[0, \alpha]$, but 'rounded' at the 'ends' in a careful way. Namely, equation (3) implies that if $f(t)$ is different from 0 and 1, then either $g(t)$ or $g(t+\alpha)$ is non-zero. Then the equation (2) says that if $g(t) \neq 0$ then $f(t) + f(t-\alpha) = 1$. Choose any $\varepsilon > 0$ such that $\varepsilon < \alpha$ and $\alpha + \varepsilon < 1/2$. On $[0, \varepsilon]$, let f be an arbitrary continuous function with values in $[0,1]$ and with $f(0) = 0, f(\varepsilon) = 1$. On $[\alpha, \alpha+\varepsilon]$ define f by $f(t) = 1 - f(t-\alpha)$, and on $[\varepsilon, \alpha]$ and $[\alpha+\varepsilon, 1]$ let f take the values 1 and 0, respectively. Finally, on $[\alpha, \alpha+\varepsilon]$ we set

$$g(t) = \sqrt{f(t)(1-f(t))},$$

and let g have value 0 elsewhere on $[0,1]$. It is a straightforward exercise to check that f and g satisfy (1)-(3) and define a projection with trace

$$\int_0^1 f(t)dt = \alpha.$$

In the general case, note that for any positive integer m the algebra $C(S^1)$ contains the subalgebra $C_m(S^1)$ of continuous $1/m$-periodic functions on $\mathbb{R}$. On $C_m(S^1)$ the shift by α looks like the shift by $\{m\alpha\} = m\alpha - [m\alpha]$. This means that $A_{\{m\alpha\}}$ is embedded as a subalgebra of A_α having the same identity element. The restriction to $A_{\{m\alpha\}}$ of the trace on A_α will be the trace on $A_{\{m\alpha\}}$. The same holds true for the projection of the trace. Moreover, the elements of $A_{\{m\alpha\}}$ which are supported on $\{-1, 0, 1\}$ will also be supported there when viewed as elements of the algebra A_α. (The case when m is a negative integer is treated similarly.) These arguments complete the proof of Lemma 8.3.1 and Theorem 8.3.1. □

Corollary 10.1.1 *The range of the trace on projections in A_α coincides with the set $(\mathbb{Z} + \alpha\mathbb{Z}) \cap [0,1]$.*

The above theorem enables us to study the isomorphism problem for the A_α. Specifically, we have the following statement.

Theorem 10.1.3 (Rieffel) *If α and β are irrational numbers in the interval $[0, 1/2]$, and if A_α and A_β are isomorphic, then $\alpha = \beta$. If α is any irrational number with fractional part $\{\alpha\}$, let $\beta = \{\alpha\}$ or $\beta = 1 - \{\alpha\}$ depending on which one lies in $[0, 1/2]$. Then A_α and A_β are isomorphic.*

Proof. If $(\mathbb{Z} + \alpha\mathbb{Z}) \cap [0,1] = (\mathbb{Z} + \beta\mathbb{Z}) \cap [0,1]$ and $\alpha, \beta \in [0,1]$, then after quick calculations one obtains $\beta = \alpha$ or $\beta = 1 - \alpha$. Since $A_\alpha \cong A_{1-\alpha}$ and, more generally, $A_\alpha \cong A_{\alpha+n}$ for all $n \in \mathbb{Z}$, we arrive at the conclusion of the theorem.

10.1.5 Morita Equivalence

Let A_α, A_β be a pair of C^*-algebras. A_α and A_β are said to be *Morita equivalent* if their categories of left modules are equivalent [251]. (As we shall see later this definition provides a useful equivalence relation on algebras which is considerably coarser than that of isomorphism.) The following theorem is a criterion for the equivalence of the irrational rotation C^*-algebras.

Theorem 10.1.4 *The algebras A_α and A_β are Morita equivalent if and only if α and β are equivalent by the action of the modular group $SL_2(\mathbb{Z})$ on the irrational numbers by linear-fractional transformations.*

Proof. Necessary conditions. First some preliminary considerations are helpful. Denote by G a locally compact group and let H and K be closed subgroups of G. Then G acts by translations on G/H and G/K. One can restrict this action to K and H respectively, so that K acts on G/H and H acts on G/K. It may be shown [252] that the corresponding transformation group C^*-algebras $C^*(K, G/H)$ and $C^*(H, G/K)$ are strongly Morita equivalent.

In the case $G = \mathbb{R}, H = \mathbb{Z}, K = \alpha\mathbb{Z}$ this means that $C^*(\alpha\mathbb{Z}, \mathbb{R}/\mathbb{Z})$ is strongly Morita equivalent to $C^*(\mathbb{Z}, \mathbb{R}/\alpha\mathbb{Z})$. The first algebra is just A_α. After we apply the homeomorphism $t \mapsto t/\alpha$ to $\mathbb{R}$ we will find that the second algebra is isomorphic to $C^*(\alpha^{-1}\mathbb{Z}, \mathbb{R}/\mathbb{Z})$. This means that A_α is strongly Morita equivalent to $A_{\alpha^{-1}}$.

It was noted above that A_α is isomorphic to $A_{\alpha+n}$ for any $n \in \mathbb{Z}$. With a slight abuse of notation we will call the group of 2×2-matrices with entries in $\mathbb{Z}$ and with the determinant ± 1 the modular group $SL_2(\mathbb{Z})$. Suppose that $SL_2(\mathbb{Z})$ acts on the set of the irrational numbers by the formula:

$$\begin{pmatrix} a & b \\ c & d \end{pmatrix} \circ \alpha \longmapsto \frac{a\alpha + b}{c\alpha + d}.$$

On the other hand it is known also that the group $SL_2(\mathbb{Z})$ is generated by the matrices

$$\begin{pmatrix} 0 & 1 \\ 1 & 0 \end{pmatrix} \quad \text{and} \quad \begin{pmatrix} 1 & 1 \\ 0 & 1 \end{pmatrix}.$$

These two matrices exactly carry the element α to the elements α^{-1} and $\alpha + 1$, respectively. It follows therefore that if α and β are irrational numbers which belong to the same orbit of the action of $SL_2(\mathbb{Z})$, then A_α and A_β are strongly Morita equivalent.

Sufficient conditions. If A and B are unital C^*-algebras which are Morita equivalent, then they are stably isomorphic, and therefore A and B will have isomorphic K_0 groups [68]. Traces of a C^*-algebra define homomorphisms from the K_0 group into $\mathbb{R}$. For C^*-algebras with unique trace which are Morita equivalent, the ranges of the corresponding homomorphisms will be isomorphic as groups. It may easily be shown [253] that the traces on $K_0(A_\alpha)$ are isomorphic as abstract groups for different α's. Therefore we need to show that for algebras which are Morita equivalent, the isomorphism which one obtains between the groups $\hat{\tau}(K_0(A_\alpha))$ are order-preserving isomorphisms under the order induced by the one on $\mathbb{R}$. (Here it is necessary to relate traces carefully to Morita equivalence. As was shown in [253] this can be done using a series of technical lemmas. We omit here further details referring the reader to [253].)

Suppose that A_α and A_β are Morita equivalent under the equivalence bimodule X [253]. Let τ be the normalized trace on A_α, and let τ_X be the corresponding non-normalized trace on A_β such that

$$\hat{\tau}_X(K_0(A_\beta)) = \hat{\tau}(K_0(A_\alpha)).$$

It may be seen that τ_X differs from the normalized trace on A_β only by a scalar multiple. From this it follows [253] that there exists a positive real number r such that $\mathbb{Z}+\beta\mathbb{Z} = r(\mathbb{Z}+\alpha\mathbb{Z})$. In particular, there are $j,k,m,n \in \mathbb{Z}$ such that $j+\beta k = r$ and $1 = r(m+\alpha n)$. Excluding r in the above equations one finds that α and β belong to the same orbit of the $SL_2(\mathbb{Z})$-actions.

Theorem is proved. □

10.1.6 Embedding of A_α

The irrational rotation algebra is known to be *not* an AF-algebra. This result follows from calculation of K-groups of A_α. Fortunately, A_α can be embedded (as C^*-subalgebra) into an AF-algebra whose K_0-group is order-isomorphic to the K_0-group of A_α.

A natural generalization of AF-algebras are $A\mathbb{T}$-*algebras* which are defined to be inductive limits of the multi-matrix algebras with entries in the space $C(S^1)$ of continuous complex-valued functions on the circle. The most amazing and deep fact established in 90s by G. A. Elliott in collaboration with D. E. Evans is that the irrational rotation algebras are $A\mathbb{T}$-algebras; see [98]. We quote this important result in below.

Lemma 10.1.7 (Elliott-Evans) *Let A_α be the irrational rotation algebra. There exists a continuous function $\mu : \mathbb{R}^+ \to \mathbb{R}^+$ such that if $\begin{pmatrix} p' & p \\ q' & q \end{pmatrix} \in SL_2(\mathbb{Z})$ is a modular matrix with $q, q' > 0$ and $\alpha \in \mathbb{R}$ such that $p/q < \alpha < p'/q'$, such that a C^*-subalgebra $A_\alpha(p/q, p'/q')$ of A_α which is*

$$A_\alpha(p/q, p'/q') \cong M_q(C(S^1)) \oplus M_{q'}(C(S^1)),$$

contains the elements within the distance $\mu(\gamma)\max(1/q, 1/q')$ of the canonical generators u, v of A_α, where γ is the quotient $\gamma = \left(\alpha - \frac{p}{q}\right)\left(\frac{p'}{q'} - \alpha\right)^{-1}$. Moreover, the subalgebra $A_\alpha(p/q, p'q')$ may be chosen to depend continuously on the parameter α.

Proof. The proof requires some preparative work which we omit completely referring the reader to [98] for the details. The proof is based on the tower construction, similar to those used in the ergodic theory by Rokhlin (Rokhlin's towers). The idea is to obtain a chain of estimates on norms (in the norm topology of the Banach algebra A_α) of the 'basic' elements in A_α. □

Theorem 10.1.5 (Elliott-Evans) *Let α be an irrational number. There exists a sequence*

$$A_1 \longrightarrow A_2 \longrightarrow \dots$$

of finite direct sums of (complex) matrix algebras over $C(S^1)$ with inductive limit isomorphic to the rotation C^*-algebra A_α. Moreover, a given sequence of such algebras has the above property if (and only if) the inductive limit is simple and unital, has a unique trace state and has K_0-group with order-unit isomorphic to $(\mathbb{Z} + \alpha\mathbb{Z}, 1)$ and K_1-group isomorphic to $\mathbb{Z}^2$.

Proof. This follows from Lemma 10.1.7. □

10.2 Artin Rotation Algebra

Let S^1 be a global cross-section of a transitive flow on a closed orientable surface of genus $g > 1$. Denote by B_α a C^* algebra, induced by the first return mapping $\phi : S^1 \to S^1$ in accordance with formula (10.1). Unlike the case of the torus, ϕ is no longer a homeomorphism, but an interval exchange transformation; see Ch. 0 for a definition. As well as the case of the torus, B_α is not an AF-algebra but can be embedded into an AF-algebra such that the respective K_0-groups are order-isomorphic; see Vershik [289], Giordano *et al.* [117]. In what follows, the algebra B_α will be called an *Artin rotation algebra.*

10.2.1 Approximationssatz

A general reference to the subject is

J. Lehner, Discontinuous Groups and Automorphic Functions, AMS Math. Surveys 7 (1964), Providence, R.I.

Let M be a compact, orientable surface of a finite genus $g > 1$. M is given by a universal covering representation $(\mathbb{H}, \Gamma)$ where

$$\mathbb{H} = \{z = x + iy \in \mathbb{C} | y \geq 0\}$$

is the Lobachevsky plane endowed with the hyperbolic metric $ds = |dz|/y$. For technical advantages we have chosen here an upper half-plane model. With minor difficulties all of the results are valid for the disc model $\Delta = \{z \in \mathbb{C} : |z| \leq 1\}$ as well. The Fuchsian group Γ is supposed to be of the first kind.

Let Γ be a Fuchsian group of signature

$$(g : m_1, \dots, m_r; s; t)$$

where the positive numbers $m_1, \ldots, m_r$ are the orders of (cyclic) elliptic subgroups of Γ, s is the number of parabolic subgroups of Γ, and, finally, t is the number of hyperbolic subgroups of Γ; see Beardon [42]. Under natural restrictions on the signature there exists a connected fundamental region for Γ which is a polygon $\widetilde{M}$ with finite, pairwise identified geodesic sides. $M = \widetilde{M}/\Gamma$ is an s-punctured Riemann surface of genus g. For simplicity we suppose $t = 0$. Γ can be chosen to be a subgroup of the modular group $SL_2(\mathbb{Z})/\{-I, I\}$; cf. the *Appendix*. Moreover, Γ is supposed to coincide with the principal congruence group $\Gamma(n)$:

$$\Gamma(n) = \pm \begin{pmatrix} 1 & 0 \\ 0 & 1 \end{pmatrix} \; Mod\; n, \qquad n \geq 0.$$

$\Gamma(n)$ is a normal subgroup of finite index. Denote by $p|n$ the set of non-trivial prime divisors of the natural number n. The genus of the Riemann surface $M = \tilde{M}/\Gamma(n)$ is calculated by the formula (see Lehner, p.359):

$$g = 1 + \frac{1}{24}\tau n^2(n-6)\prod_{p|n}\left(1 - \frac{1}{p^2}\right), \qquad \tau = \begin{cases} 2, & \text{if } n = 2, \\ 1, & \text{if } n > 2 \end{cases}.$$

(Notice that genus of M is zero if and only if $n \leq 5$.) In what follows, we suppose $g > 1$ by choosing n sufficiently big, no matter how many 'punctures' the surface M would have.

A preimage of the geodesic line on $\mathbb{H}$ is either a semi-circle or a straight line orthogonal to the axis $y = 0$. The latter fall into "equivalence classes" by the isometry $\Gamma(n)$. Each "class" represents an individual geodesic line of M. "Existence and uniqueness" conditions for geodesics involve the angles θ through the points of M; therefore such lines may have self-intersections on M. This possibility is ruled out on an auxiliary (3-dimensional) manifold $TM = (M, \theta)$ which is a unit tangent bundle to M. Clearly, the set of all geodesics on M "generates" a flow without stationary points on the manifold TM. The latter is called a *geodesic flow* on the surface M. Each trajectory of the geodesic flow corresponds to a (parametrized) geodesic line pushed through a point $x \in M$ in the direction θ. From now on, the reader is encouraged to adopt both of viewpoints.

Geodesic lines on surfaces of constant negative curvature have attracted a lot of attention in the past. It is clear, that, in contrast to the case of zero curvature, geodesic lines can have self-intersections on M; in fact, the "majority" of geodesics will have this feature. Do there exist geodesic lines without self-intersections? The answer is positive: Hadamard showed the existence of closed (periodic) geodesics and non-closed "spiral" geodesics, while Morse *et al.* established the existence of non-closed geodesics of discontinuous type [3]. Such geodesics are "rare" in the sense that their Lebesgue measure in the

[3] That is, the closure of the geodesic line on M is topologically a product of Cantor set and a linear segment.

pencil of geodesics pushed through a fixed point on M is zero. There are no other types of behaviour of geodesic lines different from those mentioned above.

There exist geodesic lines (with self-intersections) which approximate *any* geodesic line on M; this fact is due to Artin [38] who established the regional transitivity of the geodesic flow. Geodesic flows enjoy even stronger approximation properties such as ergodicity. The latter implies regional transitivity. Ergodicity of a geodesic flow means that 'almost all' geodesic lines are everywhere dense in M. Ergodicity was established by Myrberg in [208]. He proved an "Approximationssatz" for the Fuchsian groups in the unit disc.

Theorem 10.2.1 (Approximationssatz) *Let Γ be a Fuchsian of the first kind acting on the Lobachevsky plane $\mathbb{H}$. Then there exists a subset M of points of the absolute of full Lebesgue measure, which has the following property: If K is a geodesic line with the "endpoints" in M, then every geodesic line in $\mathbb{H}$ is approximated (arbitrarily close) by the images $\Gamma(K)$ of K under a certain sequence of transformations belonging to the Fuchsian group Γ. Moreover, the set (of measure zero), complementary to M at the absolute has the cardinality of the continuum.*

Remark 10.2.1 *The set K of the Approximationssatz corresponds to the geodesic lines with self-intersections; the set complementary to it at the absolute consists of closed geodesics, spiral geodesics and geodesics of discontinuous type.*

10.2.2 Artin Numbers

Let M be a closed orientable surface of constant negative curvature. If $\mathbb{H}$ is the Lobachevsky plane, we represent M as a quotient $\mathbb{H}/\Gamma$, where Γ is the isometry group isomorphic to the principal congruence group $\Gamma(n)$; cf Section 1. With a slight abuse of language a C^∞ flow ϕ^t on M will be called *minimal* if (i) every non-stationary trajectory of ϕ^t is everywhere dense in M and (ii) the set Sing ϕ^t consists of $2m$-separatrix saddles, $m \geq 2$.

Let λ be a non-stationary trajectory of ϕ^t, and let $\tilde{\lambda}$ be (a connected component of) its preimage on the Lobachevsky plane $\mathbb{H}$. The properties of $\tilde{\lambda}$ can be summarized as follows; cf. Aranson *et al* [30]:

- The extention of $\tilde{\lambda}$ to the absolute defines two distinct points, a and b, therein;
- The geodesic line through the points a, b is a geodesic line of discontinuous type;
- Up to isometries of $\mathbb{H}$, there exists a one-to-one correspondence between the triples $(\tilde{\lambda}, a, b)$ and the equivalence classes of minimal flows on M.

Let us denote by $\tilde{\gamma}$ a discontinuous-type geodesic corresponding to the triple $(\tilde{\lambda}, a, b)$. Note that, up to isometries of $\mathbb{H}$, $\tilde{\gamma}$ is uniquely defined by the flow

ϕ^t. By the remark 10.2.1 the endpoints of $\tilde{\gamma}$ belong to the set which is complementary to K at the absolute.

Denote by S_0 a fundamental polygon corresponding to the principal congruence group $\Gamma(n)$ whose closure lies in the interior of the Lobachevsky plane $\mathbb{H}$. The boundary of S_0 can be chosen to consist of the geodesic segments. It can be seen after some work that the "footpoints" of these segments at the absolute belong to the set K. We are now in position to introduce the Artin number of a minimal flow ϕ^t.

Definition 10.2.1 *Let ϕ^t be a minimal flow on a closed surface of genus $g > 1$ and let $\tilde{\gamma}$ be the geodesic line corresponding to the triple $(\tilde{\lambda}, a, b)$. Let S_0 be the fundamental polygon of the group $\Gamma(n)$ with the geodesic sides $a_1, b_1, \ldots, a_{2g}, b_{2g}$ corresponding to the generators of the group $\pi_1(M)$. Let $\varepsilon_1, \varepsilon_2, \ldots$ be a monotone sequence of positive real numbers tending to zero. Finally, let $T_0, T_1, T_2, \ldots$ be an infinite sequence of isometries $\Gamma(n)$ which approximate the geodesic $\tilde{\gamma}$ by the images of a fixed "geodesic side" of S_0 with accuracy $\varepsilon_0, \varepsilon_1, \varepsilon_2, \ldots$. By the Artin number of the flow ϕ^t one understands a real number $A(\phi) = [a_0, \ldots, a_{n_0}, a_{n_0+1}, \ldots, a_{n_1}, \ldots]$, where*

$$T_0 = \begin{pmatrix} a_0 & 1 \\ 1 & 0 \end{pmatrix} \cdots \begin{pmatrix} a_{n_0} & 1 \\ 1 & 0 \end{pmatrix}; \quad T_1 = \begin{pmatrix} a_{n_0+1} & 1 \\ 1 & 0 \end{pmatrix} \cdots \begin{pmatrix} a_{n_1} & 1 \\ 1 & 0 \end{pmatrix}; \quad \ldots$$

is the Minkowski decomposition of the positive integer matrices $T_0, T_1, T_2, \ldots$; cf. the Minkowski Lemma.

Remark 10.2.2 *The "geometric meaning" of $A(\phi)$ is an average inclination of the trajectories of ϕ^t with regard to the lattice $\mathbb{L}$ formed by the fundamental polygons $\Gamma(S_0)$. $A(\phi)$ is equal to the "angle" between $\tilde{\gamma}$ and a fixed side of S_0. Clearly, choosing different sides from the collection $a_1, b_1, \ldots, a_{2g}, b_{2g}$, one obtains $A(\phi)$'s which differ by a constant.*

Theorems 10.2.2 and 10.2.3 establish the uniqueness and invariance of Artin numbers with regard to the equivalence of minimal flows on M.

Theorem 10.2.2 *([224]) Let ϕ^t be a minimal flow on a surface of genus $g > 1$. Then the Artin number $A(\phi)$ is uniquely defined by the flow ϕ^t up to a relation on the real numbers which identifies modularly equivalent numbers.*

Let ϕ^t be a minimal flow on M. To prove that the number $A(\phi)$ is defined correctly we have to show that: (i) $A(\phi)$ is independent of the choice of the particular trajectory λ of the flow ϕ^t; (ii) $A(\phi)$ is "modularly" independent of the connected component $\tilde{\gamma}$ of the geodesic line; and, finally, (iii) $A(\phi)$ is "modularly" independent on the approximation sequence $\varepsilon_0, \varepsilon_1, \ldots$. Let us consider these items one by one.

(i) The proof is a consequence of the following lemma.

Lemma 10.2.1 *Suppose that λ and μ are two distinct trajectories of a minimal flow ϕ^t. If $\tilde{\lambda}, \tilde{\mu}$ are their respective covering trajectories on the Lobachevsky plane $\mathbb{H}$ and $\tilde{\gamma}_1 \neq \tilde{\gamma}_2$ are the geodesic lines representing $\tilde{\lambda}$ and $\tilde{\mu}$ at the absolute, then $\gamma_1 \subset$ Clos γ_2 and $\gamma_2 \subset$ Clos γ_1.*

Proof. Aranson *et al.* [30] give a detailed proof of the lemma, accompanied by the necessary figures. This fact means that the set $Clos\ \gamma_1 = Clos\ \gamma_2$ (which is a Cantor set X on M) is a full "geometric" invariant of the minimal flow ϕ^t. □

Let $X = Clos\ \gamma$ be the Cantor set mentioned in lemma 10.2.1. By the earlier discussion, to the minimal flow ϕ^t on M there corresponds a fixed-point-free minimal flow ϕ^t_X on X. (Note that every trajectory of ϕ^t_X is a geodesic line in the metric of constant negative curvature on M.) The trajectories of ϕ^t_X can be regarded as a substitute for the linear irrational flow on torus.

The flow ϕ^t_X on X is covered by a flow $\tilde{\phi}^t_X$ on the Cantor set $\tilde{X}$ of the universal covering $\mathbb{H}$. Note that the Artin number $A(\phi)$ is not affected by *any* isometry of $\mathbb{H}$. The covering geodesic $\tilde{\gamma}$ which determines $A(\phi)$ can always be thought to be passing through one (and only one, cf Lemma 10.2.4) point of the lattice $\mathbb{L} \subset \mathbb{H}$. For simplicity, let it be the lower left vertex $(0,0)$ of the fundamental polygon S_0. Clearly, if $\tilde{\gamma}' \neq \tilde{\gamma}$ is another "parallel" geodesic in $\tilde{X}$, then it can always be "shifted" so as to pass through $(0,0)$; therefore, it yields the same Artin's number $A'(\phi) = A(\phi)$. Item (i) now follows. □

(ii) Let $\tilde{\gamma}_1, \tilde{\gamma}_2, \ldots$ be a sequence of connected components which constitute the complete preimage of the geodesic γ on the Lobachevsky plane $\mathbb{H}$. For each pair $\tilde{\gamma}_n$ and $\tilde{\gamma}_{n+k}$ in this sequence there exists an isometry $g \in \Gamma$ which sends $\tilde{\gamma}_n$ to $\tilde{\gamma}_{n+k}$. If now $T_0, T_1, \ldots$ is the defining sequence for the Artin number $A(\phi)$ calculated with respect to $\tilde{\gamma}_n$, then $T_0' \ldots T_k' T_0, T_1 \ldots$ will be the defining sequence for the Artin number $A'(\phi)$ calculated with respect to the geodesic circle $\tilde{\gamma}_{n+k}$. By the fundamental property of continued fractions the numbers $A(\phi)$ and $A'(\phi)$ are modularly equivalent. Item (ii) now follows. □

(iii) To see that the Artin number $A(\gamma)$ is "modularly" independent of the sequence $\varepsilon_0, \varepsilon_1, \varepsilon_2, \ldots$ mentioned in definition 10.2.1, first let us notice the additivity property of the defining sequence of operators $T_0, T_1, \ldots$. Namely, if $\varepsilon_i > \varepsilon_{i+1}$ are two constants, and there exists ε' such that $\varepsilon_i > \varepsilon' > \varepsilon_{i+1}$, then the operator T_i in the defining sequence will decompose as $T_i' \circ T_{i+1}'$. Therefore, if $T_0, T_1, T_2, \ldots$ is the defining sequence corresponding to $\varepsilon_0, \varepsilon_1, \varepsilon_2, \ldots$ then $T_0, T_1, \ldots, T_i' T_{i+1}', \ldots$ will be the defining sequence corresponding to $\varepsilon_0, \varepsilon_1, \ldots, \varepsilon_i, \varepsilon', \varepsilon_{i+1}, \ldots$. Clearly, the continued fractions in the first and in the second case will differ only in a finite number of terms, i.e. $A(\phi)$ and $A'(\phi)$ are modularly equivalent. Such an operation we call *insertion*.

In general, a sequence $\{\varepsilon_i'\}$ is said to be *finer* than the sequence $\{\varepsilon_i\}$ if it is obtained from $\{\varepsilon_i\}$ by a finite number of operations of insertion. The set of all sequences finer than $\{\varepsilon_i\}$ is denoted by Ω. Clearly, Ω is a countable set and any sequence $\{\varepsilon_i'\} \to 0$ belongs to Ω. Item (iii) now follows. □

Theorem 10.2.3 *([224]) Let ϕ^t and ψ^t be two minimal flows on the surface M whose Artin numbers are $A(\phi)$ and $A(\psi)$, respectively. Then the flows ϕ^t and ψ^t are topologically conjugate if and only if $A(\phi)$ and $A(\psi)$ are modularly equivalent real numbers.*

Proof. Necessary conditions. Suppose that two minimal flows ϕ^t, ψ^t are topologically equivalent: $\psi^t = h\phi^t h^{-1}$. Let γ_ϕ, γ_ψ be the respective geodesics lines which define the equivalence classes of these flows; cf. Aranson *et al.* [30]. Let *Clos* γ_ϕ, *Clos* γ_ψ be the closures of the geodesic lines γ_ϕ, γ_ψ on the Lobachevsky plane $\mathbb{H}$.

Take arbitrary connected components, $\tilde{\gamma}_\phi$ and $\tilde{\gamma}_\psi$, in the preimages of γ_ϕ and γ_ψ on $\mathbb{H}$. Since the flows ϕ^t and ψ^t are equivalent, both $\tilde{\gamma}_\phi$ and $\tilde{\gamma}_\psi$ belong to the same set *Clos* $\gamma_\phi \equiv$ *Clos* γ_ψ on $\mathbb{H}$. This means that there exists an isometry $h \in \Gamma$ such that $\tilde{\gamma}_\psi = \tilde{h}\tilde{\gamma}_\phi\tilde{h}^{-1}$. By the basic property of continued fractions the rotation numbers α and β are modularly equivalent.

Sufficient conditions. Let us assume now that α and β are modularly equivalent. In constructing the conjugacy homeomorphism $h : \psi^t = h \circ \phi^t \circ h^{-1}$, the following lemma is instrumental.

Lemma 10.2.2 *Suppose that the flows ϕ^t and ψ^t are as specified before. Let $\alpha \simeq \beta$ be their modularly equivalent Artin numbers. Then ϕ^t and ψ^t have mutually homotopic transversal cycles, C_ϕ and C_ψ, intersecting the respective geodesics lines, γ_ϕ and γ_ψ.*

Proof of lemma. Let γ_ϕ and γ_ψ be as before. Since $\alpha \simeq \beta$, there exist a connected component in the preimages of γ_ϕ and γ_ψ on the universal covering such that $\tilde{\gamma}_\phi = \tilde{\gamma}_\psi = \tilde{\gamma}$. Let us denote by $\tilde{\phi}^t$ and $\tilde{\psi}^t$ the corresponding flows on the covering surface $\tilde{M}$.

Every transitive flow on M admits a transversal cycle. Take one of the flows, say ϕ^t; clearly, it is transitive on M. Denote by C_ϕ the corresponding transversal cycle. Let $\tilde{C}$ be a connected component of its lifting to the universal covering surface $\tilde{M}$. Let us show that $\tilde{C}$ is a transversal cycle to the flow $\tilde{\psi}^t$ as well.

Indeed, suppose to the contrary, that $p \in \tilde{C}$ is a contact point to the flow $\tilde{\psi}^t$. By assumption, $\tilde{\gamma}_\psi$ is an everywhere dense trajectory of the flow $\tilde{\psi}^t$. In particular, it approximates arbitrarily close the trajectory $l(p)$ of the flow $\tilde{\psi}^t$ passing through the contact point p. But $\tilde{\gamma}_\psi = \tilde{\gamma}_\phi = \tilde{\gamma}$, so that the flow $\tilde{\phi}^t$ must have a contact point with $\tilde{C}$, a contradiction. □

The idea now is to reduce the conjugacy problem on all of M to the conjugacy problem between the first return mappings on the transversal cycles C_ϕ and C_ψ. This will be done in the following steps. (Again we refer the reader to the monograph [30] for illustrations and details.)

Let $p_1, p_2, \ldots, p_{8g-8} \in \tilde{C}_\phi$ and $q_1, q_2, \ldots, q_{8g-8} \in \tilde{C}_\psi$ (where g is the genus of M) be the points where separatrices of the saddle points intersect transversal cycle. (We assume that all the saddles are Morse.) On account of the orientation of the flows $\tilde{\phi}^t$ and $\tilde{\psi}^t$, we fix the mapping $\tilde{h}$:

$$\tilde{h}(p_1) = q_1, \quad \tilde{h}(p_2) = q_2, \ldots, \tilde{h}(p_{8g-8}) = q_{8g-8}.$$

(It should be understood, that the construction takes place in a fixed fundamental polygon, S_0, of the universal covering $\tilde{M}$.) Both of the geodesic lines $\tilde{\gamma}_\phi$ and $\tilde{\gamma}_\psi$ hit the respective transversal cycles in an infinite sequence of points, and we denote the latter by $\{x_n\}_{n=-\infty}^{n=\infty}$ and $\{y_n\}_{n=-\infty}^{n=\infty}$, choosing a suitable parametrization (by time, say). Next we put

$$\tilde{h}(x_i) = y_i, \qquad \text{for} \quad i \in \mathbb{Z}.$$

Because the points x_i and y_i lie everywhere dense in the respective transversal cycles, we extend by continuity the mapping $\tilde{h}$ to the whole circle, $\tilde{h} : \tilde{C}_\phi \to \tilde{C}_\psi$. (Note that h 'respects' the interval exchange transformation on $\tilde{C}_\phi$ and $\tilde{C}_\psi$ induced by the flows ϕ^t and ψ^t.)

Now we are able to extend the mapping h to the whole fundamental polygon S_0. This can be done 'along the trajectories of flow':

$$\tilde{\phi}^t(x) = \tilde{\psi}^t(hx), \qquad x \in \tilde{C}_\phi.$$

As far as the $2g - 2$ saddle points are concerned, we must take more care; the idea is to follow the respective separatrices. When this is done we have an extended mapping $\tilde{h} : S_0 \to S_0$. Finally, we prolong it to the rest of the Lobachevky plane by the isometries from Γ. To obtain the required mapping on the surface M, it remains to take a quotient $\mathbb{H}/\Gamma$. We claim that the constructed mapping h is a homeomorphism $M \to M$.

Indeed, at 'typical' points it follows from the continuity of orbits with regard to the 'initial data'. Some precaution should be taken near the separatrix points. But this is also rather standard; see [30]. □

Let S^1 be a cross-section of the minimal flow ϕ^t on M. Let $\phi : S^1 \to S^1$ be the first return mapping induced on S^1 by ϕ^t. We have a partition of $S^1 \cong [0, 1)$ into the semi-intervals $I_i = [x_{i-1}, x_i)$ where $0 = x_0 < x_1 < \ldots < x_r = 1$ are the points where the separatrices of ϕ^t intersect first time with S^1. ϕ is known to permute the order of I_i on S^1: $(I_1, I_2, \ldots, I_r) \mapsto (I_{\pi(1)}, I_{\pi(2)}, \ldots, I_{\pi(r)})$, where π is the corresponding permutation group.

By the minimality of flow the set of points $\Lambda = \{\phi^n(x_i) | 0 \leq i < r; n \in \mathbb{Z}\}$ is everywhere dense in S^1. If we "blow-up" each separatrix of the initial flow ϕ^t, then each point in Λ will split into two points, $y_1 < y_2$, at S^1. Throwing-out all the open intervals (y_1, y_2) one comes to the Cantor set X on S^1. It follows from the discussion above (cf. also [30]) that X corresponds to the intersection of S^1 with the closure of the discontinuous geodesic line $\tilde{\gamma}$ built upon the flow ϕ^t. One defines a homeomorphism $\phi : X \to X$ so that it coincides on the set $S^1 \backslash \Lambda$ with the initial mapping $\phi : S^1 \to S^1$.

Following Putnam [249] we define a C^*-algebra generated by the minimal flow ϕ^t as a crossed product $C(X) \rtimes_\phi \mathbb{Z}$ of the C^*-algebra of continuous complex-valued functions on X by the action of powers of ϕ.

Our definition of the Artin number $A(\phi)$ suggests another C^*-algebra connected with the flow ϕ^t, this time approximately finite-dimensional. Let us denote by $\mathbb{L}$ a lattice corresponding to the abelianization $\Gamma/[\Gamma,\Gamma]$ of the Fuchsian group Γ.

Lemma 10.2.3 *The lattice $\mathbb{L}$ is isomorphic to the abelian lattice $\mathbb{Z}^{2g} \subset \mathbb{R}^{2g}$ endowed with the natural operations of addition and multiplication of the integers. The above isomorphism turns $\mathbb{L}$ into an abelian lattice.*

Proof. Taking one representative in each coset gH, where $g \in \Gamma$ and $H = [\Gamma,\Gamma]$ is the commutator of Γ, we come to the abelian sublattice $\mathbb{Z}^{2g}$ of the lattice formed by the vertices of fundamental polygons corresponding to Γ. The rest of the proof follows from the isomorphisms $\Gamma/[\Gamma,\Gamma] \simeq H^1(M,\mathbb{Z}) \simeq \mathbb{Z}^{2g}$, where M is a surface of genus g whose fundamental group is isomorphic to Γ; see [223]. □

Let $\tilde{\gamma}$ be as in the Definition 10.2.1 and let $T_0, T_1, T_2 \dots$ be the defining sequence of isometries approximating $\tilde{\gamma}$. "Lifting" $\tilde{\gamma}$ to the Euclidean space $\mathbb{R}^{2g}$, we obtain the following lemma, which is an implication of definition 10.2.1.

Lemma 10.2.4 *The geodesic line $\tilde{\gamma}$ corresponds to a hyperplane L in $\mathbb{R}^{2g}$ such that $L \cap \mathbb{Z}^{2g} = \emptyset$. The approximants $T_0, T_1, T_2, \dots$ of $\tilde{\gamma}$ are represented by the hyperplanes $L_0, L_1, L_2, \dots$ whose intersection with $\mathbb{Z}^{2g}$ consists of an infinite sequence of points.*

The two types of hyperplanes mentioned in lemma 10.2.4 may be largely viewed as a substitute for the "rational" and "irrational" directions in the lattice $\mathbb{L} \simeq \mathbb{Z}^{2g}$. The positive cones $L_0^+, L_1^+, L_2^+, \dots$ fixed in $\mathbb{Z}^{2g}$ turn the sequence $L_0, L_1, L_2, \dots$ into the sequence of ordered abelian groups (L_0, L_0^+), $(L_1, L_1^+), \dots$ converging to (L, L^+). This is the *dimension group* corresponding to the Artin number $A(\phi)$.

Using the method of "towers" one can obtain an AF-algebra $B_{A(\phi)}$ (whose K_0-group is (L, L^+)) by taking the limit of special multi-matrix C^*-algebras $M_1 \oplus \dots \oplus M_{2g}$ built upon a descending sequence $Y_0 \supset Y_1 \supset \dots$ of closed subsets in X. How do $B_{A(\phi)}$ and $C(X) \rtimes_\phi \mathbb{Z}$ relate to each other? The answer is given by the following Basic Technical Lemma:

Lemma 10.2.5 *There exists an embedding $i : C(X) \rtimes_\phi \mathbb{Z} \to B_{A(\phi)}$ such that $i_* : K_0(C(X) \rtimes_\phi \mathbb{Z}) \to K_0(B_{A(\phi)})$ is an isomorphism of ordered abelian groups.*

Proof of lemma. The proof is based on the "tower" construction; see Davidson [83] for the details. The minimal homeomorphism ϕ on the Cantor set X induces on each closed subset $Y \subset X$ a tower $\{J_1, \ldots, J_K\}$ of a finite height (see below for the definition). There is a one-to-one correspondence between towers of finite height and multi-matrix C^*-algebras. Taking a chain of inclusions $Y_0 \supset Y_1 \supset \ldots$ such that $\cap Y_i = \{y\}$ is a point in X, one gets a sequence of towers converging to an approximate finite C^*-algebra $A_{\{y\}}$. Putnam's theorem (Theorem 4.1 of [249]) says that the crossed product algebra $C(X) \rtimes_\phi \mathbb{Z}$. is embeddable into $A_{\{y\}}$, so that the K_0-groups of the two C^*-algebras are order isomorphic. Next we show that there exists an order isomorphism between $K_0(A_{\{y\}}) \stackrel{\text{def}}{=} P_{A(\phi)}$ and the dimension group (L, L^+) of the lattice $\mathbb{L} \subset \mathbb{H}$.

Definition 10.2.2 *Let $Y \subset X$ be a closed subset of the Cantor set X. Let $\phi : X \to X$ be a minimal homeomorphism on X. Define an integer-valued function $\lambda : Y \to \mathbb{Z}$ by the formula*

$$\lambda(y) = \min\{n \geq 1 \ : \ \phi^n(y) \in Y, \ y \in Y\}.$$

(λ is a well-defined function on Y and takes only a finite number of values.) We let $\lambda(Y) = \{J_1, J_2, \ldots, J_K\}$ be an ordered sequence of positive integers $J_1 < J_2 < \ldots < J_K$. Finally, for $k = 1, 2, \ldots, K$ and $j = 1, 2, \ldots, J_K$ we define the closed-open ("clopen") subsets $Y(k, j)$ as follows:

$$Y(k, j) = \phi^j(\lambda^{-1}(J_k)).$$

The collection $\{Y(k, j) | j = 1, \ldots, J_k\}$ is called a tower of height J_k.

One can observe the basic properties of the $Y(k, j)$:

- $\cup_{k=1}^K Y(k, 1) = \phi(Y)$;
- $\phi(Y(k, j)) = Y(k, j + 1)$ for $1 \leq j < J_k$;
- $\cup_{k=1}^K Y(k, J_k) = Y$.

"Towers" are adequate instruments of constructing finite-dimensional C^*-algebras starting from a closed set Y fixed on X. Every such algebra is $*$-isomorphic to a multi-matrix algebra

$$M_{J_1} \oplus M_{J_2} \oplus \ldots \oplus M_{J_K}$$

by choosing the matrix units $e_{ij}^{(k)}$ in accordance with the formula

$$e_{ij}^{(k)} = \chi_{Y(k,i)} u^{i-j}.$$

(Here u is unitary operator on the C^*-algebra $C(X)$ and $\chi_{Y(k,i)}$ is the characteristic function of the interval $Y(k, i)$.)

Fix $K = 2g$, where g is genus of the underlying surface M. The tower construction will give us the dimension group $P_{A(\phi)}$ corresponding to the AF-algebra $A_{\{y\}}$. We denote by φ_i the relevant positive integer unitary operators acting on the lattice $\mathbb{Z}^{2g}$. Consider the following commutative diagram:

$$\begin{array}{ccccccc}
\mathbb{Z}^{2g} & \xrightarrow{\varphi_0} & \mathbb{Z}^{2g} & \xrightarrow{\varphi_1} & \mathbb{Z}^{2g} & \longrightarrow & \dots \\
\downarrow i_0 & & \downarrow i_1 & & \downarrow i_2 & & \\
\mathbb{L} & \xrightarrow{T_0} & \mathbb{L} & \xrightarrow{T_1} & \mathbb{L} & \longrightarrow & \dots
\end{array}$$

where $T_0, T_1, T_2, \dots$ is the defining sequence for the Artin number $A(\phi)$. By Lemma 10.2.3 there exists an order isomorphism $i_n : \mathbb{Z}^{2g} \to \mathbb{L}$ such that $\varphi_m \circ i_{m+1} = i_m \circ T_m$. It remains to take the limit $i = \lim_{m\to\infty} i_m$ which completes the infinite commutative diagram pictured above. Therefore, $P_{A(\phi)}$ is order isomorphic to the dimension group (L, L^+) of the lattice $\mathbb{L}$.

The rest of the proof follows from Putnam's theorem (Theorem 4.1 of [249]) which says that $C(X) \rtimes_\phi \mathbb{Z}$ is embeddable into the AF-algebra $A_{\{y\}}$, so that $K_0(C(X) \rtimes_\phi \mathbb{Z}) \simeq P_{A(\phi)}$. □

Corollary 10.2.1 *Let $C(X) \rtimes_\phi \mathbb{Z}$ be a C^*-algebra generated by the minimal homeomorphism ϕ on the Cantor set X. Suppose that the rotation number of $C(X) \rtimes_\phi \mathbb{Z}$ has a continued fraction expansion $\alpha = [a_0, a_1, a_2, \dots]$. Let φ_n be a composition of isometries of the lattice $\mathbb{L} \subset \mathbb{H}$:*

$$\varphi_n = \begin{pmatrix} a_0 & 1 \\ 1 & 0 \end{pmatrix} \cdots \begin{pmatrix} a_n & 1 \\ 1 & 0 \end{pmatrix}.$$

Then the K_0-group of $C(X) \rtimes_\phi \mathbb{Z}$ is order isomorphic to a dimension group $P_\alpha = \lim_{n\to\infty}(\mathbb{L}, \varphi_n)$. Moreover, if $C(X) \rtimes_\phi \mathbb{Z}$ and $C(X) \rtimes_\psi \mathbb{Z}$ are two crossed product C^*-algebras with the rotation numbers $\alpha = [a_0, a_1, \dots]$ and $\beta = [b_0, b_1, \dots]$ respectively, then they are Morita equivalent if and only if $a_{m+k} = b_m$ for some number $k \in \mathbb{Z}$. The latter means that the irrational numbers α and β are modularly equivalent.

Proof. The first part follows from Lemma 10.2.5 and Definition 10.2.1. The necessary conditions in the second part are an implication of the basic property of continued fractions. The sufficient conditions are less evident; they follow from the Giordano-Putnam-Skau's theory; see e.g. Theorem 2.1 of [117]. □

10.2.3 Applications

There is evidence that the diophantine properties of the rotation numbers are fundamental in the analysis of dynamical systems; see Herman [144],

Gutiérrez [127]. We conclude by posing (and suggesting a solution to) problems which reveal the advantages of Artin numbers; all problems are classical and we do not comment on them.

KAM Theory Let ϕ^t be an analytic Hamiltonian flow on a surface of genus $g > 1$. Its "linearization" is represented by an interval exchange transformation (I.E.T) $x \mapsto I(x) + A(\phi)$. Under which arithmetic conditions on $A(\phi)$ is the "non-linear" flow $x \mapsto I(x) + A(\phi) + \eta(x)$ analytically (C^∞-smooth) conjugate to the linear one?

Definition 10.2.3 *An (irrational) number $\alpha \in I\!R$ is said to satisfy Herman's condition if*

$$\lim_{B \to +\infty} \limsup_{n \to \infty} \frac{\sum_{0 \le i \le n}^{a_i \ge B} \log(1 + a_i)}{\sum_{0 \le i \le n} \log(1 + a_i)} = 0, \tag{10.3}$$

where $\alpha = [a_0, a_1, \ldots]$ is the continued fraction expansion for α.

Proposition 10.2.1 *Let $\phi^t \in \mathcal{G}(M)$ be a minimal, analytic (C^∞-smooth) flow on a surface M. Let $A(\phi)$ be the Artin number of ϕ^t. If $A(\phi)$ satisfies the Herman's condition, then ϕ^t is analytically (C^∞-smooth) conjugate to its linear part.*

Proof. This is left to the reader; cf. also [223]. □

Closing Lemma Let ϕ^t be a minimal flow of class $C^r, r \ge 2$ on a surface M. Let $x \in M$ be a point on the everywhere dense trajectory $l(x)$ of the flow ϕ^t. Can we describe the set of Artin numbers $A(\phi)$ such that there exists a C^r-smooth (r as above) perturbation of ϕ^t which renders $l(x)$ periodic? The flow ϕ^t is said to be of constant type if $\sup_{i>0} a_i < \infty$, for the continued fraction expansion $A(\phi) = [a_0, a_1, \ldots]$; see Gutierrez [127].

Proposition 10.2.2 *Fix a natural number $r \ge 2$. Let ϕ^t be a minimal flow of class C^r on the surface M. Let $p \in M$ be a point lying on a non-trivially recurrent trajectory of ϕ^t. If ϕ^t is of non-constant type, then there exists an arbitrary C^r-close flow on M having a periodic trajectory through p.*

Proof. This is left to the reader; cf also [223] . □

Basic Sets of A-Diffeomorphisms Let us consider one-dimensional basic sets of A-diffeomorphisms of the surface M. If we include one of these sets in a minimal flow ϕ^t, what can be said about its Artin number $A(\phi)$? Do all irrational values arise?

Let us start with a simple example of an A-diffeomorphism of a 2-dimensional torus $T^2 = R^2/\mathbb{Z}^2$ given by the automorphism $A = \begin{pmatrix} 2 & 1 \\ 1 & 1 \end{pmatrix}$ of the lattice $\mathbb{Z}^2$. The iterates $A, A^2, A^3, \ldots$ of A converge to a 1-dimensional basic set, which is a dense curve in T^2. Let us consider the Minkowski decomposition for A: $\begin{pmatrix} 2 & 1 \\ 1 & 1 \end{pmatrix} = \begin{pmatrix} 1 & 1 \\ 1 & 0 \end{pmatrix} \begin{pmatrix} 1 & 1 \\ 1 & 0 \end{pmatrix} \begin{pmatrix} 1 & 1 \\ 1 & 0 \end{pmatrix}$. If we write the iterates of A, then we obtain the 'Artin number' for the 1-dimensional basic set γ: $A(\gamma) = [1, 1, 1, \ldots] = \frac{1+\sqrt{5}}{2}$. On surfaces of genus $g \geq 2$ the following proposition is true.

Proposition 10.2.3 *Let ϕ^t be a minimal flow on a surface M such that one trajectory of ϕ^t coincides with a 1-dimensional basic set of an A-diffeomorphism on the same surface. Then $A(\phi)$ is a quadratic surd. Conversely, if the Artin number of a minimal flow is a quadratic surd, then at least one of its trajectories is a 1-dimensional basic set of an A-diffeomorphism on M.*

Proof. This is left to the reader; cf also [223]. □

10.3 K-Theory

C^*-algebras are essentially non-commutative objects; this fact explains their universality, on the one hand, and their complexity, on the other. For practice, it is preferable to work with something commutative (like dimension groups introduced earlier), but with structure sufficiently rich for applications.

K-theory is a generalized cohomology theory (functor) which maps the category of C^*-algebras to the category of abelian groups. The basics of K-theory were laid down by Grothendieck and developed by Atiyah *et al.*; see

M. F. Atiyah, K-Theory, Benjamin, New York, 1967.

We briefly review the basic concepts of the theory, and then consider how this technique applies to the simplest problems of dynamics. The reader is referred to the excellent book of Fillmore [106], Chapter X.

***K*-Functor** As usual, by $C_0(X)$ (X is a topological space) we denote the C^*-algebra of continuous complex-valued functions $f : X \to \mathbb{C}$ vanishing at infinity with involution $f^*(x) = \bar{f}(x)$ and norm $||f|| = \max |f|$. Let A be a Banach algebra, and let $\mathbb{R}$ be the set of real numbers. A *suspension* of A is the Banach algebra $SA = C_0(A, \mathbb{R})$ with pointwise operations and supremum norm. (For the etymology of the word, see Fillmore [106].)

Definition 10.3.1 *Let A be a Banach algebra and n a non-negative integer. A K-functor of order n on A is defined to be the iterated suspension $K_n(A) = K(S^n A)$, where $S^0(A) = A$.*

Theorem 10.3.1 (Homotopy Invariance) *Denote by K_* a K-functor of order $n \geq 0$. If A and B are two Banach algebras and $\varphi : A \to C([0, 1], B)$ is a continuous homomorphism between them, then the induced mapping $\varphi_* : K_*(A) \to K_*(B)$ is an isomorphism.*

Proof. See [106]. □

Bott Periodicity Unlike the cohomology functor H_n, the K-functor enjoys the 2-periodicity property. (This fact is an important point of the whole theory.)

Theorem 10.3.2 *Let A be a Banach algebra. Then $K(S^2 A) \cong K(A)$. In other words, the functors K_n and K_{n+2} are isomorphic for all $n \geq 0$.*

Proof. This is due to Cuntz and Higson; see Fillmore [106] for the idea of a proof. □

***K*-Groups of AF-Algebras.** Bott periodicity reduces the K-theory of C^*-algebras to the calculation of the groups K_0 and K_1. What are these groups for approximately finite-dimensional C^*-algebras? The answer is simple: $K_1(A) = 0$ for any AF-algebra A (continuity argument). For $K_0(A)$ we have the following beautiful theorem.

Theorem 10.3.3 *Let A be an approximate finite C^*-algebra and let $B(A)$ be the corresponding dimension group of A. Then $K_0(A)$ and $B(A)$ are isomorphic.* [4]

Proof. Elliott [97], Fillmore [106]. □

Pimsner-Voiculescu Diagram

Theorem 10.3.4 *Let A be a C^*-algebra and let ϕ be an automorphism of α. Let $A \rtimes_\phi \mathbb{Z}$ be the crossed product of A by integers $\mathbb{Z}$. Then the following six-term diagram is commutative:*

$$
\begin{array}{ccccc}
K_0(A) & \xrightarrow{\;id - \phi_*\;} & K_0(A) & \xrightarrow{\;i_*\;} & K_0(A \times_\phi \mathbb{Z}) \\
\uparrow & & & & \downarrow \\
K_1(A \times_\phi \mathbb{Z}) & \xleftarrow{\;i_*\;} & K_1(A) & \xleftarrow{\;id - \phi_*\;} & K_1(A)
\end{array}
$$

Proof. See Pimsner-Voiculescu [244]. □

One of the important applications of the Pimsner-Voiculescu diagram involves the crossed product rotation algebra A_α defined by formula (10.1). It can be easily seen that ϕ_* in this case is the identity map. From the diagram we have $K_* \cong K_0(C(S^1)) \oplus K_1(C(S^1)) \cong \mathbb{Z}^2$. Note that the result does not depend on α's so the rotation algebras A_α cannot be discerned by their K-groups solely.

---

[4] Moreover, they are order-isomorphic if one fixes a positive cone $K_0(A)_+$ in $K_0(A)$.

10.3.1 Foliation With Reeb Components

Let T^2 be a 2-torus with a C^∞ non-singular foliation $\mathcal{F}$ on it. Two cases are possible.

(I) $\mathcal{F}$ has no closed leaves, i.e., it is conjugate to an irrational flow on T^2 (the Kronecker foliation). The respective C^*-algebra is a crossed product $C(T^2) \rtimes_\alpha \mathbb{R}$ of the space of continuous functions on T^2 by real numbers. This algebra is Morita equivalent to the irrational rotation algebra $C(S^1) \rtimes_\alpha \mathbb{Z}$.

(II) $\mathcal{F}$ has closed leaves (the rational foliation of the torus). One can split $\mathcal{F}$ into two classes: (i) there are no closed leaves with a (one-sided) trivial holonomy group (so-called 'stable' leaves in the terminology of [282]); (ii) there are 'stable' closed leaves partitioning cylindric regions on the torus. For these two classes the following statements are valid.

Theorem 10.3.5 (Torpe) *Suppose that $\mathcal{F}$ is of class (i), orientable or possibly with Reeb components. Then*

$$K_0(C^*(T^2, \mathcal{F})) = \begin{cases} \mathbb{Z} \oplus \mathbb{Z}, & \textit{when } \mathcal{F} \textit{ is orientable,} \\ \mathbb{Z}/2, & \textit{when } \mathcal{F} \textit{ is non-orientable} \end{cases}$$

$$K_1(C^*(T^2, \mathcal{F})) = \begin{cases} \mathbb{Z} \oplus \mathbb{Z}, & \textit{when } \mathcal{F} \textit{ is orientable,} \\ \mathbb{Z}/2, & \textit{when } \mathcal{F} \textit{ is non-orientable} \end{cases}$$

Theorem 10.3.6 (Torpe) *Let $\mathcal{F}(n)$ be a foliation of class (ii) with n regions without 'stable' leaves (that is, either with n orientable cylindric components, or n Reeb components), which are separated from each other by the components partitioned by 'stable' leaves. Then*

$$K_0(C^*(T^2, \mathcal{F}(n))) = \begin{cases} \mathbb{Z}, & \textit{when } \mathcal{F} \textit{ is orientable,} \\ 0, & \textit{when } \mathcal{F} \textit{ is non-orientable} \end{cases}$$

$$K_1(C^*(T^2, \mathcal{F}(n))) = \begin{cases} \mathbb{Z} \oplus \mathbb{Z}^n, & \textit{when } \mathcal{F} \textit{ is orientable,} \\ \mathbb{Z}/2 \oplus \mathbb{Z}^n, & \textit{when } \mathcal{F} \textit{ is non-orientable} \end{cases}$$

Proof of theorems 10.3.5 and 10.3.6. The idea is to decompose $\mathcal{F}$ into 'foliation boxes' containing exactly one Reeb component, or exactly one annular region with closed leaves. Next the corresponding C^*-algebras are calculated, and then using the 6-term exact sequences the groups K_0 and K_1 are determined; see [282] for the relevant technique. □

10.3.2 Baum-Connes Conjecture

As we have seen above C^*-algebras associated to foliations on surfaces generate K-theoretic groups K_0 and K_1. (Such groups are called 'analytic' homology groups.) On the other hand, there are 'geometric' K-groups associated to (measured) foliations on surfaces. It is natural to ask how they relate to each other. It was conjectured (unpublished) that such groups are isomorphic. This hypothesis is known as *Baum-Connes Conjecture.* It is still unknown whether the conjecture is true in general, but in special cases it is known to hold. Below we treat the Baum-Connes Conjecture for foliations with Reeb components on the torus.

Theorem 10.3.7 (Torpe) *Let $\mathcal{F}$ be a C^∞ foliation on the two-dimensional torus T^2 with a finite number of Reeb components. Let $K_*(C^*, \mathcal{F})$ and $K^*(T^2/\mathcal{F})$ be, respectively, the analytic and the geometric K-groups associated to the foliation $\mathcal{F}$. Then there exists a canonical isomorphism μ which maps the generators of $K_*(C^*, \mathcal{F})$ to the generators of $K^*(T^2/\mathcal{F})$:*

$$\mu : K_*(C^*, \mathcal{F}) \longrightarrow K^*(T^2/\mathcal{F}).$$

Proof. We shall prove the Baum-Connes Conjecture by the explicit construction of such generators. (For more details the reader is referred to [282].) Let $\mathcal{F}$ be a foliation of the torus which has n Reeb components and no closed leafs partitioning the annular regions.

(I) First let us consider the 'analytic' K-groups. Let $C(S^1) \oplus \kappa(L_2(I\!R))$ be an ideal in the C^*-algebra $C^*(T^2, \mathcal{F})$ that corresponds to the interior of a Reeb component; see [282]. (Here S^1 is chosen to be transversal to $\mathcal{F}$.) We consider the following two cases.
   (i) The foliation $\mathcal{F}$ is not transverse orientable. In this case the group $K^0(T^2, \mathcal{F}) \cong \mathbb{Z} \oplus \mathbb{Z}$ is generated by $[1 \otimes q]$ and the group $K^1(T^2, \mathcal{F}) \cong \mathbb{Z}/2$ is generated by $[v \otimes q + 1 \otimes (1-q)]$; see [282].
   (ii) The foliation $\mathcal{F}$ is transverse orientable. Then the group $K^0(T^2/\mathcal{F}) \cong \mathbb{Z} \oplus \mathbb{Z}$ is generated by $[1 \otimes q]$ and $[\tilde{p}]$, where $[\tilde{p}]$ is the restriction of $[p] \in K_0(C(S^1) \rtimes_\lambda I\!R$ to a compact leaf. The group $K^1(T^2/\mathcal{F}) \cong \mathbb{Z} \oplus \mathbb{Z}$ is generated by the elements $[v \otimes q + 1 \otimes (1-q)]$ and $[\tilde{u}]$, where $[\tilde{u}]$ is the restriction of $[u] \in K_1(C(S^1) \rtimes_\lambda I\!R$ to a compact leaf.

(II) For the generators of the 'geometric' group $K^*(T^2/\mathcal{F})$ the following two cases are possible:
   (i) The foliation $\mathcal{F}$ is transversely non-orientable. Then the required isomorphism between the generators is

$$\mu(S^1, \alpha) = [1 \otimes q] \in K^0(T^2/\mathcal{F})$$

   (since α may be chosen to map S^1 into the transversal used in the above description of $[1 \otimes q]$). In the complementary case one obtains:

$$\mu(pt, s) = [v \otimes q + 1 \otimes (1-q)] \in K^1(T^2/\mathcal{F}).$$

(ii) The foliation $\mathcal{F}$ is transversely orientable. Then for the required isomorphism we get

$$\mu(S^1, \beta) = [\tilde{p}] \in K^0(T^2/\mathcal{F}).$$

(Note that we may assume that all Reeb components are orientation preserving [282].) In the complementary case,

$$\mu(T^2, Id) = [\tilde{u}] \in K^1(T^2/\mathcal{F}).$$

The above formulas establish the Baum-Connes conjecture with an explicit construction of the required isomorphism μ. □

10.4 C^*-Algebras of Morse-Smale Flows

Finally let us discuss several results on the classification of the C^* algebras of (singular) flows on higher genus surfaces, obtained by X. Wang [294], [295]. Let $\mathcal{F}$ be a foliation on a compact surface M induced by a Morse-Smale flow $\phi^t : M \times \mathbb{R} \to M$ on M. In [295] the following three invariants of $\mathcal{F}$ are considered.

- $C^*(M, \mathcal{F})$ – a C^*-algebra of a flow $\mathcal{F}$ on M
- a dual graph $\widehat{G}(\mathcal{F})$ of a Morse-Smale flow $\mathcal{F}$
- a coloured dual graph $\widehat{G}^*(\mathcal{F})$ of a Morse-Smale flow $\mathcal{F}$.

The relations between these invariants and the topological conjugacy classes of Morse-Smale flows on 2-manifolds are studied. First a relevant counterexample is built, which shows that $\widehat{G}(\mathcal{F})$ cannot serve as a complete topological invariant of a Morse-Smale flow. The above fact implies that the corresponding $C^*(M, \mathcal{F})$-algebra also cannot be used for this purpose by virtue of the following classification result.

Theorem 10.4.1 *([295]) Let $\mathcal{F}$ be a foliation, induced by a Morse-Smale flow on M. Let $\widehat{G}(\mathcal{F})$ be the dual graph of $\mathcal{F}$, see §2. Then the $C^*(M, \mathcal{F})$-algebra of the flow is isomorphic to the C^*-algebra, $C^*(\widehat{G}(\mathcal{F}))$ of the dual graph. Moreover, C^*-algebras of two Morse-Smale flows are isomorphic if and only if the two dual graphs of the flow are isomorphic.*

Proof. The proof is based on the study of the combinatorial invariants of Morse-Smale flows; cf. Ch.2 where a Peixoto graph has been introduced. □

This negative result is the main reason for introducing 'colouring' on dual graphs, i.e., the so-called u and s-cycles on $\widehat{G}$, satisfying a list of axioms; see [295] for details. Roughly speaking, dual graphs lose some important information about the embedding of the 'scheme of a flow' into M. Colouring permits the restoration of this information and allows to pass from $\widehat{G}$ to the embedding $G \hookrightarrow M$. We have the following Equivalence Theorem (theorem 4.14 in [295]):

Theorem 10.4.2 *([295]) Two Morse-Smale flows* $(M, \mathcal{F})$ *and* $(M, \mathcal{F}')$ *are topologically conjugate if and only if their (general) coloured dual graphs are isomorphic.*

Proof. The proof reduces to the embedding theorem of Heffter and Edmonds; see Ch.2. □

(As an exercise we leave it to the reader to verify that $G^{R_1} \cong G^{R_2}$ implies $\widehat{G}_1^* \cong \widehat{G}_2^*$, where R_1 and R_2 are the corresponding rotations of the flow scheme, see Ch.5.) We have a counterpart to Theorem 10.4.2 (theorem 4.2 in [295]):

Theorem 10.4.3 *([295]) Let* $\widehat{G}$ *be a graph with Inc* $(E_G) = 4$ *and a partition of* E_G *into two pairs of each* $v \in V_G$*. Then* $\widehat{G}$ *is the dual graph of a Morse-Smale flow on a closed surface of genus* g *if and only if the following conditions (1), (2) and (3) hold:*

(I) The partition of E_G *decomposes* E *into a disjoint union of 4-cycles,*
(II) There is a colouring and a direction on G *which induces an orientation on each 4-cycle, u-cycle and s-cycle,*
(III) Let p, q, r *be the numbers of u-cycles, s-cycles and 4-cycles, respectively. Then* $p + q - r = 2 - 2g$*.*

Proof. In general, items (1)-(3) correspond to the basic properties of the dual coloured graphs (and therefore are necessary). The sufficient part of the theorem is again reduced to the general embedding theorem for graphs due to Heffter and Edmonds, see Ch.2. □

Bibligraphic Notes. Operator algebras were introduced by von Neumann in the 30's, see e.g.

J. von Neumann, Zur Algebra der Funktionaloperatoren und Theorie der normalen Operatoren, Math. Annalen 102 (1929), 370-427.

The theory flourished through the influence of pivotal works of I. M. Gelfand and M. Naimark, see

I. M. Gelfand and M. Naimark, On the embedding of normed rings into the ring of operators in Hilbert space, Mat. Sbornik 12 (1943), 197-213.

A. Connes idea of applying such algebras to the study of 'dynamics' of foliations has proven to be an effective tool in the area, see

A. Connes, A survey of foliations and operator algebras, Proc. Symp. Pure Math. 38 (1982), 521-627.

The structure of C^*-algebras associated to foliations on compact surfaces has been studied by X. Wang [294],[295], A. Torpe [282] and in the series of the remarkable works I. F. Putnam; see a joint paper of T. Giordano, I. F. Putnam and C. F. Skau [117] for an account. The general classification problem for C^*-algebras has been considered by G. A. Elliott; see the bibliography to [98]. Our exposition is close to the above mentioned sources.

11. Quadratic Differentials

11.1 Notations

An ample information on the quadratic differentials and quasiconformal mappings can be found in:

F. P. Gardiner, Teichmuller Theory and Quadratic Differentials, John Wiley & Sons, 1987.

(See also an *Appendix* of this book for a short review.) By a *quadratic differential* on M one understands a collection of complex-valued functions ϕ_i defined in the local charts U_i such that on the overlapping domains of local parameters z_i, z_j they transform accordingly with the 'quadratic' rule:

$$\phi_j = \phi_i \left(\frac{dz_i}{dz_j} \right)^2$$

The quadratic differential $\phi(z)dz^2$ is said to be *holomorphic (meromorphic)* if such are all of the functions ϕ_i. A point $p \in S$ is said to be *regular* if $\phi(p) \neq 0$; for the otherwise $\phi(p) = 0$ and the point p is called *critical point* of the quadratic differential ϕ. A critical point is called *finite* if it is either zero or a first order pole of quadratic differential. Critical points form a discrete set and on the compact Riemann surface there are only finitely many of them. *Trajectory* of quadratic differential $\phi(z)dz^2$ is a line where the function Arg $\phi(z)dz^2$ vanishes.

11.2 Local Theory

11.2.1 Normal Forms

Let $\phi(z)dz^2$ be a meromorphic quadratic differential given on a compact surface M. Suppose that Sing ϕ = Zer $\phi \cup$ Pol ϕ is finite, where Zer ϕ and Pol ϕ denotes the sets of zeros and poles of ϕ, respectively.

Lemma 11.2.1 *Let $z \in$ Sing ϕ be a critical point of the quadratic differential $\phi(z)dz^2$ of an odd order n. Then there exists a local parameter ζ near 0 such that the representation of ϕ in the terms of ζ is*

$$\phi(\zeta)d\zeta^2 = \left(\frac{n+2}{2}\right)^2 \zeta^n d\zeta^2.$$

Moreover, the integral $\Phi(\zeta) = \zeta^{\frac{n+2}{2}} +$ Const and the parameter ζ is uniquely determined up to a factor $C = e^{\frac{2k\pi i}{n+2}}$ where $k = 0, 1, \ldots, n+1$.

Proof. By the assumption n is an odd integer. Let us choose a local parameter in the vicinity of a point $z \in$ Sing ϕ as follows. The function ϕ in such a point admits an expansion:

$$\phi(z) = z^n(a_n + a_{n+1}z + \ldots), \qquad a_n \neq 0.$$

Since the local parameter demands the expression $\sqrt{\phi(z)}$ to be calculated, we may select a single valued branch of the square root of expression in the brackets:

$$\sqrt{a_n + a_{n+1}z + \ldots} = b_0 + b_1 z + b_2 z^2 + \ldots$$

Therefore in a small neighborhood of 0 one obtains

$$\sqrt{\phi(z)} = z^{\frac{n}{2}}(b_0 + b_1 z + b_2 z^2 + \ldots).$$

By the term-wise integration we come to

$$w = \Phi(z) = z^{\frac{n+2}{2}}(c_0 + c_1 z + \ldots), \qquad c_k = \frac{2b_k}{n + 2(k+1)}.$$

It is possible to choose a new parameter,

$$\zeta = z(d_0 + d_1 z + \ldots) = z(c_0 + c_1 z + \ldots)^{\frac{2}{n+2}},$$

by fixing a single-valued branch of the above multi-valued function. In terms of this new parameter the expression for w is especially simple:

$$w = \Phi(z) = \zeta^{\frac{n+2}{2}}.$$

The representation of the quadratic differential $\phi(z)dz^2$ in term of the parameter ζ can be found by differentiating of Φ and squaring:

$$dw = \Phi'(\zeta)d\zeta = \frac{n+2}{2}\zeta^{\frac{n}{2}}d\zeta, \qquad dw^2 = \left(\frac{n+2}{2}\right)^2 \zeta^n d\zeta^2,$$

and in this case the statement of the lemma follows.

Let ζ' be another local parameter in the above representation. This gives us that

$$\frac{n+2}{2}\zeta^{\frac{n}{2}}d\zeta = \frac{n+2}{2}(\zeta')^{\frac{n}{2}}d\zeta',$$

and after the integration we have:

$$(\zeta')^{\frac{n+2}{2}} = \zeta^{\frac{n+2}{2}} + C \quad \Rightarrow \quad \zeta' = \left[\zeta^{\frac{n+2}{2}} + C\right]^{\frac{2}{n+2}}.$$

It is clear that for $n \geq -1$ it must hold $C = 0$ (since $\zeta'(0) = 0$). For $n \leq -3$ the simple calculations imply that

$$\zeta' = C\zeta, \qquad C = e^{\frac{2k\pi i}{n+2}}, \qquad k = 0, 1, \ldots, n+1.$$

Lemma 11.2.1 is proved. □

Lemma 11.2.2 *Let z be a finite critical point (that is a critical point of the order $n \geq -1$) of the meromorphic quadratic differential $\phi(z)dz^2$. Then one can introduce a local parameter ζ in terms of which the quadratic differential has the representation:*

$$\phi(\zeta)d\zeta^2 = \left(\frac{n+2}{2}\right)^2 \zeta^n d\zeta^2.$$

Moreover, ζ is uniquely determined up to a factor $C = e^{\frac{2k\pi i}{n+2}}, k = 0, 1, \ldots, n+$ 1.

Proof. In view of Lemma 11.2.1 it remains to prove the statement for the zeros of the even order. Indeed, putting $n = 2m$ one obtains

$$\sqrt{\phi(z)} = z^m(b_0 + b_1 z + \ldots), \qquad m = 0, 1, \ldots$$

Then integrating by z we have:

$$w = \Phi(z) = z^{m+1}(c_0 + c_1 z + \ldots), \qquad c_k = \frac{b_k}{m+k+1}.$$

As before we set

$$\zeta = z(c_0 + c_1 z + \ldots)^{\frac{1}{m+1}} = z(d_0 + d_1 z + \ldots),$$

so that

$$w = \zeta^{m+1} + \ Const, \qquad dw^2 = (m+1)^2\zeta^{2m}d\zeta^2 = \left(\frac{n+2}{2}\right)^2 \zeta^n d\zeta^2.$$

The rest of the proof is similar to those of Lemma 11.2.1. □

Lemma 11.2.3 *Let $z \in Sing\ \phi$ be a second order pole $n = -2$ of the quadratic differential $\phi(z)dz^2$. Then in terms of the local parameter ζ the quadratic differential can be represented as*

$$dw^2 = \left(\frac{a_{-2}}{z^2} + \frac{a_{-1}}{z} + a_0 + a_1 z + \ldots\right)dz^2 = \frac{a_{-2}}{\zeta^2}d\zeta^2.$$

The local parameter ζ is uniquely defined up to an arbitrary factor $C \neq 0$.

Proof. In the case $n = -2$ the main term in the integral $\Phi(z)$ is logarithmic:

$$w = \Phi(z) = b_0 \log z + b_1 z + \frac{b_2}{2} z^2 + \dots$$

By setting

$$\log \zeta = \log z + \frac{b_1}{b_0} z + \frac{b_2}{2b_0} z^2 + \dots,$$

one gets $w = \log \zeta$. The representation for the quadratic differential becomes:

$$dw^2 = \left(\frac{b_0}{\zeta} d\zeta \right)^2 = \frac{a_{-2}}{\zeta^2} d\zeta^2.$$

The uniqueness of ζ may be proved as it was done before. Lemma 11.2.3 follows. □

Lemma 11.2.4 *Let $z \in Sing \;\; \phi$ be a pole of an even order $n \leq -4$ of a quadratic differential $\phi(z)dz^2$. Then there exists a local parameter ζ in terms of which the quadratic differential is given by*

$$dw^2 = \left(\frac{n+2}{2} \zeta^{\frac{n}{2}} + \frac{b}{\zeta} \right) d\zeta.$$

Proof. Suppose $n = 2m$. Let us take a single-valued branch of the square root

$$\sqrt{\phi(z)} = z^m (b_0 + b_1 z + \dots).$$

The integration gives:

$$w = \Phi(z) = b \log z + z^{m+1}(c_0 + c_1 z + \dots), \qquad b = b_{|m|-1}.$$

We choose $\zeta = z(d_0 + d_1 z + \dots) = ze^{h(z)}$ such that

$$w = \zeta^{m+1} + b \log \zeta + C.$$

Then we get a representation:

$$dw = \sqrt{\phi(z)} dz = \left((m+1)\zeta^m + \frac{b}{\zeta} \right) d\zeta = \left(\frac{n+2}{2} \zeta^{\frac{n}{2}} + \frac{b}{\zeta} \right) d\zeta.$$

Lemma 11.2.4 is proved. □

11.2.2 Finite Critical Points

Theorem 11.2.1 (Jenkins-Strebel) *Let $z \in Sing \;\; \phi$ be a singular point of the order $n \geq -1$ and let ζ by a natural parameter chosen in a small neighborhood $U = \{z : |\zeta| \leq \rho\}$ of 0. Then the leaves of $\mathcal{F}$ subdivide U by the radii*

$$Arg\ \zeta = \frac{2k\pi}{n+2}, \qquad k = 0, \ldots, n+1,$$

into the $n+2$ sectors. Each of the above sectors is of the saddle type and the trajectories in its interior are mapped by the function $w = \zeta^{\frac{n+2}{2}}$ into the upper half-plane $Re\ \zeta \geq 0$ foliated by the horizontal lines. (The corresponding phase portraits for the cases $n = -1, n = 0, n = 1$ are shown in Fig. 11.1.)

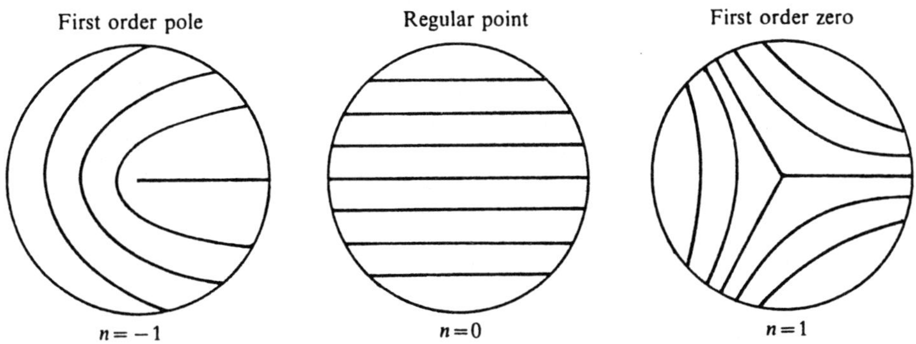

Fig. 11.1. Singular points of the orders $n = -1, n = 0, n = 1$.

Proof. Let $z \in$ Sing ϕ be a zero or a first order pole. According to Lemmas 11.2.1 and 11.2.2 the representation of $\phi(z)dz^2$ in terms of the natural parameter is

$$\left(\frac{n+2}{2}\right)^2 \zeta^n d\zeta^2.$$

If one chooses the constant of integration to be zero, the integral of the square root of ϕ becomes

$$w = \Phi(\zeta) = \zeta^{\frac{n+2}{2}}.$$

It is a straightforward exercise to check that the horizontals Im ζ = Const > 0 go by Φ into the saddle sectors

$$\frac{2k\pi}{n+2} \leq\ Arg\ \zeta \leq \frac{2(k+1)\pi}{n+2}, \qquad k = 0, 1, \ldots, n+1,$$

of the singularity in 0. The total number of such sectors is $n+2$. Theorem 11.2.1 is proved. □

11.2.3 Pole of Order 2

Theorem 11.2.2 (Jenkins-Strebel) *In the neighbourhood of a pole of order $n = -2$ the trajectories of the quadratic differential $\phi(z)dz^2$ are the images of the horizontal straight lines by the mapping*

$$\zeta = e^{\frac{w}{\sqrt{a_{-2}}}},$$

where ζ is a natural parameter. Moreover, they are the concentric circles if $a_{-2} < 0$, the radii if $a_{-2} > 0$ and the logarithmic spirals if Im $a_{-2} \neq 0$. These cases are pictured in Fig. 11.2.

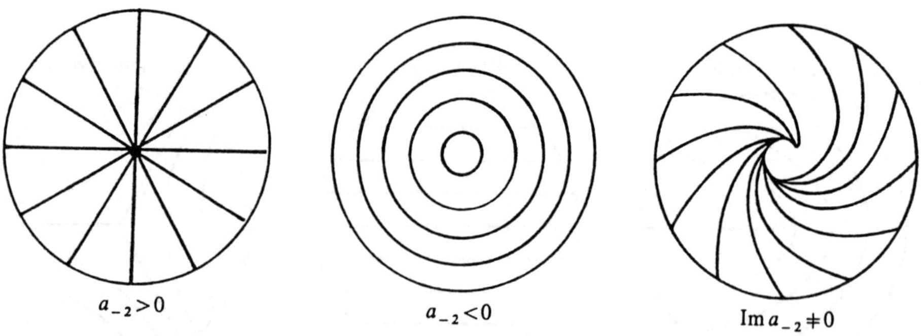

Fig. 11.2. Second order pole. Cases $a_{-2} > 0, a_{-2} < 0$, Im $a_{-2} \neq 0$.

Proof. By Lemma 11.2.3 in terms of the natural parameter ζ quadratic differential has a representation

$$\frac{a_{-2}}{\zeta^2} d\zeta^2.$$

This yields us immediately that $w = \Phi(\zeta) = b \log \zeta$, where $b = \sqrt{a_{-2}}$. There are three cases to distinguish.

(i) $a_{-2} > 0$ and b is real. The punctured disc $0 < |\zeta| \leq \rho$ is mapped onto the half-plane Re $w < b \log \rho$ $(b > 0)$. The horizontals in the w-plane go to the radii in the ζ-plane.

(ii) $a_{-2} < 0$ and b is imaginary. The punctured disc is mapped into the upper or lower half-plane. The horizontals in the w-plane correspond to the circles around $\zeta = 0$.

(iii) Im $a_{-2} \neq 0$. In this case the boundary of the punctured disc corresponds to the line in w-plane which is neither vertical nor horizontal. The trajectory arcs spirals around the point $\zeta = 0$.

All cases are considered and Theorem 11.2.2 is proved. □

11.2.4 Higher Order Poles

Theorem 11.2.3 (Jenkins-Strebel) *Let z be a pole of the order $n \leq -3$. Then there are $|n| - 2$ directions at z forming the equal angles, and the trajectories enter into z along these distinguished directions. There exists a* *neighborhood U of z such that every trajectory ray entering U tends to z. The two rays of any trajectory which stays in U tend to z in two nearby distinguished directions (thus forming an 'elliptic' sector).*

Proof. For the sake of simplicity we prove the above statement for the case when the pole has no 'logarithmic part'. (The alternative case is treated similarly, see [274] for the details.)
Let z be a pole of the odd order ≤ -3 or of the even order ≤ -4 with vanishing logarithmic term. In view of Lemmas 11.2.1 and 11.2.2, the quadratic differential is represented in terms of the natural parameter as

$$\phi(z)dz^2 = \left(\frac{n+2}{2}\right)^2 \zeta^n d\zeta^2.$$

Up to an additive constant chosen to be zero, we get $w = \zeta^{\frac{n+2}{2}}$. Let us introduce $\zeta' = w^{-1}$. Then the horizontal lines in the w-plane correspond to a family of the circles of growing diameters which are tangent to the real axis in the point $\zeta' = 0$. Using the mapping

$$\zeta = (\zeta')^{\frac{2}{|n|-2}},$$

we may recover the trajectories in terms of natural parameter ζ. (The corresponding configurations for $n = -3$, $n = -4$ and $n = -5$ are shown in Fig. 11.3.) Theorem 11.2.3 is proved. □

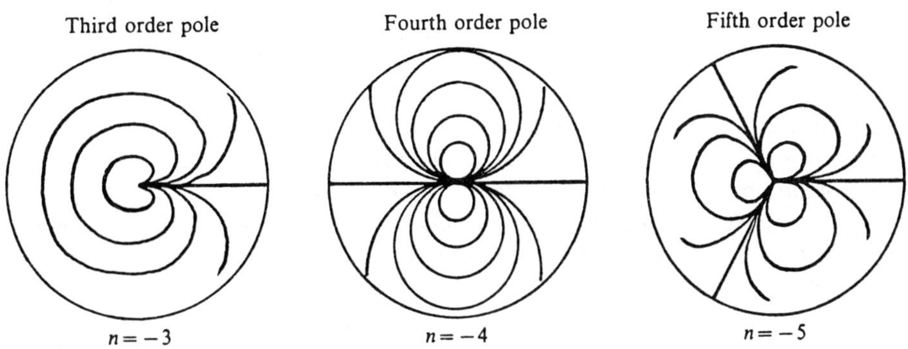

Fig. 11.3. Higher order poles. Cases $n = -3, n = -4$ and $n = -5$.

11.3 Global Behaviour of the Trajectories

Theorem 11.3.1 (Strebel) *The trajectories of a meromorphic quadratic differential $\phi(z)dz^2$ on a compact Riemann surface $\mathcal{R}$ belong to one of the following types:*

(i) *The closed trajectories filling-up the disjoint annuli (including the case of a disc punctured in the second order pole). Two annuli A_1 and A_2 coincide if and only if the trajectories $\gamma_1 \in A_1$ and $\gamma_2 \in A_2$ are freely homotopic in $\mathcal{R}\setminus Pol\,\phi$. The annulus A has a boundary ∂A degenerate to a pair of points if and only if $\mathcal{R} = S^2$ is a sphere with two second order poles with $a_{-2} < 0$;*

(ii) *The non-closed trajectories of either of the following types*
   (a) *The separatrix trajectories. At least one 'end', say γ^+, of such a trajectory γ tends to a finite $n \geq -1$ singular point. The other 'end', γ^-, tends either to another finite singular point or to an infinite $n < -1$ singular point or else is recurrent. There are only finitely many separatrix trajectories.*
   (b) *Trajectories both 'ends' of which tend to infinite singular points filling-up the simply connected regions in $\mathcal{R}$. There is a continuum of such trajectories.*
   (c) *And, finally, the trajectories both 'ends' of which are recurrent. The closure of such a trajectory, γ, is either a connected region in $\mathcal{R}$, all of $\mathcal{R} = T^2$ or a Cantor set in $\mathcal{R}$. The boundary of the region $\overline{\gamma}$ (if any) consists of the separatrix trajectories and singular points.*

Moreover, the surface $\mathcal{R}$ minus the singular points and separatrix trajectories splits into a finite number of connected regions which are either annuli, disc regions (simply connected domains) or non-trivially connected regions.

Proof. Item (i) is equivalent to the following two lemmas which we shall prove consequently.

Lemma 11.3.1 *Let $\phi(z)dz^2$ be a meromorphic quadratic differential on an arbitrary Riemann surface $\mathcal{R}$. Every closed trajectory γ of ϕ can be embedded into a maximal annular region A filled-up with closed trajectories. This region is uniquely determined except for a holomorphic quadratic differential with closed trajectories on the torus. Two annular regions A_1 and A_2 are either disjoint or identical.*

Lemma 11.3.2 *Suppose the conditions of Lemma 11.3.1 are satisfied. Let A be the maximal annular region associated to a closed trajectory γ of ϕ. If a closed trajectory γ' is freely homotopic to γ on the punctured surface $\mathcal{R}\setminus Pol\,\phi$, then $\gamma' \subset A$.*

Proof of Lemma 11.3.1. Let γ be a closed trajectory of ϕ. Then γ is the image of an interval $[0, a)$ by the mapping Φ^{-1}.There exists $\delta > 0$ such that the mapping Φ^{-1} is defined in the rectangle $\Pi = [0, a) \times (-\delta, \delta)$.

The mapping Φ^{-1} is a homeomorphism of the rectangle Π onto an annular region $A \subset \mathcal{R}$. (The horizontal lines in Π are mapped onto the closed trajectories in A.) The corresponding conformal mapping acts by the formula

$$\zeta \longmapsto \frac{a}{2\pi i} \log \zeta.$$

Extending the region A maximally into $\mathcal{R}$ we obtain the conclusion of the Lemma. (The only exceptional case is the torus $\mathcal{R} = T^2$ with a holomorphic differential on it. Here several extensions are possible, see [274] for details.) By maximality, regions A_1 and A_2 are either disjoint in $\mathcal{R}$ or coincide. Lemma 11.3.1 is proved. □

Proof of Lemma 11.3.2. If $\mathcal{R}$ is a torus and ϕ is holomorphic then if γ is closed then all the trajectories are closed, and there is one annular region.

For the other surfaces, let γ be a closed curve freely homotopic to a point in the punctured surface $\mathcal{R}\setminus$ Pol ϕ. Then γ bounds a disc D which can be mapped conformally onto the unit disc $|z| \leq 1$. Function $\phi(z)$ is analytic on $|z| \leq 1$ and the trajectories Arg $\phi(z)dz^2$ = Const must define at least one pole. This contradicts the choice of the punctured surface. Thus γ cannot be homotopic to a point.

If γ is freely homotopic to a closed trajectory γ', then γ, γ' form a boundary of annular region A which can be mapped conformally onto the annulus $r_0 \leq |z| \leq r_1$. Therefore the trajectory γ' belongs to the same annulus A as trajectory γ. Lemma 11.3.2 is proved. □

(ii) In view of the structure theorems (Theorems 11.2.1-11.2.3) for trajectories of ϕ in the vicinity of finite singular points there exists a finite number of the separatrix leaves and the claim (a) follows. Claim (b) may be deduced from the following lemma.

Lemma 11.3.3 *Let $\phi(z)dz^2$ be a meromorphic quadratic differential given on an arbitrary Riemann surface $\mathcal{R}$. Then, every trajectory γ whose ends tend to the infinite singular points, is embedded into a horizontal strip.*

Proof. Consider a semi-trajectory γ^+ which tends to a pole z of order -2. In terms of the natural parameter the quadratic differential near z has the representation:

$$dw^2 = \frac{a_{-2}}{\zeta^2} d\zeta^2, \qquad |\zeta| < \rho,$$

and therefore $w = \sqrt{a_{-2}} \log \zeta +$ Const where a_{-2} is not a negative real. We choose a constant of integration such that γ^+ corresponds to a subinterval $t > t_0$ of the real axis. The mapping Φ^{-1} is analytic on the disc around $w = u$, and it is easy to see that both in the case of the node $a_{-2} < 0$ and focus Im $a_{-2} \neq 0$ Φ^{-1} maps onto a horizontal strip of w-plane.

Let now z be a pole of the order $n \leq -3$. Then there is a point $z_1 \in \gamma^+$ such that the vertical trajectory through z_1 bounds a half-plane of w. Therefore Φ^{-1} is a conformal mapping of the disc $|w - u| < u - u_0$ onto a rectangle $u_1 < u < u_2, |v| \leq \delta$. for some $u_1 < u_2$ and $\delta > 0$. Lemma 11.3.3 is proved. □

Finally, let us consider the case (c). The existence of the recurrent trajectories of a quadratic differential follows readily from the simplest examples on torus, see [274] for details. On the surfaces of higher genus it may be shown similarly using the general methods elaborated earlier. Theorem 11.3.1 is proved. □

Bibliographic Notes. The equivalence relation between the holomorphic quadratic differentials and measures foliations on closed Riemann surfaces has been established by J. Hubbard and H. Masur [147]. They proved that every holomorphic quadratic differential defines an equivalence class of measured foliations by horizontal trajectories of the differential. The converse of this statement is also true. A simpler proof of this fact is due to M. Wolf [304]. The same author establishes a connection between the measured foliations and the harmonic maps on the Riemann surface, see [305].

12. Flat Structures

Flat metric is a metric of zero curvature with a finite number of degenerate points. Each of degenerate points is a cone of the angle θ. Flat metric taken together with cone singularities defines a flat structure on M. Flat structures are connected in many ways with other objects such as quadratic differential, measured foliations, interval exchange transformations, principal curvature lines and billiards in the rational polygons.

12.1 Flat Metric With Cone Singularities

12.1.1 Notations

By a *standard cone* of total angle θ one understands a topological space $V_\theta = \{(r,t) : r \geq 0, t \in \mathbb{R}/\theta\mathbb{Z}\}$ endowed with the metric

$$ds^2 = dr^2 + r^2 dt^2.$$

(What the total angle means geometrically, is shown in Fig. 12.1.) The total angle is a unique invariant of V_θ. Another frequently used quantities are the *point curvature* $k = 2\pi - \theta$ and the *weight* $\beta = (\theta/2\pi) - 1$.

Lemma 12.1.1 *Let $\mathbb{C}$ be a complex plane of parameter z. Then $\mathbb{C}$ endowed with the metric $ds^2 = |z|^{2\beta}|dz|^2$ is isometric to V_θ.*

Proof. If $z = x + iy$ then the isometry is given by the formula

$$x = ar^{\frac{2\pi}{\theta}} \cos\left(\frac{2\pi}{\theta}t\right), \qquad y = ar^{\frac{2\pi}{\theta}} \sin\left(\frac{2\pi}{\theta}t\right),$$

where $a = (\theta/2\pi)^{\frac{2\pi}{\theta}}$. □

A converse of Lemma 12.1.1 is true as well. Let us introduce the relevant notations. Function $h : U \to \mathbb{R}$ on a domain $U \subset \mathbb{C}$ is said to be *harmonic with logarithmic singularity of weight* β in point $p \in U$ if the function $z \mapsto h(z) - \beta \log|z - p|$ is harmonic.

Lemma 12.1.2 *Let U be an open region in $\mathbb{C}$ such that $h : U \to \mathbb{R}$ is a harmonic function with logarithmic singularity of weight $\beta > -1$ in a point $p \in U$. If one endows U with the metric $ds^2 = e^{2h}|dz|^2$, then U is isometric to a cone V_θ with $\theta = 2\pi(\beta + 1)$.*

Proof. For simplicity, we let $p = 0$. There exists a holomorphic function $g(z)$ in a neighbourhood of 0 such that

$$Re\ g(z) = h(z) - \beta \log |z|.$$

One can unfold $e^{g(z)} = a_0 + a_1 z + \ldots$ where $a_0 \neq 0$. Let us set

$$b_k = \frac{\beta + 1}{\beta + k + 1} a_k.$$

The series $\sum b_k z^k$ converges in a neighbourhood of 0 to an analytic function $f(z)$ such that $f(0) \neq 0$. Finally, we let $w(z) = z[f(z)]^{\frac{1}{\beta}+1}$. It is verified directly, that $w^\beta dw = e^{g(z)+\beta \log z} dz$. Taking the absolute values, we get $ds^2 = e^{2h(z)}|dz|^2 = |w|^{2\beta}|dw|^2$. In view of Lemma 12.1.1, we are done. □

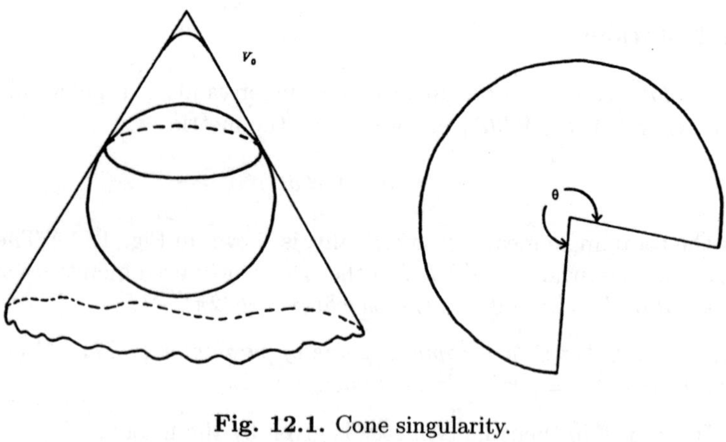

Fig. 12.1. Cone singularity.

Two riemannian metrics ds_0 and ds_1 on the manifold M are said to be *conformally equivalent* if there exists a function $h : M \to \mathbb{R}$ such that $ds_1^2 = e^{2h} ds_0^2$. The equivalence class of conformal metrics is called a *conformal structure* on M. If M is an orientable surface, then conformal structures coincide with the complex structures on M. The following lemma says that in singular points conformally equivalent flat metrics are "parametrized" by the total angle θ.

Lemma 12.1.3 *Let M be a surface with two conformal metrics ds_0 and ds_1: i.e $ds_1 = e^h ds_0$. Let ds_0 be a flat metric. Then ds_1 is flat if and only if h*

is a harmonic function. Moreover, $p \in M$ is a cone of angle θ_0 for ds_0 and of angle θ_1 for ds_1 if and only if h has a logarithmic singularity of weight $\beta_1 - \beta_0$ in p.

Proof. If p is a non-singular point both of ds_0 and ds_1 and ds_0 is flat, then Lemma 12.1.2 implies that ds_1 is flat in p. Conversely, if both ds_0 and ds_1 are flat then one can choose the coordinates (x, y) and (u, v) near p such that $ds_0^2 = dx^2 + dy^2$ and $ds_1^2 = du^2 + dv^2$. Denote $z = x + iy$ and $w = u + iv$ and let $w = g(z)$ be a unital mapping in these coordinates. Then g is holomorphic and $ds_1^2 = |dw|^2 = |g'(z)|^2|dz|^2 = e^{2h} ds_0$, where $h(z) = \log |g'(z)|$ is a harmonic function.

If p is a cone singularity of angle θ_0 for ds_0 and θ_1 for ds_1, then there exist coordinates z and w in the neighbourhood of p such that $z = w = 0$ at p and

$$ds_0^2 = |z|^{2\beta_0}|dz|^2, \qquad ds_1^2 = |w|^{2\beta_1}|dw|^2.$$

Again, let $w = g(z)$ be unital; g is holomorphic and $g'(z)$ does not vanish at the origin (simple zero). It means that $g(z) = zg_1(z)$ where $g_1(0) \neq 0$. Thus $ds_1^2 = |w|^{2\beta_1}|dw|^2 = |zg_1(z)|^{2\beta_1}|g'(z)|^2|dz|^2$. But $ds_1^2 = e^{2h} ds_0^2 = e^{2h}|z|^{2\beta_0}|dz|^2$. Therefore $h(z) = (\beta_1 - \beta_0)\log|z| + \beta_1 \log|g_1(z)| + \log|g'(z)|$ is a harmonic function with logarithmic singularity of weight $\beta_1 - \beta_0$, since $g_1(0) \neq 0$ and $g'(0) \neq 0$. □

12.1.2 Classification of Closed Flat Surfaces

By a closed *flat* surface one means the surface endowed with the metric of zero curvature and having a finite number of singular points $p_1, \ldots, p_n$ where it is isometric to a cone of angles $\theta_1, \ldots, \theta_n$, respectively. Every (closed) flat surface admits geodesic triangulation with vertices being singular points and edges which are geodesic in the chosen metric; cf Troyanov [283]. Such a triangulation is called *normal*. With this one can prove a Gauss–Bonnet formula for the flat surfaces.

Lemma 12.1.4 *If M is a closed flat surface with singularities $p_1, \ldots, p_n$ of angle $\theta_1, \ldots, \theta_n$, respectively, then*

$$\sum_{i=1}^{n} k_i = 2\pi\chi(M),$$

where $k_i = 2\pi - \theta_i$ is the point curvature in p_i and $\chi(M)$ is the Euler characteristic of M.

Proof. Fix a normal triangulation T of M. If q is a vertex of T, we set

$$\theta_q = \begin{cases} \theta_i & \text{if} \quad q = x_i \\ 2\pi & \text{for otherwise} \end{cases}$$

Denote by a, b, c the number of vertices, edges and faces of triangulation T, respectively. Because T is a triangulation, $2b = 3c$ and according to the

Euler equality, $2\pi(a-b+c) = \pi(2a-c) = 2\pi\chi(M)$. Since every triangle of T is Euclidean, $\pi c = \sum \theta_q$. Inserting this to the Euler equality we get $2\pi\chi(M) = 2\pi a - \sum\theta_q = \sum(2\pi - \theta_i) = \sum k_i$. □

Finally, we are able to classify flat surfaces. Recall that metric ds_1^2 is conformally equivalent to the metric ds^2 fixed on the Riemann surface (M, ds^2) if there exists function $h : M \to \mathbb{R}$ such that $ds_1^2 = e^{2h}ds^2$. (Note that h can take infinite values of logarithmic growth, if point is singular.)

Theorem 12.1.1 (Troyanov) *Let M be a closed orientable surface. Let $p_1, \ldots, p_n \in M$ be points coming with the positive numbers $\theta_1, \ldots, \theta_n$ such that*

$$\sum_{i=1}^{n}(2\pi - \theta_i) = 2\pi\chi(M).$$

Then in each equivalence class of conformal structures on M there exists a flat metric with cone singularities $\theta_1, \ldots, \theta_n$ in points $p_1, \ldots, p_n$. Moreover, this metric is unique after a normalization, say, by the condition area of M is equal to 1.

Proof. First, let us prove the uniqueness. If, to the contrary, ds and ds' are two such metrics, then by conformal equivalence there exists a function h such that $ds' = e^h ds$. Because metrics are flat, h must be harmonic on a compact surface M. Therefore h is a constant which by normalization of the area is 0.

To prove the existence part, the following lemma will be used.

Lemma 12.1.5 *Let M be a closed Riemann surface and let $x_1, \ldots, x_n \in M$ be a set of points coming with the real numbers $\alpha_1, \ldots, \alpha_n$ such that*

$$\sum_{i=1}^{n}\alpha_i = 0.$$

Then there exists a harmonic function $h : M \to \mathbb{R}$ with logarithmic singularities of weight $\alpha_1, \ldots, \alpha_n$ in the points $x_1, \ldots, x_n$. Two such functions h and h' may differ only by a constant.

Proof. Basically follows from the divisor theorem for the meromorphic functions, see the Appendix. □

Suppose first that the genus of M is $g > 1$. Then M can carry a non-trivial holomorphic quadratic differential ω. Suppose that $y_1, \ldots, y_k$ are zeros of ω whose order is $m_1, \ldots, m_k$, respectively. Then $ds_0^2 = |\omega|$ defines [1] a flat metric with cone singularities $y_1, \ldots, y_k$ of weight $m_1/2, \ldots, m_k/2$. By the

[1] This construction is given in more details in the next Section. On metrics, connected with the quadratic differentials, see Appendix.

Gauss-Bonnet formula, to the one hand, and the assumption of theorem, to the other, we have

$$\sum_{i=1}^{k}\left(\frac{m_i}{2}\right)=\sum_{j=1}^{n}\beta_j=2g-2,$$

where $\beta_j=\frac{\theta_j}{2\pi}-1$. In particular, $\frac{1}{2}\sum m_i-\sum\beta_j=0$. By Lemma 12.1.5, there exists a harmonic function $h:M\to\mathbb{R}$ with logarithmic singularities of weight β_j in x_i and weight $-m_i/2$ in y_i. Therefore, $ds^2=e^{2h}ds_0^2$ is the required metric.

In the case $g=0$, proof is more elaborate; see Troyanov [283] for the idea.

□

12.2 Connection With Quadratic Differentials and measured foliations

Special class of flat metrics on M is given by the holomorphic quadratic differentials. This link provides us with an analytic method which will be used in the classification of orbits in rational billiards. Yet another way to think of flat metrics is as of measured foliations and, in the case cone angle is a multiply of 2π, as of an interval exchange transformation.

Lemma 12.2.1 *Let $\omega:TM\to\mathbb{C}$ be a holomorphic quadratic differential of order m in the singular point $p\in M$. Then there exists a coordinate z of p such that $z(p)=0$ and $\omega=z^m dz^2$. Moreover, this coordinate is unique, up to a rotation through the angle $2k\pi/(m+2), k\in\mathbb{Z}$.*

Proof. Existence follows form Lemma 12.1.2; for an alternative proof see Chapter on quadratic differentials. Let $z^m dz^2=w^m dz^2$ be two such coordinates. Then we have

$$\frac{dw}{dz}=\pm\left(\frac{z}{w}\right)^{\frac{m}{2}}. \tag{12.1}$$

The solution is given by the equation $w=e^{id}z$. After the substitution into (12.1), one readily gets $d=2k\pi/(m+2)$. □

Theorem 12.2.1 *Let M be a closed Riemann surface and let ω be a non-trivial holomorphic quadratic differential on M. Let $p_1,\dots,p_n$ be zeros of ω of order $m_1,\dots,m_n$, respectively. Then ω defines a flat metric on M with cone singularities of angle $(m_1+2)\pi,\dots,(m_n+2)\pi$ in the points $p_1,\dots,p_n$.*

Proof. The metric is given accordingly with the formula $ds^2=|\omega|$. The rest of the theorem follows from Lemma 12.1.2. □

Another way of defining flat structures is via (pair of) measured foliations. The vertical foliation and the perpendicular horizontal foliation of a quadratic differential define a pair of transverse measured foliations. Conversely, suppose $\mathcal{F}_1$ and $\mathcal{F}_2$ are a pair of transverse measured foliations. One may find local coordinates (x, y) for which $\mathcal{F}_1$ and $\mathcal{F}_2$ are locally given by dx and dy, respectively. Then $\mathcal{F}_1$ and $\mathcal{F}_2$ are the vertical and horizontal foliation of a quadratic differential $dz^2 = (dx + idy)^2$. At this point we refer the interested reader to the classical paper of Hubbard and Masur [147]. Clearly, in view of Theorem 12.2.1, foliations $\mathcal{F}_1$ and $\mathcal{F}_2$ define a flat metric on M.

Further we are interested in a 1-parameter family of foliations $\mathcal{F}_\theta$ given by the horizontal trajectories of quadratic differential $e^{i\theta}\omega$. If we impose yet another restriction – that all zeros of ω are of even order – then foliation $\mathcal{F}_\theta$ is orientable.

Let I be a closed transversal to $\mathcal{F}_\theta$ for a fixed θ. There exist [2] $\theta \in S^1$ such that a first return mapping on $T : I \to I$ is correctly defined. (Mapping T is defined along the leaves of $\mathcal{F}_\theta$.) The following lemma specifies the properties of T; cf Vorobets [291].

Lemma 12.2.2 *Mapping T is an interval exchange transformation so that the number of the intervals of continuity is bounded by a constant which depends only on the flat structure. Each trajectory in the direction θ issued from point $x \in I$ either returns to I or goes to a singular point in a time which is less or equal to a constant independent of x.*

Proof. Let $x_1, \ldots, x_m$ be such points of I that the trajectories in the direction θ issued from x_i go to the singular points. Note that the number of such points does not exceed the sum of multiplicities of all singular points of the flat structure. One should add to these points at most two points which are transformed by T into the ends of the interval I. Let J be an arbitrary interval of those defined by points $x_1, \ldots, x_m$. By theorem on returnability of Poincaré, the trajectories issued form the points of J come back to I in one and the same time which does not exceed $S/|J|$ where S is the area of surface M and $|J|$ is the length of J. Evidently, the restriction of T to J is a shift. Lemma follows. □

Note that with the suspension construction, a partial converse of Lemma 12.2.2 is true. However, not every flat structure can be obtained from the suspension over interval exchange transformation, see Vorobets [291] for the reference.

12.3 Rational Billiards

An amazing application of flat structures is given by the billiards in polygons whose angles are rational multiples of π. The study of dynamics of

[2] This is a rather delicate question whose complete answer is unknown to the author. H. Masur proved that for "almost all" θ's there exist trajectories coming back arbitrary close to the initial point.

such billiards reduces to the study of geodesic flow on a compact surface M endowed with a flat structure; cf Zemlyakov and Katok [310]. In this Section we introduce notations needed for interpretation of the Veech dichotomy; cf next Section.

Suppose one has a point mass moving at unit speed in straight lines in a polygon Δ in the plane. At a side the angle of reflection equals the angle of incidence. At a vertex the reflection is undefined. This gives a flow defined on the set of unit tangent vectors to Δ except for for those vectors which determine orbits that hit a vertex. This flow is called the *billiard flow.*

Definition 12.3.1 *The billiard is called rational if all angles of the polygon are rational multiples of* π.

For each side s of Δ let $r_s \in O_2$ be the linear part of the reflection operator in s. In other words, r_s is the reflection in the line parallel to s passing through the origin. Let Γ denote a subgroup of $O(2)$ generated by r_s.

Lemma 12.3.1 *Group Γ is finite if and only if the billiard is rational.*

Proof. Follows from the fact that for given initial angle of the billiard ball only finitely many angle can appear after all possible reflections. □

There exists a compact surface corresponding to the rational billiard. Let us index copies of Δ by the elements of Γ. We glue copy Δ_{γ_1} to copy Δ_{γ_2} isometrically along the side s, if and only if $\gamma_1 \circ r_s = \gamma_2$. Instead of reflecting the orbit in Δ_{γ_1} in side s, the orbit is continued in Δ_{γ_2} in the same direction. After all possible gluings are done, a compact surface M is obtained. This surface is locally isometric to Euclidean space, except at points corresponding to vertices. At these points a certain number of copies of Δ are glued together and they form an angle $2\pi c$ around the vertex, where c is a positive integer. (Of course, number c depends only on the group Γ.) The resulting metric is a flat metric with cone singularities. At the singularity the metric is given by the formula

$$ds^2 = dr^2 + (crd\theta)^2.$$

Clearly, the geodesic flow on M fibers over S^1 on the invariant surfaces which are homeomorphic to M; on every fiber M a billiard flow is given.

Typical question asked about dynamics of billiards is existence of the periodic orbits. Peroidic orbit means that after bouncing several times, the point mass returns to its initial position and direction. Periodic orbits are known to exist for the rational billiards and they lie densely in the unit circle of parameter θ; cf [80].

By *minimal* billiard flow one understands a flow on M having an orbit whose closure is M itself. Existence of such billiard flows means that the point mass comes arbitrary close to any point in any direction of the billiard. The following theorem was proved in [54].

Theorem 12.3.1 (Boldrighini-Keane-Marchetti) *For all but a countable number of values of θ the interval exchange transformation generated by the billiard flow is minimal.*

Proof. Let $T : I \to I$ be first return mapping on a closed transversal I to the billiard flow. First, we exclude all directions $\theta \in S^1$ such that billiard flow in that directions has an orbit connecting two or more singularities. There is at most countable number of such directions, called exceptional. We are to show that if θ is not exceptional, then T is minimal.

Indeed, T is aperiodic, i.e. for each $y \in I$ the orbit $Orb\ (y) = \{T^n y : n \in \mathbb{Z}\}$ is infinite. By $q_i, i = 0, \ldots, n-1$ we denote a separation point, i.e. point of discontinuity of mapping T. Let us set

$$D_\infty = \bigcup_{i=0}^{n-1} Orb\ (q_i) \cup \{1\}.$$

Let F be a finite union of right open intervals with endpoints belonging to the countable set D_∞. Take $x \in \partial F$ and suppose that $x = T^k q_j$ for some $k \geq 0$ and some j. Then either $T^{-1}x$ is a boundary point of F or it is in $D = \{q_0, q_1, \ldots, q_{n-1}\}$. Since F has only a finite number of boundary points, there must be a positive integer s such that $T^{-s}x = y \in D$. If $y = q_i$, this would mean a periodic orbit for q_i (this case we excluded). Thus y must be a separation point, $s = 1$ and x must be a vertex. It means that T would leave invariant a subset $\Omega \in I$, so that T is minimal. □

A complete description of "phase portraits" of rational billiards is unknown. We single out an interesting result of

G. A. Galperin, Non-periodic and not everywhere dense billiard trajectories in convex polygons and polyhedrons, Commun. Math. Phys. 91 (1983), 187-211.

who proves that rational billiards may have both periodic and minimal components corresponding to one and the same direction $\theta \in S^1$.

12.4 Veech Dichotomy

We conclude this Chapter with a result of Veech on the classification of billiard trajectories in the isosceles triangles; cf [288]. The idea of Veech is to study "symmetries" in flat structure induced by the rational billiard. Roughly speaking, if a billiard flow (in fixed direction) subdivides M into the "annuli" filled-up with the periodic trajectories or regions filled-up with regular trajectories joining singular points, then there is an automorphism of flat structure which fixes singular points and sends billiard orbits to the billiard orbits. The group $\Gamma(u)$ of all such automorphisms must form a lattice (discrete subgroup of cofinite volume) in the group $SL_2(\mathbb{R})/\pm I$. If we know *a priori* that $\Gamma(u)$ is a lattice (for example by an explicit construction, see Veech [288]) then rather complete information can be obtained about the billiard on M. Note that this approach is akin to the idea of Cayley graph of a flow, see Chapter 4.

Let us fix integers $g, n > 0$ and let M_g be a closed orientable surface of genus g. By $H(g,n)$ we denote the group of orientation preserving homeomorphisms $\phi : M_g \to M_g$ which fix n distinct points of M_g. Recall that a flat structure on M_g is given by an atlas u such that:

(i) coordinate translations are locally of the form $z \to \pm z + c$,
(ii) u is compatible with the given orientation, and
(iii) u is maximal with respect to (i) and (ii).

By $\Omega(g,n)$ we denote a set of flat structures on M_g with cone singularities in the points $p_1, \dots, p_n \in M_g$. Fix $u \in \Omega(g,n)$. We define a group $Aff\ (u)$ as the set of homeomorphisms $\phi \in H(g,n)$ such that ϕ is *affine linear* in every chart of atlas u. Note that if $\phi \in Aff(u)$ then the derivative of ϕ in every u-chart is a constant matrix A defined up to a factor $\pm I$. Because ϕ is orientation preserving, $det\ A > 0$.

It will be always assumed further that flat structures $u \in \Omega(g,n)$ are *positive*, i.e. quadratic differential ω corresponding to u is the square of a holomorphic 1-form on M. If one fixes a direction vector $\xi \in \mathbb{R}^2$, then the foliation of the plane $\mathbb{R}^2$ by the lines parallel to ξ can be glued together to a foliation $\mathcal{F}_\xi(u)$ on the surface M_g, for every $u \in \Omega(g,n)$. The singularities $p_1, \dots, p_n$ of $\mathcal{F}_\xi(u)$ are $l-prong$ saddles, non-orientable in general. But after we imposed positivity on u, $\mathcal{F}_\xi(u)$ is orientable. Notice also that $\mathcal{F}_\xi(u) = \mathcal{F}_{\alpha\xi}(u)$ for all real $\alpha \neq 0$.

Lemma 12.4.1 *Let $u \in \Omega(g,n), \phi \in Aff(u)$ and $A \in SL_2(\mathbb{R})/\pm I$. If a non-zero vector $\xi \in \mathbb{R}^2$ is such that $A\xi = \pm\xi$, then either*

(i) $A \equiv Id$ or
(ii) every leaf of $\mathcal{F}_\xi(u)$ is closed or a saddle connection.

In case (ii) surface M_g decomposes into maximal cylinders $C_1, \dots, C_r$ of closed leaves such that each cylinder C_j is bounded by saddle connections. Moreover, there exists $k > 0$ such that $\phi^k C_j = C_j$ and $\phi^k|_{\partial C_j} = Id$.

Proof. Assume $A \not\equiv Id$, else there is nothing to prove. It is not hard to see that $\phi\mathcal{F}_\xi(u) = \mathcal{F}_\xi(u)$, that is ϕ is a "symmetry" of $\mathcal{F}_\xi(u)$. We may choose $k > 0$ so that ϕ^k maps each separatrix of $\mathcal{F}_\xi(u)$ to itself. Note that ϕ^k is a translation along each separatrix with a fixed point at at least one end. Therefore, $\phi^k = Id$ on each separatrix.

It is known that a separatrix is either dense in an open set or is a saddle connection. Former case is excluded, since it would follow that $\phi^k = Id$ and $A(\phi^k) = Id$. However, $A(\phi)$ is a non-trivial unipotent and therefore $A(\phi)^k \neq Id$. Thus every separatrix of $\mathcal{F}_\xi(u)$ is a saddle connection so that collection of all saddle connections bounds cylinders $C_1, \ldots, C_r, r > 0$ on M_g. □

Recall that an assignment $\phi \mapsto D\phi$ where $D\phi$ is "linear part" of ϕ, determines a homomorphism $a : Aff(u) \to SL_2(\mathbb{R})/\pm I$ from affine linear homeomorphisms of M_g to the group of projective linear transformations. In what follows, we denote by $\Gamma(u)$ the image of $Aff\ (u)$ under a.

Lemma 12.4.2 *If $u \in \Omega(g,n)$ then $\Gamma(u)$ is a discrete subgroup of $SL_2(\mathbb{R})/\pm I$.*

Proof. Follows from the fact that directions of vectors $\xi \in \mathbb{R}^2$ are discrete (and dense) in the unit circle S^1. □

In general, $\Gamma(u)$ is *not* a lattice (discrete subgroup of cofinite volume) in $SL_2(\mathbb{R})/\pm I$. But in the case when it does, rather strong classification results can be obtained, see [288]. It occurs, in particular, when one considers a rational billiard in the isosceles triangles.

Definition 12.4.1 *Fix $n > 2$. By $T(n)$ we understand an isosceles triangle with angles $\pi/n, \pi/n$ and $(n-2)\pi/n$.*

Theorem 12.4.1 (Dichotomy of Veech) *Let $\xi \in \mathbb{R}^2$ be a non-zero vector. Consider a billiard in the triangle $T(n)$. There are two possibilities for billiard trajectories which are somewhere parallel to ξ:*

(i) Either for every $x \in T(n)$ the billiard trajectory through x in the direction ξ is periodic or begins and ends in a vertex;
(ii) Or for every $x \in T(n)$ the billiard trajectory through x in the direction ξ begins and ends at distinct vertices of $T(n)$ or it is minimal.

Proof. Difficult part of the proof is to show that for the billiard in question the corresponding flat structure u admits a discrete symmetry group $\Gamma(u)$ which is a lattice. The following lemma yields an explicit representation to such a symmetry.

Lemma 12.4.3 (Veech) *Let $g \geq 1$ and $\varepsilon = 1, 2$ such that $n = 2g + \varepsilon \neq 4$. Let C_n be the smooth affine complex curve* [3] *given by the equation*

$$y^2 + x^n = 1.$$

Let $q_n = c_n \frac{dx^2}{y^2}, c_n > 0$ be the holomorphic quadratic differential on C_n chosen so that $||q_n|| = 1$. The pair (C_n, q_n) determines a point $x_n \in \mathcal{M}(g)$ in the

[3] See the *Appendix* for the relevant definitions and basic facts.

moduli space $\mathcal{M}(g) = \Omega(g,n)/H(g,n)$ *such that* $\Gamma(x_n)$ *is a lattice with the generators*

$$\sigma_n = \begin{pmatrix} 1 & 2\cot\frac{\pi}{n} \\ 0 & 1 \end{pmatrix}, \qquad \beta_n = \begin{pmatrix} \cos\frac{2\pi}{n} & -\sin\frac{2\pi}{n} \\ \sin\frac{2\pi}{n} & \cos\frac{2\pi}{n} \end{pmatrix}.$$

Proof. We refer the interested reader to the original proof due to Veech [288]. □

Suppose now that $u \in \Omega(g,n)$ is a flat structure corresponding to our billiard. By Lemma 12.4.3 $\Gamma(u)$ is a lattice. If $\xi \in \mathbb{R}^2$ is a direction vector such that the foliation $\mathcal{F}_\xi(u)$ admits a saddle connection, then there exists $\phi \in Aff(u)$ such that $a(\phi) \neq Id$ and $\phi\mathcal{F}_\xi(u) = \mathcal{F}_\xi(u)$. In particular, every leaf of $\mathcal{F}_\xi(u)$ is either closed or a saddle connection. To finish the proof of theorem, it remains to apply Lemma 12.4.1. □

Bibliographic Notes. Flat metric is a typical example of a Liouville metric. The latter is characterized by complete integrability of the geodesic lines. The geometric study of flat metrics with cone singularities has been undertaken by Troyanov; see for example [283] and other works of the same author. An application of flat structures to the billiards is due to Zemlyakov and Katok [310]. An interaction between the billiards, the Teichmuller theory and flat metrics was studied by Veech [288]; see also a survey of Vorobets [291].

13. Principal Curvature Lines

13.1 Local Theory

Theory of surface foliations grew from the classical problems of differential geometry. By the end of the 19th century due to the works of Cayley, Darboux and Picard an important class of foliations given by the lines of principal curvature was singled out. To this date theory of such foliations is the hardest one both locally (umbilic points, Carathéodory conjecture) and globally (periodic and non-trivially recurrent leaves). An excellent introduction to the area is the monograph:

C. Gutiérrez & J. Sotomayor, Lines of Curvature and Umbilical Points on Surfaces, Instituto de Matemática Pura e Aplicada, Rio de Janeiro, 1994, ISBN 85-244-0057-9.

13.1.1 Notations

The space of C^∞ immersions of the surface S into E^3 is denoted by $I(S, \mathrm{E}^3)$. It is endowed with the uniform C^r topology, in which two mappings $i_1, i_2 : S \to \mathrm{E}^3$ are ε-close if for all $k \le r$ and k-jets J^k, $|J_1^k(x) - J_2^k(x)| < \varepsilon$ uniformly in $x \in S$. Fix an immersion $i \in I$ and a point $x \in S$. Further we consider the local immersions and for such we fix a *Monge chart* (u, v) connected with the point x. In this chart the immersion acts by the formula $(u, v) \mapsto (u, v, f(u, v))$, where f is of the class C^∞ at $(0, 0)$.

Consider the first and the second fundamental forms of the immersed surface,

$$\begin{aligned} \Omega^1 &= E(u,v)du^2 + 2F(u,v)dudv + G(u,v)dv^2, \\ \Omega^2 &= L(u,v)du^2 + 2M(u,v)dudv + N(u,v)dv^2. \end{aligned}$$

Here we have

$$\begin{aligned} &E = 1 + f_u^2, \quad F = f_u f_v, \quad G = 1 + f_v^2, \\ &L = \frac{f_{uu}}{\sqrt{1 + f_u^2 + f_v^2}}, \quad M = \frac{f_{uv}}{\sqrt{1 + f_u^2 + f_v^2}}, \quad N = \frac{f_{vv}}{\sqrt{1 + f_u^2 + f_v^2}}. \end{aligned}$$

The normal curvature is defined by the quotient Ω^2/Ω^1. The maximal and the minimal values of the normal curvature at x can be found from the equation

$$(M^2 - LN)k^2 + (GL + EN - 2FM)k + F^2 - EG = 0.$$

The corresponding principal directions must obey the following differential equation,

$$(EM - FL)du^2 + (EN - GL)dudv + (FN - GM)dv^2 = 0.$$

If $(0,0)$ is an umbilical point then the quadratic equation for the curvature k has a double root $k_1 = k_2$. Since in this point the Gaussian curvature $K = k_1k_2$ is positive, it can always be made minimum (maximum) of the function f (that is $f_u(0,0) = f_v(0,0) = 0$). This fact especially simplifies the calculations, and for the coefficients of the 'curvature lines' equation one obtains $A(u,v) = f_{uv}, B(u,v) = \frac{1}{2}(f_{vv} - f_{uu}), C(u,v) = -f_{uv}$. Now we have the following differential equation of the umbilical point:

$$f_{uv}du^2 + (f_{vv} - f_{uu})dudv - f_{uv}dv^2 = 0, \quad f_{uv}(0,0) = (f_{vv} - f_{uu})(0,0) = 0.$$

Definition 13.1.1 *An isolated umbilical point given by the above differential equation, is called simple if the curves $A(u,v) \equiv f_{uv} = 0$ and $B(u,v) \equiv f_{vv} - f_{uu} = 0$ intersect transversely at the point $(0,0)$. In other words, the Jacobian*

$$\begin{vmatrix} A_u & A_v \\ B_u & B_v \end{vmatrix} = A_uB_v - A_vB_u$$

does not vanish at $(0,0)$.

13.1.2 Invariants of the 2-Jets

Let $w(0)$ be a germ of the $\mathbb{Z}_2$-symmetric vector field and let $P_2(u,v)$ and $Q_2(u,v)$ be the quadratic polynomials in the right hand sides of $w(0)$. Let us study the action of the general linear group $GL_2(\mathbb{R})$ on the 2-jets of $w(0)$:

$$\frac{du}{dt} = a^1_{11}u^2 + 2a^1_{12}uv + a^1_{22}v^2, \quad \frac{dv}{dt} = a^2_{11}u^2 + 2a^2_{12}uv + a^2_{22}v^2.$$

For the sake of brevity we let

$$\begin{aligned}
\mu &= (a^1_{11}a^2_{22} - a^1_{22}a^2_{12})^2 - 4(a^1_{11}a^2_{12} - a^1_{12}a^2_{11})(a^1_{12}a^2_{22} - a^1_{22}a^2_{12}),\\
\eta &= -27(a^1_{22})^2(a^2_{11})^2 - 4a^1_{22}(a^1_{11} - 2a^2_{12})^3 + 18a^1_{22}a^2_{11}(a^1_{11} - 2a^2_{12})(a^2_{22} - 2a^1_{12}) -\\
&- 4a^2_{11}(a^2_{22} - 2a^1_{12})^3 + (a^1_{11} - 2a^2_{12})^2(a^2_{22} - 2a^1_{12})^2,\\
R &= [3a^2_{11}(a^2_{22} - 2a^1_{12}) - (a^1_{11} - 2a^2_{12})^2]x^2 + [(a^1_{11} - 2a^2_{12})(a^2_{22} - 2a^1_{12}) - 9a^1_{22}a^2_{11}]xy +\\
&+ [3a^1_{22}(a^1_{11} - 2a^2_{12}) - (a^2_{22} - 2a^1_{12})^2]y^2,\\
\theta &= (a^1_{11}a^2_{12} - a^2_{11}a^1_{12})[a^1_{11}a^1_{22} - (a^1_{12})^2 - 2a^2_{12}a^1_{22}] + (a^1_{12}a^2_{22} - a^2_{12}a^1_{22})[a^2_{11}a^2_{22} -\\
&- (a^2_{12})^2 - 2a^1_{12}a^2_{11}] + (a^1_{11}a^2_{22} - a^2_{11}a^1_{22})(a^1_{22}a^2_{11} + a^1_{12}a^2_{12}).
\end{aligned}$$

Denote by $J_0 \subset J$ a subspace $J_0 = E_0 \cup E_2$ of 2-jets, such that $E_0 = \{\mu \neq 0\}$ and $E_2 = \{\mu = \theta = 0, \eta < 0\}$. The set E_0 of codimension 0 consists of pairs (P_2, Q_2) which are relatively prime. [1] The set E_2 of codimension 2 consists of (P_2, Q_2) having a quadratic common factor $M(u,v)$, which is undecomposable into the linear factors.

Lemma 13.1.1 *Quadratic system of the class J_0 can be brought by linear non-degenerate transformations $q \in GL_2(\mathbb{R})$ to one of the k-parametric normal forms, $k \leq 2$. They are:*

(I) *In case $\eta < 0$:*

$$\frac{du}{dt} = \alpha u^2 + \beta uv - v^2, \quad \frac{dv}{dt} = (\alpha + 1)uv + \beta v^2, \quad \alpha \neq 0;$$

(II) *In case $\eta > 0$:*

$$\frac{du}{dt} = \alpha u^2 + (\beta - 1)uv, \quad \frac{dv}{dt} = (\alpha - 1)uv + \beta v^2, \quad \alpha\beta(\alpha + \beta - 1) \neq 0;$$

(III) *In case $\eta = 0, R \neq 0$:*

$$\frac{du}{dt} = \alpha u^2 + uv, \quad \frac{dv}{dt} = (\alpha - 1)uv + v^2 \quad \alpha \neq 0;$$

(IV) *In case $\eta = R = 0$:*

$$\frac{du}{dt} = uv, \quad \frac{dv}{dt} = \pm u^2 + v^2.$$

Proof. Let us consider a differential 1-form

$$\begin{aligned}\omega &= (a_{11}^2 u^2 + 2a_{12}^2 uv + a_{22}^2 v^2)du - (a_{11}^1 u^2 + 2a_{12}^1 uv + a_{22}^1 v^2)dv = \\ &= a_{11}^2 u^2 du + (2a_{12}^2 uvdu - a_{11}^1 u^2 dv) + (a_{22}^2 v^2 du - 2a_{12}^1 uvdv) - a_{22}^1 v^2 dv,\end{aligned}$$

which vanishes on $w(0)$. There is a natural isomorphism $V \otimes V \otimes dV \cong V \otimes V \otimes V$ between the differential forms ω and the space of cubic forms. Therefore it is sufficient to study $GL_2(\mathbb{R})$-normal forms of the following cubic:

$$F = a_{11}^2 u^3 + (2a_{12}^2 - a_{11}^1)u^2 v + (a_{22}^2 - 2a_{12}^1)uv^2 - a_{22}^1 v^3.$$

(Cubic form F is called a *comitant form* of system $w(0)$. It does not vanish identically, since otherwise we have $\mu = \eta = \theta = 0$. Note also that the discriminant of F coincides with η.) It is classical that by the linear transformations $q \in GL_2(\mathbb{R})$ cubic form can be brought to either of the following normal forms

[1] In other words, have neither linear nor quadratic common factors.

$$\text{(I)}\quad v(u^2+v^2),\quad \text{(II)}\quad uv(v-u),\quad \text{(III)}\quad u^2v,\quad \text{(IV)}\quad u^3.$$

(The above normal forms correspond to the possible distribution of the roots of F: it is either $(1_{I\!R}, 1_{\mathbb{C}}, 1_{\mathbb{C}})$ – one simple real root and two complex conjugate roots, or $(1_{I\!R}, 1_{I\!R}, 1_{I\!R})$ – three simple real roots, or $(2_{I\!R}, 1_{I\!R})$ – one double and one simple real, or else $(3_{I\!R})$ – one triple real root.) Comparing (I)-(IV) with F, one obtains immediately

$$\begin{aligned}
&\text{(I)} && a_{11}^2 = a_{22}^2 - 2a_{12}^1 = 0, \quad 2a_{12}^2 - a_{11}^1 = 1, \quad a_{22}^1 = -1,\\
&\text{(II)} && a_{11}^2 = a_{22}^1 = 0, \quad a_{22}^2 - 2a_{12}^1 = 1, \quad 2a_{11}^2 - a_{11}^1 = -1,\\
&\text{(III)} && a_{11}^2 = a_{22}^2 - 2a_{12}^1 = a_{22}^1 = 0, \quad 2a_{12}^2 - a_{11}^1 = 1,\\
&\text{(IV)} && a_{11}^2 = 1, \quad 2a_{12}^2 - a_{11}^1 = a_{22}^2 - 2a_{12}^1 = a_{22}^1 = 0.
\end{aligned}$$

To obtain normal forms (I)-(IV) mentioned in the lemma, it remains to denote a_{11}^1 by α and a_{22}^2 by β, and substitute it in $w(0)$. In the case of the normal forms (III) and (IV) an additional simplification is available. Namely, substituting $v \mapsto \beta^{-1}v$ in the case of (III) and $u\sqrt{|\beta|} \mapsto u, \alpha u + \beta v \mapsto v$ in the case of (IV), one comes to its final form, mentioned in the lemma. (In the last case, sgn β corresponds to $\pm$ in the normal form (IV)). The restrictions on the parameters α and β follow from the condition $\mu \neq 0$, what can be checked directly. In case (I) the condition $\alpha \neq 0$ is equivalent to $\{\mu \neq 0\} \cup \{\mu = \theta = 0\}$, where $\mu = -\alpha[(\alpha+1)^2 + \beta^2], \theta = (\alpha+1)[(\alpha-1)^2 + \beta^2]$.

Also a direct computation of η and R for (I)-(IV) gives us an $GL_2(I\!R)$-invariant description of the orbits of the normal forms obtained. Lemma 13.1.1 is proved. □

13.1.3 Stability Lemma

Lemma 13.1.2 *A 2-jet of the germ $w(0)$ is topologically sufficient if and only if it belongs to the class J_0.*

Proof. (A) Sufficient conditions. We suppose that the 2-jet J_0 has been brought to one of the normal forms (I)-(IV) of Lemma 13.1.1. Our proof will consist in the step-by-step resolving of the germ $w(0)$ at the critical point 0. The relevant technique was described earlier.

Normal form (I). The blowing-up process of resolving (unfolding) the singularity at the point (0,0), works as follows. After the first step of blowing-up, the germ w in the local chart (u, η), where $\eta = v/u$, gets the form

$$\frac{du}{dt} = \alpha u + \beta u\eta - u\eta^2 + \frac{1}{u}\phi(u, u\eta),$$
$$\frac{d\eta}{dt} = \eta + \eta^3 - \frac{\eta}{u^2}\phi(u, u\eta) + \frac{1}{u^2}\psi(u, u\eta),$$

and in the local chart (θ, v), where $\theta = u/v$, it can be written as:

$$\frac{d\theta}{dt} = -1 - \theta^2 - \frac{\theta}{v^2}\psi(\theta v, v) + \frac{1}{v^2}\phi(\theta v, v),$$
$$\frac{dv}{dt} = \beta v + (\alpha + 1)\theta v + \frac{1}{v}\psi(\theta v, v).$$

First, let us notice that there exists a characteristic orbit $u = 0$, which enters the origin of (u, η). It implies that $(0,0)$ of w also admits such. This guarantees (along with the Lojasiewicz conditions, see §1.2.2) that blowing-up will be over after a finite number of steps. (Although it will be checked immediately in below).

At the axis $u = 0$ system (u, η) has a unique singular point $(0,0)$, which is, by the condition $\alpha \neq 0$, elementary (a node when $\alpha > 0$ and a saddle when $\alpha < 0$). At the axis $v = 0$ system (θ, v) has no singular points and therefore no trajectory of w enters the origin in the direction $u = 0$. Thus, after the first step of the blowing-up process, we obtain hyperbolic singular points. It follows readily from the above equations that the perturbation terms ϕ and ψ do not change the scheme of elementary points of the singularity. (Since the coefficients of the higher degree monomials of w do not enter the linear parts of (u, η) and (θ, v).) It means that the 2-jet of w is C^0-sufficient and defines the topology of orbits near the origin of w. Note that the conjugating homeomorphism h satisfies the equality $h(-x, -y) = -h(x, y)$, and therefore h commutes with the involution θ.

Normal form (II). Blowing-up the singularity at the origin of w with the 2-jet of the form (II), we rewrite it in the local charts (u, η) and (θ, v) as follows:

$$\frac{du}{dt} = \alpha u + (\beta - 1)u\eta + \frac{1}{u}\phi(u, u\eta),$$
$$\frac{d\eta}{dt} = -\eta + \eta^2 - \frac{\eta}{u^2}\phi(u, u\eta) + \frac{1}{u^2}\psi(u, u\eta);$$

$$\frac{d\theta}{dt} = -\theta + \theta^2 - \frac{\theta}{v^2}\psi(\theta v, v) + \frac{1}{v^2}\phi(\theta v, v),$$
$$\frac{dv}{dt} = \beta v + (\alpha - 1)\theta v + \frac{1}{v}\psi(\theta v, v).$$

System (u, η) has two singular points at the axis $u = 0$: $M_1(0,0)$ and $M_2(0,1)$. At M_1 the eigenvalues are: $\lambda_1 = \alpha, \lambda_2 = -1$ and at $M_2 : \lambda_1 = \alpha + \beta - 1, \lambda_2 = 1$. By the condition $\alpha\beta(\alpha + \beta - 1) \neq 0$, the points M_1 and M_2 are elementary.

System (θ, v) has a unique singular point $M_3(0,0)$ which lies on the axis $v = 0$. M_3 is an elementary singularity since at this point $\lambda_1 = -1, \lambda_2 = \beta \neq 0$. Thus, the scheme of elementary points of the singularity of w with the 2-jet of form (II) does not depend on terms of the order higher than 2. As before, it means that the 2-jet of the germ w is C^0-sufficient, and defines completely the topology of orbits near the origin of w. Here again, the conjugating homeomorphism h commutes with the involution θ, $h(-x, -y) = -h(x, y)$.

Normal form (III). Unfolding the singularity of w with the 2-jet in the normal form (III), let us rewrite it in the local charts (u, η) and (θ, v),

$$\frac{du}{dt} = \alpha u + u\eta + \frac{1}{u}\phi(u, u\eta),$$
$$\frac{d\eta}{dt} = -\eta - \frac{\eta}{u^2}\phi(u, u\eta) + \frac{1}{u^2}\psi(u, u\eta),$$

$$\frac{d\theta}{dt} = \theta^2 + \frac{1}{v^2}\phi(\theta v, v) - \frac{\theta}{v^2}\phi(\theta v, v),$$
$$\frac{dv}{dt} = v + (\alpha - 1)\theta v + \frac{1}{v}\psi(\theta v, v).$$

At the axis $u = 0$ the system (u, η) has a unique singularity $(0, 0)$ with the eigenvalues $\lambda_1 = \alpha, \lambda_2 = -1$. By the condition $\alpha \neq 0$, this singularity is elementary hyperbolic, being a saddle if $\alpha > 0$ and a node if $\alpha < 0$.

The unique singularity $(0, 0)$ of system (θ, v) is also elementary (non-hyperbolic). It can be easily seen, that it is a saddle-node of the multiplicity 2. Therefore, the singularity scheme is completely defined by the above two singularities, and does not depend on the higher degree monomials of the germ w. It means that the topology of w is defined by its 2-jet.

Normal form (IV). Finally, let us unfold the singularity of the germ w with the 2-jet in the form (IV). In the local charts (u, η) and (θ, v) it writes

$$\frac{du}{dt} = u\eta + \frac{1}{u}\phi(u, u\eta)$$
$$\frac{d\eta}{dt} = \pm 1 - \frac{\eta}{u^2}\phi(u, u\eta) + \frac{1}{u^2}\psi(u, u\eta),$$

$$\frac{d\theta}{dt} = \mp\theta^3 + \frac{1}{v^2}\phi(\theta v, v) - \frac{\theta}{v^2}\psi(\theta v, v)$$
$$\frac{dv}{dt} = v \pm \theta^2 v + \frac{1}{v}\psi(\theta v, v).$$

There are no singular points of the system (u, η) at the axis $u = 0$. Taking another chart given by (θ, v) we find that there is a unique singularity $(0, 0)$ which lie at $v = 0$. It is again elementary non-hyperbolic. According to the sign $\mp$ it is either a saddle $(-)$ or a node $(+)$ of the multiplicity 3. Thus, the singularity scheme is not influenced by the higher-order monomials of w, and we are done. It implies that the 2-jet of w is C^0-sufficient. Since all cases are considered, (A) follows.

(B) Necessary conditions. Indeed, suppose to the contrary, that the 2-jet of w is C^0-sufficient, but $w(0) \notin J_0 = E_0 \cup E_2$. It means that either $w(0) \in \{\mu =$

$0, \theta \neq 0\}$ or $w(0) \in \{\mu = \theta = 0, \eta \geq 0\}$. Let us study consequently the both cases.

(i) $w(0) \in \{\mu = 0, \theta \neq 0\}$. In this case the components $P_2(u,v), Q_2(u,v)$ have a linear common factor, $L(u,v)$. The regular curve $L(u,v) = 0$ passes through the point $(0,0)$ and is filled up with the singular points. (The case of a non-isolated singularity.)

(ii) $w(0) \in \{\mu = \theta = 0, \eta \geq 0\}$. The components $P_2(u,v), Q_2(u,v)$ have a quadratic common factor, $M(u,v)$, which is decomposable into the linear factors, $M(u,v) = L_1(u,v)L_2(u,v)$. In this case we have a pair of straight lines (case $L_1 \equiv L_2$ is not excluded), $L_1(u,v) = 0, L_2(u,v) = 0$ through $(0,0)$, filled up with the singularities. (Again the case of a non-isolated singular point.)

In both cases, (i) and (ii), it is easy to construct a perturbation $\phi(u,v), \psi(u,v)$ by the higher degree monomials, such that the singularity curves $L_i(u,v) = 0$ vanish. (E.g. one can take monomials $A_{2k}u^{2k}, B_{2k}v^{2k}, k \geq 2$ for this purpose.) Thus $(0,0)$ becomes an isolated singularity. This is a contradiction with the assumption that the 2-jet is topologically sufficient.

Lemma 13.1.1 is proven. □

Let $\mathcal{O}^+ (\mathcal{O}^-; \mathcal{O}^\pm)$ denote a separatrix of singular point p, whose nearby orbits tend to p (do not tend to p; tend to p from one side). Clearly, each singularity can be encoded by a finite ordered sequence of such separatrices. (Example: $\mathcal{O}^+\mathcal{O}^+$ gives a singularity with two elliptic sectors of the Euler-Poincaré index +2.)

Corollary 13.1.1 *Let the 2-jet of the germ $w(0)$ be of the class J_0. Then the configurations of the orbits near 0 are C^0-conjugate to one of the topological types, given by Table 13.1.*

Singularity	Index	Normal form	Conditions	$GL_2(\mathbb{R})$-invariants
$\mathcal{O}^+\mathcal{O}^+$	+2	(I) (IV)	$\alpha > 0$ $-$	$\eta < 0, \mu < 0$ $\eta = R = 0, \mu < 0$
$\mathcal{O}^-\mathcal{O}^-$	0	(I) (I) (IV)	$\alpha < 0$ $\alpha = -1$ $+$	$\eta < 0, \mu > 0$ $\eta < 0, \mu = \theta = 0$ $\eta = R = 0, \mu > 0$
$\mathcal{O}^+\mathcal{O}^-\mathcal{O}^-\mathcal{O}^+\mathcal{O}^-\mathcal{O}^-$	0	(II)	$\alpha\beta(\alpha+\beta-1) > 0$	$\eta > 0, \mu > 0$
$\mathcal{O}^+\mathcal{O}^-\mathcal{O}^+\mathcal{O}^+\mathcal{O}^-\mathcal{O}^+$	+2	(II)	$\alpha\beta(\alpha+\beta-1) < 0$, $(\alpha-1)(\beta-1)(\alpha+\beta) < 0$	$\eta > 0, \mu < 0, \theta < 0$
$\mathcal{O}^-\mathcal{O}^-\mathcal{O}^-\mathcal{O}^-\mathcal{O}^-\mathcal{O}^-$	-2	(II)	$\alpha\beta(\alpha+\beta-1) < 0$, $(\alpha-1)(\beta-1)(\alpha+\beta) > 0$	$\eta > 0, \mu < 0, \theta > 0$
$\mathcal{O}^\pm\mathcal{O}^+\mathcal{O}^\pm\mathcal{O}^+$	+2	(III)	$\alpha < 0$	$\eta = 0, R \neq 0, \mu < 0$
$\mathcal{O}^\pm\mathcal{O}^-\mathcal{O}^\pm\mathcal{O}^-$	0	(III)	$\alpha > 0$	$\eta = 0, R \neq 0, \mu > 0$

Table 13.1. Invariants of the germs of class J_0

Proof. Follows from Lemma 13.1.1. It remains to notice only that the condition $\alpha\beta(\alpha+\beta-1) < 0$ implies $(\alpha-1)(\beta-1)(\alpha+\beta) \neq 0$. (The last expression is defined by the invariant θ, introduced in previous section.) □.

13.1.4 Classification of Simple Umbilics

Theorem 13.1.1 *Let $x \in S$ be an isolated simple umbilical point of a given C^∞ immersion $i \in I$ of the surface S into E^3. Then the 3-jet of the germ $i(x)$ at the point x is topologically sufficient. Moreover, the configurations of lines of the principal curvature near a simple umbilical point are exhausted by a finite number of topological classes represented by Fig. 13.1.*

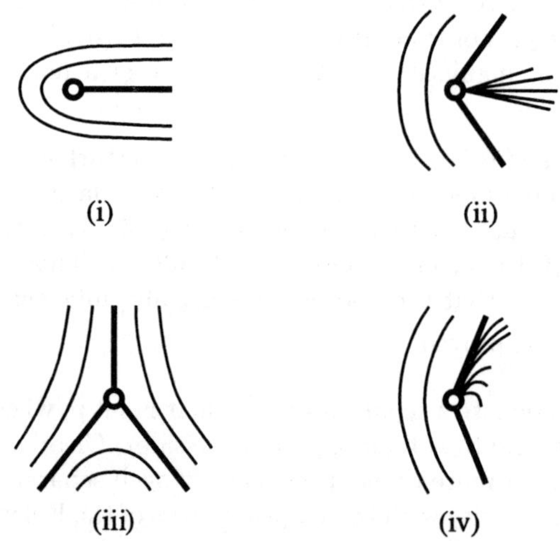

Fig. 13.1. The simple umbilical points.

Proof. Roughly, proof will consist in the reduction of the vector field given by the curvature 2-form to a 'good' vector field with Z_2-symmetry, studied in previous sections. We shall see that the simplicity of an umbilic is equivalent to the condition $w(0) \in J_0$. (It will allow us to apply the Stability Lemma.) However, not all orbit configurations of Table 13.1 are realized as principal line configurations near an umbilical point. Simple umbilics 'avoid' elliptic sectors, forming a subclass of the class J_0.

Proof is split into two parts. First the differential 2-form which defines the principal line configurations is reduced by the proper diffeomorphisms to the vector field with Z_2-symmetry. Next for the normal form obtained we check the $GL_2(\mathbb{R})$-invariants of the respective 2-jets.

Part I. Formally, the vector field which defines curvature lines writes as

$$w^{\pm} = [-B(u,v) \pm \sqrt{A^2(u,v) + B^2(u,v)}]\frac{\partial}{\partial u} + A(u,v)\frac{\partial}{\partial v}.$$

(Further we fix one of the signs and consider w^+.) The above vector field is not a 'good' vector field. At the umbilical point $A(0,0) = B(0,0) = 0$ it is not analytic, and the price of this is the singular curve $A(u,v) = 0$ which goes to $(0,0)$ from the one side. In other words, Sing $w^+ = \{(u,v)|A(u,v) = 0, B(u,v) \geq 0\}$. Therefore the origin of w^+ is a non-isolated equilibrium point.

Let us introduce a chain of the coordinate transformations,

$$(u,v) \xrightarrow{\phi} (x,y) \xrightarrow{\psi} (\xi,\zeta),$$

where ϕ and ψ are diffeomorphisms of $U \subset \mathbb{R}^2$ everywhere, except, may be, the origin of w. Let us study the action of the both diffeomorphisms consequently.

(i) ϕ acts by the formula $(u,v) \mapsto (A(u,v), B(u,v))$. By simplicity condition ϕ is a diffeomorphism everywhere in U. The elementary calculations in the local chart (x,y) yields us

$$\frac{dx}{dt} = \Phi_2(x,y)x - \Phi_1(x,y)y + \Phi_1(x,y)\sqrt{x^2+y^2},$$
$$\frac{dy}{dt} = \Psi_2(x,y)x - \Psi_1(x,y)y + \Psi_1(x,y)\sqrt{x^2+y^2}.$$

Here

$$\Phi_1(x,y) = A_u(u,v), \qquad \Phi_2(x,y) = A_v(u,v),$$
$$\Psi_1(x,y) = B_u(u,v), \qquad \Psi_2(x,y) = B_v(u,v),$$

are C^∞ functions of x, y after the substitution

$$u = \frac{1}{J}[B_v(0)x - A_v(0)y + O(x^2+y^2)], \quad v = \frac{1}{J}[-B_u(0)x + A_u(0)y + O(x^2+y^2)],$$

where $J = A_u(0)B_v(0) - A_v(0)B_u(0)$ is a Jacobian of the transformation in 0. (Geometrically, ϕ is a transition to the nonlinear coordinates, defined by the curves $A(u,v) = B(u,v) = 0$.)

(ii) ψ acts by the formula $(x,y) \mapsto (2\xi\zeta, \xi^2 - \zeta^2)$. In a complex representation ψ acts by the formula $z \mapsto z^2$. The Jacobian $J(\psi) = 2(\xi^2 + \zeta^2)$ of ψ does not vanish in U, except at the origin where ψ fails to be a diffeomorphism. Making necessary substitutions and time-scaling, in the local chart (ξ,ζ) it takes the form,

$$\frac{d\xi}{dt} = \Psi_2\xi^2 + (\Phi_2 + \Psi_1)\xi\zeta + \Phi_1\zeta^2,$$
$$\frac{d\zeta}{dt} = \Phi_2\xi^2 + (\Phi_1 - \Psi_2)\xi\zeta - \Psi_1\zeta^2.$$

where $\Phi_1, \Phi_2, \Psi_1, \Psi_2$ are C^∞ Z_2-symmetric functions of ξ and ζ: $\Phi_i(\xi,\zeta) = \Phi_1(-\xi,-\zeta), \Psi_i(\xi,\zeta) = \Psi_i(-\xi,-\zeta)$. The 'main parts' of the above functions in $(0,0)$ can be written as

$$\Phi_1(\xi,\zeta) = A_u(0,0) + O(\xi^2+\zeta^2), \quad \Phi_2(\xi,\zeta) = A_v(0,0) + O(\xi^2+\zeta^2),$$
$$\Psi_1(\xi,\zeta) = B_u(0,0) + O(\xi^2+\zeta^2), \quad \Psi_2(\xi,\zeta) = B_v(0,0) + O(\xi^2+\zeta^2).$$

Finally, one gets a 'good' system of differential equations given by a vector field with Z_2-symmetry,

$$\frac{d\xi}{dt} = B_v(0,0)\xi^2 + [A_v(0,0) + B_u(0,0)]\xi\zeta + A_u(0,0)\zeta^2 + O(\xi^4 + \psi^4),$$
$$\frac{d\zeta}{dt} = A_v(0,0)\xi^2 + [A_u(0,0) - B_v(0,0)]\xi\zeta - B_u(0,0)\zeta^2 + O(\xi^4 + \zeta^4).$$

Clearly, the above germ is Z_2-symmetric in $(0,0)$.

Part II. Let $f : \mathbb{R}^2 \to \mathbb{R}$ be a C^∞ function, which defines in the Monge chart U a local immersion $i \in I$ of the same class. It was already notified that we can always assume $f_u(0,0) = f_v(0,0) = 0$ in an umbilical point $(0,0)$. Imposing the umbilic conditions $A(0,0) = B(0,0) = 0$ yields us $f_{vv} - f_{uu} = f_{uv} = 0$ at the point $(0,0)$. The germ of the function f at an isolated umbilical point must take the form

$$f(u,v) = \gamma(u^2+v^2) + pu^3 + qu^2v + ruv^2 + sv^3 + O(u^4+v^4).$$

Moreover, one can always assume $q = 0$, otherwise the Monge chart U should be rotated [2] (as show the appropriate calculations which we omit here).

Now let us write down, keeping notation u and v, the equation (ξ,ζ) in terms of the local immersion, given by the above function f:

$$\frac{du}{dt} = 3su^2 + 3(r-p)uv + \phi(u,v),$$
$$\frac{dv}{dt} = 2ru^2 - 3suv + (3p-r)v^2 + \psi(u,v),$$

where $\phi(u,v)$ and $\psi(u,v)$ are higher $2k$-order monomials, $k \geq 2$. The $GL_2(\mathbb{R})$-invariants of the 2-jet of the system (u,v) are:

$$\mu = 18r(3p-r)[s^2+(r-p)^2], \quad \eta = 16(3p-2r)^2[9s^2 - 12pr + 8r^2],$$
$$\theta = (r-p)(25r^3 + 81p^2r + 27rs^2 - 90pr^2 - 54ps^2),$$
$$R = [3r(3p-2r) - 9s^2]x^2 + 3s(3p-2r)xy - (3p-2r)^2y^2.$$

Lemma 13.1.3 *The simplicity condition, $J(0) = 2r(3p-r) \neq 0$, written for the umbilical point given by the function f, is equivalent to that of the 2-jet of w to belong to the class J_0.*

[2] A transformation, which keeps the Riemannian structure of S.

Proof of Lemma 13.1.3. (i) ($\Rightarrow$) Indeed, let $(0,0)$ be a simple umbilic, $r(3p-r) \neq 0$. If $\mu \neq 0$ we are done. Let $\mu = 0$, then $s = r - p = 0$. This implies immediately that $\theta = 0, \eta < 0$. Therefore, simple umbilics are in the class $J_0 = \{\mu \neq 0\} \cup \{\mu = \theta = 0, \eta < 0\}$.

(ii)($\Leftarrow$) Conversely, let the equation (u, v) be of the class J_0. Then if $\mu \neq 0$, $r(3p - r) \neq 0$ and we are done. Alternatively, let $\mu = \theta = 0, \eta < 0$. Suppose to the contrary, that either $r = 0$ or $r = 3p$. It can be easily seen, that both conditions imply $\eta \geq 0$. A contradiction. □

The first part of Theorem 13.1.1 follows from the Stability Lemma and Lemma 13.1.3. (It remains to notice solely that the 2-jet of the vector field (u, v) involves the coefficients of the 3-jet of the immersion f, and only them.)

Now, by the Stability Lemma, we can confine ourselves by the study of the 'shorted' system (u, v) with $\phi(u, v) = \psi(u, v) = 0$. Let us show that the simple umbilical points 'avoid' principal line configurations with an elliptic sector. Namely, on account of the Table 13.1, we shall prove that

(I) condition $\eta < 0$ implies $\mu > 0$;

(II) conditions $\eta > 0, \mu < 0$ imply $\theta > 0$;

(III) conditions $\eta = 0, R \neq 0$ imply $\mu > 0$;

(IV) conditions $\eta = R = 0$ imply $\mu > 0$.

Indeed, let us do it step-by-step using the invariant expressions obtained above. Hereby one can always assume $r = 1$, since the simplicity condition and the appropriate time-scaling imply this. Then the invariants are transformed into:

$$\begin{aligned}
&\mu = 18(3p-1)[s^2 + (1-p)^2], \quad \eta = 16(3p-2)^2[9s^2 - 12p + 8],\\
&\theta = -81p^3 + 171p^2 + 54p^2s^2 - 115p - 81ps^2 + 27s^2 + 25,\\
&R = [3(3p-2) - 9s^2]x^2 + 3s(3p-2)xy - (3p-2)^2y^2.
\end{aligned}$$

(I) By $\eta < 0$ one obtains $4(3p - 2) > 9s^2 \geq 0$. In other words, $p > 2/3$. The restriction $3p - 1 > 0$ follows. It means that $\mu > 0$.

(II) Suppose that $\eta > 0, \mu < 0$. It yields the following inequalities

$$9s^2 - 12p + 8 > 0, \quad 1 - 3p > 0.$$

The invariant θ can be represented as

$$\theta = 3(9s^2 - 12p + 8)(1 - 3p) - 81p^3 + 63p^2 - 7p + 1 + 54p^2s^2.$$

Since $9s^2 > 12p - 8$, the invariant θ can be evaluated from below,

$$\theta > 3(9s^2 - 12p + 8)(1 - 3p) - 9p^3 + 15p^2 - 7p + 1 =$$
$$= (1 - 3p)[3(9s^2 - 12p + 8) + \frac{1}{3}(1 - 3p)^2 + \frac{2}{3}(1 - 3p)] > 0.$$

(III) Suppose that $\eta = 0$. Then either $p = 2/3$ or $p = (3/4)s^2 + 2/3$. In both cases it is a straightforward exercise to check that μ is positive.

(IV) Let finally, $\eta = R = 0$. It implies immediately, that $s = 3p - 2 = 0$. The direct substitution yields us $\mu = 2 > 0$. We are done.

Now it remains to compare the obtained restrictions on the $GL_2(\mathbb{R})$-invariants with those given by Table 13.1. This yields us the principal line configurations, shown in Fig. 13.1. (Where the corresponding non-orientable singularities π^t/θ are pictured.)

Theorem 13.1.1 is proven. □

13.2 Carathéodory Conjecture

Such a conjecture suggests that the index of an isolated umbilical point of a convex immersion of 2-dimensional sphere cannot exceed +1, cf e.g. [1]. The Carahéodory conjecture is true for the simple umbilical points; cf. Fig 13.1 and Theorem 13.1.1. Below we propose a proof of this conjecture in general situatuion following an unpublished preprint [221].

13.2.1 Notations

Constant mean curvature (CMC–) surfaces in E^3 are known to admit continuous family of local, non–trivial, isometric deformations preserving mean curvature (H–deformations). In the case surface is compact Umehara showed that converse is also true.

The lines of maximal and minimal curvature of CMC–surface form an orthogonal net called *réseau de Bonnet*, cf Cartan [71]. Bonnet Theorem says that if CMC–surface is simply connected and umbilic–free, then under H–deformations the orthogonal net "rotates" through a constant angle which can be taken as parameter of deformation.

If CMC–surface is not simply connected or umbilic–free, Cartan seems to be the first to ask about possible scenario of evolution of *réseau de Bonnet* under H–deformations.

We study evolution of orthogonal nets in case CMC–surface is simply connected with single umbilic or, equivalently, doubly connected and umbilic–free. Namely, if we "pinch" the umbilic, CMC–surface becomes an annulus whose points undergo H–deformations accordingly with the Bonnet Theorem. In general the rotation angle is no longer constant through all points because annulus cannot be covered by a single chart.

However, Bonnet Theorem implies that every curve of the orthogonal net is a φ–geodesic line whatever H–deformations are applied to the CMC–surface. Metric φ is given by the linear element $ds = |\varphi||dz|$, where φdz^2 is a holomorphic quadratic "differential" associated to CMC–surface. Of course, $\varphi(0) = 0$ at the umbilical point.

This observation is crucial, because φ–geodesics near n–th order zeros of holomorphic quadratic form are something well–understood due to the works of Strebel [274]. Roughly speaking, φ–geodesics fill–up annulus either by "hyperbolas" or "radii". Thus far, possible configurations of *réseau de Bonnet* near umbilic looks like singularity with a finite number of hyperbolic and parabolic sectors.

Despite independent interest, the orthogonal nets are auxiliary for us. We postulate different fact here: H–deformations of orthogonal nets give an amazingly simple proof to a *Caratheodory'sche Vermutung.*

Theorem 13.2.1 *Let S^2 be a C^∞ surface which bounds a convex compact body in the Euclidean space E^3. Then S^2 has at least two umbilical points. In other words, the Euler-Poincaré index of isolated umbilical point is at most* $+1$.

(Some historical comments on the conjecture can be found in [1]; see also the end of this chapter.)

13.2.2 φ–Geodesics

Until further indications, M is a simple domain of complex parameter z. Holomorphic functions $\varphi(z)$ are supposed to vanish at the unique point of M which we identify with 0. An order $n \geq 1$ is assigned to 0, if there exists complex constant $a \neq 0$ such that $\varphi(z) = az^n + O(|z|^{n+1})$.

Flat metric φ with cone of angle $(n+2)\pi$ is given by

$$|ds| = |\varphi||dz|,$$

provided $\varphi(z)dz^2$ is a quadratic form on M. [3] By a *φ-geodesic line* in M one understands line conisting of shortest arcs relatively metric φ. Any two points in M (including 0) may be joined by unique φ–geodesic line. Strebel classified possible types of φ–geodesics in the neighborhood of n–th order zero by proving the following lemma.

Lemma 13.2.1 *([274]) Any two points in a neighborhood M of n–th order zero of holomorphic 2–form $\varphi(z)dz^2$ can be joined by a unique φ–geodesic. Moreover, each φ–geodesic is either an arc defined by equation $Arg\ \varphi(z)dz^2 =$ Const, or is composed of two radii centered in 0 with the minimal angle* $\geq 2\pi/(n+2)$.

[3] For a systematic exposition of flat structures the reader is refered to Chapter 12. We also recommend paragraphs 3.5 and 3.6 of Chapter 17 for additional information on the quadratic differentials.

Foliation $\mathcal{F}$ on $M\backslash 0$ is said to be *geodesic* if every leaf of $\mathcal{F}$ is a φ–geodesic line. Before we state general lemma on structure of geodesic foliations, let us consider an example when all $\mathcal{F}$'s can be obtained by "brute force".

If 0 is a double zero, then φ–metric is given by the linear element $ds^2 = (u^2 + v^2)(du^2 + dv^2)$ where $u + iv$ is a natural parameter. Metric $|ds|$ is Liouville's and geodesic lines in this metric are completely integrable. General integral is known to be of the form

$$\int \frac{du}{\sqrt{u^2 - a}} \pm \int \frac{dv}{\sqrt{v^2 + a}} = a',$$

where a, a' are two independent constants. Easy calculations show that no information will be lost if we suppose $a = 0$. The integral takes the form $\ln|u| \pm \ln|v| = a'$. Geodesic foliation is described by two "families of curves": $v = Cu$ and $v = C/u$, where C is an arbitrary constant. Thus, $\mathcal{F}$ near double zero is either a "node" with geodesics radii tending to 0, or a "saddle" with four sectors filled–up with geodesic "hyperbolas".

Let w be a finite "word" on the alphabet consisting of two symbols h and p. We introduce elementary operations on w:

(i) cyclic permutation of symbols in w, and
(ii) contraction of p–symbol: $p^2 = p$.

The words are *equivalent* $w_1 \sim w_2$ if and only if w_2 can be obtained from w_1 by elementary operations. Equivalence class of word w is denoted by $[w]$.

Fix an integer number $n \geq 1$. To every symbol h in w we assign a *weight* $|h| = 2\pi/(n+2)$. To every symbol p we assign weight $|p| = \alpha_i$, where α_i is a positive real. Weight of w is an additive function equal to the sum of weights of symbols entering w. Equivalence class $[w]$ is called *normalized* if $|w| = 2\pi$ for all $w \in [w]$. (Note that weight of w is one and the same for all $w \in [w]$.)

Lemma 13.2.2 *Let h and p stay for hyperbolic and parabolic sectors, respectively. We encode singular point w by sequence of symbols h and p in the order h– and p–sectors occur when turning clockwise around the singularity. Then:*

(i) each φ–geodesic foliation $\mathcal{F}$ is topologically equivalent to singularity w of a normalized equivalence class $[w]$;
(ii) each normalized equivalence class $[w]$ can be realized as φ–geodesic foliation $\mathcal{F}$ with singularity $w \in [w]$ in a neighborhood of n–th order zero of φ for some $n \geq 1$.

Proof. Denote by M a neighborhood of n–th order zero of φ. Let us introduce partial order for points $x, y \in M$: $x \leq y$ if and only if $Arg\ x \leq Arg\ y$. If $x \in M$ is an arbitrary point, then by Lemma 13.2.1 φ–geodesic line through

x is either (i) a hyperbola $Arg\ \varphi dz^2 = Const$ or (ii) a radius Ox. Let us consider first possibility.

(i) Hyperbola $Arg\ \varphi dz^2 = Const$ must tend to the asymptotic rays Oz_1, Oz_2 with $z_1 < x < z_2$, enclosing the angle $2\pi/(n+2)$. Clearly, the only possibility to geodesic foliation $\mathcal{F}$ is to form a hyperbolic sector z_1Oz_2. Of course, along Oz_1 and Oz_2 $Arg\ \varphi dz^2$ is constant.

(ii) Let Ox be geodesic radius through x, distinct from the boundary radii of hyperbolic sector. Then through nearby points $|x - y| < \varepsilon$ one can draw geodesic radii Oy. Denote by y_1Oy_2 a maximal connected parabolic sector filled-up with geodesic radii. Clearly, $y_1 < x < y_2$. An angle enclosed between two boundary radii, we denote by α. In general, $0 \leq \alpha \leq 2\pi$.

If hyperbolic sector h is followed by another hyperbolic sector h, we write this as hh. If h is followed by a parabolic sector, we put it as hp. Parabolic sector p followed by parabolic sector p, gives a larger parabolic sector $p = pp$ and contraction rule (ii) follows. Of course, "weights" of sectors are equal to angles swept by these sectors.

Finally, according to definition of normalized equivalence class each singularity consists of sequence of parabolic and hyperbolic sectors; every curve in these sectors is a geodesic arc.

Part (ii) of Lemma 13.2.2 is proved by similar argument. □

13.2.3 CMC–Surfaces

Every smooth immersion $f : M \to E^3$ of orientable surface M into the Euclidean space E^3 induces a Riemann structure on M; let $z = u + iv$ be the corresponding local parameter. With respect to z the first fundamental form can be written as $ds^2 = e^{2\lambda}|dz|^2$.

If $ldu^2 + 2mdudv + ndv^2$ is the second fundamental form, we consider a complex quadratic form φdz^2, such that $\varphi(z) = \frac{1}{2}(l - n) - im$. Mainardi–Codazzi equations imply that φ is holomorphic on M if and only if $f(M)$ is a CMC–surface. Locally, along the lines of minimal and maximal curvature $Arg\ \varphi dz^2 = 0$ and $\varphi(0) = 0$ at the umbilic points.

A continuous deformation f_t of the immersion $f = f_0$ is an isometry of surface M such that $M \times [0,1] \to E^3$ is a continuous mapping. Deformation f_t is called an *H–deformation* if $H_t = H$ for all $t \in [0,1]$, where $H : M \to \mathbb{R}$ is mean curvature function.

CMC–surfaces are known to admit non–trivial H–deformations and in case of compact surfaces, they are the only ones with such a property. Of course, there are known many examples of compact CMC–surfaces of genus $g > 0$.

What happens with the lines of principal (i.e. minimal or maximal) curvature of CMC–surface during an H–deformation? If M is a local CMC–surface without umbilics, principal curvature lines of $f_0(M)$ and $f_t(M)$ form two families of parallels intersecting each other with the constant angle proportional to parameter t (Bonnet Theorem, see e.g. [76]). Note, that if we fix φ–metric on M corresponding to $f_0(M)$, then principal curvature lines of $f_t(M)$ are φ–geodesic lines of inclination t. If umbilical points are allowed, then the law is given by following lemma.

Lemma 13.2.3 *Suppose that $M_0 = f_0(M)$ is a canonical CMC–surface with quadratic function $\varphi = z^n, n \geq 1$. Let φ be metric on M corresponding to M_0. If $M_t = f_t(M)$ is an H–deformation of M_0, then one of two principal curvature lines of M_t are φ–geodesic lines on M for any $t \geq 0$.*

Proof. In the polar coordinate system the coefficients of the second fundamental form of surface M_t are given by

$$\begin{aligned} l &= He^{\lambda} + |z|^n \cos(2t - n\ Arg\ z), \\ m &= |z|^n \sin(2t - n\ Arg\ z), \\ n &= He^{\lambda} - |z|^n \cos(2t - n\ Arg\ z), \end{aligned} \tag{13.1}$$

where t is parameter of H–deformation, cf [285]. Two cases are possible.

(i) An H–deformation, such that t is constant on M. It can be immediately seen that in new coordinates $\tilde{u} = \cos t\ u + \sin t\ v$, $\tilde{v} = -\sin t\ u + \cos t\ v$ the first and the second forms of surfaces M_0 and M_t are the same. By the fundamental theorem, surfaces M_0 and M_t may differ only by a rigid motion in E^3. Thus, H–deformation is trivial.

(ii) A non–trivial H–deformation. By item (i), t varies for the points of M. Thus far, associated to every $z \in M \backslash 0$, there is a chart in which the second fundamental form of surface $M_t(z)$ writes as

$$l = He^{\lambda} + \cos 2t, \quad m = \sin 2t, \quad n = He^{\lambda} - \cos 2t,$$

where t is the deformation parameter, cf [303]. Straightforward calculations show that principal curvature lines of surface $M_t(z)$ are φ–geodesic lines of slope t on M. (This fact follows also from the Bonnet Theorem.) Since every regular point $z \in M$ can be endowed with such a chart, Lemma 13.2.3 is proved. □

13.2.4 Proof of Theorem 13.2.1

Take a convex C^∞ immersion $f_0 : S^2 \to E^3$ of the 2–sphere into the Euclidean space E^3 which is not totally umbilic (i.e. there are no $U \subseteq S^2$ such that $f_0(U)$ is a part of the round sphere). In other words, umbilics are supposed isolated and their number is finite. Denote by ds_0 a Riemann metric on S^2 induced by the immersion f_0 and by $H : S^2 \to I\!R$ the corresponding mean curvature function.

Definition 13.2.1 *By a Hopf spheroid in E^3 we understand a convex C^∞ immersion $f : S^2 \to E^3$ such that there exists at least one umbilical point p and a small closed disc $D \ni p$ such that $H(D) = Const$.*

Lemma 13.2.4 *There exist infinitely many Hopf spheroids in* E^3.

Proof. By the results of Wente and Kapouleas any compact orientable surface S_g of genus $g > 0$ admits an immersion into E^3 which is a CMC-surface with $H > 0$; cf. [159], [299]. Fix $g \geq 2$ and consider the lines of principal curvature of any such immersion. By the index argument, there exists an umbilic $p \in S_g$ and a small closed disc $D \ni p$ which is a convex local surface in E^3. We separate this local surface from S_g. To obtain a Hopf spheroid, it remains to complete this piece of CMC-surface to a C^∞ immersion $S^2 \to E^3$. By Urysohn's lemma this can be done in an infinite number of ways. □

Lemma 13.2.5 *For the Hopf spheroids the Caratheodory conjecture is true.*

Proof. Without loss of generality we can assume that the umbilic point p of Hopf spheroid is unique. (For otherwise, if there are more than one umbilic then we are done.) Since a Hopf spheroid is locally CMC, we apply Lemma 13.2.3 to identify the curvature lines in the disc $D \ni p$ with φ-geodesic lines in the vicinity of a singularity w.

Let $w \in [w]$ be a word of the minimal length in the normalized equivalence class $[w]$. According to Lemma 13.2.2, there exists a singularity of order n whose topological type is encoded by the sequence w of symbols h and p. Let w admit $\langle h \rangle$ symbols of type h and $\langle p \rangle$ symbols of type p. By the normalization axiom, $\langle h \rangle \leq n + 2$.

To estimate the Euler–Poincaré index of singularity w, note that the parabolic sectors make no contribution to the index value and the number $\langle p \rangle$ can be neglected. To the contrary, if there are no hyperbolic sectors (i.e. $w = p$) we necessarily have one parabolic sector. The general formula is true:

$$Ind\ w = \begin{cases} 1 - \frac{\langle h \rangle}{2} & \text{if} \quad w \neq p, \\ +1 & \text{if} \quad w = p. \end{cases}$$

In either case $Ind\ w \leq 1$ and by the index argument the conjecture follows. □

Now we are ready to finish the proof of Theorem 13.2.3. But first we wish to outline the main idea. To every convex C^∞ immersion $f_0 : S^2 \to E^3$ one can relate a Hopf spheroid. This spheroid is uniquely defined by f_0 and is a 'modification' of f_0 which has an interesting 'mechanical' interpretation.

Suppose that f_0 is a convex steel ball filled-up with a gas under a pressure. Let p be an isolated umbilic of f_0. We drill a small hole in p and glue-up a soap film D into this hole maintaining a pressure [4] inside the ball. We also 'deform' slightly the 'edges' of the cut in order to keep the modified surface $f : S^2 \to E^3$ in the class C^∞. We claim that f is a Hopf spheroid.

[4] The absolute value of the pressure depends on how 'flat' is the surface at the point p. Of course, by 'pressure' we understand difference of pressures inside and outside the steel ball.

Indeed, $f(D)$ is a local CMC-surface with an umbilic point $p \in D$. Moreover, the index of umbilic on the Hopf spheroid is equal to the index of p on f_0. (This is because the foliation by principal curvature lines at the 'steel part' of ball remains intact.) In general, if $\mathcal{F}_0$ and $\mathcal{F}$ are foliations by the principal curvature lines on f_0 and f, respectively, then $\mathcal{F}$ is obtained from $\mathcal{F}_0$ by a *homotopy of openning of a leaf*; cf. Chapter 3.

Let f_0 be as above. If p is an isolated umbilic of f_0 then we take a closed disc $|D| \leq r$ centred at the point p. We are going to define a local CMC-surface $f(D)$. Let $z = u + iv$ be a local parameter which corresponds to a part of CMC-surface with an umbilic; see the beginning of this section. By the results of Umehara [285] (see also [71], [76]) there exists a family of isometric H-deformations depending on a real parameter t:

$$I = e^{2\lambda}|dz|^2, \qquad II_t = l du^2 + 2m\, dudv + n dv^2, \tag{13.2}$$

with l, m and n given by equations (13.1). The Mainardi-Codazzi and Gauss equations for I, II_t:

$$\frac{\partial \varphi}{\overline{\partial} z} = \frac{\partial H}{\partial z}, \qquad |\varphi|^2 = e^{4\lambda}(H^2 - K), \tag{13.3}$$

where $\varphi = e^{it} z^n$ is a complex quadratic form φdz^2, are satisfied for any real t. (Indeed, the first equation is true since $H = Const$ and φ is holomorphic; the second equation follows from $|\varphi e^{it}| = |\varphi|$ and the fact that H-deformation is an isometry.) Therefore, the fundamental forms (13.2) are realized by a concrete local CMC-surface for each real number t.

Let $f_t(D)$ be a family of local CMC-surfaces described above. Denote by A an annular region which surrounds disc D:

$$A = \{z = u + iv |\ r \leq |z| \leq r + \varepsilon\}. \tag{13.4}$$

To glue-up $f_t(D)$ properly, we fix the metric λ so that $\lambda|_{\partial A_{r+\varepsilon}} = \lambda|_{\partial A_r}$, where the left part denotes a metric on the exterior boundary of A which is induced by metric of the surface f_0. The boundary condition $\lambda|_{\partial A_r}$ gives a unique solution $f_{t=t^*}(D)$ to the Gauss equation, so that a representative in the family $f_t(D)$ is fixed.

To obtain a C^∞ Hopf spheroid it remains to conjugate $f_{t^*}(D)$ with the rest of the sphere:

$$f(S^2) = \begin{cases} f_{t^*}(D), & \text{if } \ z \in D \subset Int\ D_{r+\varepsilon}, \\ f_0(S^2), & \text{if } \ z \in S^2 \backslash D_{r+\varepsilon}. \end{cases} \tag{13.5}$$

By the Urysohn Lemma, function f in formula (13.5) can be chosen C^∞ for an arbitrary small ε, see formula (13.4). Moreover, taking r sufficiently small we can fix number n (see (13.1)) equal to the order of quadratic form φ at point p of the surface f_0. (Such an order is correctly defined for any φ, not necessary holomorphic.)

Thus, the surface f given by equation (13.5) is a Hopf spheroid. By the Lawson-Tribuzy theorem f is uniquely defined up to a rigid motion in E^3; see [177]. To finish the proof of Caratheodory conjecture, it remains to notice that passage from f_0 to f gives us a homotopy $h(\mathcal{F}_0) = \mathcal{F}$ between foliations induced by curvature lines. In particular, $Ind\, p_0 = Ind\, p$. By Lemma 13.2.5, the Caratheodory conjecture follows. □

13.3 Elements of Global Theory

13.3.1 Structural Stability

Structural stability of the foliations by lines of the principal curvature are defined similarly as for the general foliations. Namely, immersion i of class C^r of a compact orientable surface S into $\mathbb{R}^3$ is *structurally stable* if for any immersion i' sufficiently C^s-close with $r > s \geq 3$, there exists a homeomorphism h of S which maps lines of $\mathcal{F}_i$ to the lines of $\mathcal{F}_{i'}$.

Let us denote by D_1, D_2 and D_3 the local configurations of lines of principal curvature near simple umbilical points, given by Theorem 13.1.1. (Note that subscript corresponds to the number of separatrices at the respective umbilic.)

A compact (periodic) leaf of foliation $\mathcal{F}$ defined by immersion i is called a *principal* cycle of the immersion i. The first return mapping (holonomy) is defined between any two small transversal segments to the principal cycle C. Denote by π such a mapping on the same transversal defined along the leaves of foliation $\mathcal{F}$. Principal cycle C is said to be *hyperbolic* if $\pi' \neq 1$ in the points belonging to C.

The line of principal curvature is called an *umbilical connection* if it 'starts' and 'finishes' at the umbilical points D_i being separatrix of such points. (Case of one and the same point D_i is not excluded and then line is called a *loop*.)

Definition 13.3.1 (Gutierrez-Sotomayor) *Denote by $I_0^r \subset I^r$ a set of C^r-smooth immersions $i \in I^r$ of compact surface S such that:*

(i) all the umbilical points of i are of type D_i;
(ii) all the principal cycles of i are hyperbolic;
(iii) i has no umbilical connections;
(iv) the limit set of every line of principal curvature is a union of umbilical points and principal cycles.

Theorem 13.3.1 (Gutierrez-Sotomayor) *The set I_0^r is open and dense in the space I^r of C^r-smooth immersions for all $r \geq 4$. Moreover, every $i \in I_0^r$ is C^3-structurally stable.*

Proof. We refer the interested reader to the original proof given in [130]. In the outline, to prove this result one should go through all stages of the proof of to stability of Morse-Smale foliations. First, a filtration is constructed. Next, one establishes a homeomorphism between two "nearby" principal line configurations. The density property is more delicate. One should demonstrate how to "destroy" the non-trivial recurrent curvature lines; however, this can also be done with a success, see [131]. □

13.3.2 Bifurcation of Umbilical Connections

As far as the principal curvature foliations of the first degree of stability is concerned, Gutierrez & Sotomayor [133] considered foliations of Definition 13.3.1 when condition (iii) is no longer present. Namely, let $i, j \in \{1, 2, 3\}$. A C_{ij} umbilical connection is one which joins umbilical points p and q of types D_i and D_j, respectively. Umbilical connection C_{ij} is called *simple* if one of the following three conditions is satisfied:

- C_{ij} is neither C_{11} nor a C_{kk} umbilical loop with $k \in \{2, 3\}$;
- For C_{11} with $p \neq q$ the first return mapping T on a segment transversal to C_{11} it holds $\lim_{x \to 0}(T(x)/x) \neq 1$;
- For $i = j \in \{2, 3\}$ and $p = q$ the first return mapping T on segment transversal to C_{11} it holds $\lim_{x \to 0}(\log\ T(x)/\log x) \neq 1$.

Let us denote by I_1^r a subspace of the space I_0^r consisting of immersions satisfying all the conditions of Definition 13.3.1, except condition (iii), and such that there exists exactly one simple umbilical connection. If an immersion $i \in I_1^r$ is such that its umbilical connection is not contained in the limit set of other umbilical separatrix, then i is said to belong to the space $\tilde{I}_1^r$.

Theorem 13.3.2 (Gutierrez-Sotomayor) *Let $r \geq 6$. Then the following is true:*

(i) Set I_1^r ($\tilde{I}_1^r$) is a one-to-one immersed (embedded) Banach submanifold of I^r of codimension 1 and of class C^{r-4}.
(ii) $\tilde{I}_1^r$ is an open subset of I_1^r.
(iii) Every $i \in I_1^r$ ($i \in \tilde{I}_1^r$) is C^4-structurally stable in $I^r \backslash I_0^r$ (in $I^r \backslash (I_0^r \cup I_1^r)$).

Moreover, the set $\tilde{I}_1^r$ is dense in the subspace $I_1^r \cup \hat{I}^r$, where $\hat{I}^r$ is a space of immersions with violated condition (i) of Definition 13.3.1.

Proof. The idea of proof is to establish that in the bifurcation of the above described umbilic connections, exactly one hyperbolic principal cycle can be born. Here the method of "normal forms" of simple umbilics is applied, and then the transition (Dulac) functions are calculated. We refer the interested reader to the original work [133]. □

Bibliographic Notes. Simple umbilics attracted attention of A. Cayley [72] and G. Darboux [82]. The lines of the principal curvature have been intensively studied at the IMPA (Brazil); see [130]–[134]. The Carathéodory conjecture has been considered by H. Hamburger [136],[137], G. Bol [53], T. Klotz [164] and other mathematicians whose names are not mentioned here with an apology from the author. The global theory of lines of principal curvature has been elaborated in a series of remarkable works of C. Gutiérrez and J. Sotomayor; cf. [130]–[134]. Our exposition is close to the above papers.

14. Differential Equations

Ordinary differential equations (ODE) correspond to the orientable foliations. Allowing the ODE to be unsolved with respect to the derivative of highest order gives us an interesting class of non–orientable foliations. A method of integration of such differential equations has been suggested by Cayley and Darboux in the context of principal curvature lines on surfaces. Later Hartman and Wintner in the series of works [138] — [140] developed a general method of the integration of such ODE's. In this chapter we introduce the reader to the theory of Hartman and Wintner as well as to the later (geometric) method of A. G. Kuzmin.

14.1 Characteristic Curve

14.1.1 Background and Notations

Motivated by the differential geometry of surfaces, Hatrman and Wintner studied solutions of the following differential equation:

$$adx^2 + 2bdxdy + cdy^2 = 0, \tag{14.1}$$

where a, b, c are continuous functions of (x, y). Their main achievement was a theorem on existence of a characteristic curve of the singular point $a = b = c = 0$ which is a solution to the above differential equation. (In fact, in each of two families of solutions there exists such a curve.) Below we prove the theorem of Hartman and Wintner.

In a vicinity of the singular point $(x, y) = (0, 0)$ let the coefficients of (14.1) be continuous functions of the form

$$a = \alpha_1 x + \beta_1 y + f_1, \qquad b = \alpha_2 x + \beta_2 y + f_2, \qquad c = \alpha_3 x + \beta_3 y + f_3,$$

where α_i, β_i are six constants and each of the three functions $f_i = f_i(x, y)$ satisfy the condition $f_i(x, y) = O(\sqrt{x^2 + y^2})$. We assume that $b^2 - ac \geq 0$ vanishes only at the point $(0, 0)$.

Definition 14.1.1 *By a solution path of (14.1) we mean a set of points* (x, y) *which can be represented in the form* $x = x(\tau), y = y(\tau)$ *with* $x(\tau), y(\tau)$ *differentiable functions satisfying (14.1) on a* τ*–interval. Moreover, neither* $(x(\tau), y(\tau))$ *nor* $(dx/d\tau, dy/d\tau)$ *is different from* $(0,0)$ *at any point of that interval.*

(The set of solution paths of (14.1) splits into two distinguishable systems, S_1 and S_2, and they represent *all* the solution paths of (14.1).)

Definition 14.1.2 *By a solution path of (14.1) reaching the origin we mean a solution path* $x = x(\tau), y = y(\tau)$ *defined on an interval* $0 \leq \tau < \tau_0$ *and such that* $(x(\tau), y(\tau)) \to (0,0)$ *as* $\tau \to \tau_0$.

If $x = x(\tau), y = y(\tau)$ is a solution path, then in the polar coordinates $r = \sqrt{x^2 + y^2}, \theta = \arctan y/x$ there are two continuously differentable functions $r = r(\tau) > 0, \theta = \theta(\tau)$. Let us introduce three linear trigonometric forms

$$L_i(\theta) = \alpha_i \cos\theta + \beta_i \sin\theta,$$

where α_i, β_i are constants defined earlier. Finally, we introduce the following (quadratic and cubic) trigonometric polynomials:

$$Q(\theta) = L_2^2(\theta) - L_1(\theta)L_3(\theta),$$
$$M(\theta) = L_(\theta)\cos^2\theta + 2L_2(\theta)\sin\theta\cos\theta + L_3(\theta)\sin^2\theta.$$

14.1.2 Theorem of Hartman and Wintner

Theorem 14.1.1 (Hartman–Wintner) *Suppose that* $a(x, y), b(x, y), c(x, y)$ *are continuous functions in the vicinity of the point* $(0,0)$ *as specified before. Let also the following three conditions hold:*

(i) $L_1(\theta), L_2(\theta), L_3(\theta)$ *have no common zero;*
(ii) $Q(\theta) \not\equiv 0$;
(iii) $M(\theta) \not\equiv 0$.

Then:

(I) Each of the systems S_1 *and* S_2 *of solution paths of (14.1) contains at least one solution path reaching the origin, cf Definition 14.1.2.*
(II) To every solution path reaching the origin as specified in (I), there exists an angle θ_0 *such that:*

(a) $\theta \to \theta_0$, *where* $\theta = \arctan y/x$ *and*
(b) $\phi \to \theta_0$ *(mod* π*), where* $\phi = \arctan dy/dx$,

as $(x, y) \to (0, 0)$.

Proof. First, we introduce the following notation:

$$F(\theta) = L_3(\theta), \quad G_j(\theta) = -L_2(\theta) + (-1)^j \sqrt{Q(\theta)},$$

where the square root takes non–negative values for $j = 1, 2$. It can be seen, that in the polar coordinates the solution paths tend uniformly to the limit $G_j(\theta)/F(\theta)$. If, in addition, we set

$$J_j(\theta) = G_j(\theta)\cos\theta - L_3(\theta)\sin\theta,$$

then the following identity is true:

$$M(\theta)L_3(\theta) = J_1(\theta)J_2(\theta).$$

In order to prove the theorem, one has to show that for $j = 1, 2$ function J_j must change the sign at some $\theta = \theta_0^j$ while the polynomial $L_3(\theta_0^j) \neq 0$.

Indeed, there exist values θ_0 such that $L_3(\theta_0) \neq 0$ if $M(\theta_0) = 0$. This means that the product $M(\theta)L_3(\theta)$ must change the sign at θ_0. Therefore, the same is true for either $J_1(\theta)$ or $J_2(\theta)$. It follows from $J_1(\theta) = J_2(\theta + \pi)$ that each of the functions $J_j(\theta)$ has a zero (θ_0 or $\theta_0 + \pi$) at which it changes the sign.

Theorem of Hartman and Wintner is a consequence of the following lemma.

Lemma 14.1.1 ([138]) *Let $J(\theta)$ be as before. Then:*

(i) If $J(\theta)$ changes the sign at a point $\theta = \theta_0$ then there exists at least one solution path reaching the origin in a definite direction;
(ii) If $J(\theta)$ assumes both positive and negative values then every solution path reaching the origin does it in a definite direction.

Proof. Let us consider the following limits

$$F(\theta) = \lim_{r\to+0} f(r\cos\theta, r\sin\theta)/r, \quad G(\theta) = \lim_{r\to+0} g(r\cos\theta, r\sin\theta)/r,$$

for the solution path of the differential equation $dx/dt = f(x,y), dy/dt = g(x,y)$. Consider also a function $K(\theta) = G(\theta)\sin\theta + F(\theta)\cos\theta$. If θ is near θ_0, then $K(\theta) \neq 0$ so that we have a differential equation (see [138]):

$$r\frac{d\theta}{dr} = \frac{J(\theta) + O(1)}{K(\theta) + O(1)},$$

where the right-hand side is a continuous function, $R(r,\theta)$. Moreover, if $\delta > 0$ is small, then $R(r, \theta_0 + \delta)$ and $R(r, \theta_0 - \delta)$ are of opposite sign and $R(0, \theta)$ is not 0 on the interval $|\theta - \theta_0| \leq \delta$. There exists a solution $\theta = \theta(r)$ of the above differential equation, defined on the interval $0 < r \leq \varepsilon$ and satisfying condition $\theta(r) \to \theta_0$ as $r \to 0$. Clearly, after introducing a parameter t (see

[138] for the details), the curve $(r(t), \theta(r(t))$ gives a solution path reaching the origin with a definite direction. Item (i) follows.

To prove item (ii) of the lemma we suppose the contrary. Then, as $t \to b$ there are no finite limit value for the function $\theta = \theta(t)$. It can be seen ([138]) that $\theta(t)$ is bounded, but by the assumption has no limit. Hence $\theta(t)$ is not monotone. Consequently, $\theta'(t)$ changes the sign infinitely often. Thus there is an interval I with a non-empty set of points where function $J(\theta)$ changes the sign. On the other hand, if $J(\theta) \not\equiv 0$ then $\theta(t) = \theta$ can happen only for a unique value of t near b. (The latter follows from the assumption that $\theta(t)$ has the same sign as $J(\theta)$ if t is close to b.) This contradiction proves item (ii) of the lemma. □

To finish the proof of the Hartman–Wintner theorem, let us apply Lemma 14.1.1. We have the following consequences:

(i) For every $j = 1, 2$ there exist solution paths reaching the origin with the definite direction $\theta_0 = \theta_0^j$;
(ii) To every solution path reaching the origin, there is a number θ_0 equal to the slope of the path in the origin and such that $J_j(\theta_0) = 0$.

This argument finishes the proof of the theorem. □

14.2 Classification

Motivated by the study of characteristics of partial differential equations, A. G. Kuzmin suggested in 1982 a geometric classification of the singular points of (14.1); cf [173], [174]. Along with Hartman and Wintner, Kuzmin assumes that the coefficients of differential equation are in a general position. In this case a complete qualitative picture of the solution paths near singular point is possible. In the later paper [174] Kuzmin studies a "global behaviour" of the solution paths; there also a lot of examples of the phase portraits of particular differential equations are given. Below we expose a "local" theory of A. G. Kuzmin.

14.2.1 Generic Singularities

Let us consider a differential equation

$$a(x, y)dy^2 - 2b(x, y)dxdy + c(x, y)dx^2 = 0 \tag{14.2}$$

in a domain D of the plane xOy. Let $a, b, c \in C^1(D)$ be C^1 functions of (x, y). Suppose also that $a = b = c = 0$ if and only if $x = y = 0$ and

$$\Delta(x, y) = b^2 - ac > 0 \quad \text{for} \quad r = \sqrt{x^2 + y^2} > 0. \tag{14.3}$$

The integral curves of the first and the second families of equation (14.2) are solutions of the equation

$$\frac{dy}{dx} = \frac{b - (-1)^i\sqrt{\Delta}}{a}, \tag{14.4}$$

for $a(x,y) \neq 0$ and $i = 1, 2$. Note that the points where $a = 0$ and $b \leq 0$ are singular for the equation (14.4) if $i = 1$ and the points where $a = 0$ and $b \geq 0$ are singular for the same equation if $i = 2$. At these singular points, except point 0, the direction field can be recovered from a "complementary" equation

$$\frac{dy}{dx} = \frac{c}{b + (-1)^i\sqrt{\Delta}}. \tag{14.5}$$

It was shown in the previous section that each integral curve which reaches the point O does it in a definite direction so that dy/dx has a limit. Moreover, there exists at least one such integral curve for the both families determined by equation (14.2).

In what follows we impose even stronger condition than the one of (14.3), namely, that

$$\Delta_0(x,y) \equiv (b_1x + b_2y)^2 - (a_1x + a_2y)(c_1x + c_2y) > 0 \quad \text{for} \quad r > 0, \tag{14.6}$$

where $a_1 = \partial a/\partial x(0,0), a_2 = \partial a/\partial y(0,0), \ldots, c_2 = \partial c/\partial y(0,0)$ is a "linearization" of (14.3). Let us introduce the following notation:

$$\begin{aligned} \theta &= \arctan(y/x), \qquad\qquad (14.7)\\ M(\theta) &= (a_1\cos\theta + a_2\sin\theta)\sin^2\theta - 2(b_1\cos\theta + b_2\sin\theta)\sin\theta\cos\theta \\ &\quad + (c_1\cos\theta + c_2\sin\theta)\cos^2\theta, \\ R_i(\theta) &= -(a_1\cos\theta + a_2\sin\theta)\sin\theta + (b_1\cos\theta + b_2\sin\theta)\cos\theta \\ &\quad - (-1)^i\sqrt{\Delta_0(\cos\theta, \sin\theta)}\cos\theta. \end{aligned}$$

14.2.2 Theorem of A. G. Kuzmin

Theorem 14.2.1 (Kuzmin) *Let $a, b, c \in C^2(D)$ so that $a = b = c = 0$ for $x = y = 0$ and let inequality (14.6) hold. Let the polynomial $M(\theta)$ have (a) one root; (b) two roots; (c) three roots $\theta_1 < \theta_2 < \theta_3$ where θ_1 and θ_2 or θ_2 and θ_3 correspond to one and the same family of integral curves; or (d) three roots $\theta_1 < \theta_2 < \theta_3$, where θ_1 and θ_3 correspond to one family and θ_2 to the other. Let, finally, H and P denote the hyperbolic and parabolic sectors of the singularity, respectively. Then, for both families of integral curves the local schemes of the point O have the form $\{H\}$, $\{H, P\}$, $\{H, P\}$, or $\{H, H, H\}$, respectively.*

Proof. It follows from (14.6) that

$$a_1^2 + a_2^2 \neq 0, \qquad b_1^2 + b_2^2 \neq 0, \qquad c_1^2 + c_2^2 \neq 0. \tag{14.8}$$

We assume that the direction $\theta = \pi/2$ is exceptional for the first family of integral curves; this can always be obtained by rotation of coordinate system in a suitable manner. From (14.4) and (14.7) we obtain

$$a_2 = 0, \quad b_2 > 0, \tag{14.9}$$

and $a_1 \neq 0$ by virtue of (14.8). The direction $\theta = \pi/2$ is not exceptional for the second family, as it follows from (14.8) and (14.9).

The remaining exceptional directions for $-\pi/2 < \theta \leq \pi/2$ are determined by the equation

$$L(\lambda) = (a_1 - 2b_2)\lambda^2 + (c_1 - 2b_1)\lambda + c_1 = 0,$$

where $\lambda = \tan\theta$. (This follows from $M(\theta) = 0$ and (14.9).) From $R_i(\theta) = 0$ we find an equation which determines the exceptional directions for the first and second families separately:

$$P_i(\lambda) = b_1 + (b_2 - a_1)\lambda - (-1)^i\sqrt{(b_1 + b_2\lambda)^2 - a_1(c_1 + c_2\lambda)} = 0. \tag{14.10}$$

It is not hard to see that

$$P_1(\lambda)P_2(\lambda) = a_1 L(\lambda), \tag{14.11}$$

$$P_1(\lambda) > P_2(\lambda) \quad \text{for} \quad -\infty < \lambda < +\infty. \tag{14.12}$$

(Inequality (14.12) follows from (14.6).) Note also some useful consequences of (14.10):

$$P_1(\lambda) = -a_1\lambda + O(1),\ P_2(\lambda) = (2b_2 - a_1)\lambda + O(1) \text{ as } \lambda \to -\infty$$
$$P_1(\lambda) = (2b_2 - a_1)\lambda + O(1),\ P_2(\lambda) = -a_1\lambda + O(1) \text{ as } \lambda \to +\infty \tag{14.13}$$

Let us prove the statement of the theorem in case (a). Suppose that $\theta = \pi/2$ is a simple root of the polynomial $M(\theta)$. Then by (14.7) we have

$$a_1 - 2b_2 \neq 0. \tag{14.14}$$

$M(\theta)$ does not have roots for $-\pi/2 < \theta < \pi/2$ according to the assumption. Then, due to (14.11), $P_1(\lambda)$ and $P_2(\lambda)$ do not have roots either and from (14.13) we find that $0 < a_1 < 2b_2$. From (14.4) we obtain

$$\frac{dx}{dy} - \frac{x}{y} = -\frac{\mu Q_i(\mu)}{2b_2} + O(\mu Q_i) + O(1), \tag{14.15}$$

where $\mu = x/y$, $Q_i = b_1\mu + b_2 - a_1 - (-1)^i\sqrt{(b_1\mu + b_2)^2 - a_1(c_1\mu + c_2)}$ and $O(1) \to 0$ as $r \to 0$. Since $Q_1(0) = 2b_2 - a_1 > 0$, the direction $\theta = \pi/2$ is contained in a normal domain of the hyperbolic type (Nemytskii & Stepanov [209]). Thus the local scheme will be $\{H\}$. The uniqueness of the integral curve which enters the point O is demonstrated in the same way as in the case of the equation (14.4) with the right-hand side whose numerator and denominator are representable as a sum of homogeneous polynomials and the higher order terms.

Now let $\theta = \pi/2$ be a triple root of the polynomial $M(\theta)$. Then, instead of (14.14) we obtain from (14.7) that $a_1 = 2b_2, c_2 = 2b_1$ and $c_1 \neq 0$ and on account of (14.9) we find from (14.6) that $\Delta_0 = (c_2x - a_1y)^2/4 - a_1c_1x^2 > 0$ for $r > 0$. Thus $c_1 < 0$. It is easy to verify that $Q_1Q_2 = a_1c_1\mu^2$ and $Q_2(0) = -2b_2$ in this case. Thus $Q_1(\mu) = -\mu^2a_1c_1/(2b_2) + O(\mu^2)$. Then we find from (14.15) that the direction $\theta = \pi/2$ is contained in a normal domain of the hyperbolic type. The local scheme is again $\{H\}$.

To prove case (b) of the theorem, let us suppose that $\theta = \pi/2$ is a simple root of the polynomial $M(\theta)$. Then (14.14) is fulfilled and for $-\infty < \lambda < +\infty$ one of the functions $P_1(\lambda)$ or $P_2(\lambda)$ has a double root. Thus, from (14.13) we find that $0 < a_1 < 2b_2$.

We assume that $-2b_1b_2 + c_2(a_1 - b_2) \geq 0$, for otherwise we substitute $x \to -x$ in (14.4). Then $P_1(\lambda)$ has a root in the interval $-\infty < \lambda < +\infty$. Namely, $L(\lambda)$ has a double root at the point

$$\lambda_0 = \frac{2b_1 - c_2}{2(a_1 - 2b_2)},$$

where $P_1 + P_2 = 2[-2b_1b_2 + c_2(a_1 - b_2)]/(a_1 - 2b_2) \leq 0$, so that $P_1(\lambda_0) < 0$ and $P_2(\lambda_0) < 0$ due to (14.11) and (14.12).

Since $Q_1(0) = 2b_2 - a_1 > 0$, it follows from (14.15) that the direction $\theta = \pi/2$ can be contained in a normal domain of the hyperbolic type. From the equality

$$\frac{dy}{dx} - \frac{y}{x} = P_i(\lambda) + O(1) \tag{14.16}$$

we find that the direction of $y = \lambda_0 x$ is contained in a normal domain of mixed (parabolic and hyperbolic) type. The smoothness of the coefficients a, b and c claimed in the theorem ensures that there exists an infinite set of integral curves entering the point O in this direction. In this case the local scheme is $\{H, P\}$ as stated in the theorem.

In the case (c) of theorem, suppose that the sequence of roots of $M(\theta)$ is arranged in an increasing order; triples of roots corresponding to the first and second families alternate. We assume also that along with (14.9) the following is satisfied: $\theta = \pi/2$ corresponds to the second of the three consecutive exceptional directions of the first family. $L(\lambda)$ has two roots in the interval $-\infty < \lambda < +\infty$. If $0 < a_1 < 2b_2$, then by (14.13) $P_1(\lambda)$ or $P_2(\lambda)$ has two roots for $-\infty < \lambda < +\infty$ what contradicts the assumption. If $a_1 < 0$, then

$P_1(\lambda)$ and $P_2(\lambda)$ have two roots λ_1 and λ_2, respectively, where $\lambda_1 < \lambda_2$ what contradicts the assumption as well. Thus, $a_1 > 2b_2$. From (14.15) it follows that the direction $\theta = \pi/2$ is contained in a normal domain of the parabolic type. Since the signs of $P_1(\lambda)$ and $P_2(\lambda)$ change from positive to negative when λ grows, and since $a_1 > 0$ it follows from (14.16) that the directions of $y = \lambda_1 x$ and $y = \lambda_2 x$ are contained in the normal domains of hyperbolic type. The scheme $\{H, P\}$ follows.

Finally, to prove case (d) of the theorem let us note that the roots of the first and the second families alternate in the sequence of roots of $M(\theta)$. In view of what has been said above, we conclude that $a_1 < 0$. Similarly, from (14.15) and (14.16) we find that the directions $\theta = \pi/2, y = \lambda_1 x$ and $y = \lambda_2 x$ are contained in the normal domains of hyperbolic type. The local scheme $\{H, H, H\}$ follows. Theorem of Kuzmin is proved. □

Bibliographic Notes. The singularities of differential equations unsolved relatively the highest order derivative attracted a lot of attention because of a strong link to the differential geometry of surfaces. After the works of Poincaré on the curves defined by differential equations É. Picard [239] studied singularities of "quadratic" equation whose derivatives are involved in a quadratic relation. He proved that in a generic (general position) case there are no solution paths which "spiral" around the singularity. Hartman and Wintner [138], [139], [140] proved the existence of a "characteristic curve" which enters the singularity with definite direction. (Such a curve exists in each of the two families of solution paths.) In 1982 Kuzmin [173] classified "generic" singularities of the quadratic equations using the method of normal domains of Nemytskii & Stepanov [209]. The same author made an effort to develop the global theory of foliations given by quadratic equations in a simple domain and on the sphere; see [174]. In this work one finds a collection of instructive examples of such foliations.

15. Positive Differential 2–Forms

15.1 Notations

Let $U \subseteq M$ be a local chart on M and let $\omega(u,v)$ be a C^r-quadratic differential form:

$$a(u,v)du^2 + b(u,v)dudv + c(u,v)dv^2,$$

where a, b and c are real-valued functions of class C^r. By a *positive C^r differential 2-form* on M one understands a C^r-quadratic differential form ω such that for every point $x \in M$, the set $\omega^{-1}(x)(0)$ is either

- A union of two transversal lines (such a point is called *regular*),or
- An isolated point (such a point is called *singular*).

The space of all positive differential 2-forms endowed with C^r Whitney topology is denoted by $Q^r(M)$.

Remark 15.1.1 *Let ω be a C^r-quadratic differential form on M. Then the following properties are pairwise equivalent:*

(i) $\omega \in Q^r(M)$;
(ii) For each point $x \in M$ there exists a local chart U such that if

$$\omega(u,v) = a\ du^2 + b\ dudv + c\ dv^2 \quad \textit{then} \quad b^2 - 4ac \geq 0$$

where the lower bound is attained if and only if x is a singular point.

Definition 15.1.1 *A C^r-configuration on M is a triple $(\mathcal{F}_1, \mathcal{F}_2, S)$ where S is a closed subset Sing ω and $\mathcal{F}_1, \mathcal{F}_2$ are two transversal one-dimensional C^r foliations on $M \backslash S$.*

If $\omega \in Q^r(M)$ is a positive differential 2-form written in the local chart (u,v) as $a(u,v)dv^2 + b(u,v)dudv + c(u,v)dv^2$ then foliations $\mathcal{F}_1$ and $\mathcal{F}_2$ are tangent to the 'vector field'

$$\frac{du}{dt} = -b(u,v) \pm \sqrt{(b^2 - 4ac)(u,v)}, \quad \frac{dv}{dt} = 2a(u,v),$$

except possibly along the curve $\{a(u,v) = 0 | u \leq 0\}$. (Note that the above differential equations are by no means 'good' since the singularity $(0,0)$ is non-isolated.)

Definition 15.1.2 *Two forms $\omega_1, \omega_2 \in Q^r(M)$ are said to be equivalent if there exists a homeomorphism $h : M \to M$ such that $h(S_1) = S_2$ and $h(\mathcal{N}_1) = \mathcal{N}_2$, where $\mathcal{N}_1$ and $\mathcal{N}_2$ are the transversal nets on $M \backslash S$ formed by the pairs $\mathcal{F}_1, \mathcal{F}_2$.*

Definition 15.1.3 *A positive C^r differential 2-form $\omega \in Q^r(M)$ is said to be structurally stable if there exists a neighborhood $\mathcal{U}$ of ω in $Q^r(M)$ such that every $\omega' \in \mathcal{U}$ is equivalent to ω.*

Let $\omega \in Q^r(M)$ be a positive differential 2-form on M and let $a(u,v), b(u,v), c(u,v)$ be its coefficients in a chart U. Let $x \in M$ be a point on the surface M. By a derivative form, $D\omega(x)$, in the point x one understands the form

$$D\omega(x) = Da(x)du^2 + Db(x)dudv + Dc(x)dv^2,$$

where $Da(x), Db(x), Dc(x)$ are usual derivative of the real-valued functions a, b, c taken at the point x.

Definition 15.1.4 (Guíñez) *A singular point $x \in M$ of the positive differential 2-form ω is said to be simple if $D\omega(x)$ is a positive differential 2-form in x. The subspace of positive differential 2-forms $\omega \in Q^r(M)$ such that all its singular points are simple is denoted by $\mathcal{G}^r(M)$.*

15.2 Local Theory

15.2.1 Normal Forms

Lemma 15.2.1 (Guíñez) *Let $x \in M$ be a singular point of a positive differential 2-form $\omega \in Q^r(M)$. The following statements are pairwise equivalent:*

(i) The point x is a simple singular point of ω;

(ii) There exists a local chart U near x such that if ω in this chart has the form

$$\omega(u,v) = (a_1u + a_2v + M_1(u,v))\, dv^2 + (b_1u + b_2v + M_2(u,v))\, dudv + (c_1u + c_2v + M_1(u,v))\, du^2,$$

where $M_i = O(u^2+v^2)$ then the following inequalities hold:

$$b_1^2 - 4a_1c_1 > 0, \qquad (b_1^2 - 4a_1c_1)(b_2^2 - 4a_2c_2) - (b_1b_2 - 2a_1c_2 - 2a_2c_1)^2 > 0.$$

(iii) The statement (ii) holds true for any local chart in x;

(iv) There exists a local chart U such that ω writes as

$$\omega(u,v) = (v + M_1(u,v))\, dv^2 + (b_1u + b_2v + M_2(u,v))\, dudv + (-v + M_3(u,v))\, du^2,$$

where $M_i(u,v) = O(u^2+v^2)$ and $b_1 \neq 0$.

Proof. The only hard parts are (i) ⇒ (ii) and (ii) ⇒ (iv). Let us consider them consequently.

(i) ⇒ *(ii)*. Let x be a simple point of positive C^r differential 2-form ω which admits the expansion of item (ii). Then

$$D\omega(x) = (a_1u + a_2v)dv^2 + (b_1u + b_2v)dudv + (c_1u + c_2v)du^2.$$

Since $D\omega(x)$ must be positively defined for all $(u, v) \in U$ the inequalities of item (ii) follow.

(ii) ⇒ (iv). Suppose that the form ω is as in item (ii). Let us consider the action of the linear group $GL_2(\mathbb{R})$ on ω defined by the substitutions $u' = \alpha u + \beta v, v' = \gamma u + \delta v$. The relevant coefficients of the form ω' may be written as

$$\begin{aligned}
a_1' &= \delta^2(\alpha a_1 + \gamma a_2) + \beta\delta(\alpha b_1 + \gamma b_2) + \beta^2(\alpha c_1 + \gamma c_2),\\
a_2' &= \delta^2(\beta a_1 + \delta a_2) + \beta\delta(\beta b_1 + \delta b_2) + \beta^2(\beta c_1 + \delta c_2),\\
c_1' &= \gamma^2(\alpha a_1 + \gamma a_2) + \alpha\gamma(\alpha b_1 + \gamma b_2) + \alpha^2(\alpha c_1 + \gamma c_2),\\
c_2' &= \gamma^2(\beta a_1 + \delta a_2) + \alpha\gamma(\beta b_1 + \delta b_2) + \alpha^2(\beta c_1 + \delta c_2).
\end{aligned}$$

Let us consider the following possibilities:

(i) Suppose that $a_1 = c_1 = 0$. Then we set $\beta = \gamma = 0$ and check that also $a_1' = c_1' = 0$. The rest of the equations writes as $a_2' = \delta^3 a_2, c_2' = \alpha^2\delta c_2$. Since by (ii) $a_2c_2 < 0$ it is easy to choose α and δ such that $a_2' = -c_2' = 1$.

(ii) Suppose that $a_1 \neq 0, c_1 = 0$. Fixing $\gamma = 0$ we always have $c_1' = 0$. Therefore $a_1' = \alpha\delta(\delta a_1 + \beta b_1)$ with $b_1 \neq 0$. Setting $\beta = -\delta a_1/b_1$ we get $a_1' = 0$. One is back to the case (i).

(iii) Let, finally, $a_1 \neq 0, c_1 \neq 0$. We put $\alpha = \lambda\gamma$ and search for a real solution of the cubic equation

$$c_1\lambda^3 + (b_1 + c_2)\lambda^2 + (a_1 + b_2)\lambda + a_2 = 0.$$

(Clearly, since $c_1 \neq 0$ such a solution always exists.) It remains to check that the solvability of the above cubic equation implies $c_1' = 0$. Therefore we are back to the case (ii).

Since all possibilities are considered Lemma 15.2.1 is proved. □

Definition 15.2.1 (Guíñez) *Let $x \in M$ be a simple singular point of the positive differential 2-form ω. Point x is said to be hyperbolic simple singular point of ω if the homogeneous cubic polynomial $D\omega(x)(u, v)$ obtained from $D\omega(x)$ by the substitution $du = u, dv = v$:*

$$v\left[v^2 + b_2uv + (b_1 - 1)u^2\right] = 0,$$

has only simple roots.

It is clear that the hyperbolicity conditions are equivalent to the conditions

$$(b_2/2)^2 + 1 - b_1 \neq 0, \quad \text{and} \quad b_1 \neq 1.$$

Since b_1 does not vanish we have to consider the following three cases:

$$D_1 : (b_2/2)^2 + 1 < b_1;$$
$$D_2 : (b_2/2)^2 + 1 > b_1 > 0;$$
$$D_3 : b_1 < 0.$$

15.2.2 Classification

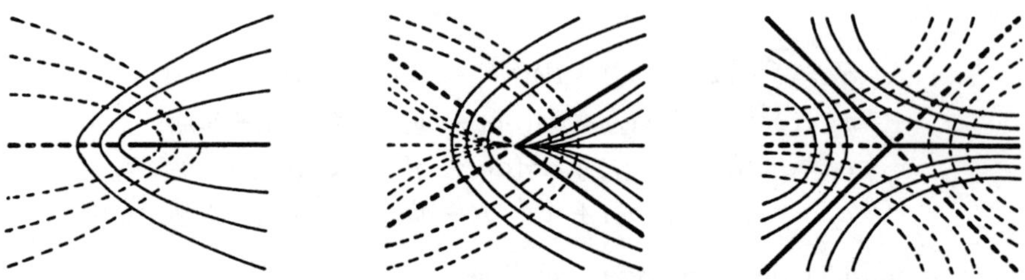

Case $\mathbf{D_1}$ Case $\mathbf{D_2}$ Case $\mathbf{D_3}$

Fig. 15.1. The local configurations near the hyperbolic simple singular point.

Theorem 15.2.1 (Guíñez) *Let $x \in M$ be a hyperbolic simple singular point of the positive differential 2-form $\omega \in \mathcal{G}^r(M)$. Then ω is locally stable at the point x. Moreover, the corresponding configurations of leaves near x are topologically conjugate to those shown in Fig. 15.1.*

Proof. Let $\omega \in \mathcal{G}^r(M)$ be the positive differential 2-form with a simple hyperbolic singularity at $x \in M$. The vector field $Y^{\pm} = P\partial/\partial u + Q^{\pm}\partial/\partial v$, where

$$P(u,v) = 2a(u,v), \qquad Q^{\pm}(u,v) = -b(u,v) \pm \sqrt{(b^2 - 4ac)(u,v)},$$

is tangent to one of the foliations $\mathcal{F}_1, \mathcal{F}_2$, except possibly when $a(u,v) = 0$. Making a 'blow-up' of the singularity at the point $(0,0)$:

$$u = s, \qquad v = st + y(s),$$

where $y = y(s)$ is the unique solution satisfying the equation $a(s, y(s)) = 0$ in the point $y(0) = 0$, we get the vector field:

$$Z^{\pm} = S\frac{\partial}{\partial s} + T^{\pm}\frac{\partial}{\partial t}.$$

Here we have:

$$\begin{aligned}
S(s,t) &= 2st\,[1+U(s,t)]\,, \qquad U(0,t)=0,\\
T^{\pm}(s,t) &= s^{-1}\left[-(t+y'(s))S + S_1 \pm \sqrt{S_1^2 + SS_1}\right],\\
S_1(s,t) &= -s\,[b_1 + b_2 t + U_1(s,t)]\,, \quad U_1(0,t)=0,\\
S_2(s,t) &= -2s\,[-t + U_2(s,t)]\,, \qquad U_2(0,t)=0.
\end{aligned}$$

We 'rescale' the vector field $Z^{\pm}$ by $r^{\pm}(s,t)$:

$$\begin{aligned}
\widetilde{Z}^{\pm}(s,t) &= r^{\pm}(s,t)Z^{\pm} = \widetilde{S}_i\frac{\partial}{\partial s} + \widetilde{T}\frac{\partial}{\partial t},\\
r^{\pm}(s,t) &= (st)^{-1}\left[-(t+y'(s))S + S_1 \pm \sqrt{S_1^2 + SS_2}\right].
\end{aligned}$$

For $i = 1, 2$ after the direct computations one obtains:

$$\begin{aligned}
\widetilde{S}_i(s,t) &= \begin{cases} 2s\,[1+U(s,t)]\, f_i(s,t), & \text{if } s>0,\\ 2s\,[1+U(s,t)]\, f_{3-i}(s,t) & \text{if } s<0,\end{cases}\\
f_i(s,t) &= 2t(t+y'(s))(1+U(s,t)) - (b_1 + b_2 t + U_1(s,t)) +\\
&\quad + (-1)^i\sqrt{(b_1+b_2t+U_1(s,t))^2 - 4t(1+U(s,t))(-t+U_2(s,t))},\\
\widetilde{T}(s,t) &= 4(1+U(s,t))[t(t+y'(s))^2(1+U(s,t)) +\\
&\quad + (t+y'(s))(b_1+b_2t+U_1(s,t)) + (-t+U_2(s,t))].
\end{aligned}$$

(The above formulas will be used in the local analysis of singular points of types $D_1 - D_3$ to which we proceed now.)

Case D_1. We have $(b_2/2)^2 + 1 < b_1$. In this case $b_1 > 0$ and the following estimates are valid:

$$\begin{aligned}
\frac{\partial S_1}{\partial s^+}(0,0) &= -2b_1 < 0, \qquad \frac{\partial S_1}{\partial s^-}(0,0) = 0,\\
\frac{\partial \widetilde{S}_1}{\partial t}(0,0) &= 0, \qquad \frac{\partial \widetilde{T}}{\partial t}(0,0) = 4(b_1 - 1) > 0.
\end{aligned}$$

Cases D_2 and D_3. We have $(b_2/2)^2 + 1 > b_1$. For the sake of simplicity we suppose that $1 - b_1 > 0$ (other cases are treated similarly). One obtains easily that $f_1(0,t) < 0$ and $t_2 < 0 < t_1$, where

$$t_{1,2} = -\frac{b_2}{2} \pm \sqrt{\left(\frac{b_2}{2}\right)^2 + 1 - b_1}.$$

In the case D_2 we get

$$\frac{\partial \widetilde{S}_1}{\partial s^+}(0,0) = -b_1 - |b_1|, \quad \frac{\partial \widetilde{S}_1}{\partial s^-}(0,0) = -b_1 + |b_1|,$$
$$\frac{\partial \widetilde{S}_1}{\partial t}(0,0) = 0, \quad \frac{\partial \widetilde{T}}{\partial t}(0,0) = 4(b_1 - 1) < 0,$$

and in the case D_3 for $i = 1, 2$:

$$\frac{\partial \widetilde{S}_1}{\partial s^+}(0,t_i) = 2f_1(0,t_i) < 0, \quad \frac{\partial \widetilde{S}_1}{\partial s^-}(0,t_i) = 0,$$
$$\frac{\partial \widetilde{S}_1}{\partial t}(0,t_i) = 0, \quad \frac{\partial \widetilde{T}}{\partial t}(0,t_i) = (-1)^{i+1} 8t_i \sqrt{\left(\frac{b_2}{2}\right)^2 + 1 - b_1} > 0.$$

The construction of the topological equivalence in each of the above cases is straightforward. Theorem 15.2.1 is proven. □

15.3 Space of Forms

15.3.1 Stable Subspace

Let $S^r(M) \subset \mathcal{G}^r(M)$ be a subset in the space $\mathcal{G}^r(M)$ of the positive differential 2-forms with simple singular point such that:

- $\omega \in S^r(M)$ has a finite number of singular points and periodic leaves all of which are hyperbolic;
- The limit set of every leaf of $\mathcal{F}_1, \mathcal{F}_2$ defined by ω is a union of singular points and periodic leaves;
- $\mathcal{F}_1, \mathcal{F}_2$ have no double separatrices. (It means that there are no separatrix of the singular point which is a separatrix of two different singular points or twice a separatrix of the same singular point.)

15.3.2 Theorem of V. Guinez

Theorem 15.3.1 (Guíñez) *For each $1 \le r \le \infty$ the set $S^r(M) \subset \mathcal{G}^r(M)$ is non-empty and*

(i) All elements in $S^r(M)$ are structurally stable and the set $S^r(M)$ is open;
(ii) $S^r(M)$ is dense in $\mathcal{G}^r(M)$ in the C^r Whitney topology;

Proof. (i) The openness of $S^r(M)$ follows from the local stability of the hyperbolic singular points and periodic leaves and stability of the separatrices.

On the other hand, the structural stability of $\omega \in S^r(M)$ may be proved using the method of canonical regions (filtration decomposition) similar to those applied to the Morse-Smale foliations.

(ii) Density of $S^r(M)$. We split the proof into a series of lemmas. Consider the following subsets in $Q^r(M)$: H_1 is the subset in $\mathcal{G}^r(M)$ whose singular points are hyperbolic; H_2 is the subset in H_1 which generates foliations without non-trivial recurrent leaves; H_3 is the subset in H_2 which generates foliations without double separatrices. We shall prove that:

(i) H_1 is dense in $\mathcal{G}^r(M)$;
(ii) H_2 is dense in H_1;
(iii) H_3 is dense in H_2;
(iv) And, finally, $S^r(M)$ is dense in H_3.

Lemma 15.3.1 *The set H_1 is dense in the space $\mathcal{G}^r(M)$ with regard to the C^r Whitney topology.*

Proof. Let $\omega \in \mathcal{G}^r(M)$ and let x be a singularity of ω. Take a local chart U in which ω writes as:

$$\omega(u,v) = (v+M_1(u,v))dv^2+(b_1u+b_2v+M_2(u,v))dudv+(-v+M_3(u,v))du^2,$$

with $b_1 \neq 0$ and $M_i(u,v) = O(u^2+v^2)$. Let V_1, V_2 be neighbourhoods of $(0,0)$ such that $\overline{V}_1 \subset V_2 \subset \overline{V}_2 \subseteq U$. Take a smooth function $\phi : \mathbb{R}^2 \to [0,1]$ such that $\phi^{-1}(1) = \overline{V}_1$ and $\phi^{-1}(0) = \mathbb{R}^2 \backslash V_2$. Fix $\delta \geq 0$ and define

$$\begin{aligned}\omega_\delta(u,v) &= (v+M_1(u,v))dv^2 + (1+2\delta\phi(u,v))(b_1u+b_2v+M_2(u,v))dudv \\ &\quad + (-v+M_3(u,v))du^2.\end{aligned}$$

Since functions $g_1(\delta) = (1+2\delta)^2(b_2/2)^2+1-(1+2\delta)b_1$ and $g_2(\delta) = 1-(1+2\delta)b_1$ are non-zero for $\delta \neq 0$, there exists $\omega' \in \mathcal{G}^r(M)$ arbitrary C^r-close to ω such that Sing ω' = Sing ω and x is a hyperbolic singular point of ω'. Since ω has only finitely many singular points, Lemma 15.3.1 is proved. □

Lemma 15.3.2 *The set H_2 is dense in the set H_1 with regard to the C^r Whitney topology.*

Proof. We shall explain how to eliminate the non-trivial recurrent leaves of $\omega \in H_1$. (It is done in a similar way to one described for the Morse-Smale foliations.)

Let R be a small square lying in the set $M\backslash$ Sing ω with the horizontal sides ac and bd along $\mathcal{F}_1$ and the vertical sides ab and cd along $\mathcal{F}_2$. Denote by (X_1, X_2) the C^r unit vector fields tangent to $(\mathcal{F}_1, \mathcal{F}_2)$.

Let x be an interior point of R and let γ_1, γ_2 be the leaves of $\mathcal{F}_1, \mathcal{F}_2$ passing through x. Suppose that there exist points p_1, p_2 in $\gamma_1 \cap \gamma_2 \cap$ Int R such that the oriented segments pp_1 (resp. pp_2) of γ_1 (resp. γ_2) have the first return point with ∂R at $[a,b]$ (resp. $[c,d]$), and the last return point at $[c,d]$ (resp $[a,b]$) . (The above situation always happens if we have recurrent leaves.)

Consider a family of C^r vector fields

$$X_1^\alpha = X_1(\omega) + \delta\alpha Y, \qquad \delta > 0, \qquad 0 \le \alpha \le 1,$$

where Y is a vector field transversal to $X_1(\omega)$ in the points of $Int\ R$ going upwards and vanishing outside R. If δ is chosen small enough, then X_1^α lies near X_1 and is transversal to X_2. We may consider a family of positive differential 2-forms $\{\omega_\alpha\} \subset Q^r(M)$ given by

$$\omega_\alpha = \begin{cases} \lambda\omega(X_1^\alpha(\omega), X_2(\omega)) & \text{on } R, \\ \omega & \text{on } M\backslash R, \end{cases}$$

where λ is a positive function on R. Since the surface M is orientable, the first $\xi \mapsto p_1(\xi)$ and the last $\xi \mapsto p_2(\xi)$ return mapping are monotone, one being increasing and other decreasing on γ_2. Therefore we may choose α for which the above mapping is periodic, and there are no recurrent leaves. Lemma 15.3.2 is proved. □

Lemma 15.3.3 *The set H_3 is dense in the set H_2 with regard to the C^r Whitney topology.*

Proof. Since any $\omega \in H_2(M)$ has a finite number of singular points, the same holds true for the double separatrices. The double separatrices can be broken by small perturbations which create new periodic leaves. Lemma 15.3.3 is proved. □

To finish the proof of Theorem 15.3.1 it remains to show that the set $S^r(M)$ is dense in $H_3(M)$. For this it is enough to establish that any $\omega \in H_3(M)$ can be approximated by ω' with a finite number of periodic hyperbolic leaves. But this is an easy property of the first return mappings in the vicinity of a periodic leaf. (Here an extra argument is needed since the Reeb components are possible. However, it does not change considerably the basic idea of the proof, see [122] for details.) Theorem 15.3.1 is proven. □

Bibliographic Notes. Foliations given by the positive differential 2-form have been a focus of several works of mathematicians affiliated to IMPA (Brazil); see C. Gutierrez, J. Sotomayor [130] — [134] and V. Guíñez [122],[123]. An interest to such foliations can be explained by a certain relation to the principal curvature foliations and Carathéodory conjecture.

16. Control Theory (by B. Piccoli)

This chapter is dedicated to the optimal control. We give a brief and simple introduction to the optimal control problems. Our aim is not a complete treatise of the subject but quite a not too technical introduction to understand the solution to a natural geometric problem on manifolds we want to study.

Problem 16.0.1 *Consider a two dimensional manifold M and two smooth vector fields in a generic position. Given any two points p and q on M we want to reach q from p in minimum time using only integral curves of the two given vector fields X and Y.*

After reformulating this question as an optimal control problem it is possible to study it in details and give a satisfactory combinatorial classification of "optimal flows" using techniques of geometric control theory – cf Piccoli [242], and Bressan-Piccoli [60]. Such a classification is similar to the classification of Morse-Smale flows via the Peixoto graphs and is another example of the increasing two sided relationship it is developing between control theory and differential geometry and topology of manifolds, see also

A. A. Agrachev, Methods of Control Theory, in Mathematic Geometry, Proc. ICM 94, Dekker, New York, 1998.

V. Jurdjevic, Geometric Control Theory, Cambridge University Press, 1997.

16.1 Introduction

In every day life it happens usually to face systems that are evolving following some physical laws but at the same time being affected by an external agent that modifies the evolution and in many cases is due to the actions of a human being. We experience this when we are driving a car or simply when we are walking. In the latter case the gravitation is the main physical force involved, while through the use of our massless we can manage to move without falling (that would be the natural consequences of the forced involved if there is no control on them). In some cases it is possible to model the evolution by a controlled dynamical system whose evolution depends upon some parameters called "controls" that are chosen by a "controller". The controller designs controls in a suitable way to achieve a given goal, e.g. in the case of the car or of the man walking to reach a final position without accidents and possibly in minimum time. The goal is described mathematically by an optimization problem. Optimal control theory is the theory that studies optimization problems for control systems.

16.2 Optimal Control

A control on an n-dimensional manifold M is given assigning a family of smooth vector fields:

$$\mathcal{F} = \{X_u : u \in U\} \tag{16.1}$$

where U is a set called the control set and each X_u is a smooth vector field on M.

Remark 16.2.1 *Sometime a control system is introduced as a controlled equation:*

$$\dot{x} = f(x, u), \qquad x \in M, \quad u \in U,$$

where $f : \mathbb{R}^n \times U \to \mathbb{R}^n$, and U is the control set. In this case the family of vector fields is clearly given by

$$\mathcal{F} = \{f(\cdot, u) : u \in U\}.$$

A trajectory of the control system $\mathcal{F}$ is a piecewise smooth curve $x : [a, b] \to M$ for which there exist $a = t_0 < t_1 < \cdots < t_N = b$ and $u_1, \ldots u_N \in U$ such that on each (t_{i-1}, t_i), $i = 1, \ldots, N$, x is differentiable and $dx/dt(t) = X_{u_i}(t)$.

An *optimization problem* is defined giving the *boundary conditions*

$$x(0) = x_0, \qquad x(T) \in S \tag{16.2}$$

where $x_0 \in M$, $S \subset M$, and a *cost functional*

$$J(x, u) = \int_0^T L(x, u)\, dt + \psi(x(0), x(T)). \tag{16.3}$$

The aim is to minimize the cost among the trajectories of (16.1) that satisfy (16.2) (notice that T is not fixed). The function L is called *running cost* and ψ the *final cost.* In the case $L \equiv 1$ and $\psi \equiv 0$ we have a minimum time problem indeed a solution is a trajectory that steers x_0 to S in minimum time. While if $\psi \equiv 0$, S is a point and $\mathcal{F}$ is the set of all vector fields, then we are in the case of the classical calculus of variation problem.

We refer the reader to the above cited monograph of Jurdjevic, for the general definitions and results. The theory of optimal trajectories is a wide topic, but we only observe that in the case of smooth systems on $\mathbb{R}^n$, if the set $\mathcal{F}(x) = \{X_u(x) : u \in U\}$ is a convex set for all x then the existence of optimal trajectories is granted.

Let us call $\mathcal{P}(x_1)$, $x_1 \in M$, the optimization problem (16.2), (16.3), with $S = \{x_1\}$. One can try to solve the family of problems $\mathcal{P} = \{\mathcal{P}(x_1) : x_1 \in M\}$ at the same time looking for a feedback control, that is a function $u : M \to U$, such that the solutions of

$$\dot{x} = f(x, u(x)), \qquad x(0) = x_0 \tag{16.4}$$

are optimal. The feedback $u(x)$ is in general discontinuous and (16.4) has not forward uniqueness. Indeed for any $x_1 \in M$ we want to find a trajectory γ_{x_1}, solution of (16.4), that steers x_0 to x_1. The set of trajectories $\{\gamma_{x_1} : x_1 \in M\}$ is called a synthesis, see [243], for general theory of syntheses, and can be thought as a forward multivalued optimal feedback flow, briefly an "optimal flow".

16.3 Optimal Flows

Let us now restate our geometric problem in terms of control theory. The problem is precisely the minimum time problem on M with boundary conditions $x(0) = p$ and $x(T) \in S = \{q\}$, for the control system $\mathcal{F} = \{X, Y\}$. This control system can also be rewritten as $\dot{x} = F(x) + uG(x)$, $u \in \{-1, +1\}$, where

$$F = \frac{X+Y}{2}, \qquad G = \frac{Y-X}{2}.$$

We start considering the problem in the case $M = \mathbb{R}^2$. If we fix $p = 0$ we can try to solve all the minimum time problems for every $q \in \mathbb{R}^2$ giving a synthesis. To have existence of solutions it is necessary to guarantee that the set of velocities is convex, that is to consider all controls u such that $|u| \leq 1$. Indeed there are some optimal trajectories, called turnpikes, for which $u \notin \{-1, 1\}$. If we reverse time we obtain the problem of minimum time stabilization at the origin and in this case one usually assumes $F(0) = 0$. Hence we assume $F(0) = 0$ (the case $F(0) \neq 0$ can be treated using the same techniques).

Finally we obtain the following problem. Let (F, G) be a pair of smooth vector fields on the plane such that $F(0) = 0$ in a fixed coordinate system. Consider the problem of reaching all the points of the plane from the origin in minimum time for the control system

$$\frac{dx}{dt} = F(x) + G(x)u, \qquad |u(t)| \leq 1. \tag{16.5}$$

Under generic assumptions on F and G, listed in [241], [59], the problem admits a solution. Namely, denote by $R(\tau)$ the set of points in the plane which can be reached from the origin within time τ, also called *reachable set in time* τ. For any $\tau > 0$ one can define a *feedback control* $u = \varphi(x)$ such that: (i) the set $R(\tau)$ splits into a finite number of submanifolds V_i such that the restriction $\varphi|_{V_i}$ is a smooth function and (ii) every trajectory of the equation

$$\frac{dx}{dt} = F(x) + G(x)\varphi(x) \tag{16.6}$$

starting from the origin reaches points of $R(\tau)$ in minimum time. For generic vector fields $F, G \in C^3$ the optimal feedback $\varphi(x)$ is essentially unique, and the corresponding "optimal flow" is unique (with possibly two distinct trajectories for points on special curves some of which are called overlaps, see below), so that one regards (16.6) as a differential equation with discontinuous right-hand side defined uniquely by the vector fields F and G.

It is not restrictive to assume that $\tau = 1$ and to study the set $R = R(1)$ of points *reachable* from the origin within unit time.

Define the vector fields (that are our original vector fields)

$$X = F - G, \qquad Y = F + G,$$

corresponding to the constant controls $u \equiv -1$ and $u \equiv 1$ of system (16.5), respectively. Every solution to (16.5) with such a control function is called an *X-trajectory* and *Y-trajectory*, respectively. The set $R = R(1)$ of reachable points splits into a finite number of open regions filled-up with the X- and Y-trajectories of system (16.5). The boundaries between these regions are called *frame curves.* The intersection points of frame curves are called *frame points.*

A C^2-curve in the plane is called a regular *turnpike* if it is defined by the equation

$$G \wedge [F, G] = 0,$$

where $[F, G] = (\nabla G)F - (\nabla F)G$ is the Lie bracket of the two vector fields. For every turnpike S one can define a control $\varphi_S(x)$ such that S becomes a trajectory of the feedback equation (16.6). Such a trajectory is called a *Z-trajectory.* Each time–optimal trajectory is a finite concatenation [1] of X-, Y- and Z-trajectories.

If we do not bother about degenerate cases, there exists only finite number of types of frame curves and frame points. It has been proved by Piccoli [242] that there are five types of generic frame curves, namely:

- The X- and Y-trajectories coming from the origin;
- The topological frontier of the reachable set;

[1] Concatenation $\gamma_2 \star \gamma_1$ of two trajectories $\gamma_1 : [a, b] \to I\!R^2, \gamma_2 : [b, c] \to I\!R^2$ is a trajectory $[a, c] \to I\!R^2$ such that its restriction to $[a, b]$ and $[b, c]$ coincides with γ_1 and γ_2, respectively. In other words, point moves along trajectory γ_1 and then along trajectory γ_2.

- A *switching* curve, i.e. a curve on which optimal trajectories change control from $+1$ to -1 or *vice versa*;
- The regular turnpikes; and
- An *overlap* curve, i.e. a C^2-curve whose points admit a double time–optimal connection with the origin.

The above curves are dubbed by the letters X, Y, F, C, Z and K, respectively. Generic frame points are obtained as intersection of different types of frame curves. However, not all combinatorial possibilities can be realized, see [242]. There exist the following types of frame points:

$$(X,Y),\quad (X,F),\quad (X,C),\quad (X,Z),\quad (X,K),\quad (Y,F),\quad (Y,C),$$
$$(Y,Z),\quad (Y,K),\quad (F,C),\quad (F,Z),\quad (F,K),\quad (C,C),\quad (C,Z),$$
$$(C,K),\quad (Z,K)\quad \text{and}\quad (K,K).$$

16.4 Generic Optimal Flows on the Plane

Denote by Σ an optimal flow defined by differential equation (16.6). Let R be a reachable set of Σ, except internal (i.e. not boundary) points of the overlap curves.

Definition 16.4.1 (Bressan–Piccoli) *Optimal flows Σ_1 on R_1 and Σ_2 on R_2 are equivalent, or $\Sigma_1 \sim \Sigma_2$, if there exists a homeomorphism $h : R_1 \to R_2$ such that the following holds:*

(i) h maps the arcs of optimal trajectories for Σ_1 onto the arcs of optimal trajectories for Σ_2;
(ii) h induces a bijection on the frame curves that are not the overlap curves, i.e. h sends frame curves of Σ_1 to frame curves of Σ_2 keeping the types of frame curves; and
(iii) if $U \subseteq R_1$ is an open region of R_1 bounded by frame curves and filled-up with X- or Y-trajectories of Σ_1, then $h(U)$ is bounded by the corresponding (see item (ii)) frame curves and is filled-up by X- or Y-trajectories of Σ_2, respectively.

The space Ω of control systems (16.5) falls into the equivalence classes under the relation $\sim$ of Definition 16.4.1. The rest of this section is devoted to description of these equivalence classes via a discrete object which is a planar graph with some extra data assigned to it. Recall that graph is *planar* if it has a proper embedding into the sphere. Fix $\Sigma \in \Omega$ a feedback flow in the plane.

Definition 16.4.2 *By a graph $\mathcal{G}$ associated to Σ one understands a planar finite graph whose edges correspond to the frame curves and whose points correspond to the frame points of Σ. Every edge can be one of the following type: X, Y, F, C, Z and K corresponding to the types of frame curves. An edge of type X, Y and Z has an orientation defined by the orientation of the flow Σ on it. The edges of type C have a positive side what reflects the fact that the trajectories cross C-frame curve passing from one side to another. For a fixed rotation system on $\mathcal{G}$ (see Chapter 2) corresponding to the planar embedding, the faces of $\mathcal{G}$ which are not bounded by the curves of F-type, have a sign $+$ or $-$ depending on the X- or Y-trajectories fill-up the respective reachable set. Finally, to each frame point x it is associated a set of frame curves. These correspond to curves reached from x via the optimal flow excluding X, Y and Z frame curves, see [60] for details.*

Given two graphs $\mathcal{G}_1, \mathcal{G}_2$ as in Definition 16.4.2, we say that they are *isomorphic*, $\mathcal{G}_1 \sim \mathcal{G}_2$, if and only if they are:

- isomorphic as graphs;
- the isomorphism preserves the orientation of the oriented edges, the positive sides of edges of type C and the set of frame curves associated to each frame point;
- endowed with equivalent rotation systems so that the corresponding faces are of the same sign.

Theorem 16.4.1 (Bressan–Piccoli) *If Σ_1, Σ_2 are two optimal flows in the plane and $\mathcal{G}_1, \mathcal{G}_2$ are their graphs, then $\Sigma_1 \sim \Sigma_2$ if and only if $\mathcal{G}_1 \sim \mathcal{G}_2$.*

Proof. Assume first that $\Sigma_1 \sim \Sigma_2$ and let h be homeomorphism which realizes the equivalence. It can be verified directly that graphs $\mathcal{G}_1$ and $\mathcal{G}_2$ are isomorphic and all axioms of Definition 16.4.2 are satisfied.

To prove the converse, suppose that $\mathcal{G}_1 \sim \mathcal{G}_2$. Because of equivalence of rotation systems we have the same system of faces associated to $\mathcal{G}_1$ and $\mathcal{G}_2$. One can fill-up these faces with the X- or Y-trajectories accordingly to the sign of each particular face. We extend a homeomorphism h between $\mathcal{G}_1$ and $\mathcal{G}_2$ regarded as topological spaces (1-dimensional CW-complex) to the whole plane along the X- and Y-trajectories. It remains to check that the axioms of equivalence $\sim$ on the set of graphs transform into the axioms of Definition 16.4.1. □

It is natural to look at how large is the set of graphs $\mathcal{G}$ among all planar graphs. In other words, which graphs are admissible in the sense that they are graphs of concrete optimal flows in the plane? We shall formulate admissibility conditions $G1 - G5$ for the graphs.

Condition G1. Consider the set of graphs $\mathcal{E}$ whose elements are defined locally so that each one corresponds to a type of frame point. A point x of graph $\mathcal{G}$ is said to be *admissible* if there exists graph $\mathcal{G}' \in \mathcal{E}$ such that $\mathcal{G}$

contains a copy of $\mathcal{G}'$ to which x belongs. Condition $G1$ consists in that all points of graph $\mathcal{G}$ are admissible.

Condition G2. Consider the set of graphs that contain exactly one point of type (X, Y) called the *origin* of the graph. Assume also that condition $G1$ holds. Denote by e an Y-edge and let x be the initial point of e. If x is not the origin, there exists an Y-edge e_1 for which x is a terminal point. We take the initial point x_1 of E_1 and proceed in the same way. It yields us a finite collection $e_1, \ldots, e_n$ of edges such that edge e_i has a common point with edge e_{i+1}. A subsequence $e_1, \ldots, e_m$ of this collection is formed by the Y-edges. By η^+ we denote the union of such edges. The similar collection of X-edges is denoted by η^-. The $\eta^\pm$ is called *admissible* if it is corresponds to a $\gamma^\pm$ trajectory of a concrete feedback flow, here $\gamma^\pm$ are the constant control $u = \pm 1$ trajectories issued from the origin. (For a classification of local behaviors around $\gamma^\pm$ via a finite word from a finite alphabet, see [240].) Condition $G2$ is that $\mathcal{G}$ has exactly one (X, Y) point which is the origin and collections of edges $\eta^\pm$ are admissible.

Condition G3. If e is an C-edge, we denote by A_1 and A_2 two faces of $\mathcal{G}$ such that $A_1 \cap A_2$ contains e. Then e is *admissible* if it is possible to define functions $F \wedge G$ and $G \wedge [F, G]$ that take the same sign on A_1 and A_2. Condition $G3$ claims that all C-edges of $\mathcal{G}$ are admissible.

Condition G4. Condition $G4$ requires that there is only one closed curve formed by F-edges which bounds a region with frame points and edges inside, and no such points and edges outside the region.

Condition G5. If we construct a curve for each frame point connecting the frame curves associated to it, then it is possible to choose them not crossing each other. Moreover they split the graph in regions that resemble "flow box regions" for an optimal flow, see [60] for details.

Theorem 16.4.2 (Bressan-Piccoli) *If a planar graph $\mathcal{G}$ satisfies the conditions G1-G5, then there exists an optimal flow Σ in the plane whose graph is isomorphic to $\mathcal{G}$.*

Proof. In outline, proof consists in the construction of the flow Σ directly on $\mathcal{G}$. This is done covering step by step "flow box regions" and using the admissibility conditions. The classification of frame points and curves with explicit examples is here crucial. See [60] for details.

16.5 Optimal Flows on 2-Manifolds

The notion of control system (16.5) and feedback flow (16.6) can be correctly redefined for more complicated topological spaces, such as closed surfaces and surfaces with boundary. Unlike the planar case, the graph $\mathcal{G}$ formed by the frame curves and points is not planar and the embedding of $\mathcal{G}$ into surface is not always cellular, see Chapter 2. However, the analogues of Theorems 16.4.1 and 16.4.2 are true. (The work in this direction is under the way.) We quote two theorems from an unpublished preprint [227].

Theorem 16.5.1 *Let Σ_1 and Σ_2 be structurally stable optimal flows given on a compact 2-dimensional manifold M. Then Σ_1 is equivalent to Σ_2 if and only if their complexes O_{w1} and O_{w2} are isomorphic.*

Theorem 16.5.2 *Let O_w be arbitrary complex isomorphic to an admissible complex $\mathfrak{D}$. Then there exists a structurally stable optimal flow on a compact 2-dimensional manifold whose complex $O_w{}'$ coincides with O_w.*

Bibliographic Notes. The fundamentals of control theory have been laid by Lyapunov and Pontryagin. The ideas of global analysis and differential geometry changed the outlook and content of the theory. Structurally stable optimal flows on the two-dimensional manifolds have been studied by A. Bressan and B. Piccoli; see e.g. [60] and [242].

Part III

Appendix

17. Riemann Surfaces

For a profound understanding of the theory, we recommend:

R. Fricke & F. Klein, Vorlesungen uber die Theorie der automorphen Funktionen. Vol. I and II, Teubner, Leipzig, 1897 and 1901.

17.1 Uniformization Theorem

Definition 17.1.1 *By a Riemann surface one understands a 1-dimensional complex connected manifold. The set of charts*

$$z_\alpha : U_\alpha \to \mathsf{C}$$

is called a set of analytic coordinate charts. Traditionally, if a Riemann surface is compact it is said to be closed; if it is non-compact, it is referred to as being open.

(Immediate examples of Riemann surfaces are given by the complex plane C endowed with a unique analytic coordinate chart $(\mathsf{C},\ Id)$, and by the Riemannian sphere $\mathsf{C} \cup (\infty)$ with two analytic coordinate charts $U_1 = \mathsf{C}$ and $U_2 = (\mathsf{C}\backslash 0) \cup (\infty)$.) A continuous mapping

$$f : M \to N$$

between two Riemann surfaces is said to be *holomorphic* if for every local coordinate $\{U, z\}$ of M and the corresponding local coordinate $\{V, w\}$ of N the mapping $w \circ f \circ z^{-1}$ is a holomorphic mapping from C to C. The holomorphic mapping f is called a *conformal mapping* if it establishes a homeomorphism between M and N. The mappings

$$f : M \to \mathsf{C}, \qquad \mathcal{M} \to \mathsf{C} \cup (\infty)$$

are called the *holomorphic function* and *meromorphic function* on M, respectively. The mapping f is a *constant mapping* if it sends M to a point $z \in \mathsf{C}$.

Proposition 17.1.1 *Let M and N be Riemann surfaces and suppose M be compact. Suppose that $f : M \to N$ is a holomorphic mapping. Then f is either a constant mapping or a surjective mapping. In particular N is also compact.*

Proof. If f is not a constant mapping then $f(M)$ is open and compact. Thus $f(M)$ is a closed subset in N. Since M and N are connected, $f(M) = N$. □

'Existence theorems' classify surfaces on which there exists a complex analytic structure. Briefly, the question is on what M there exists a real-valued function $f : M \to \mathbb{R}$ for which

$$\frac{\partial^2 f}{\partial x^2} + \frac{\partial^2 f}{\partial y^2} = 0 \tag{17.1}$$

in every local chart (x, y) chosen on M. It turns out that the answer for 2-dimensional (real) orientable manifolds is always positive. For the simply connected manifolds (such as the 2-sphere S^2 and 2-disc D^2) the solution of (17.1) is given explicitly by the Poisson formula [101], while for the n-connected domains (which are models for surfaces of higher genus) the proof of existence uses Green's and subharmonic functions which admit singular points. The following theorem is true.

Theorem 17.1.1 (Riemann Mapping Theorem) *Let M be a simply connected Riemann surface. Then M is conformally equivalent to one and only one of the following three surfaces:*

(i) Riemann sphere $\mathsf{C}^* = \mathsf{C} \cup \{\infty\}$;
(ii) Complex plane C;
(iii) The unit disc $\Delta = \{z \in \mathsf{C} : |z| < 1\}$ (or, what is conformally the same, the upper half-plane $\mathsf{H} = \{z \in \mathsf{C} : Im\ z > 0\}$.

All the above surfaces are conformally distinct, and are called, respectively, elliptic, parabolic and hyperbolic Riemann surfaces.

Proof. The above theorem is a rather deep result and requires considerable background to be proved [101]. We give the proof in outline so that the reader can see essential links existing between the Riemann surfaces and the theory of harmonic functions.

First we introduce the following 'harmonic' classification of Riemann surfaces. A non-compact Riemann surface S is parabolic if every negative subharmonic function [101] on S is constant; otherwise S is called hyperbolic.

Using subharmonic functions and Perron families, one can define Green's functions for the Riemann surface just as can be done for the disc domain by the Poisson formula. The Green's function, G_p, on the surface S has a singularity in the point $p \in S$, and is positive and harmonic on $S \backslash p$. To describe the singularity, we consider a local parameter z which maps a neighbourhood of p onto the unit disc such that $z(p) = 0$. Then it is required that $G_p + \log |z|$

be harmonic at p (this is a definition which does not depend on the choice of the local parameter). The Green function is characterized by the property that among all positive and harmonic functions on $S\backslash p$, and possessing the same singularity, the function G_p is the smallest. If a Green's function exists for some $p \in S$, then it exists for every point in S. Moreover, such function exists if and only if S is hyperbolic.

In the case S is parabolic or compact, Green's functions do not exist but it is possible to prove the existence of a function $u_{p,q}$ with the following properties:

- $u_{p,q}$ is harmonic in $S\backslash(p \cup q)$;
- If $z(p) = 0$, then $u_{p,q} - \log|z|$ is harmonic at p, and if $z(q) = 0$, then $u_{p,q} + \log|z|$ is harmonic at q;
- Outside parameter discs containing p and q the function $u_{p,q}$ is bounded.

Suppose now that S is simply connected. If S is hyperbolic, one takes the Green's function G_p and forms a conjugate G_p^*. The function

$$f(z) = e^{-(G_p + iG_p^*)}$$

is extended by the analytic continuation to S. Using the Monodromy Theorem [101], we conclude that the extended function is single-valued on S. Finally, applying the Maximum Principle, it can be shown that $f(z)$ maps S conformally to the unit disc.

If S is parabolic or compact, the conformal mapping of S into the extended plane can be constructed with the aid of the function $u_{p,q}$. In case S is parabolic the boundary of the image consists of one point, whereas the boundary is empty if S is compact. □

The above mentioned simply connected Riemann surfaces allows one to classify all Riemann surfaces. The idea is that C^*, C and H can be regarded as a universal covering for any Riemann surface. The construction of the covering is intimately connected with the study of various subgroups of the automorphism group Aut U acting on the respective simply connected surfaces $U = \mathsf{C}^*, \mathsf{C}, \mathsf{H}$. (Such subgroups are known as Kleinian, Fuchsian, modular, etc. groups acting on U.) From this point of view each Riemann surface can be represented as a factor-surface U/Γ, where $\Gamma \subset$ Aut U.

17.2 Discrete Groups

17.2.1 Möbius Transformations

Denote by C a 2-dimensional Euclidean (x, y)-plane representing the complex numbers $z = x + iy$. (For brevity it is further referred to as a 'complex plane' C.)

Definition 17.2.1 *Let $z \in \mathsf{C}$ be a point lying in a complex plane C. The transformation*

$$z \longmapsto \frac{az+b}{cz+d}$$

where a, b, c and d are complex numbers such that $ad - bc \neq 0$, is called a Möbius (or, linear-fractional) transformation of C. Since the nominator and the denominator of the Möbius transformation can be multiplied by a complex number, one can always suppose that $ad - bc = 1$ so that the Möbius transformations are naturally identified with the group $PSL(2, \mathsf{C}) \simeq SL(2, \mathsf{C})/\pm I$.

The Möbius transformations form a group by the composition with an inverse element defined by the formula

$$z \longmapsto \frac{-dz+b}{cz-a}$$

Let $\mu : \mathsf{C} \to \mathsf{C}$ be a Möbius transformation $\mu \in PSL(2, \mathsf{C})$ of a complex plane C. Clearly μ acts one-to-one and sends the 'extended complex plane' $\mathsf{C}^* = \mathsf{C} \cup (\infty)$ to itself. (Clearly, the point $-d/c$ is sent under μ to the point ∞, and vice versa.)

There are several notable objects in the complex plane which are in a sense 'invariant' under the μ-actions. The first such is the set of fixed points of $\mu : \mathsf{C} \to \mathsf{C}$. To calculate them we need to solve the quadratic equation

$$cz^2 + (d-a)z - b = 0.$$

(i) Suppose additionally that $c \neq 0$. Then the above equation has the roots

$$z_1, z_2 = \frac{a - d \pm \sqrt{(a+d)^2 - 4}}{2c}$$

(Recall that we always assume $ad - bc = 1$.) It has either two distinct roots, if $(a+d)^2 - 4 \neq 0$, or a unique root

$$w = \frac{a-d}{2c}$$

otherwise. Therefore, a Möbius transformation $\mu \in PSL(2, \mathsf{C})$ different from the identity, has either one or two fixed points lying in the complex plane C.

(ii) If $c = 0$ then we must suppose $a \neq 0, d \neq 0$ since otherwise the determinant $ad - bc$ will vanish. In this case the fixed points of μ are

$$z_1 = \frac{b}{d-a}, \qquad z_2 = \infty.$$

We sum up the above considerations in the following statement.

Proposition 17.2.1 *The only Möbius transformation $\mu \in PSL(2, \mathsf{C})$ with more than 2 fixed points is the identity transformation $\mu \equiv Id_{\mathsf{C}}$.*

Another geometrical object which behaves 'nicely' with regard to the actions of the group $PSL(2, \mathbb{C})$ are the circles in the complex plane $\mathbb{C}$. (As a 'generalized circle' in $\mathbb{C}$ one includes also the straight lines in $\mathbb{C}$, which represent a degenerate case.)

Proposition 17.2.2 *Denote by $A(x^2+y^2)+b_1x+b_2y+C=0$ a 'generalized circle' in the complex plane $z = x+iy$. Then it is sent by the Möbius transformation $\mu \in PSL(2, \mathbb{C})$ to a 'generalized circle' $A'(x^2+y^2)+b_1'x+b_2'y+C'=0$ in $\mathbb{C}$.*

Proof. Let us reparametrize the equation of 'generalized circle', introducing complex variables z and $\bar{z}$, where $\bar{z}$ is complex conjugate of z. Clearly,

$$x = \frac{1}{2}(z+\bar{z}), \qquad y = \frac{i}{2}(\bar{z}-z), \qquad z\bar{z} = x^2+y^2.$$

The substitution of x, y in the equation of 'generalized circle' yields the equation

$$Az\bar{z} + \frac{1}{2}(b_1 - ib_2)z + \frac{1}{2}(b_1+ib_2)\bar{z} + C = 0.$$

Denoting $B = \frac{1}{2}(b_1 - ib_2)$ one obtains the following canonical form of the 'generalized circle' equation:

$$Az\bar{z} + Bz + \bar{B}\bar{z} + C = 0,$$

where A and C are real. We apply the following substitution (which is a Möbius transformation relative to the new parameters):

$$z = \frac{-dz'+b}{cz'-a}, \qquad \bar{z} = \frac{-\bar{d}\bar{z}'+\bar{b}}{\bar{c}\bar{z}-\bar{a}}.$$

After z and $\bar{z}$ are inserted into the equation of a 'generalized circle', the latter has the form:

$$\begin{aligned}[][Ad\bar{d} - B\bar{c}d - \bar{B}c\bar{d} + Cc\bar{c}]z'\bar{z}' + [-A\bar{b}d + B\bar{a}d + \bar{B}\bar{b}c - C\bar{a}c]z' + \\ + [-Ab\bar{d} + Bb\bar{c} + \bar{B}a\bar{d} - Ca\bar{c}]\bar{z}' + Ab\bar{b} - B\bar{a}b - \bar{B}a\bar{b} + Ca\bar{a} = 0.\end{aligned}$$

It can be easily checked that the coefficient before the term $z'\bar{z}'$ is real; also the constant term is real. The coefficients before z' and $\bar{z}'$ are complex conjugate. Thus we come again to the canonical representation of the 'generalized circle.' □

In the remainder of this subsection our purpose will be the classification of 'typical' Möbius transformations. We have seen earlier that such a classification can be based on the number of fixed points which Möbius transformation $\mu \in PSL(2, \mathbb{C})$ may have, and also the behaviour of μ with regard to these points. As a 'quantitive' characteristic of the 'qualitative' behaviour of μ one considers the so-called *multiplier*, M, of the transformation μ.

Definition 17.2.2 *Let $\mu \in PSL(2, \mathbb{C})$ be a Möbius transformation having a pair of fixed points z_1, z_2 (not necessary distinct). By a multiplier of μ one understands a complex number M which satisfies the equation*

$$\frac{z' - z_1}{z' - z_2} = M \frac{z - z_1}{z - z_2}.$$

where $z' = \mu(z)$ is the image of z under the transformation μ.

(It can be easily calculated that M is expressed through parameters of μ as $M + M^{-1} = a + d$.) The advantage of using M is that the n-th iteration $\mu \circ \ldots \circ \mu$ of the Möbius transformation μ is given by the general formula:

$$\frac{z' - z_1}{z' - z_2} = M^n \frac{z - z_1}{z - z_2},$$

where $z' = \mu^n(z)$.

Definition 17.2.3 *Let $M = Ae^{i\theta}$ be a complex polar representation of the multiplier M of the Möbius transformation $\mu \in PSL(2, \mathbb{C})$. We divide the class of Möbius transformations into the following types:*

Transformation	*Parameters of μ*	*Number of fixed points*	*Multiplier M*
Hyperbolic	$a + d$ *is real and* $\lvert a + d\rvert > 2$	2	$M = A$
Elliptic	$a + d$ *is real and* $\lvert a + d\rvert < 2$	2	$M = e^{i\theta}$
Loxodromic	$a + d$ *is complex*	2	$M = Ae^{i\theta}$
Parabolic	$a + d = \pm 2$	1	—

17.2.2 Fuchsian Group

We shall be interested in those subgroups of $PSL(2, \mathbb{C})$ which leave invariant certain geometrical objects in $\mathbb{C}$, such as the unit circle $x^2 + y^2 = 1$, or the straight line $y = 0$ of the complex plane $\{z \in \mathbb{C} | z = x + iy\}$. The motivation for such choice is actually item (iii) of the Riemann Mapping Theorem, which classifies all conformally distinct simply connected Riemann surfaces. First, let us establish the following

Proposition 17.2.3 *The most general Möbius transformation sending the unit circle $S^1 = \{(x, y) | x^2 + y^2 = 1\}$ to itself and sending the interior of S^1 into itself has the form*

$$z \longmapsto \frac{az + \bar{c}}{cz + \bar{a}}, \qquad a\bar{a} - c\bar{c} = 1. \tag{17.2}$$

Proof. The equation of the unit circle is $z\bar{z} - 1 = 0$ and it maps under the Möbius transformation into the circle

$$(d\bar{d} - c\bar{c})z'\bar{z}' + (\bar{a}c - \bar{b}d)z' + (a\bar{c} - b\bar{d})\bar{z}' + b\bar{b} - a\bar{a} = 0.$$

Clearly, necessary and sufficient conditions for the invariance of S^1 are furnished by the equations:

$$\bar{a}c - \bar{b}d = a\bar{c} - b\bar{d} = 0, \quad d\bar{d} - c\bar{c} = a\bar{a} - b\bar{b} \neq 0.$$

Denote by $\lambda = b/\bar{c} = a/\bar{d}$ in the first equation. Then it has the form $b = \lambda\bar{a}, a = \lambda\bar{d}$. Substituting it in the second equation one arrives at

$$d\bar{d} - c\bar{c} = \lambda\bar{\lambda}(d\bar{d} - c\bar{c}) \neq 0,$$

whence we conclude that $\lambda\bar{\lambda} = 1$. The equality $ad - bc = 1$ implies

$$\lambda(d\bar{d} - c\bar{c}) = 1,$$

and therefore λ must be real number equal to ± 1. (The sign of λ depends upon the sign of $d\bar{d} - c\bar{c}$. In case the interior of S^1 is transformed into itself, it is easy to see that λ is equal to 1.) Finally we obtain

$$b = \bar{c}, \qquad d = \bar{a},$$

which proves the proposition. □

Corollary 17.2.1 *A Möbius transformation which leaves the unit circle invariant cannot be loxodromic.*

It can be verified directly that the Möbius transformations given by (17.2) form a non-commutative subgroup of the group $PSL(2, \mathbb{C})$. As we shall see later, this group plays an essential rôle in the whole theory of Riemann surfaces. The transformations $\mu \in PSL(2, \mathbb{C})$ form, in general, a continuous 4-complex parameter [1] group of transformations of $\mathbb{C}$. Such a group acts 'continuously' on $\mathbb{C}$ in the sense of the following definition.

Definition 17.2.4 *A subgroup $G \subset PSL(2, \mathbb{C})$ is said to contain infinitesimal (or, continuous) transformations, if for some region $U \subset \mathbb{C}$ and any given $\varepsilon > o$ there is a transformation $\mu \in G$ such that, for all $z \in U$, $|\mu(z) - z| < \varepsilon$.*

It is not hard to see that the subgroup of Möbius transformations is infinitesimal if and only if its 'free parameters' vary continuously as complex numbers in the space of parameters. (In other words, such a subgroup is a Lie group acting on $\mathbb{C}$.) We study those subgroups of Möbius transformations which are, in a sense, opposite to those which are infinitesimal. Such subgroups, called *discontinuous subgroups*, will be of the major interest to us.

Definition 17.2.5 *A subgroup of $PSL(2, \mathbb{C})$ is called (properly) discontinuous in the complex plane $\mathbb{C}$ if there exists a point z_0 and a region U enclosing z_0 such that all the transformations from this subgroup, different from identity, send z_0 outside U: $\mu(z_0) \in \mathbb{C} \backslash U$.*

[1] On account of the 'normalized' Möbius transformations with $ad - bc = 1$, $PSL(2, \mathbb{C})$ has only 3 independent complex parameters.

Immediate examples of discontinuous groups are given by finite groups acting in the complex plane C (rotation through rational angles, reflections with respect to a finite number of axes in C, etc). There exist examples of *infinite* discontinuous groups. We introduce the following definition.

Definition 17.2.6 *Transformations $\mu \in PSL(2, \mathsf{C})$ which leave invariant the unit circle sending its interior into itself and which act discontinuously on* C *are called Fuchsian, and the respective group of transformations of the complex plane* C *is called a Fuchsian group.*

The Fuchsian groups fall into three major classes.

- Hyperbolic transformations with two fixed points on the invariant unit circle.
- Elliptic transformations with two fixed points lying at the opposite sides of the invariant unit circle.
- Parabolic transformations with the unique fixed point on the invariant unit circle.

17.2.3 Limit Set of Fuchsian Groups

There exists another important characteristic of Fuchsian groups acting on C which is connected with the limit behaviour of points $z \in \mathsf{C}$ under the action of such groups. (In fact, this is a 'dynamical' characteristic of the Fuchsian group, which can be regarded as a diffeomorphism $\mu : \mathsf{C} \to \mathsf{C}$ of the complex plane C.) First let us establish the following elementary property related with the 'dynamics' of a Fuchsian group.

Proposition 17.2.4 *All limit points of the Fuchsian group G belong to the invariant unit circle S^1.*

Proof. This is a consequence of the following two statements.

(i) In the neighbourhood of a limit point z^* there is an infinite number of distinct points which are G-equivalent to any point of C, with the possible exception of z^* itself and one other point of C;
(ii) If a closed subset $S \subset \mathsf{C}$ of points of C consisting of more than one point is transformed into itself by all $g \in G$, then S contains all the limit points of G.

For the proof of item (i) we refer the reader to [108]. To prove item (ii), suppose to the contrary that there is a limit point z^* which does not belong to S. Then there exists a small neighbourhood of z^* which contains no points of S. Denote by z_1 and z_2 two points in S. By item (i) at least one of them must be G-equivalent to a point in the neighbourhood of z^*. This is in contradiction with the choice a G-invariant S.

Proposition 17.2.4 follows now from the item (ii) supposing $S = S^1$ is G-invariant unit circle of the Fuchsian group G. □

Theorem 17.2.1 *Let G be a Fuchsian group acting on the complex plane* $\mathbb{C}$*. Suppose that the set of limit points S of G consists of more than two points. Then*

(i) either $S = S^1$*, that is, the set S coincides with the invariant unit circle* S^1*,*
(ii) or the set S is a Cantor set in S^1*, that is, S is nowhere dense on the invariant unit circle* S^1*.*

Proof. From the above argument it follows that if the limit set S consists of more than two points, then S is a perfect set. [2] It must be shown now that unless every point of the invariant unit circle S^1 is a limit point, then the limit points form a set which does not contain all the points of any arc lying in the invariant circle S^1.

Let z_0 be a point lying on the invariant circle S^1 and which is not a limit point (such points are called *ordinary points*). Then the points in the neighbourhood of z_0 are also ordinary points. In particular, the points on a sufficiently small arc h of the invariant circle S^1 which passes through z_0 are also ordinary points.

Let z^* be a limit point. Then by item (i) of the proof of Proposition 17.2.4, there are points in the neighbourhood of z^* which are G-equivalent to points of h. These points are ordinary and lie on the invariant circle S^1. Since there are ordinary points on S^1 in the neighbourhood of limit point, this implies that the set of limit points is nowhere dense on the invariant unit circle S^1. □

Theorem 17.2.1 allows us to classify Fuchsian groups with regard to their 'dynamical' properties. We adopt the following terminology.

Definition 17.2.7 *A Fuchsian group G acting on the complex plane* $\mathbb{C}$ *is called a Fuchsian group of the first kind if every point of the G-invariant unit circle* S^1 *is a limit point of the group G. The group G is a Fuchsian group of the second kind if the limit points of G are nowhere dense on the invariant unit circle* S^1*.*

To give a criterion for whether the given Fuchsian group is of the first or second kind, one needs a notion of the fundamental region (polygon) associated with a Fuchsian group G. Denote by R a subset of the interior of the invariant unit circle of a Fuchsian group G which 'consists of' G-equivalent points $z \in \mathbb{C}$, where z runs through the whole domain $\mathbb{C}$. Let $z \in R$ be a point in R and let $O(z) = \{g(z)|g \in G\}$ be the orbit of the point z under the action of the group G.

Definition 17.2.8 *By a fundamental region* R_0 *of the Fuchsian group G one understands a simply connected open region which is a maximal cross-section*

[2] A set is perfect if it is closed and dense in itself.

of orbits of all points $z \in R$. In other words, R_0 is a region containing exactly one element of each orbit in $\cup_{z \in R} O(z)$.

Clearly, the fundamental region R_0 is not uniquely defined on a given Fuchsian group G. We choose R_0 to be a polygon region bounded by $4n$-sided polygon ∂R_0. (This is always possible to do; choose the sides of the polygon to be the arcs of circles in C, see [108].) For the fundamental region R_0 associated with the Fuchsian group G the following holds:

$$R = \bigcup_{g \in G} g(R_0).$$

(This property can be used as an alternative definition of the region R.) A general criterion for G to be Fuchsian group of the first or second kind is given by the following statement.

Proposition 17.2.5 *If R contains in its interior a point of the invariant unit circle S^1, then the Fuchsian group G is of the second kind. Otherwise, G is a Fuchsian group of the first kind.*

Proof. The first claim of the proposition is evident. (If a point on the invariant circle belongs to the interior of the set R, then it is an ordinary point. Therefore not all points of S^1 are limit points. Hence the group G is of the second kind.)

To complete the proof, let us establish the converse statement: If G is a Fuchsian group of the second kind, then R contains a point of the invariant circle S^1 in its interior. Indeed, let z_0 be an ordinary point on the invariant circle. Then one can draw a small circle around z_0 such that it bounds a disc D, all points of which can be covered by a finite number of transforms of region R. Denote these iterations by $R_1, \ldots, R_n$. (The sides of these regions are always orthogonal to S^1.) As before, let h be an arc of the invariant circle lying in D. With the exception of a finite number of points (which are 'vertices' of polygons) each point of h is interior to one of the above regions. Let $z_k \in \text{Int } R_k$ be one such point. Since each R_k was produced from R by iterations of the G-action, the point z_k has its preimage in the interior of R. □

17.2.4 Modular Group

Definition 17.2.9 *The group of Fuchsian transformations of the complex plane C given by the formula*

$$z \longmapsto \frac{az+b}{cz+d}, \qquad ad - bc = 1, \qquad a, b, c, d \in \mathsf{Z}$$

is called a modular group.

Note that the modular group is indeed Fuchsian, since it fixes the real axis as an invariant circle, and maps the upper half-plane to itself. The upper half-plane is mapped conformally to the unit disc with the help of the substitution

$$w = \frac{z+i}{iz+1}$$

which carries the invariant real axis to the invariant unit circle S^1. Note, that modular group leaves invariant the lattice of (complex) rational points in the complex plane.

17.2.5 Examples

Below we consider three examples of modular groups acting on the upper half-plane of the complex plane and establish the corresponding fundamental regions R_0.

Example 17.2.1 *Suppose that parameter c in the modular transformation vanishes, so that we have $ad = 1$. Then necessarily $a = d = \pm 1$, and b takes any integer value. The modular transformation in this case is a translation with period n along the real axis which acts by the formula*

$$z \longmapsto z + n, \qquad n \in \mathsf{Z}.$$

As a fundamental region we take a strip enclosed by the two lines orthogonal to the real axis passing through the points $\pm 1/2$. The covering subgroup acts by the formula $z \mapsto z + 1$.

Example 17.2.2 *A more complicated example occurs in the case $c = \pm 1$. We then have $ad \mp b = 1$. For any integer values of a and d there is an integer value of b so that the modular group is given by*

$$z \longmapsto \frac{az + 1 \pm ad}{\pm z + d}, \qquad a, d \in \mathsf{Z}.$$

The fundamental region R_0 in this case is formed by the parallel lines through $\pm 1/2$, and from below R_0 is bounded by arc of the unit circle through the origin of complex plane C.

Example 17.2.3 *Consider, finally, a more general class of modular groups acting on the upper complex half-plane. Let n be an integer greater than 1, and consider those transformations in the modular group for which the parameters b and c are divisible by n. We are therefore considering the transformations of the form*

$$z \longmapsto \frac{az + nb'}{nc'z + d'}, \qquad ad - n^2 b'c' = 1, \qquad a, b', c', d' \in \mathsf{Z}.$$

Let us first check that such transformations form a group. Clearly, the inverse transformation

$$z \longmapsto \frac{-dz + nb'}{nc'z - a},$$

is of the same form. Denote by $(\alpha z + n\beta')/(n\gamma' z + \delta)$ *the second transformation in the set, then the composition of two transformations takes the form*

$$z \longmapsto \frac{(\alpha a + n^2\beta' c')z + n(\alpha b' + \beta' d)}{n(\gamma' a + \delta c')z + n^2\gamma' b' + \delta d}$$

Since $n(\alpha b' + \beta' d)$ *and* $n(\gamma' a + \delta c')$ *are divisible by* n*, the above transformations form a group. Setting* $c' = 0$*, we get the subgroup of translations* $z \mapsto z + mn$ *considered in Example 17.2.1, generated by translation* $U_n(z) = z + n$*. Let us turn now to the case* $n = 2$*. The modular group is generated by two transformations*

$$U(z) = z + 2, \qquad T = \frac{z}{2z+1}.$$

For an arbitrary n *such modular groups are called the principal congruence groups and are denoted by* $\Gamma(n)$*.*

It is not hard to see that the modular groups of Examples 17.2.1 - 17.2.3 are Fuchsian groups of the first kind.

17.3 Teichmuller Space

Quadratic differentials have formed a massive part of the theory of Riemann surfaces since Teichmuller's discovery of their pertinence to the moduli problem. Teichmuller's idea was put on a solid foundation by Ahlfors and Bers; Ahlfors proved the existence of 'global' solution to the Beltrami equation in the class of quasiconformal mappings. Such mappings may be viewed as 'paths' in the Teichmuller (moduli) space. To each quasiconformal mapping one can associate a holomorphic quadratic differential. Quadratic differentials have rich geometry since they induce measured foliations on Riemann surfaces.

17.3.1 Conformal Invariants

A conformal invariant of the Riemann surface M is a quantity which remains constant after any conformal (holomorphic) transformations of M. A conformal invariant is called a *module* of the Riemann surface M. Below we give expressions for moduli in the cases of 'quadrilaterals' and annuli.

A quadrilateral $Q(z_1, z_2, z_3, z_4)$ is a simple domain in C whose points z_1, z_2, z_3, z_4 belong to the boundary ∂Q so that their ordering corresponds to the positive orientation of ∂Q versus Q. If a conformal mapping f sends Q to a Euclidean rectangle $R = \{x+iy \mid 0 < x < a, 0 < y < b\}$ so that z_1, z_2, z_3, z_4 are sent to the vertices of R, then f is said to be a *canonical* mapping. Every quadrilateral possesses a canonical mapping and the canonical mapping is uniquely determined by Q.

A number $M(Q(z_1, z_2, z_3, z_4)) = a/b$ is called a *module* of the quadrilateral $Q(z_1, z_2, z_3, z_4)$. Note that it does not depend on the particular choice of the canonical rectangle and $M(Q(z_1, z_2, z_3, z_4)) = 1/M(Q(z_2, z_3, z_4, z_1))$. Finally, a module is a conformal invariant of quadrilaterals: if f is a conformal mapping of a domain A and $Q(z_1, z_2, z_3, z_4) \subset A$ is a quadrilateral such that $f(Q)$ is a simple domain, then $M(Q(z_1, z_2, z_3, z_4)) = M(f(Q)(f(z_1), f(z_2), f(z_3), f(z_4)))$.

The module depends continuously on the quadrilateral. Namely, suppose that $Q_n(z_1^n, z_2^n, z_3^n, z_4^n)$ is a sequence of quadrilaterals converging to $Q(z_1, z_2, z_3, z_4)$ from inside: $\overline{Q}_n \subset Q$ for every n and for every $\varepsilon > 0$ there is $n > n_\varepsilon$ such that each point of the sides $Q_n(z_1^n, z_2^n, z_3^n, z_4^n)$ has a distance $< \varepsilon$ from the sides of Q. Then the continuity means:

$$\lim_{n\to\infty} M(Q_n(z_1^n, z_2^n, z_3^n, z_4^n)) = M(Q(z_1, z_2, z_3, z_4)).$$

The module of a ring domain (annulus) is more complicated. Any doubly connected domain A maps conformally to an annulus $r_1 < |z| < r_2$ where $r_1 < r_2$ may take values 0 and ∞. The module of A is calculated according to the formula:

$$M(A) = \log \frac{r_2}{r_1}.$$

(The module is assumed infinite when either $r_1 = 0$ or $r_2 = \infty$.) As for quadrilaterals, the module of a ring domain is a continuous invariant. It is not hard to check that $M(A)$ is a conformal invariant of the domain A.

Modules of domains B of higher connectivity are calculated using a path family method. This method associates to each path in B a non-negative Borel measurable function ρ: one then takes an infimum of a certain integral of ρ over B.

17.3.2 Quasiconformal Mappings

Let $f : X \to X$ be an orientation preserving homeomorphism of a simple domain X. The number

$$\mu(f) = \sup_Q \frac{M(f(Q)(f(z_1), f(z_2), f(z_3), f(z_4)))}{M(Q(z_1, z_2, z_3, z_4))}$$

is called a *maximal dilatation* of f, provided $Q(z_1, z_2, z_3, z_4)$ are quadrilaterals such that $\overline{Q} \subset X$. For maximal dilatation one always has $\mu(f) \geq 1$. Since a module is a conformal invariant, the maximal dilatation of a conformal mapping is 1. The converse is also true: if the maximal dilatation of f is equal to 1, then f is conformal.

Definition 17.3.1 *An orientation preserving homeomorphism f with a finite maximal dilatation is called a quasiconformal mapping. If $\mu(f)$ is bounded by a positive constant K, the mapping f is called K-quasiconformal.*

Clearly, f is 1-quasiconformal if and only if f is conformal. If mapping f is K-quasiconformal, then its inverse f^{-1} is also K-quasiconformal. If $f : X \to Y$ is K_1-quasiconformal and $g : Y \to Z$ is K_2-quasiconformal, then $g \circ f$ is K_1K_2-quasiconformal.

Along with conformal mappings, the quasiconformal mappings enjoy the compactness property: a sequence $\{f_n\}$ of K-quasiconformal mappings of domain X uniformly converges in X either to a constant or to a K-quasiconformal mapping f.

17.3.3 Beltrami Equation

Until now it has never been assumed that the quasiconformal mapping is differentiable. In this section we do so by assuming that f is an orientation preserving diffeomorphism. Let us consider complex derivatives:

$$\partial f = \frac{1}{2}(f_x - if_y), \qquad \bar{\partial} f = \frac{1}{2}(f_x + if_y).$$

The 'directional' derivative then has the form $\partial_\alpha f = \partial f + \bar{\partial} f e^{-2i\alpha}$; easy argument shows that $\max_\alpha |\partial_\alpha f(z)| = |\partial f(z)| + |\bar{\partial} f(z)|$ and $\min_\alpha |\partial_\alpha f(z)| = |\partial f(z)| - |\bar{\partial} f(z)|$. A *dilatation quotient* is set to be

$$D_f = \frac{|\partial f| + |\bar{\partial} f|}{|\partial f| - |\bar{\partial} f|}.$$

The dilatation quotient is finite and is equal to 1 if and only if f is conformal. There exists a remarkable characterization of the quasiconformal diffeomorphisms.

Lemma 17.3.1 *The orientation preserving diffeomorphism f is K-quasiconformal if and only if the dilatation condition $D_f(z) \leq K$ holds everywhere in X.*

Proof. To prove the sufficient conditions, let us take a quadrilateral $Q \subset X$. Denote by w a mapping induced on the canonical rectangles $R(0, M, M+i, i)$ of Q and $R'(0, M', M' + i, i)$ of $f(Q)$. By the conformal invariance, the dilatation quotient D_w is majorized by K. The desired result $M' \leq KM$ follows from the length-area argument. The necessary conditions follow from the lower estimate for $M(f(Q))$ furnished by Rengel's inequality. □

The class of K-quasiconformal diffeomorphisms no longer enjoys the compactness property: that is why we cannot introduce the quasiconformality directly, via the dilatation quotient (although historically it was done by Grotzsch). However, the quasiconformal mappings are differentable 'almost everywhere' in the domain X. To put it more precisely, one needs the following notion. A continuous real-valued function u is said to be *absolutely continuous on lines* if for each rectangle $\{x+iy \mid a \leq x \leq b, c \leq y \leq d\} \subset X$, the function $x \mapsto u(x+iy)$ is absolutely continuous on $[a,b]$ for almost all $y \in [c,d]$ and $y \mapsto u(x+iy)$ is absolutely continuous on $[c,d]$ for almost all $x \in [a,b]$. With this, one gets an analytic characterization of quasiconformality.

Lemma 17.3.2 *An orientation preserving homeomorphism f of a domain X is K-quasiconformal if f is absolutely continuous on the lines in X and* $\max_\alpha |\partial_\alpha f(z)| \leq K \min_\alpha |\partial_\alpha f(z)|$ *almost everywhere in X.*

Proof. Since f is absolutely continuous on the lines, it has partial derivatives almost everywhere; moreover, these derivatives are of class L^2. The rest of the proof follows from the argument of Lemma 17.3.1. □

Let $f : X \to X'$ be a K-quasiconformal mapping and let $z \in X$ be a point of differentiability of f. The dilatation conditions may be rewritten in the following form:

$$|\bar{\partial} f(z)| \leq \frac{K-1}{K+1} |\partial f(z)|.$$

Quotient $\mu(z) = \bar{\partial} f(z)/\partial f(z)$ defined almost everywhere in X is called a *complex dilatation* of f. Since f is continuous, μ is a Borel measurable function. It can be easily seen that $|\mu(z)| < 1$.

Definition 17.3.2 *The differential equation $\bar{\partial} f = \mu \partial f$ where μ is measurable function such that $||\mu||_\infty < 1$, is called a Beltrami equation.*

Theorem 17.3.1 *A homeomorphism f is K-quasiconformal if and only if f is an L^2-solution to the Beltrami equation where μ is such that $|\mu| < 1$ for almost all z.*

Proof. The necessary conditions are evident. The sufficient conditions follow from Lemma 17.3.2 if we note that $||\mu||_\infty < 1$ implies that f is an orientation preserving homeomorphism. □

17.3.4 Ahlfors-Bers' Theorem

The study of the conformal modules of Riemann surfaces is known as the *moduli problem.* The quasiconformal mappings allow one to 'move' from point to point in the 'moduli space'. Let us be more precise.

A homeomorphism f between two Riemann surfaces M_1, M_2 is said to be K-quasiconformal if, for any local parameter $h_i, i = 1,2$ in the atlas for M_i, the mapping $h_2 \circ f \circ h_1^{-1}$ is K-quasiconformal. The mapping will be called *quasiconformal* if it is K-quasiconformal for some $K \geq 1$. Suppose that f_1, f_2 are quasiconformal mappings of a Riemann surface M. According with the uniformization theorem, the universal covering of M is either the extended plane, the complex plane, or the disc with the Fuchsian group of the first kind. Then f_1 and f_2 are said to be *equivalent* if $f_2 \circ f_1^{-1}$ is homotopic to a conformal mapping of $f_1(M)$ onto $f_2(M)$.

Definition 17.3.3 *The set of all equivalence classes of quasiconformal mappings of M is called a Teichmuller space T_M of the Riemann surface M.*

It was shown earlier that there exists a duality between the quasiconformal mappings and the dilatation functions given 'in the small'. By the *Beltrami differential*, μ, one understands a differential which transforms from one local parameter to the other by the rule:

$$\mu = (\mu \circ g)\frac{\partial \bar{g}}{\partial g},$$

where g is a covering transformation. (For simply connected domains μ coincides with the complex dilatation function.) The following statement is a simplified version of the Ahlfors-Bers' theorem, which establishes the duality between the quasiconformal mappings and the Beltrami differentials 'in the large'.

Theorem 17.3.2 (Ahlfors-Bers) *Let μ be a Beltrami differential on the Riemann surface M. Then there is a quasiconformal mapping of M onto another Riemann surface with the complex dilatation μ. This mapping is uniquely defined up to a conformal mapping.*

Proof. Let μ be the Beltrami differential for the covering group G on the universal cover H. There exists a quasiconformal mapping $f : \mathsf{H} \to \mathsf{H}$ with the complex dilatation μ. By the transformation rule for the Beltrami differential, f and $f \circ g$ have the same complex dilatation for every $g \in G$. Therefore $f \circ g \circ f^{-1}$ is conformal and f induces an isomorphism of G onto the Fuchsian group $G' = \{f \circ g \circ f^{-1} | \, g \in G\}$. Denote by π and π' the canonical projections of H onto M and $M' = \mathsf{H}/G'$. The equation $\varphi \circ \pi = \pi' \circ f$ defines a quasiconformal mapping of $\varphi : S \to S'$ with the complex dilatation μ. □

17.3.5 Geometry of Quadratic Differentials

The following 'naive' observation establishes an important link between the quasiconformal mappings and measured foliations on the surfaces. If S^1 is a small circle centered in the point x of the surface M, then after the quasiconformal transformation $f : M \to M$ it maps to an ellipse E^1 centered at the point $f(x)$. The big and the small axes of E^1 generate two orthogonal fields of tangent elements on M. The induced pair of foliations will be measured foliations with (common) singularities. Later we shall see that the converse is true as well.

By the *quadratic differential* on M one understands a collection of the complex-valued functions ϕ_i defined in the local charts U_i such that on the overlapping domains of local parameters z_i, z_j they transform according to the 'quadratic' rule:

$$\phi_j = \phi_i \left(\frac{dz_i}{dz_j} \right)^2$$

The quadratic differential $\phi(z)dz^2$ is said to be *holomorphic (meromorphic)* if all of the functions ϕ_i are holomorphic (meromorphic). A point $p \in S$ is said to be *regular* if $\phi(p) \neq 0$; otherwise, p is called a *singular point* of the quadratic differential ϕ. Singular points form a discrete set and on a compact Riemann surface there are only finitely many of them.

If M is a Riemann surface we define a deformation of the complex structure of M as follows. For a fixed holomorphic quadratic differential q on M and a positive number $k \in [0, 1)$ let $z = \xi + i\eta$ and $z' = \xi' + i\eta'$ be two local parameters on M such that ξ and η correspond to the lines $Arg\ q = 0$ and $Arg\ q = \pi$, while for ξ' and η' we have:

$$\xi' = K\xi, \qquad \eta' = \eta, \qquad K = \frac{1+k}{1-k}.$$

Quadratic differential $q' = (dz')^2$ is said to be *induced* by the K-quasiconformal deformation (M, q, k). The transformation (M, q, k) leaves invariant the length of vertical lines, but stretches horizontal lines by a factor K. Note that for $k = 0$ we have an identical (conformal) transformation.

Below we introduce a 'natural parameter' connected with the holomorphic differentials on S and study its geometry at the regular and singular points. Let p be a regular point of the quadratic differential, and let $q \to h(q) = z$ be a local parameter near p which sends p to the origin. Since $\phi(0) \neq 0$, there is a simply connected domain containing the origin where the two branches of $z \to \sqrt{\phi(z)}$ are single-valued. For a fixed branch of $\sqrt{\phi}$ every integral function

$$z \longrightarrow \Phi(z) = \int \sqrt{\phi(z)}dz$$

is also single-valued, and is determined uniquely up to an additive constant. From the choice of w it follows that, after the transformation $z \to \Phi(z)$, the holomorphic quadratic differential assumes the form

$$dw^2 = \phi(z)dz^2,$$

with ϕ equal to 1. We call the function $w = \Phi(z)$ a *natural parameter* at p. Thus it is proved that

Proposition 17.3.1 *Near the regular point the local representation of holomorphic quadratic differential ϕ in terms of the natural parameter is the constant function 1.*

The notion of natural parameter can be extended to the set of singular points. Suppose $p \in S$ is a zero of ϕ of the order n. As before let $q \to h(q) = z$ be a local parameter which sends p to the origin. Then there exists a disc $D(r)$ of 'radius' r centered at 0 in which $\phi(z) = z^n\psi(z)$ with $\psi(0) \neq 0$. Let us fix a single-valued branch of $\sqrt{\psi}$ in $D(r)$ along its positive radius $I = \{x|0 \leq x \leq r\}$ and also fix the branch of $z \to z^{n/2}$ in $D(r)\backslash I$. (If n is even, no such cut is needed.) If n is odd, we come to the function

$$z \longmapsto \Phi(z) = \int_0^z \sqrt{\phi(z)}dz = z^{\frac{n-2}{2}}(c_0 + c_1 z + \ldots), \qquad c_0 \neq 0,$$

which is single-valued on $D(r)\backslash I$. Now we fix the local parameter $q \to \zeta = [\Phi(z)]^{\frac{2}{n+2}}$ near the singularity p. Further ζ is called a *natural parameter* at p. From $\phi = (\Phi')^2$ one derives

$$\phi(z)dz^2 = \left(\frac{n+2}{2}\right)^2 \zeta^n d\zeta^2.$$

Thus we have shown the following

Proposition 17.3.2 *Near the singular point of the order n the holomorphic quadratic differential has the representation* $\zeta \mapsto (1+n/2)^2\zeta^n$ *in terms of the natural parameter* ζ.

By a *regular path* in S one understands continuously differentiable monotone mapping $\gamma : I \to S$, where I is the unit segment of the real line. If $\phi(z)$ is a holomorphic quadratic differential on S, then the function

$$t \longmapsto Arg[\phi(z(t))(z'(t))^2], \qquad t \in I,$$

is well defined modulo 2π. By a *straight line* in the geometry induced by ϕ we mean the argument-constant mapping γ, which we write in the form

$$Arg(\phi(z)dz^2) = Const.$$

At the regular point $p \in S$ the straight line through p is locally a usual straight line taken in the Euclidean sense (this is a consequence of the representation of ϕ in the 'natural parameter' w as a constant). Near the critical point $p \in S$ the behaviour of 'straight lines' through p is more complicated. Namely, if the singularity has order n then $n+2$ rays emerging from p represent the same straight line with the prescribed 'slope'. (The behaviour of 'parallel' straight lines near p looks like foliation near the $n+2$-prong saddle singularity, see the earlier chapters.)

17.3.6 Associated Metric

Let $\phi(z)$ be a holomorphic quadratic differential given on the Riemann surface S. The invariant differential

$$|\phi(z)|^{\frac{1}{2}}|dz|,$$

is called a *line element* of the metric induced by ϕ. An arbitrary path $\gamma : I \to S$ which does not pass through the singular points of holomorphic differential ϕ can be subdivided into small parts, and its length $l(\gamma)$ is the sum of the lengths of its parts given by the formula:

$$l(\gamma) = \int_\gamma |\phi(z)|^{1/2}|dz|.$$

(Clearly, this definition is invariant with respect to a reparametrization of γ and the choice of the local parameters.) The following statement is the existence theorem for the local geodesics (the shortest curves joining two arbitrary points of small disc region in S) on Riemann surfaces.

Proposition 17.3.3 *Every point of a Riemann surface has a neighbourhood in which any two points can be joined by a unique shortest curve.*

Proof. Suppose first that $p \in S$ is a regular point of quadratic differential ϕ. Denote by V the maximal disc around p and let $w(V) = \{w : |w| < r\}$ be its image under a natural parameter $w = \Phi(z)$. Let $V_0 \subset V$ be the preimage of $|w| < r/2$, and let p_1, p_2 be arbitrary points in V_0. Then the preimage γ_0 of the line segment connecting $\Phi(p_1)$ and $\Phi(p_2)$ is the unique shortest curve which joins p_1 and p_2 on S. If γ is another curve in S joining p_1 and p_2, then clearly $l(\gamma_0) < l(\gamma)$.

Suppose now that p is a zero of ϕ of order n. If ζ is a natural parameter near p, then the holomorphic differential has the form

$$\phi(\zeta) = \left(\frac{n+2}{2}\right)^2 \zeta^n, \qquad w = \Phi(\zeta) = \zeta^{\frac{n+2}{2}},$$

and is given in a disc $|\zeta| < r$. Let now V_0 be the preimage of the disc $|\zeta| < 2^{-2/(n+2)}r$ on S. Then any two points p_1 and p_2 in V_0 can be joined by a unique shortest curve. This is either a straight line segment in w-plane, or it is composed of two radii in the ζ-plane emerging from the origin. In the former case $|\operatorname{Arg}\zeta_1 - \operatorname{Arg}\zeta_2| < 2\pi/(n+2)$, where ζ_1 and ζ_2 are ζ-images of p_1 and p_2. □

Similarly, one can establish the existence of 'global geodesics' on S. Such geodesics also satisfy 'uniqueness conditions', with respect to the relevant homotopy class of curves on S. Namely, the following is valid.

Proposition 17.3.4 *Let S be a compact Riemann surface and $p, q \in S$ be the arbitrary points on S. Then every homotopy class of curves joining p and q on S contains a unique geodesic arc.*

Proof. Let us replace S by its universal covering surface $\tilde{S}$. Let z_1, z_2 be two points on $\tilde{S}$. Since the distance from both of these points to $\partial\tilde{S}$ is infinite, we can find a Jordan domain $G, \bar{G} \subset \tilde{S}$, such that $z_1, z_2 \in G$ and any arc connecting z_1 and z_2 in $\tilde{S}$ and leaving G cannot be minimizing in length.

Denote by a the infimum of the length of the curves in $\tilde{S}$ which join z_1 and z_2.

Let (γ_i) be a minimizing sequence of curves in G from z_1 to z_2, that is, $l(\gamma_i) \to a$. Subdivide the parameter interval $[0, l(\gamma_i)]$ into n equal parts and take n so large that the endpoints of the resulting subarcs of γ_i can be joined by a unique shortest arc in $\tilde{S}$. For a subsequence (γ_{i_k}) these $n+1$ endpoints converge. By joining the limit points with shortest arcs in $\tilde{S}$ we obtain a shortest arc γ from z_1 to z_2. □

17.3.7 Isothermal Coordinates

The most important case of differentials arises when $m = 2, n = 0$. In this case ϕ is called a *quadratic differential.* (The close case of $m = 1, n = 0$ is called an *Abelian differential*, and its square yields a quadratic differential.) The importance of quadratic and Abelian differentials already appears from the following natural differential-geometric construction, called *isothermal coordinates.*

Let S be an orientable C^1 surface embedded into $\mathbb{R}^3$, given by the local Monge chart $f = f(u,v)$. The induced metric on S is defined by a line element ds, where

$$ds^2 = Edx^2 + 2Fdxdy + Gdy^2,$$

where

$$E = 1 + f_u^2, \qquad F = f_u f_v, \qquad G = 1 + f_v^2,$$

are the coefficient of the first 'fundamental' (or Gaussian) form of the surface S. The first fundamental form is invariant with regard to the choice of local parameters. Using the complex notation $dz = dx + idy, d\bar{z} = dx - idy$, we obtain the Abelian differential

$$ds = \lambda |dz + \mu d\bar{z}|,$$

where

$$\lambda^2 = \frac{E + G + 2\sqrt{EG - F^2}}{4}, \qquad \mu = \frac{E - G + 2iF}{E + G + 2\sqrt{EG - F^2}}.$$

It is a classical result, which is not hard to prove by direct calculations, that the function f is harmonic (or the mapping f is conformal) if and only if $E = G, F = 0$. This implies immediately that $\mu = 0$, so that we have

$$ds = \lambda |dz|.$$

The local coordinates, z, of S which have such a property are called *isothermal.* (Such coordinates exist in all points of the surface, except possibly a finite number of points. As local coordinates one may choose lines of minimal and maximal curvature of S; see Chapter 13.)

Let us consider another local parameter of S which defines the local coordinates w. If the coordinates z and w are both isothermal (that is for w it holds $ds = \lambda_1|dw|$), and if the induced mapping $z \to w$ is defined in some non-empty open set of the plane, then

$$\lambda_1|dw| = \lambda|dz|.$$

Provided S is orientable, this means that the mapping $z \to w$ is conformal. Therefore, we have proved the following

Proposition 17.3.5 *The isothermal coordinates define a natural conformal structure for an orientable C^1 surface, which thus becomes a Riemann surface.*

17.4 Complex Curves

The complex algebraic curve in C^2 is a subset C of $\mathsf{C}^2 = \mathsf{C} \times \mathsf{C}$ of the form

$$C = \{(x, y) \in \mathsf{C}^2 : P(x, y) = 0\}, \tag{17.3}$$

where $P(x, y)$ is a polynomial in two variables with complex coefficients. After suitable compactification, every complex algebraic curve becomes a Riemann surface. This gives us another way to look at Riemann surfaces: as complex algebraic curves in projective space. Using methods of algebraic geometry (Bezout theorem, divisors, etc) we are able to establish the important degree–genus formula and Riemann–Roch theorems for non–singular projective curves. We culminate with the proof of the Teichmuller formula for the dimension of the space of conformal structures on the Riemann surface – the result which we announced earlier.

17.4.1 Projective Curves

Let $P(x, y)$ be a non-constant polynomial in two variables with the complex coefficients. $P(x, y)$ is said to have *no repeated factors* if we cannot write

$$P(x, y) = Q^2(x, y)R(x, y)$$

for some non-constant polynomials $Q(x, y)$ and $R(x, y)$. If $P(x, y)$ is a non-constant polynomial with complex coefficients and no repeated factors, then a *complex algebraic curve* in C^2 defined by $P(x, y)$ is the set C defined by formula (17.3). The answer to the question of when two complex polynomials define the same complex algebraic curve is given by the following theorem.

Theorem 17.4.1 (Hilbert's Nullstellensatz) *If $P(x, y), Q(x, y)$ are polynomials with complex coefficients then*

$$\{(x,y) \in \mathbb{C}^2 : P(x,y) = 0\} = \{(x,y) \in \mathbb{C}^2 : Q(x,y) = 0\}$$

if and only if there exist positive integers m and n such that $P(x,y)$ divides $Q^n(x,y)$ and $Q(x,y)$ divides $P^m(x,y)$. In other words, $P(x,y)$ and $Q(x,y)$ have the same irreducible factors, possibly occurring with different multiplicities.

Proof. We refer the reader to:

M. F. Atiyah and I. G. Macdonald, Commutative algebra, Addison-Wesley, 1969.

□

Recall that a non-zero polynomial $P(x,y)$ is *homogeneous* of degree d if $P(\lambda x, \lambda y) = \lambda^d P(x,y)$ for all $\lambda \in \mathbb{C}$.

Lemma 17.4.1 *If $P(x,y)$ is a non-zero homogeneous polynomial of degree d in two variables with complex coefficients then it factors as a product of linear polynomials*

$$P(x,y) = \prod_{i=1}^{d} (\alpha_i x + \beta_i y)$$

for some $\alpha_i, \beta_i \in \mathbb{C}$.

Proof. One divides $P(x,y)$ by the highest order monomial y^d. Now the polynomial of one variably x/y is considered. Since the field $\mathbb{C}$ is algebraically closed, the lemma follows. □

A complex algebraic curve C in $\mathbb{C}^2$ is never compact. For many purposes it is useful to compactify C by adding "points at infinity". For this the concept of a projective space is used. The idea is to identify each $(x,y) \in \mathbb{C}^2$ with a one-dimensional complex linear subspace of $\mathbb{C}^3$. Every one-dimensional linear subspace of $\mathbb{C}^3$ which does not belong to the plane $\{(x,y,z) \in \mathbb{C}^3 : z = 0\}$ contains a unique point of the form $(x,y,1)$. The one-dimensional subspaces of $\{(x,y,z) \in \mathbb{C}^3 : z = 0\}$ correspond to the "points at infinity". In general, the following definition is adopted.

Definition 17.4.1 *Complex projective space $\mathbb{P}_n$ of dimension n is the set of complex one-dimensional subspaces of the complex vector space $\mathbb{C}^{n+1}$. In particular, when $n = 1$ we have the complex projective line $\mathbb{P}_1$ and when $n = 2$ we have the complex projective plane $\mathbb{P}_2$.*

By introducing homogeneous coordinates and the quotient topology, each $\mathbb{P}_n$ can be eventually compactified, see e.g.

F. Kirwan, Complex Algebraic Curves, Cambridge Univ. Press 1992.

If $P(x,y,z)$ is a non-constant homogeneous polynomial with complex coefficients with no repeated factors, then by a *projective curve* C defined by $P(x,y,z)$ one understands a set

$$C = \{[x,y,z] \in \mathsf{P}_2 : P(x,y,z) = 0\},$$

where $[x,y,z]$ are homogeneous coordinates in P_2. Note that this definition is correct, since the condition $P(x,y,z) = 0$ is independent of the choice of homogeneous coordinates $[x,y,z]$ because of the homogeneity of P itself: $P(\lambda x, \lambda y, \lambda z) = 0 \Leftrightarrow P(x,y,z) = 0$. A *degree* of projective curve C defined by a homogeneous polynomial $P(x,y,z)$ is defined to be the degree d of $P(x,y,z)$. The projective curve C is called *irreducible* if polynomial $P(x,y,z)$ is irreducible, i.e. has no non-constant polynomial factors other than scalar multiplies of itself. An irreducible projective curve D defined by a homogeneous polynomial $Q(x,y,z)$ is called a *component of* C if and only if $Q(x,y,z)$ divides $P(x,y,z)$. A projective curve defined by a linear equation $\alpha x + \beta y + \gamma z = 0$ is called a *projective line* provided α, β, γ are not all zero. By a *tangent line* to a projective curve C in P_2 defined by a homogeneous polynomial $P(x,y,z)$ at a non-singular point $[a,b,c] \in \mathsf{C}^3$ one understands a projective line

$$\frac{\partial P}{\partial x}(a,b,c)x + \frac{\partial P}{\partial y}(a,b,c)y + \frac{\partial P}{\partial z}(a,b,c)z = 0$$

The main advantage of complex projective curves over complex curves is given by the following lemma.

Lemma 17.4.2 *Any projective curve C in P_2 is compact and Hausdorff.*

Proof. Since P_2 is compact and Hausdorff it is enough to show that C is a closed subset of P_2. This happens if and only if a set $\{(x,y,z) \in \mathsf{C}^3 \backslash \{0\} : P(x,y,z) = 0\}$ is closed. But this is true since polynomials are continuous functions. □

If C and D are two complex projective curves in P_2 how many intersection points can they have in P_2? The answer is given by the famous *Bézout theorem*, but before stating it we need some notations. An intersection multiplicity $I_p(C,D)$ of two complex projective curves C and D in the point p is defined to be ∞ if p lies on a common component of C and D; otherwise it is a non-negative integer equal to 0 when p does not belong to $C \cap D$ and 1 if and only if p is a non-singular point of both C and D and the tangent lines to C and D at p are distinct. (In other cases, which are not generic, the definition of I_p is more elaborate; see the above cited book of F. Kirwan.)

Theorem 17.4.2 (Bézout) *If C and D are two projective curves of degrees n and m in P_2 which have no common component, then they have precisely nm points of intersection counting multiplicities:*

$$\sum_{p \in C \cap D} I_p(C, D) = mn.$$

Proof. The proof is based on a concept of resultant $R_{P,Q}$ of two polynomials in one variable P(x) and Q(x). (The resultant is a determinant of an $(m+n)$ by $(m+n)$ matrix with enties being complex coefficients of P and Q.) The resultant vanishes if and only if $P(x)$ and $Q(x)$ have a non-constant common factor.

An important observation is that for two homogeneous polynomials $P(x, y, z), Q(x, y, z)$ the polynomial $R_{P,Q}(y, z)$ in y, z is identically zero, if and only if $P(x, y, z)$ and $Q(x, y, z)$ have a non-constant homogeneous common factor. It is not hard to see that the resultant $R_{P,Q}(y, z)$ is a homogeneous polynomial of degree mn provided n and m are the degrees of $P(x, y, z)$ and $Q(x, y, z)$. On account of the Main Theorem of Algebra (and some technical details which are omitted), Bézout's theorem follows. □

17.4.2 Degree–Genus Formula

Our interest towards (non-singular) complex projective curves is motivated by their topological properties – each of them is a (compact) orientable surface. Below we shall see how the genus of such a surface is affected by the degree of irreducible complex projective curve which represents this surface. First, let us prove the following elementary lemma.

Lemma 17.4.3 *A complex projective line L in P_2 is homeomorphic to the two-dimensional unit sphere:*

$$S^2 = \{(u, v, w) \in \mathsf{R}^3 : u^2 + v^2 + w^2 = 1\}.$$

Proof. After a proper projective transformation we may assume that L is defined by the equation $z = 0$. It is not hard to see that the mapping $\phi : S^2 \to L$ given by the formula $\phi(u, v, w) = [u + iv, 1 - w, 0]$ establishes the required bijection. □

Theorem 17.4.3 (degree-genus formula) *A non-singular complex projective curve of degree $d > 0$ in the complex projective plane P_2 is homeomorphic to a closed orientable surface of genus*

$$g = \frac{(d-1)(d-2)}{2}.$$

Proof. Let us consider a singular complex projective curve C_0 which is a union of d projective lines in P_2 taken in "general position". The first observation is that C_0 has exactly $\frac{1}{2}d(d-1)$ singular points, which are the intersection points of these lines.

Indeed, after checking that formula is true for $d = 1$ and 2, let us suppose that it is so for $d = n - 1$. The total number of singular points in this case is $\frac{1}{2}(n-1)(n-2)$. Let us add an extra projective line L to the curve C_0. By the Bézout theorem the number of intersection points $L \cap C_0$ is equal to $(n-1)$. Now we add two numbers:

$$\frac{(n-1)(n-2)}{2} + n - 1 = \frac{n(n-1)}{2},$$

which proves the required formula for the number of intersections.

By Lemma 17.4.3, every projective line in C_0 is topologically a sphere. It is a little step from here to see that if we "perturb" slightly C_0 in P_2 every singular point of C_0 will give rise to a "handle" between pairs of spheres, as shown in Fig. 17.1. One thus obtains a non-singular curve C which is topologically a sphere with glued $\frac{1}{2}(d-1)(d-2)$ "handles". □

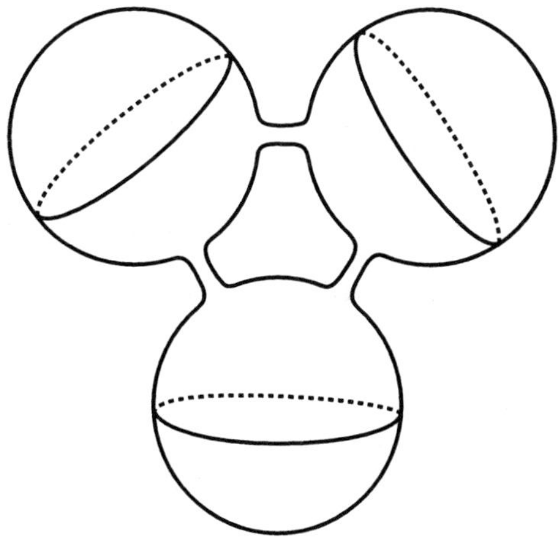

Fig. 17.1. Deformation of the singular curve C_0.

17.4.3 Elliptic Curves

Non-singular complex projective curves carry more information than only topological information. One may readily consider such curves as Riemann surfaces and formulate the moduli problem. An especially interesting interaction between the "algebraic" and "complex" approaches arises out in the case $d = 3$, or elliptic curves. Topologically, an elliptic curve is a torus.

Let Λ be a lattice in the complex plane C defined by the formula

$$\Lambda = \{n\omega_1 + m\omega_2 : n, m \in \mathsf{Z}\},$$

where ω_1 and ω_2 are non-zero complex numbers linearly independent over R (i.e., the quotient ω_1/ω_2 is not real). Note that Λ is an abelian subgroup of C isomorphic to Z × Z. Define a *Weierstrass* $\wp$*-function* on C by the formula

$$\wp(z) = \frac{1}{z^2} + \sum_{\omega \in \Lambda \setminus 0} \left(\frac{1}{(z-\omega)^2} - \frac{1}{\omega^2} \right).$$

$\wp(z)$ is a meromorphic doubly periodic function on C and its derivative is given by the series $\wp'(z) = -2\sum_{\omega\in\Lambda} 1/(z-\omega)^3$. The significance of the $\wp$-function in complex analysis is that it "parametrizes" complex projective curves of degree 3 (elliptic curves); such a function is also known as an elementary elliptic function because of its historic role in the problem of rectifying arcs of an ellipse. These (and other) beautiful properties may be deduced from the following important identity.

Lemma 17.4.4 *Let g_2 and g_3 be two complex numbers defined by the convergent series*

$$g_2 = g_2(\Lambda) = 60 \sum_{\omega \in \Lambda \setminus 0} \frac{1}{\omega^4}, \qquad g_3 = g_3(\Lambda) = 140 \sum_{\omega \in \Lambda \setminus 0} \frac{1}{\omega^6}.$$

Then the derivative of the Weierstrass $\wp$-function satisfies the equality

$$\wp'(z) = 4\wp^3(z) - g_2\wp(z) - g_3.$$

Proof. Denote by $s_m = s_m(\Lambda) = \sum_{\omega\neq 0} \omega^{-m}$. Then the Laurent series for the $\wp$-function write as

$$\wp(z) = \frac{1}{z^2} + 3s_4 z^2 + 5s_6 z^4 + \ldots$$

The differentiation and term-by-term cancellation give the required identity. □

Note that in view of Lemma 17.4.4 the points $(\wp(z), \wp'(z))$ belong to the cubic curve $y^2 = 4x^3 - g_2 x - g_3$ which may be viewed as "parametrized" by a complex variable z. In homogeneous coordinates $x' = x/z, y' = y/z$ the same curve is denoted by C_Λ according to the following definition.

Definition 17.4.2 *Denote by C_Λ a complex projective curve in P_2 defined by the polynomial*

$$Q_\Lambda(x, y, z) = y^2 z - 4x^3 + g_2 x z^2 + g_3 z^3,$$

where $g_2 = g_2(\Lambda)$ and $g_3 = g_3(\Lambda)$ are as specified in Lemma 17.4.4.

It takes a time to see that C_Λ is a non-singular complex projective curve and the mapping $u : \mathsf{C}/\Lambda \to C_\Lambda$ which acts from the torus to the elliptic curve according to the formula

$$u(z+\Lambda) \longmapsto \begin{cases} [\wp(z), \wp'(z), 1] & \text{if} \quad z \notin \Lambda \\ [0,1,0] & \text{if} \quad z \in \Lambda \end{cases} \tag{17.4}$$

is a homeomorphism. In fact much more is true.

Theorem 17.4.4 *Let $u : \mathsf{C}/\Lambda \to C_\Lambda$ be a mapping 17.4. Then u and its inverse $u^{-1} : C_\Lambda \to \mathsf{C}/\Lambda$ are holomorphic functions.*

Proof. This follows from the meromorphic properties of $\wp(z)$ and $\wp'(z)$. A good treatise of this and other properties of elliptic functions is given in

R. Fricke, Die elliptishen Funktionen und ihre Amwendungen, I and II, Teubner 1916 and 1922.

Theorem 17.4.4 has two important implications.

Corollary 17.4.1 *Complex structures on the two-dimensional torus C/Λ are in one-to-one correspondence with the coefficient space of elliptic curves of Definition 17.4.2. In particular, the Teichmuller space of torus has complex dimension 1. Every conformal structure in this space is parametrized by the number ω_1/ω_2 of lattice Λ.*

Corollary 17.4.2 *The elliptic curve C_Λ carries structure of an abelian group induced by the lattice Λ.*

17.4.4 Divisors and the Riemann–Roch Theorem

We have seen that complex projective curves are Riemann surfaces. Thus we can look at holomorphic functions and differentials on such curves. This gives us yet another insight to the complex curves – those coming from the classical Riemann–Roch theory for meromorphic functions with prescribed singularity data. The point is that in a proper interpretation, the Riemann–Roch theorem is true for arbitrary algebraic curves and varieties.

Let C be a non-singular projective curve in P_2. A *divisor* on C is a formal sum

$$D = \sum_{p \in C} n_p p$$

such that $n_p \in \mathsf{Z}$ for every $p \in C$ and $n_p = 0$ for all but finitely many $p \in C$. The *degree* of D is defined to be the number

$$deg\ D = \sum_{p \in C} n_p.$$

We add and subtract divisors and multiply them by integers in the obvious way. The set of all divisors on C is an abelian group, denoted by *Div* C. The degree is a homomorphism from *Div* C to Z. If $n_p \geq 0$ for all $p \in C$ we write this fact as $D \geq 0$. Hence, we have a partial order on the set of divisors: $D \geq D'$ if and only if $D - D' \geq 0$. Note that the homomorphism *deg* is order-preserving since $deg\ D \geq deg\ D'$ if $D \geq D'$.

Let f be a meromorphic function on C such that every singular point p of f goes with the negative integer n_p if p is n_p-th order pole of f and with the positive integer n_p if p is an n_p-th order zero of f. The divisor of f is a sum

$$(f) = \sum_{p \in C} n_p p.$$

Note that $(fg) = (f) + (g)$ and $(f/g) = (f) - (g)$ for any pair of meromorphic functions on C. The divisor is called *principal* if it is a divisor of some meromorphic function on C. Two divisors D and D' are said to be *equivalent* if they differ by a principal divisor. Because the number of poles of meromorphic function on C always is equal to the number of zeros, counted with multiplicities, the degree of the principal divisor is 0. In particular, equivalent divisors must have the same degree.

If ω is a meromorphic differential on C, we define the divisor (ω) exactly in the same fashion as for a meromorphic function. Divisor is called *canonical* if it is a divisor of some meromorphic differential on C. Because any two meromorphic differentials ω and η on C can be written as

$$\eta = f\omega$$

for some meromorphic function f on C, for the respective divisors we have: $(\eta) = (f) + (\omega) \sim (\omega)$. Thus, any two canonical divisors on C are equivalent. Therefore, if C is fixed, all canonical divisors have the same degree. (In fact, this degree is a topological invariant of C equal to $2g - 2$.)

A classical problem for the meromorphic functions consists in the evaluation of the "number" of such functions taking prescribed singularity data at the fixed points of C. Let D be a divisor on C. Denote by $L(D)$ a set of meromorphic functions $\{f\}$ with singular points coinciding with *Sing* D and such that (i) if n_p is negative, then f has a zero at p of order greater or equal to $|n_p|$; (ii) if n_p is positive, then f has a pole at p of order greater or equal to $-n_p$. Formally, functions in $L(D)$ must satisfy the inequality

$$(f) + D \geq 0.$$

It is easy to check that $L(D)$ is a complex vector space, whose dimension we denote by $l(D) = dim\ L(D)$. Like the degree function, $l(D)$ is an invariant of the equivalence relation on the set of divisors: If $D \sim D'$ then $l(D) = l(D')$. Even more, $l(D)$ is a good topological invariant of the complex curve C as it we have the following

Theorem 17.4.5 (Riemann–Roch) *Let D be a divisor and κ a canonical divisor on a non-singular complex projective curve of genus g. Then*

$$l(D) - l(\kappa - D) = deg\ D - g + 1. \tag{17.5}$$

Proof. See, e.g., [270]. □

17.4.5 Application: Dimension of the Teichmuller Space

Let M be a Riemann surface of genus g. In case $g = 1$ we have estimated the (complex) dimension of the Teichmuller space T_g associated to M. Below we use our knowledge of the Riemann–Roch theorem to estimate $dim_{\mathbb{C}} T_g$ in the case $g > 1$. We have to prove the following corollary.

Corollary 17.4.3 (Teichmuller) *Let M be a closed Riemann surface of genus $g > 1$. Then the complex dimension of its Teichmuller space is equal to $3g - 3$.*

Proof. We have shown earlier that the space of conformal structures T_g on M is in a one-to-one correspondence with the space of holomorphic quadratic differentials $Q(M)$ on M. Thus we have to calculate $l(\kappa_Q)$ where κ_Q is a canonical divisor associated to quadratic differential Q.

For this let us substitute $\kappa = \kappa_Q$ and $D = \kappa_Q$ in formula (17.5). We get

$$l(\kappa_Q) - l(0) = deg\ \kappa_Q - g + 1.$$

We have mentioned earlier that for an abelian (holomorphic) differential $deg\ \kappa = 2g-2$, where g is the genus of the supporting surface. For a quadratic (holomorphic) differential we have to double the degree: $deg\ \kappa_Q = 4g - 4$. Now in view of this formula and the identity $l(0) = 0$ we obtain $l(\kappa_Q) = 4g - 4 - g + 1 = 3g - 3$. Thus the complex dimension of the quadratic canonical divisor is $3g - 3$. □

Bibliographic Notes. Riemann surfaces have been introduced in [254] and further developed by F. Klein, H. Poincaré, É. Picard, P. Koebe and others; see the bibliography in [108]. The general theory can be found in [108], [178], [179], [270]. The theory of Riemann surfaces is an example of Algebra, Analysis and Geometry being brought together in a bouquet of unforgetable scent.

Bibliography

1. **Alekseevskij, D. V., Vinogradov, A. M. and Lychagin, V. V.**
1991 *Basic Ideas and Concepts of Differential Geometry,* Encycl. of Math. Sci. Vol. 28, Geometry I, Springer.
2. **Andronov, A. A. and Leontovich, E. A.**
1965 *Dynamical systems of the first order of stability in the plane,* Mat. Sbornik 68, 328-372, MR 33# 2866,
3. **Andronov, A. A., Leontovich, E. A., Gordon, I. I. and Maier, A. G.**
1966 *Qualitative theory of dynamical systems of the second order,* Moscow, MR 33# 7650, Translation: Halstead Press, New York-Toronto, Israel Program for Sci. Translations, Jerusalem-London, 1973.
4. 1973 *Theory of Bifurcations of Dynamical Systems on a Plane,* Israel Program for Sci. Transl., Wiley, New York.
5. **Andronov, A. A. and Pontryagin, L. S.**
1937 *Systèmes grossiers,* Comptes Rendus de l'Académie des Sciences de l'URSS 14(5), 247-250.
6. **Anjos, Dos**
1986 *Polynomial vector fields on the torus,* Bol. Soc. Bras. Mat. 17(2), 1-22.
7. **Anosov, D. V.**
1967 *Geodesic flows on closed Riemannian manifolds with negative curvature,* Trudy Mat. Inst. Steklov 90, Translation: Proc. Steklov Inst. Math. 90.
8. 1987 *On the behavior in the Euclidean or Lobachevsky plane of the trajectories that cover trajectories of flows on closed surfaces. I,* Math. USSR Izvestiya 51, 16-43.
9. 1988 *On the behavior in the Euclidean or Lobachevsky plane of the trajectories that cover trajectories of flows on closed surfaces. II,* Math. USSR Izvestiya 52, 451-478.
10. 1989 *On infinite curves on the Klein bottle,* Mat. Sb. 180, 39-56, Translation: Math. USSR, Sbornik 66, MR 90f:58145.
11. 1990 *Infinite curves on the torus and on closed surfaces of negative Euler characteristic,* Proc. Steklov Math. Inst. 185, 33-58.
12. 1992 *How curves on the universal covering plane that cover non-self-intersecting curves on a closed surface can go to infinity,* Proc. Steklov Math. Inst. 191, 35-45.
13. 1995 *Flows on closed surfaces and behavior of trajectories lifted to the universal covering plane,* J. of Dynamical and Control Syst. 1, 125-138.
14. 1995 *On the behavior of trajectories, in the Euclidean and Lobachevsky plane, covering the trajectories of flows on closed surfaces. III,* Izvestiya Ross. Akad. Nauk, Ser. Mat. 59, 63-96, MR 97c:58134.
15. **Aranson, S. H.**
1964 *Rotation of a field in the theorem of the theory of structurally stable dy-*

namical systems on torus, Dokl. Akad. Nauk SSSR 156, 995-998, Translation: Soviet Math. Dokl. 1964, 5, 733-736, MR 29# 337.

16. 1969 *Trajectories on non-orientable two-dimensional manifolds*, Mat. Sbornik 80, 314-333.
17. 1975 *Certain arithmetical properties of dynamical systems on two-dimensional manifolds*, Dokl. Akad. Nauk SSSR 222, no 2, 265-268, Translation: Soviet Math. Dokl. 16, 605-609, MR 52# 9292.
18. **Aranson, S. H. and Grines V. Z.**
1973 *On certain invariants of dynamical systems on two-dimensional manifolds (necessary and sufficient conditions of topological equivalence of transitive systems)*, Mat. Sb. 90, 372-402.
19. 1978 *On the representation of minimal sets of flows on 2-dimensional manifolds with geodesic curves*, Izv. Akad. Nauk SSSR 42, 104-129.
20. 1986 *Topological classification of flows on closed two-dimensional manifolds*, Uspekhi Mat. Nauk 41, 149-169, Translation: Russian Math. Surveys.
21. 1990 *Topological classification of cascades on closed two-dimensional manifolds*, Uspekhi Mat. Nauk 45, 3-32, Translation: Russian Math. Surveys.
22. **Aranson, S. H., Grines, V. Z. and Zhuzhoma, E. V**
1995 *On the geometry and topology of surface flows and foliations and Anosov problem*, Mat. Sb. 186, 25-66, Translation: Mat. Sbornik 186, 1107-1146, MR 96h:58149.
23. 1996 *On Anosov–Weil problem*, Preprint.
24. **Aranson, S. H., Malkin, M. I. and Zhuzhoma, E. V.**
1993 *Local structure and smoothness preventing quasiminimality for flows on the torus*, Differents. Uravneniya, Translation: Differential Equations 29, 789-791.
25. **Aranson, S., Telnyh, I. and Zhuzhoma, E.**
1998 *Transitive and highly transitive flows on closed nonorientable surfaces*, Mat. Zametki 63, no 4, 625-627, Translation: Mathematical Notes.
26. **Aranson, S. H. and Zhuzhoma, E. V.**
1976 *The topological classification of singular dynamical systems on the torus*, Izv. Vys. Uchebn. Zaved. Mat. 20, 104-107, Translation: Izv. VUZ'ov.
27. 1988 *C^1-Cherry flows with gray cells*, Methods of Qual. Th. Diff. Eq., Gorky, 5-10, MR 88f:57043.
28. 1994 *About structure of quasiminimal sets of foliations on surfaces*, Mat. Sbornik 185, no.8, 31-62, Translation: Mat. Sbornik.
29. 1996 *Maier's theorem and geodesic laminations of surface flows*, J. Dynam. Control. Systems. 2, 557-582.
30. **Aranson, S. Kh, Belitsky, G. R. and Zhuzhoma, E. V.**
1996 *Introduction to the Qualitative Theory of Dynamical Systems on Surfaces*, Translations of Mathematical Monographs 153, AMS.
31. **Aranson, S. H, Bronstein, I. U, Zhuzhoma, E. V and Nikolaev, I. V.**
1998 *Qualitative theory of foliations on the closed surfaces*, Itogi nauki i techniki, VINITI, Dinamicheskie Sistemy-5, Translation: J. Math. Sci. 90, 2111-2149.
32. **Arnold, V. I.**
1961 *Small denominators. I*, Izv. Akad. Nauk SSSR 25, 21-86, Translation: Translations Amer. Math. Soc., 2nd Series, 46, 213-284.
33. 1980 *Chapitres Supplémentaires de la Thèorie des Équations Différentielles Ordinaires*, Édition Mir, Moscou.
34. 1989 *Mathematical Methods of Classical Mechanics*, Second edition, Springer.
35. 1992 *Ordinary Differential Equations*, Springer.

36. **Arnold, V. I., Afraimovich, V. S., Shilnikov, L. P. and Ilyashenko, Yu. S.**
1994 *Theory of bifurcations*, Encyclopedia of Mathematical Scinces, Dynamical Systems V, 5-218, Springer.
37. **Arnold, V. I. and Ilyashenko, Yu. S.**
1988 *Ordinary differential equations*, Encyclopedia of Mathematical Sciences, Dynamical Systems I, 11-147, Springer.
38. **Artin, E.**
1924 *Ein mechanishes System mit quasiergodishen Bahnen*, Abh. Math. Sem. Hamburg, 170-175, Collected papers: 499-504, Springer.
39. **Aschbacher, M.**
1988 *Finite Group Theory*, Cambridge Univ. Press.
40. **Atiyah, M. F.**
1967 *K-Theory*, W. A. Benjamin Inc., New York-Amsterdam.
41. **Axler, S., Bourdon, P. and Ramey, W.**
1992 *Harmonic Function Theory*, Graduate Texts in Mathematics 137, Springer.
42. **Beardon, A. F.**
1983 *The Geometry of Discrete Groups*, Graduate Texts in Mathematics 91, Springer.
43. **Beck, A.**
1974 *Continuous Flows in the Plane*, Springer.
44. **G. Belitskii, G. and Bykov, N.**
1998 *Spaces of cohomologies associated with linear functional equations*, Ergod. Th. and Dynam. Syst. 18, 343-356.
45. **Bendixson, I.**
1901 *Sur les courbes définies par les équations différentielles*, Acta Math. 24, 1-88.
46. **Beniere, J.-Ch.**
1997 *Feuilletage minimaux et sans minimaux sur les surfaces non compactes*, Thèse, Lyon.
47. **Beniere, J.-Ch., and Meigniez, G.**
1997 *Flows without minimal set*, Private communication.
48. **Bers, L.**
1977 *Quasiconformal mappings, with applications to differential equations, function theory and topology*, Bull. Amer. Math. Soc. 83, 1083-1100.
49. **Bhatia, N. P. and Szego, G. P.**
1979 *Stability Theory of Dynamical Systems*, Springer.
50. **Birkhoff, G. D.**
1927 *Dynamical Systems*, New York, AMS Colloquium Publications 9, First Edition.
51. **Blohin, A.**
1972 *Smooth ergodic flows on surfaces*, Trudy Mosk. Mat. Obsh. 27, 113-128, MR 51#6882.
52. **Bogdanov. R. I.**
1979 *Local orbital normal forms of vector fields in the plane*, Trudy Sem. Petrovsk. 5, 51-84, MR# 80j:58056.
53. **Bol, G.**
1943-44 *Über Nabelpunkte auf einer Eiflache*, Math. Z. 49, 389-410.
54. **Boldrighini, C., Keane, M. and Marchetti, F.**
1978 *Billiards in polygons*, Annals of Probability 6, 532-540.

55. **Bolsinov, A. F. and Fomenko, A. T.**
1995 *Orbital equivalence of integrable Hamiltonian systems with two degrees of freedom. A classification theorem. I,* Russian Acad. of Sci. Sbornik, Mathematics 81, no 2, 421-465.
56. 1995 *A criterion for the topological conjugacy of Hamiltonian flows on two-dimensional compact surfaces,* Russian Math. Surveys. 50, no 1, 193-194.
57. **Boothby, W. M.**
1951 *The topology of regular curve families with multiple saddle points,* Amer. J. Math. 73, 405-438.
58. 1951 *The topology of the level curves of harmonic functions with critical points,* Amer. J. Math. 73, 512-538.
59. **Bressan, A. and Piccoli, B.**
1997 *Structural stability for time–optimal planar syntheses,* Dynamics of Continuous, Discrete and Impulsive Systems 3, 335-371.
60. 1998 *A generic classification of time-optimal planar stabilizing feedbacks,* SIAM J. Control Optim 36, 12-32.
61. **Broer, H. W., Dumortier, F., van Strien, S, J. and Takens, F.**
1991 *Local study of planar vector fields,* Structures in Dynamics, 161-241, Studies in Mathematical Physics, Volume 2, North-Holland.
62. **Bronstein, I. and Nikolaev I.**
1992 *Singular points of the line element fields in the plane,* Izv. Akad. Nauk Resp. Moldova, Matematika 3(9), 23-29, MR# 94f:34060.
63. 1994 *On Peixoto graphs of Morse-Smale foliations,* CRM-2214, Univ. de Montréal, 23p, Preprint.
64. 1994 *Smooth orbital equivalence in a neighborhood of a critical point on a plane,* Differential Equations 30, 1357-1360; Differents. Uravn. 30, 1462-1464.
65. 1995 *Orbital normal forms and bifurcations of fields of linear elements on the plane,* Differential Equations 31, 874-878; Differents. Uravn. 31, 934-938.
66. 1997 *Peixoto graphs of Morse-Smale foliations on surfaces,* Topology and its Appl. 77, 19-36.
67. 1998 *Structurally stable fields of line elements on surfaces,* Nonlinear Analysis 34, 461-477.
68. **Brown, L. G., Green, P. and Rieffel, M. A.**
1977 *Stable isomorphism and strong Morita equivalence of C^*-algebras,* Pacific J. Math. 71, 349-363.
69. **Burnside, W.**
1911 *The Theory of Groups of Finite Order,* Second Edition, Cambridge, England.
70. **Cartan, E**
1922 *Lecons Sur Les Invariants Intégraux,* Paris, Hermann.
71. 1942 *Sur les couples de surfaces applicables avec conservations des courbures principales,* Bull. Sci. Math. 66, 55-85.
72. **Cayley, A.**
1863 *On differential equations and umbilici,* Philos. Mag., 26, 373-379 and 441-452, Coll. Works: V. 6.
73. 1878 *On the theory of groups,* Proc. London Math. Soc. (1), 9, 126-133.
74. 1878 *The theory of groups: graphical representation,* Amer. J. Math. 1, 174-176.
75. **Cherdak, B. M.**
1967 *Asymptotical properties of curves defined by its curvatures,* Tr. Tomsk Universiteta 191, 150-157.

76. **Chern S. S.**
1985 *Deformations of surfaces preserving principal curvatures*, Differential Geometry and Complex Analysis, H. E. Rauch memorial volume, Springer, 155-163.
77. **Cherry, T.**
1937 *Topological properties of solutions of ordinary differential equations*, Amer. J. Math. 59, 957-982.
78. 1938 *Analytic quasi-periodic curves of discontinuous type on torus*, Proc. London Math. Soc. 44, 175-215.
79. **Connes, A.**
1982 *A survey of foliations and operator algebras*, Proc. of Symp. in Pure Math. 38, Part I, 521-628.
80. **Cornfeld, I. P., Fomin, S. V. and Sinai, Ya. G.**
1982 *Ergodic Theory*, Grundlehren der Mathematischen Wissenschaften 245, Springer.
81. **Coxeter, H. S. M. and Moser, W. O. J.**
1972 *Generators and Relations for Discrete Groups*, Third Edition, Ergebnisse der Mathematik und ihrer Grenzgebiete, Bd. 14, Springer.
82. **Darboux, G.**
1896 *Sur la forme des lignes de courbure dans la voisinage d'un ombilic*, Lecons sur la Theorie des Surfaces, IV, Note 7, Gauthier Villars, Paris.
83. **Davidson, K. R.**
1996 *C^*-Algebras by Example*, Fields Institute Monographs 6, AMS.
84. **Dehn, M.**
1938 *Die Gruppe der Abbildungsklassen*, Acta Math. 69, 135-206.
85. 1987 *Lectures on group theory*, Max Dehn, Papers on Group Theory and Topology translated by John Stillwell, 5-46, Springer.
86. **de Melo, W. and van Strien, S.**
1988 *One-dimensional dynamics: The Schwarzian derivative and beyond*, Bull. Amer. Math. Soc. 18, 159-162.
87. **de Rezende, K. A. and Franzosa, R. D.**
1993 *Lyapunov graphs and flows on surfaces*, Trans. Amer. Math. Soc. 340, 767-784.
88. **Denjoy, A.**
1932 *Sur les courbes définies par les équations différentielles à la surface du tore*, J. Math. Pures et Appl. (ser 9) 11, 333-375.
89. 1958 *Le phenomene ergodique et les trajectoires sur le tore*, C.R. Acad. Sci. Paris. 247, 15, 1072-1078.
90. **Detchenya, L. V. and Sadovskii, A. P.**
1999 *Orbital normal forms of two-dimensional analytic systems with zero linear part and nonzero quadratic part*, Differential Equations 35, 50-56.
91. **Dubrovin, B. A., Fomenko, A. T. and Fucks, D. B**
1986 *Modern Geometry – Methods and Applications*, Moscow, Nauka, Translation: Modern Geometry – Methods and Applications, Part I, the geometry of surfaces, transformation groups and fields, Graduate Texts in Mathematics 93, Springer.
92. **Dumortier, F.**
1977 *Singularities of vector fields on the plane*, J. Differential Equations 23, 53-106.
93. **Eberlein, P. and O'Neil, B**
1973 *Visibility manifolds*, Pacific J. Math 46, 45-109.

94. **Edmonds, J. R.**
1960 *A combinatorial representation for polyhedral surfaces (abstract)*, Notices Amer. Math. Soc. 7, 648; 948.

95. **Effros, E. G.**
1981 *Dimensions and C^*-algebras*, Regional Conf. Series in Math. No. 46, AMS.

96. **Effros, E. G. and Shen, C. L.**
1980 *Approximately finite C^*-algebras and continued fractions*, Indiana J. Math. 29, 191-204.

97. **Elliott, G. A.**
1995 *The classification problem for amenable C^*-algebras*, Proc. of the International Congress of Mathematics, Zurich 1994, 922-932, Birkhauser Verlag, Basel, Switzerland.

98. **Elliott, G. A. and Evans, D. E.**
1993 *The structure of the irrational rotation C^*-algebra*, Ann. of Math. 138, 477-501.

99. **Ezell, C. L.**
1978 *Branch point structure of covering maps onto nonorientable surfaces*, Trans. Amer. Math. Soc. 243, 123-133.

100. **Fathi, A., Laudenbach, F., and Poénaru, Z.**
1979 *Travaux de Thurston sur les surfaces*, Séminaire Orsay, France, Astérisque 66-67.

101. **Farkas, H. M. and Kra, I.**
1980 *Riemann Surfaces*, Graduate texts in Mathematics 71, Springer.

102. **Feldman, E. A.**
1964 *Geometry of immersions. II*, Bull. Amer. Math. Soc. 70, 600-607.

103. 1965 and 1966 *Geometry of immersions. I and II*, Trans. Amer. Math. Soc. 120 (1965), 185-224; and 125 (1966), 181-215.

104. 1967 *On parabolic and umbilic points of immersed hypersurfaces*, Trans. Amer. Math. Soc. 127, 1-28.

105. **Field, M.**
1976 *Une version équivariant de théorème de le transversalité de Thom*, C.R. Acad. Sci. Paris 282, 379-381.

106. **Fillmore, P. A.**
1996 *A User's Guide to Operator Algebras*, CMS Monographs, John Wiley & Sons.

107. **Fomenko, A. T.**
1991 *Topological classification of integrable systems*, The Geometry of Hamiltonian Systems ed. by Ratiu, Springer, 131-339.

108. **Ford, L. R.**
1951 *Automorphic Functions*, Chelsea Publishing Company, New York.

109. **Forni, G.**
1997 *Solutions of the cohomological equation for area-preserving flows on compact surfaces of higher genus*, Ann. of Math. 146, 295-344.

110. **Franks. J.**
1979 *Morse-Smale flows and homotopy theory*, Topology 18, 199-215.

111. **Franks, J. and Misiurewicz, M.**
1990 *Rotation sets of toral flows*, Proc. Amer. Math. Soc. 109, 243-249.

112. **Fricke, R. and Klein F.**
1897 and 1901 *Vorlesungen uber die Theorie der automorphen Funktionen. Vol. I and II*, Teubner, Leipzig.

113. **Fucks, D. B. and Rokhlin, V. A.**
1977 *Introduction to Topology, Geometric Chapters*, Moscow, Nauka.

114. **Gardiner, C. J.**
1985 *The structure of flows exhibiting nontrivial recurrence on two-dimensional manifolds*, J. Differential Equations 57, 138-158.

115. **Gardiner, F. P.**
1987 *Teichmuller Theory and Quadratic Differentials*, John Wiley & Sons.

116. **Gelfand, I. M. and Naimark, M.**
1943 *On the embedding of normed rings into the ring of operators in Hilbert space*, Mat. Sbornik 12, 197-213.

117. **Giordano, T., Putnam, I. F. and Skau, C. F.**
1995 *Topological orbit equivalence and C^*-crossed products*, J. reine angew. Math. 469, 51-111.

118. **Godbillon, C.**
1969 *Géométrie Différentielle at Mécanique Analytique*, Hermann, Paris.

119. 1983 *Dynamical Systems on Surfaces*, Springer.

120. **Golab, S., Kuszma, M. and Opial, Z.**
1958-59 *La courbure d'une courbe plane et l'existence d'une asymptote*, Ann. Polon. Math. 5, 275-283.

121. **Gross, J. L. and Tucker, T. W.**
1987 *Topological Graph Theory*, John Wiley & Sons.

122. **Guínez, V.**
1988 *Positive quadratic differential forms and foliations with singularities on surfaces*, Trans. Amer. Math. Soc. 309, 477-502.

123. 1990 *Nonorientable polynomial foliations on the plane*, J. of Differential Equations 87, 391-411.

124. **Guttiérrez, C.**
1975 *Structural stability for line fields with singularities on two-dimensional manifolds*, Lecture Notes in Math. 468, 17-19, Springer.

125. 1979 *Smoothing continuous flows and the converse of Denjoy-Schwartz Theorem*, An. Acad. Brasil Cienc. 51, 581-589.

126. 1986 *Smoothing continuous flows on 2-manifolds and recurrences*, Ergod. Th. and Dyn. Syst. 6, 17-44.

127. 1986 *On the C^r-closing lemma for flows on the torus T^2*, Ergod. Th. and Dyn. Sys. 6, 45-56.

128. 1988 *Foliations on surfaces having exceptional leaves*, Lecture Notes in Math. 1331, 73-85, Springer.

129. **Gutiérrez, C. and de Melo, W.**
1977 *The connected components of Morse–Smale vector fields on 2-manifolds*, Lecture Notes in Math. 597, 230-251.

130. **Gutiérrez, C. and Sotomayor, J.**
1982 *Structurally stable configurations of lines of principal curvature*, Asterisque 98-99.

131. 1983 *An approximation theorem for immersions with structurally stable configurations of lines of principal curvature*, Lecture Notes in Math. 1007, Springer.

132. 1986 *Principal lines on surfaces immersed with constant mean curvature*, Trans. Amer. Math. Soc. 293, 751-766.

133. 1990 *Periodic lines of curvature from Darbouxian umbilical connections*, Lecture Notes in Math. 1455, 196-229, Springer.

134. 1994 *Lines of Curvature and Umbilical Points on Surfaces*, Instituto de Matemática Pura e Aplicada, Rio de Janeiro, ISBN 85-244-0057-9.

135. **Halmos, P. R.**
1950 *Measure Theory*, Van Nostrand.

136. **Hamburger, H.**
1940 *Beweis einer Caratheodoryschen Vermutung, I,* Ann. of Math. 41, 63-86.
137. 1941 *Beweis einer Caratheodoryschen Vermutung, II und III,* Acta Mathematica 73, 175-332.
138. **Hartman, P. and Wintner, A.**
1953 *On the behavior of the solutions of real binary differential systems at singular points,* Amer. J. of Math. 75, 117-126.
139. 1953 *Envelopes and discriminant curves,* Amer. J. of Math. 75, 142-158.
140. 1953 *On the singularities in nets of curves defined by differential equations,* Amer. J. of Math. 75, 277-297.
141. **Hector, G. and Hirsch, U.**
1981 *Introduction to the Geometry of Foliations, Part A and B,* Vieweg & Sohn.
142. **Hedlund, G.**
1936 *Two-dimensional manifolds and transitivity,* Ann. of Math. 37, 3, 534-542.
143. **Heffter, L.**
1891 *Ueber das Problem der Nachbargebiete,* Math. Annalen 38, 477-508.
144. **Herman, M. R.**
1979 *Sur la conjugaison différentiable des difféomorphismes du cercle à des rotations,* Publ. Math. IHES 49, 2-233.
145. **Hilbert, D. and Cohn-Vossen, S.**
1956 *Geometry and the Imagination,* Chelsea Publishing Company. New York.
146. **Hopf, H.**
1989 *Differential Geometry in the Large,* Lecture Notes in Math. 1000, Springer.
147. **Hubbard, J. and Masur, H.**
1979 *Quadratic differentials and foliations,* Acta Math. 142, 221-274.
148. **Hurwitz, A.**
1891 *Ueber Riemann'sche Flachen mit gegebenen Verzweigungspunkten,* Math. Annalen 39, 1-61.
149. **Hurwitz, A. and Courant, R.**
1964 *Vorlesungen uber Algemeine Funktionentheorie und Elliptische Funktionen,* Springer.
150. **Imanishi, H.**
1979 *On codimension one foliations defined by closed one forms with singularities,* J. Math. Kyoto Univ. 19-2, 285-291.
151. **Inaba, T.**
1999 *An example of a flow on a noncompact surface without minimal set,* Ergod. Th. and Dynam. Syst. (*to appear*)
152. **Jenkins, J.**
1972 *A toplogical three pole theorem,* Indiana Univ. Mathematics J. 21, No.11, 1013-1018.
153. **Kadyrov, A. A.**
1983 *Critical points of differential equations with unoriented trajectories in the plane,* Differents. Uravn. 19, 2039-2048, Translation: Differential Equations 19, 1473-1483.
154. **Kaplan, W.**
1940 *Regular curve-families filling the plane, I,* Duke Math. J. 7, 154-185.
155. 1941 *Regular curve-families filling the plane, II,* Duke Math. J. 8, 11-46.

156. 1942 *The structure of a curve-family on a surface in the neighborhood of an isolated singularity*, Amer. J. Math. 64, 1-35.
157. 1948 *Topology of level curves of harmonic functions*, Trans. Amer. Math. Soc. 63, 514-522.
158. 1976 *On the Three Pole Theorem*, Math. Nachr. 75, 299-309.
159. **Kapouleas, N.**
1991 *Compact constant mean curvature surfaces in Euclidean three-space*, J. Diff. Geom. 33, 683-715.
160. **Katok, A. B.**
1973 *Invariant measures of flows on oriented surfaces*, Soviet Math. Doklady 14, (4), 1104-1108; DAN 211, No.4, 775-778.
161. 1980 *Interval exchange transformations and some special flows are not mixing*, Israel J. Math. 35, 301-310.
162. **Katok, A. and Hasselblatt, B.**
1995 *Introduction to the Modern Theory of Dynamical Systems*, Encyclopedia of Mathematics and its Applications, Cambridge Univ. Press.
163. **Keane, M.**
1975 *Interval exchange transformations*, Math. Z. 141, 25-31.
164. **Klotz, T**
1959 *On G. Bol's proof of Carathéodory's conjecture*, Comm. Pure and Appl. Math. 12, 277-311.
165. **Kobayashi, S. and Nomizu, K.**
1963 and 1969 *Foundations of Differential Geometry, 2 vols*, New-York: John Wiley & Sons.
166. **Kochergin, A. V.**
1972 *On absence of mixing for the special flows over the circle rotation and flows on two-dimensional torus*, Dokl. Akad. Nauk SSSR 205, 515-518, MR 46#5751.
167. 1973 *Change of time in flows and mixing*, Izv. Akad. Nauk SSSR, Ser. Mat. 37, 1275-1298, MR 49#10855.
168. 1975 *On mixing in special flows over a shifting of segments and in smooth flows on surfaces*, Mat. Sb. 96, Translation: Mathematics of the USSR - Sbornik 25, 441-469.
169. 1976 *Nonsingular saddle points and the absence of mixing*, Mat. Zametki 19, no 3, 453-468, Translation: Math. Notes, 19, 277-286.
170. **Kolmogorov, A. N.**
1953 *On dynamical systems with an integral invariant on a torus*, Dokl. Akad. Nauk SSSR 93, 5, 763-766, MR 16–36g.
171. **Kontsevich, M. and Zorich, A.**
1997 *Lyapunov exponents and Hodge theory*, Preprint of IHES 13, 1-16.
172. **Kotus, J., Krych, M. and Nitecki, Z.**
1982 *Global structural stability of flows on open surfaces*, Mem. Amer. Math. Soc. 261, V.37.
173. **Kuzmin, A. G.**
1982 *The behavior of the integral curves in a neighborhood of a branch point of the discriminant curve*, Vestnik Leningrad Univ. Math 14, 143-149.
174. 1982 *Indices of singular and irregular points of direction fields in the plane*, Vestnik Leningrad Univ. Math 14, 285-292.
175. **Langevin R. and Possani C.**
1989 *Courbure totale de feuilletages et enveloppes*, C.R. Acad. Sci. Paris. Ser.I Math. 309, no.13, 821-824.

176. **Langevin R and Sifre J-C.**
1993 *Asymptotes des courbes tracees sur les varietes de Hadamard*, Ann. Fac. Sci. de Toulouse, 11, 3, 375-427.

177. **Lawson H. B and Tribuzy R. A.**
1981 *On the mean curvature function for compact surfaces*, J. Diff. Geom. 16, 179-183.

178. **Lehner, J.**
1964 *Discontinuous Groups and Automorphic Functions*, Math. Surveys. V. 8, Amer. Math. Soc., Providence, Rhode Island.

179. **Lehto, O.**
1986 *Univalent Functions and Teichmuller Spaces*, Graduate texts in mathematics 109, Springer.

180. **Leontovich, E. and Maier, A.**
1937 *On trajectories defining the qualitative structure of the decomposition of a sphere into trajectories*, Dokl Akad. Nauk SSSR 14, 251-254.

181. 1955 *On a scheme defining the topological structure of decomposition into trajectories*, Dokl. Acad. Nauk SSSR 103, 557-560. MR 17-262.

182. **Levitt, G.**
1982 *Feuilletages des surfaces*, Ann. Inst. Fourier 32, 179-217.

183. 1982 *Pantalons et feuilletages des surfaces*, Topology 21, 9-33.

184. 1983 *Foliations and laminations on hyperbolic surfaces*, Topology 22, 119-135.

185. 1983 *Flots topologiquement transitifs sur les surfaces compactes sans bord: contrexemple à une conjecture de Katok*, Ergod. Th. and Dynam. Sys. 3, 241-249.

186. 1983 *Feuilletages des surfaces*, Thèse, Paris.

187. 1987 *La décomposition dynamique et la différentiabilité des feuilletages des surfaces*, Ann. Inst. Fourier 37, 85-116.

188. **Levitt, G. and Rosenberg, H.**
1987 *Differentiability and topology of labyrinths in the disc and annulus*, Topology 26, 173-186.

189. **Lima, E. L.**
1964-65 *Common singularities of commuting vector fields on 2-manifolds*, Comm. Math. Helvetici 39, 97-110.

190. **Liousse, I.**
1995 *Dynamique générique des feuilletages transversalement affines des surfaces*, Bull. Soc. math. France 123, 493-516.

191. **Lyndon, R. C. and Schupp, P. E.**
1977 *Combinatorial Group Theory*, Ergebnisse der Mathematik und ihrer Grenzgebiete, Bd. 89, Springer.

192. **Magnus, W., Karras, A. and Solitar, D.**
1976 *Combinatorial Group Theory: presentations of groups in terms of generators and relations*, Dover Publications, New York.

193. **Maier, A. G.**
1939 *De trajectoires sur les surfaces orientées*, C.R. Acad. Sci. URSS 24, 673-675.

194. 1943 *About trajectories on orientable surfaces*, Mat. Sbornik 12, no.1, 71-84.

195. **Markley, N.**
1969 *The Poincaré-Bendixson theorem for the Klein bottle*, Trans. Amer. Math. Soc. 135, 159-165.

196. 1970 *On the number of recurrent orbit closures*, Proc. Amer. Math. Soc. 25, 413-416.

197. 1970 *Homeomorphisms of the circle without periodic points*, Proc. London Math. Soc. 20, 688-698.
198. **Markus, L.**
1954 *Global structure of ordinary differential equations in the plane*, Trans. Amer. Math. Soc. 76, 127-148.
199. **Martens M., Strien van S., Melo de W. and Mendes P.**
1990 *On Cherry flows*, Ergod. Th. and Dynam. Sys. 10, 531-554.
200. **Masur, H.**
1982 *Interval exchange transformations and measured foliations*, Ann. of Math. 115, 169-200.
201. **Masur, H. and Smillie, J.**
1993 *Quadratic differentials with prescribed singularities and pseudo-Anosov diffeomorphisms*, Comment. Math. Helvetici 68, 289-307.
202. **Meyer, K. R.**
1968 *Energy functions for Morse-Smale systems*, Amer. J. of Math. 90. 1031-1040.
203. **Michel, M.-F.**
1995 *Formes normales de certaines équations différentieles quadratiques*, C.R. Acad. Sci. Paris. 321, Sér. I, 283-286.
204. **Minkowski, H.**
1910 *Geometrie der Zahlen*, Leipzig.
205. **Morse, M. and Jenkins, J.**
1952 *The existence of pseudoconjugates on Riemann surfaces*, Fund. Math. 39, 269-287.
206. **Moser, J. K.**
1968 *Lectures on Hamiltonian Systems*, Mem. Amer. Math. Soc. 81, Providence, Rhode Island.
207. **Moussu, R.**
1970 *Feuilletage sans holonomie d'une varieté fermée*, C.R. Acad. Sci. Paris 270, 1308-1311.
208. **Myrberg, P. J.**
1931 *Ein Approximationssatz fur die Fuchsschen Gruppen*, Acta Math. 57, 389-409.
209. **Nemytskii V. V. and Stepanov V. V.**
1960 *Qualitative Theory of Differential Equations*, Princeton Univ. Press.
210. **Neumann, D.**
1978 *Smoothing continuous flows on 2-manifolds*, J. Differential Equations 28, 327-344.
211. 1978 *Central sequences in dynamical systems*, Amer. J. of Math. 100, 1-18.
212. **Neumann, D. and O'Brien, T.**
1976 *Global structure of continuous flows on 2-manifolds*, J. Differential Equations 22, 89-110.
213. **Nielsen, J.**
1924 *Über topologische Abbildungen geschlosener Flächen*, Abh. Math. Sem. Hamburg Univ., Heft 3, 1, 246-260.
214. 1927 *Untersuchungen zur Topologie der geshlossener zweiseitige Flachen. I*, Acta Math. 50, 189-358.
215. 1929 *Untersuchungen zur Topologie der geshlossener zweiseitige Flachen. II*, Acta Math. 53, 1-76.
216. 1932 *Untersuchungen zur Topologie der geshlossener zweiseitige Flachen. III*, Acta Math. 58, 87-167.
217. 1944 *Surface transformation classes of algebraically finite type*, Danske Vid. Selsk. Math.-Phys. Medd. 68.

218. **Nikolaev, I.**
1994 *Foliations with singularities of semi-integer index*, CRM-2206, Université de Montréal, 33p, Preprint.
219. 1996 *The Poincaré-Bendixson theorem and arational foliations on the sphere*, Ann. Inst. Fourier 46, 1159-1181.
220. 1998 *Finite groups and flows on surfaces*, Nonlinearity 11, 213-231.
221. 1998 *A note on the Carathéodory conjecture*, CRM-2535, Université de Montréal, 10p, Preprint.
222. 1998 *Graphs and flows on surfaces*, Ergod. Th. and Dyn. Syst. 18, 207-220.
223. 2000 *An invariant of minimal flows coming from the K_0-group of a crossed product C^*-algebra*, Ergod. Th. and Dyn. Syst. 20, 1449-1468.
224. 1999 *Geometry of a crossed product*, Proc. Amer. Math. Soc. 128, 1177-1183.
225. 2000 *Simple umbilics revisited*, Differential Geom. Appl. 13, 235-249.
226. 2000 *Non-wandering flows on the 2-manifolds*, J. of Differential Equations (to appear)
227. **Nikolaev, I. and Piccoli, B.**
1998 *Classification of stable time–optimal controls on 2-manifolds*, Preprint.
228. **Nikolaev, I. and Vulpe, N.**
1997 *Topological classification of quadratic systems at infinity*, J. London Math. Soc. 55, 473-488.
229. **Nikolaev, I. and Zhuzhoma, E.**
1999 *Flows on 2-Dimensional Manifolds*, Lect. Notes in Math. 1705, Springer.
230. **Nitecki, Z.**
1971 *Differentiable Dynamics. An introduction to the orbit structure of diffeomorphisms*, MIT Press. Cambridge, Massachusets, London.
231. **Novikov, S. P.**
1965 *Topology of foliations*, Trudy Mosk. Mat. Obshch. 14, 248-278.
232. **Olver, P.**
1986 *Applications of Lie groups to differential equations*, Graduate texts in Math. 107, Springer.
233. **Oxtoby, J. C.**
1953 *Stepanoff flows on the torus*, Proc. Amer. Math. Soc. 4, 982-987.
234. **Palis, J. and de Melo, W.**
1982 *Geometric Theory of Dynamical Systems. An introduction*, Springer.
235. **Pedersen, G. K.**
1979 *C^*-Algebras and their Automorphism Groups*, Acad. Press, London-New York-San Francisco.
236. **Peixoto, M. M.**
1962 *Structural stability on two-dimensional manifolds*, Topology 1, 101-120.
237. 1973 *On the classification of flows on 2-manifolds*, Dynamical Systems ed. by Peixoto, Proc. of Symp. held at Univ. of Bahia. Salvador, Brazil 1971, Acad. Press, 389-419.
238. **Peixoto, M. M. and Pugh, C. C.**
1968 *Structurally stable systems on open manifolds are never dense*, Amer. J. Math. 87, 423-430.
239. **Picard, É**
1895 *Sur les points singuliers des équations différentielles du premier ordre*, Math. Annalen. 46, 521-528.
240. **Piccoli, B.**
1994 PhD Thesis, SISSA-ISAS.

241. 1996 *Regular time-optimal syntheses for smooth planar systems*, Rend. Sem Mat. Univ. Padova 95, 59-79.

242. 1996 *Classifications of generic singularities for the planar time-optimal synthesis*, SIAM J. Control and Optimization 34, 1914-1946.

243. **Piccoli, B. and Sussmann, H. J.**
1999 *Regular synthesis and sufficiency conditions for optimality*, SIAM J. Control and Optimization, (to appear)

244. **Pimsner, M. and Voiculescu, D.**
1980 *Exact sequence for K-groups and Ext-groups of certain cross-product C^*-algebras*, J. Operator Theory 4, 93-118.

245. **Plykin, R. V.**
1974 *Sourses and sinks of A-diffeomorphisms on surfaces*, Mat. Sbornik 94, 243-264, Translation: Mat. Sbornik.

246. 1984 *On the geometry of hyperbolic attractors of smooth cascades*, Russian Math. Surveys 39, no. 6, 85-131.

247. **Poincaré, H.**
1881 and 1882 *Memoire sur les courbes definies par une equation differentielle*, J. math. pures et appl. 3-e sér. 7, 375-422; and 8, 251-296.

248. **Pupko, V. I.**
1967 *Non-self-intersecting curves on closed surfaces*, Dokl. Akad. Nauk SSSR 177, 272-274.

249. **Putnam, I. F.**
1989 *The C^*-algebras associated with minimal homeomorphisms of the Cantor set*, Pacific J. of Math. 136, 329-353.

250. **Richards, J.**
1963 *On the classification of non-compact surfaces*, Trans. Amer. Math. Soc. 106, 259-269.

251. **Rieffel, M. A.**
1974 *Morita equivalence for C^*-algebras and W^*-algebras*, J. of Pure and Appl. Algebra 5, 51-96.

252. 1976 *Strong, Morita equivalence of certain transformation group C^*-algebras*, Math. Ann. 222, 7-22.

253. 1981 *C^*-algebras associated with irrational rotations*, Pacific J. of Mathematics 93, 415-429.

254. **Riemann, B.**
1851 *Grundlagen fur eine allgemeine Theorie der Functionen einer veranderlichen complexen Grosse*, Inauguraldissertation, Gottingen, Bernhard Riemann, Collected papers, 35-80, Springer, 1990.

255. **Ringel, G.**
1974 *Map Colour Theorem*, Die Grundlehren der mathematischen Wissenschaften in Einzeldarstellungen, Bd.209.

256. **Rosenberg, H**
1983 *Labyrinths in the disc and surfaces*, Ann. of Math. 117, 1-33.

257. **Sabidussi, G.**
1958 *On a class of fixed-point-free graphs*, Proc. Amer. Math. Soc. 9, 800-804.

258. **Saito, T.**
1952 *On the measure-preserving flow on torus*, J. Math. Soc. Japan, 3, 279-284 and 4, 338.

259. **Sataev, E. A.**
1975 *On the number of invariant measures for flows on orientable surfaces*, Izv. Akad. Nauk SSSR Ser. Mat. 39, no.4, 860-878, Translation: Math. USSR Izv. 9, 813-830.

260. **Schoeneberg, B.**
1974 *Elliptic Modular Functions*, Die Grundlehren der mathematischen Wissenschaften, Bd. 203. Springer.

261. **Schwartz, A.**
1963 *A generalization of the Poincaré-Bendixson theorem to closed two-dimensional manifolds*, Amer. J. Math. 85, 453-458.

262. **Schwartzman, S.**
1957 *Asymptotic cycles*, Annals of Math. 66, 270-284.

263. **Schweitzer, P.**
1978 *Some problems in foliation theory and related areas*, Lecture Notes in Math. 652, 240-252.

264. **Sibirsky, K. S.**
1988 *Introduction to the Algebraic Theory of Invariants of Differential Equations*, Manchester University.

265. **Sinai, Ya. G. and Khanin, K. M.**
1989 *Smoothness of conjugacies of diffeomorphisms of the circle with rotations*, Russian Math. Surveys 44, 69-99.

266. 1992 *Mixing for some classes of special flows over rotation of the circle*, Funkt. Anal. i Pril. 26, no 3, 1-21, Translation: Funct. Anal. and its Applic., 1992, 26, 155-169.

267. **Smale, S.**
1967 *Differentiable dynamical systems*, Bull. Amer. Math. Soc. 73, 747-811.

268. **Sotomayor, J.**
1968 *Generic one-parameter families of vector fields on two-dimensional manifolds*, Bull. Amer. Math. Soc. 74, 722-726.

269. 1973 *Generic one-parameter families of vector fields on two-dimensional manifolds*, Publ. Math. IHES 43, 5-46.

270. **Springer, G.**
1957 *Introduction to Riemann surfaces*, Addison-Wesley.

271. **Sternberg, S.**
1957 *On differential equations on the torus*, Amer. J. of Math. 79, 397-402.

272. 1964 *Lectures on Differential Geometry*, Englewood Cliffs, N.J., Prentice-Hall.

273. **Stowe, D.**
1986 *Linearization in two dimensions*, J. of Differential Equations 63, 183-126.

274. **Strebel, K.**
1984 *Quadratic Differentials*, Ergebnisse der Math. and ihrer Grenzgebiete, Bd.5, Springer.

275. **Struik, D. J.**
1950 *Lectures on Classical Differential Geometry*, Addison Wesley, 1950.

276. **Takens, F.**
1974 *Singularities of vector fields*, Publ. Math. IHES 43, 47-100.

277. **Tamura, I.**
1976 *Topology of Foliations*, Moscow, Mir Publisher.

278. **Teichmuller, O.**
1939 *Extremale quasikonforme Abbildungen und quadratische Differentiale*, Abh. Preus. Akad. Wiss., math-naturw. Kl. 22, 197, Collected Papers, Springer (1982), 337-531.

279. **Thurston W. P.**
Preprint. *On the geometry and dynamics of diffeomorphisms of surfaces.*

280. 1988 *On the geometry and dynamics of diffeomorphisms of surfaces*, Bull. Amer Math. Soc. 19, 417-431.

281. **Tishler, D.**
1970 *On fibering certain foliated manifolds over* S^1, Topology 9, 153-154.
282. **Torpe, A. M.**
1985 *K-theory for the leaf space of foliations by Reeb components*, J. Func. Anal. 61, 15-71.
283. **Troyanov, M.**
1986 *Les surfaces euclidiennes a singularites coniques*, Enseign. Math. 32, 79-94.
284. **Tucker, T. W.**
1983 *Finite groups acting on surfaces and the genus of a group*, J. Combin. Theory Ser. B 34, 82-98.
285. **Umehara M.**
1990 *A characterization of compact surfaces with constant mean curvature*, Proc. Amer. Math. Soc. 108, 483-489.
286. **Veech, W. A.**
1978 *Interval exchange transformations*, J. D'Analyse Math. 33, 222-278.
287. 1982 *Gauss measures for transformations on the space of interval exchange maps*, Ann. of Math. 115, 201-242.
288. 1989 *Teichmuller curves in moduli space, Eisenstein series and an application to triangular billiards*, Invent. Math. 97, 553-583.
289. **Vershik, A. M.**
1981 *Uniform algebraic approximation of shift and multiplication operators*, Dokl. Akad. Nauk SSSR 259, no 3, 526-529. Translation: Soviet Math. Dokl. 24, 97-100.
290. **von Neumann, J.**
1929 *Zur Algebra der Funktionaloperatoren und Theorie der normalen Operatoren*, Math. Annalen 102, 370-427.
291. **Vorobets, Ya. B.**
1996 *Plane structures and billiards in rational polygons: the Veech alternative*, Uspekhi Mat. Nauk 51, no. 5, 3-42.
292. **Walsh, J. A.**
1995 *Rotation vectors for toral maps and flows: a tutorial*, Intern. J. Bif. Chaos 5, 321-348.
293. 1995 *Directions for structurally stable flows on surfaces via rotation vectors*, Topology and Appl. 66, 49-65.
294. **Wang, X.**
1987 *On the* C^**-algebras of foliations in the plane*, Lect. Notes Math. 1257, Springer.
295. 1990 *The* C^**-algebras of Morse-Smale flows on two-manifolds*, Ergod. Th. and Dynam. Sys. 10, 565-597.
296. **Warner, F. W.**
1983 *Foundations of Differentiable Manifolds and Lie Groups*, Graduate Texts in Mathematics 94, Springer.
297. **Weil, A.**
1931 *On systems of curves on a ring-shaped surface*, J. of the Indian Math. Soc. 19, No.5, 109-112; No.6, 113-114.
298. 1936 *Les familles de courbes sur le tore*, Mat. Sbornik 1, No 5, 779-781.
299. **Wente H. C.**
1986 *Counterexample to a conjecture of H. Hopf*, Pacific J. Math. 121, 193–243.
300. **White, A. T.**
1973 *Graphs, Groups and Surfaces*, North-Holland Publishing Company, Amsterdam.

301. **Whitney, H.**
1933 *Regular families of curves*, Ann. of Math. 34, 244-270.
302. **Williams, R. F.**
1967 *One dimensional non-wandering sets*, Topology 6, 473-478.
303. **Wolf, J. A.**
1966 *Surfaces of constant mean curvature*, Proc. Amer. Math. Soc. 17, 1103-1111.
304. **Wolf, M.**
1996 *On realizing measured foliations via quadratic differentials of harmonic maps to R-trees*, J. Anal. Math. 68, 107-120.
305. 1998 *Measured foliations and harmonic maps of surfaces*, J. of Differential Geom. 49, 437-467.
306. **Yano, K.**
1985 *Asymptotic cycles on two-dimensional manifolds*, Adv. St in Pure Math. 5, 359-377, Foliations, Proc. Symp. Tokyo, 1983.
307. **Yoccoz, J.-C.**
1984 *Il n'y a pas de contre-exemple de Denjoy analytique*, C.R. Acad. Sc. Paris, t.298, Série I, no.7, 141-144.
308. 1984 *Conjuguaison differentiable des diffeomorphismes du circle dont le nombre de rotation verifie une condition Diophantienne*, Annales Sci. de l'Ecole Norm. Sup. 17, 333-361.
309. **Young, L. S.**
1977 *Entropy of continuous flows on compact manifolds*, Topology 16, 469-471.
310. **Zemlyakov, A and Katok, A.**
1975 *Topological transitivity of billiards*, Mat. Zametki 18, 291-300.
311. **Zhirov, A. Yu.**
1995 *Hyperbolic attractors of diffeomorphisms of orientable surfaces. I*, Russian Acad. Sci. Sb. Math. 82, 135-174.
312. 1995 *Hyperbolic attractors of diffeomorphisms of orientable surfaces. II*, Russian Acad. Sci. Sb. Math. 83, 23-65.
313. **Zieschang, H., Vogt, E. and Coldwey, H.-D.**
1980 *Surfaces and Planar Discontinuous Groups*, Springer.
314. **Zorich, A.**
1987 *Quasiperiodic structure of the level surfaces of a Morse-Smale 1-form, close to rational. Problem of S. P. Novikov*, Izv. Akad. Nauk SSSR 51, 1322-1344.
315. 1994 *Hamiltonian flows of multivalued Hamiltonian on closed orientable surfaces*, MPI 94-29, Max-Planck Institut fur Mathematik, Bonn, Preprint.
316. 1996 *Finite Gauss measure on the space of interval exchange transformations. Lyapunov exponents*, Annals Inst. Fourier 46, 325-370.

Index